Springer Water

The book series Springer Water comprises a broad portfolio of multi- and interdisciplinary scientific books, aiming at researchers, students, and everyone interested in water-related science. The series includes peer-reviewed monographs, edited volumes, textbooks, and conference proceedings. Its volumes combine all kinds of water-related research areas, such as: the movement, distribution and quality of freshwater; water resources; the quality and pollution of water and its influence on health; the water industry including drinking water, wastewater, and desalination services and technologies; water history; as well as water management and the governmental, political, developmental, and ethical aspects of water.

Sunny Dhiman · Gunjan Mukherjee
Editors

Resource Resurgence

Mitigating Wastewater, Smart Recycling and Novel Technologies

Springer

Editors
Sunny Dhiman
Department of Biotechnology, University Institute of Biotechnology
Chandigarh University
Mohali, Punjab, India

Gunjan Mukherjee
University Institute of Biotechnology
Chandigarh University
Mohali, Punjab, India

ISSN 2364-6934 ISSN 2364-8198 (clcctronic)
Springer Water
ISBN 978-3-032-02749-8 ISBN 978-3-032-02750-4 (eBook)
https://doi.org/10.1007/978-3-032-02750-4

This Springer imprint is published by the registered company Springer Nature Switzerland AG
The registered company address is: Gewerbestrasse 11, 6330 Cham, Switzerland

Preface

Water scarcity, environmental degradation, and the pressing need for sustainable development have brought the management of wastewater to the forefront of global environmental discourse. What was once perceived as a waste burden is now increasingly recognized as a potential resource—rich in nutrients, energy, and recoverable water. *Resource Resurgence: Mitigating Wastewater, Smart Recycling and Novel Technologies* emerges as a timely contribution to this transformative narrative, presenting an in-depth exploration of cutting-edge advancements that redefine how we perceive and utilize wastewater.

This book brings together interdisciplinary research and practical insights that span the continuum from fundamental concepts to futuristic innovations. The opening chapter sets the tone by discussing the evolving paradigm that views wastewater not merely as effluent but as a reservoir of recoverable resources. Building on this shift in perspective, subsequent chapters delve into technological advances that are shaping the new era of sustainable water management.

From membrane bioreactors and nanofiltration systems to microbial fuel cells, this book highlights a range of technologies that are revolutionizing wastewater treatment and resource recovery. Chapters on nutrient recovery and anaerobic digestion reveal the dual benefit of environmental conservation and energy generation, while the integration of microalgae as biofactories points towards synergistic strategies for renewable energy and wastewater remediation.

The volume also showcases real-world applications of smart recycling technologies and the role of advanced sensing tools in ensuring process efficiency and optimization. Technological integration—often the linchpin of successful implementation—is examined for its potential to create holistic and scalable wastewater management solutions.

However, the journey toward sustainable wastewater reuse is not without its challenges. Infrastructure gaps, policy limitations, and socio-economic concerns, particularly in the Indian context, are critically analysed. These discussions are complemented by a thorough evaluation of the economic and social impacts of innovative wastewater solutions, underscoring the necessity of inclusive and cost-effective strategies.

The final chapter projects into the future, surveying emerging technologies and anticipating potential breakthroughs that promise to elevate wastewater management to new heights.

As the global community moves towards circular economy models and climate resilience, *Resource Resurgence* aspires to be a catalyst for change—illuminating pathways that transform waste into wealth, and challenges into opportunities.

This book is intended for environmental scientists, researchers, policy-makers, industry professionals, and students who are passionate about sustainability and resource optimization. By bridging the gap between scientific research, practical implementation, and policy insight, *Resource Resurgence* hopes to inspire a new era of wastewater valorization that contributes meaningfully to water security, energy recovery, and environmental conservation.

We invite readers to explore these transformative ideas and technologies and to join the global movement towards smarter, cleaner, and more resource-efficient water management systems.

Mohali, India

Sunny Dhiman
Gunjan Mukherjee

Contents

About the Editors

Dr. Sunny Dhiman currently serves as Associate Professor at the Department of Biotechnology, University Institute of Biotechnology, Chandigarh University, India. He has been actively engaged in research, education, and scientific writing within his field for over 14 years. With extensive experience exceeding a decade in academic teaching and research, Dr. Dhiman has contributed significantly to various reputed international journals such as *Food Production, Processing and Nutrition, International Microbiology, Journal of Food Process Engineering, Biologia, Journal of Scientific and Industrial Research, Mycology, Fermentation, Discover Sustainability, Indian Journal of Microbiology* and *Research Journal of Biotechnology*, among others. His research pursuits encompass a diverse array of interests, including Microbial Fermentation, Waste Management, Bioremediation, and Waste Valorization. With a publication record comprising over 55 peer-reviewed articles including 18 book chapters, and two patents. Currently, Dr. Dhiman focuses on investigating microbial metabolic capabilities for waste valorization, Functional Food Development, purification, and optimization studies of microbial enzymes and pigments. Beyond his research endeavors, Dr. Dhiman actively participates as a guest editor and reviewer for several international journals. He has guided several graduate and post graduate dissertation projects and is currently guiding two doctorate students. He is a member of numerous scientific/professional societies including the Microbiologists Society of India, American Society of Microbiology, IAENG and IFERP.

Prof. (Dr.) Gunjan Mukherjee currently serves as a Professor and Director at University Institute of Biotechnology, Chandigarh University, India. With over two decades of international recognition in the field of microbial fermentation, bioprocess development, reactor design, and lignocellulosic biomass to generation of value-added chemicals and biofuel. Dr. Mukherjee embodies the epitome of expertise in industrial biotechnology. His professional trajectory took him to prestigious positions at leading organizations, including Senior Scientist roles at Praj Industries Limited, Pune, India, and TERI-New Delhi, India. His global reach extends to a visiting Professorship at Deakin University, Australia. A hallmark of his career is the culmination of international patents and numerous research publications in esteemed journals. Dr. Mukherjee actively serves as an Editor, Associate Editor, and honorary member of editorial boards for various reputed journals. His exemplary contributions have garnered him numerous honours and awards, such as the PARIS Award from Praj Industries Limited, Pune, India, and the Rising Star award from TERI, India.

Chapter 1
Waste Water as a Resource: Shifting Paradigms Towards Resource Recovery

Nishtha Paul, Narayan Singh, Jaswini Gedam, and Sunny Dhiman

Abstract With the increasing global population, particularly in urban areas, there is continuous pressure on natural water sources to meet rising demands. These sources are shrinking yearly due to global warming and erratic rainfall patterns, leading to physical water scarcity. Therefore, exploring new options for a sustainable water supply is crucial. Wastewater treatment presents a viable solution. Cities generate large volumes of wastewater from various household and industrial activities. As urban populations grow, the volume of wastewater increases, posing significant health risks. Many countries have been establishing wastewater treatment plants to repurpose treated water for various uses to address this. However, only a small percentage of wastewater is currently treated, with a larger portion remaining untreated and unused. Emerging paradigms now focus on resource recovery from wastewater, such as extracting nutrients, energy, and materials (metals or organic compounds) for industrial reuse. This shift in perspective transforms wastewater from being a burden into a valuable resource, contributing to sustainable water management. It also reduces economic expenses and alleviates pressure on other resources. While the processes involved in resource recovery are complex, they are achievable. In this chapter, we discuss the various types of resources that can be recovered from wastewater, the methodologies developed over time, the challenges faced, and potential solutions to overcome these challenges. The chapter also emphasizes the need for global policy frameworks to achieve the common goal of the safe and productive use of these resources.

N. Paul (✉) · J. Gedam
Department of Biosciences, University Institute of Biotechnology, Chandigarh University, Mohali 140413, Punjab, India
e-mail: p.nish8@gmail.com

N. Singh (✉)
Department of Botany, Sai School of Biosciences, Sri Sai University, Palampur, Himachal Pradesh, India
e-mail: narayan.thakur90@gmail.com

S. Dhiman
Department of Biotechnology, University Institute of Biotechnology, Chandigarh University, Mohali 140413, Punjab, India

S. Dhiman and G. Mukherjee (eds.), *Resource Resurgence*, Springer Water,
https://doi.org/10.1007/978-3-032-02750-4_1

Keywords Wastewater · Resource recovery · Sustainable water supply · Wastewater reuse · Water recycling · Nutrient recovery

1.1 Introduction

Water is widely used to sustain living forms, human health, and the prosperity of the economy. Although less than 1% of water resources is directly suitable for human use, it remains the planet's most abundant natural resource, covering 71% of its surface. However, human-induced issues such as urbanization, population growth, and rapid industrialization are putting increasing pressure on available resources (Kamali et al., 2019). The majority of water pollution is caused by human activities, including the release of chemicals and pharmaceuticals (Choudri et al., 2020) such as medicines, personal care products, detergents, pesticides, dyes, preservatives (Lin et al., 2022) and food additives, oil and plastic pollution (León et al., 2019), although some pollution can occur naturally.

Every day, we use water in our households and businesses for various purposes (Larsen et al., 2013). Lakes, rivers, and subterranean groundwater are the primary sources of the water we consume, and the majority of it returns to these areas after we have used and contaminated it. This leads to the eutrophication of contaminated water resources due to high levels of sulfate, nitrogen, and phosphate, as well as high amounts of chemical and biochemical oxygen demand values. Industrial effluent also causes acidification and soil contamination due to hydrocarbon emissions. Serious pollution occurs if not treated before being released into water bodies.

Wastewater needs to be treated before it is released into natural water bodies, however, ~80% of the wastewater remains untreated when released into these bodies, resulting in a reduction in water quality (Sonune & Ghate, 2004). About 70% of the wastewater produced in high-income countries is treated; in middle-income countries, this percentage drops to 38%, and in lower-middle-income countries, it decreases to 28% (Devda et al., 2021). Over the past century, water use has grown at a rate twice as fast as population growth. Due to the high population density in cities and growing urbanization, urban water resources are at risk. According to estimates, ~685 million people living in more than 570 cities will experience a 10% reduction in freshwater supplies by 2050 due to climate change (Devda et al., 2021). More than 600 million people already struggle with water shortages, which are exacerbated by declining groundwater levels and poor water quality. Since the discovery of additional water sources is unlikely, wastewater recycling can help address these shortages. The most effective and immediate solution to water shortages is wastewater treatment and reuse, even though efforts are being made to explore new water sources such as rainwater harvesting, seawater desalination, and dew collection (Sushmitha et al., 2019). To ensure equitable water distribution across all sectors, the Sustainable Development Goals (SDG 2030) Agenda emphasizes water security, water reuse, and recycling potential (Batool & Shahzad, 2021).

The term "wastewater" refers to "utilized water" and can be defined as a mixture of groundwater, surface water, stormwater, and liquid or water-carried waste removed from homes, businesses, and industrial facilities. It typically contains a high concentration of waste products that require oxygen, pathogenic (disease-causing) substances, organic materials, plant growth-promoting nutrients, inorganic compounds, minerals, and sediments. Additionally, it may include hazardous substances. Billions of tons of wastewater are generated worldwide each year. However, data on global wastewater generation remains limited, though organizations such as the Food and Agriculture Organization (FAO) and Global Water Intelligence provide valuable insights and recommendations (Batool & Shahzad, 2021). The majority of the key stages in a Wastewater Treatment Plant (WWTP)—including wastewater collection and transportation, physical and chemical treatment, biological treatment, sludge treatment, and discharge—require significant energy consumption. Various treatment technologies are being developed to effectively remove contaminants from wastewater.

1.2 Types of Waste Water

Wastewater is generated from various sources, and it can be categorized into the following types:

- **Domestic wastewater (Municipal wastewater)**

This type of wastewater originates from homes and residences, including water from toilets, sinks, washing machines, showers, dishwashers, and other household activities. It contains oils, soaps, detergents, food particles, and other organic matter. It primarily consists of:

- **Greywater**

Greywater refers to wastewater generated from domestic activities such as laundry, dishwashing, and bathing. It may contain food particles, detergents, and soap residues. In urban areas like Bangalore, approximately 810–1000 million litres per day (MLD) of greywater is produced, with an average per capita contribution of 70 litres per day (lpcd), accounting for about 67% of daily water consumption. On-site treatment using multistage processes (physical, chemical, and biological) enables its reuse for irrigation, vehicle washing, and toilet flushing (Sushmitha et al., 2019).

- **Blackwater**

Blackwater is wastewater mainly from toilets, containing human waste products such as urine and faeces. It is highly polluted and requires thorough treatment before being discharged or reused. In many rural areas worldwide, untreated blackwater is often released directly into the environment through pits, leading to soil and water contamination due to a lack of awareness about its consequences. Electrochemical-based treatment methods are among the most versatile and efficient options for on-site blackwater treatment. Under the Swachh Bharat Abhiyan, India has aimed to eliminate open defecation by constructing lavatories in rural areas and seeking awareness about hygiene and sanitation (Anusuyadevi et al., 2023).

- **Non-domesticated wastewater**

Non-domestic wastewater originates from sources other than municipal wastewater, including industrial facilities, agricultural fields, and stormwater runoff. It may contain chemicals, heavy metals, pesticides, and other pollutants, requiring specialized treatment before disposal or reuse. The main types of non-domestic wastewater include:

1. **Agricultural wastewater**
   Runoff water from agricultural fields contains herbicides, fertilizers, and pesticides, often carrying high levels of nutrients such as nitrogen and phosphorus. Additionally, this wastewater is sometimes reused in livestock farming (Kumar et al., 2023).
2. **Industrial wastewater**
   Factories, industries, and manufacturing processes generate industrial effluents, which mainly contain metals, oils, and other hazardous compounds. The composition of industrial wastewater varies significantly depending on the industry and the materials used. Examples include wastewater from mining, textile production, food processing, and chemical manufacturing (Singh et al., 2023).
3. **Rainwater runoff (stormwater)**
   Rainfall or snowmelt runoff collects contaminants such as chemicals, garbage, and oil as it flows over impermeable surfaces like parking lots and roadways. This runoff originates from various sources, including nurseries, greenhouses, urban areas, highways, airports, and agricultural fields (Rodak et al., 2020).
4. **Leachate**
   Leachate is water that percolates through landfill waste, collecting hazardous materials such as **metals, chemicals, and organic matter**. Its chemical composition varies **seasonally within a single landfill** and significantly **between different landfills**. Due to its high toxicity and resistance to biodegradation, landfill leachate requires **specialized treatment** to ensure safe disposal. Its low biodegradability leads to an **increased COD/BOD ratio**, posing environmental risks (Naveen et al., 2018) (Fig. 3.1).

1.3 Current Scenario of Waste Water Treatment

Wastewater treatment aims to safely dispose of industrial and human effluents without harming public health or the environment. While wastewater can be repurposed for irrigation, industries, and agriculture, it requires varying levels of treatment to remove contaminants. In certain treatment systems, further treatment is required to eliminate

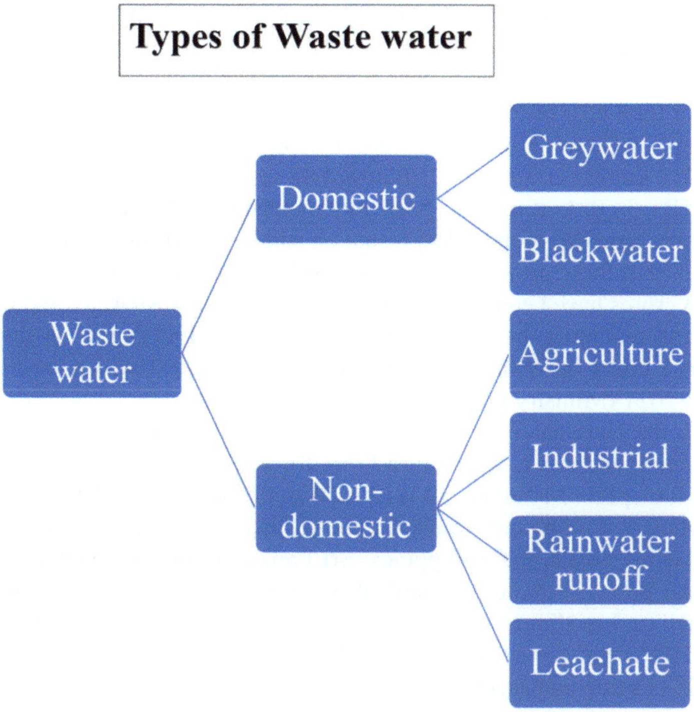

Fig. 3.1 Different forms of waste water generated from domestic and non-domestic sources

nutrients from the wastewater. Advanced wastewater treatment facilities make use of highly developed machinery and procedures (Dong et al., 2022).

In India, pollution control responsibilities are shared by the Ministry of Housing and Urban Affairs (MoHUA), the Ministry of Environment, Forest, and Climate Change (MoEF&CC), and the Ministry of Jal Shakti, with the Central Pollution Control Board (CPCB) overseeing regulatory compliance. The MoEF&CC issued the **Water Purification System (Regulation of Use) Rules, 2023**, providing guidelines for wastewater management. Government initiatives such as **Namami Gange** and the **National River Conservation Plan (NRCP)** support wastewater treatment and river conservation, with projects worth thousands of crores allocated for sewage treatment plants (STPs), sewer networks, and faecal sludge treatment plants (FSTPs). Under **AMRUT and Swachh Bharat Mission (Urban) 2.0**, significant funds have been directed toward wastewater management. Additionally, the **Water (Prevention & Control of Pollution) Act, 1974**, and the **Environment (Protection) Act, 1986,** mandate industries to install Effluent Treatment Plants (ETPs) to meet discharge standards before releasing wastewater into natural water bodies. The CPCB and State Pollution Control Boards (SPCBs) monitor compliance and take necessary action against violations, ensuring adherence to environmental regulations and sustainable water management.

With a project-approved cost of Rs. 5961.75 crore, 34 rivers in 77 towns throughout 16 States in the nation that are contaminated are cleaned up by NRCP. A sewage treatment capacity of 2677 million litres per day (m/d) has been established. At a total cost of Rs. 30,458 crore, 353 projects have been approved under the Namami Gange initiative, comprising 157 projects for the treatment of 4952 m/d and a 5212 km sewer network. Till now, 883 sewerage and septage management projects of Rs. 34,081 crores have been undertaken under the Ministry of Housing & Urban Affairs' Atal Mission for Rejuvenation and Urban Transformation (AMRUT) initiative; 370 of these projects, costing Rs. 8258 crores, have been finished thus far Rs. 15,883 crores have been given to States and UTs for wastewater and used water management, including the establishment of STPs and FSTPs (faecal sludge treatment plants), under the Swachh Bharat Mission (Urban) 2.0, which was introduced in October 2021 (Ministry of Jal Shakti, 2021).

In 2012, the National Water Policy (NWP) was released in India. It encourages water recycling and reuse, with return flows for demand management and efficient water use. Except for Myanmar, all WEPA (Water Environment Partnership in Asia) partner nations—Cambodia, China, Indonesia, Japan, the Republic of Korea, Lao PDR, Malaysia, Nepal, the Philippines, Sri Lanka, Thailand, and Vietnam—have laws in place and wastewater standards that industrial zones must adhere to.

For small and medium-sized enterprises (SMEs), Common Effluent Treatment Plants (CETPs) are considered one of the most practical wastewater treatment options. CETPs treat industrial effluents collectively, helping to reduce pollution and improve environmental quality. As per the Central Pollution Control Board report for 2011–12, a total of 193 CETPs have been established in various Indian states (https://cpcb.nic.in/status-of-cetps).

There are primarily two types of wastewater treatment and collection systems (Devda et al., 2021).

Off-site systems, where wastewater is collected in a septic tank and transported to a different location for treatment.

On-site systems, where wastewater is directly delivered to a treatment plant via a sewage network.

Only a small percentage of water pollutants, which are less harmful to the environment and public health, can be recovered as resources from wastewater. Apart from this, only a limited portion of wastewater is safely and systematically reused after resource recovery. India's current sewage treatment capacity of 11,786 million litres per day (MLD) can treat only 20% of the country's wastewater generation (Kaur et al., 2012). Similarly, while 60% of industrial wastewater is properly treated, the remaining 40%, especially from small-scale industries, is discharged into freshwater bodies without regulation by Common Effluent Treatment Plants (CETPs). In Delhi, 10 CETPs discharge effluents into the Yamuna River, and along the riverbank, 234 wastewater treatment plants (WWTPs) covering only 5% of the city have been established (Batool & Shahzad, 2021).

Earlier wastewater treatment methods primarily included activated sludge processes, secondary treatments, and biological treatments. Since its adoption as the global standard in 1930, activated sludge technology remains highly effective today. However, new technologies such as membrane bioreactors (MBRs) have evolved from the same principles, while other treatment techniques have also gained significance (Al-Asheh et al., 2021). Although treating discharged water remains challenging with existing methods, researchers continue to develop efficient wastewater treatment strategies (Al-Asheh et al., 2021).

Currently, wastewater treatment relies on a combined process of biological, chemical, and physical methods to ensure effective purification. Methods such as adsorption, coagulation, membrane filtration, oxidation, and biological treatments are widely used for wastewater treatment (Crini & Lichtfouse, 2019).

1.4 Waste Water Treatment Methods and Resource Recovery

The following headings are used to categorize water treatment and recycling methods. Primary, secondary, and tertiary water treatment technologies.

1.4.1 Primary Water Treatment Technologies

Screening, filtration, centrifugation, sedimentation, coagulation, gravity separation, and flotation techniques are used for water treatment at the primary stage (Pang & Abdullah, 2013; Sravan et al., 2024). These techniques are often applied to highly contaminated water and are classified as physical methods due to their high efficiency, simplicity, recyclability, and flexibility in removing pollutants. Materials such as sticks, rust particles, organic debris, suspended solids, charcoal or ash, micro plastics, textile fibres, and grease are removed to prevent interference with subsequent processes. Another advantage of this approach is the reduced need for chemical treatment. Since the physical treatment method is independent of biological processes, it is often considered more reliable than other treatment methods (Sravan et al., 2024).

Primary treatment is a crucial step in wastewater management; however, relying solely on primary treatment has limitations. Studies have shown that it can remove approximately 25–50% of the incoming biochemical oxygen demand (BOD), 50–70% of total suspended solids (including 72–78% of microplastics and 92–93% of textile fibres), and 65% of oil and grease (Iyare et al., 2020). While colloidal and dissolved elements remain largely unaffected, some organic nitrogen, organic phosphorus, and heavy metals associated with solids are also eliminated during primary sedimentation (Burton, 2018). Secondary treatment methods are implemented to remove the unaffected substances.

1.4.2 Secondary Water Treatment Technologies

Biological pathways for the microbial removal of both soluble and insoluble contaminants are an essential part of secondary water treatment (Iyare et al., 2020). Wastewater undergoes secondary treatment after the completion of primary treatment. In this process, a reactor sustaining a high concentration of microorganisms facilitates water circulation. Microorganisms, primarily bacterial and fungal strains, break down organic matter, producing ammonia gas, carbon dioxide, and water. In some cases, organic materials are transformed into substances such as nitrate, glucose, and alcohol, thereby reducing the biochemical oxygen demand (BOD) of the effluents (Zhou et al., 2023). Additionally, microorganisms help detoxify harmful inorganic substances. Recent advancements in secondary treatment have enhanced the recovery of chemical energy as biogas (Ugrina & Milojković, 2024) and improved treatment efficiency by integrating biological methods with physical or chemical processes (Sravan et al., 2024).

Secondary treatment typically involves both aerobic and anaerobic digestion to effectively treat wastewater.

(a) **Aerobic digestion**

This process is commonly divided into suspended growth systems (such as the activated sludge process and aerated lagoons) and attached growth systems [such as trickling filters and rotary biological contactors (RBCs)], both of which require oxygen for the breakdown of organic matter (Shen et al., 2019; Sravan et al., 2024). This method is effective in eliminating the dissolved and suspended organics, volatile organics, nitrates, phosphorus, biochemical oxygen demand (BOD), and chemical oxygen demand (COD). It can reduce the amount of biodegradable organics by up to 90%. However, one major drawback is the significant production of biosolids, which require costly management and treatment. Microbial molecular ecology methods have been widely used in wastewater treatment to study flocs and biofilms that develop in aerobic treatment systems, such as trickling filters (Sanz & Köchling, 2007). Further advancements in aerobic digestion include:

- Thermophilic Aerobic Digestion (TAD): This process operates at high temperatures, effectively inactivating pathogens through the rapid breakdown of organic matter. TAD is a promising pretreatment method for anaerobic digestion of organic waste, as studies have demonstrated its effectiveness in accelerating the decomposition of sewage sludge (Chun et al., 2024).
- Autothermal Thermophilic Aerobic Digestion (ATAD): An advanced form of TAD, ATAD operates at temperatures between 45 °C and 70 °C, ensuring complete pathogen destruction.
- Post-Aerobic Digestion (PAD): This method acts as a solid-side stream nutrient removal process. Research has shown that this step-by-step approach enhances nutrient removal and stabilizes biosolids, thereby improving municipal sewage sludge treatment performance (McNamara et al., 2022).

- Oxygen-Induced Thiosulfate Production: This method has demonstrated high thiosulfate production rates and low sludge yields, contributing to the development of secondary wastewater treatment systems that are more energy- and space-efficient (Jiang et al., 2023).

(b) **Anaerobic digestion**

In anaerobic treatment, the absence of oxygen allows usage of anaerobic digesters and upflow anaerobic sludge blanket (UASB) reactors. Complex organic matter is broken down into simpler organic molecules containing nitrogen, carbon, and sulphur by facultative and anaerobic bacteria. Methane, hydrogen sulphides, ammonia, and nitrogen are the key gases produced during this process, particularly from high-strength wastewater or sludge digestion (Abdelsalam et al., 2017). This technique significantly reduces the biological load of wastewater. However, one drawback of anaerobic digestion is the production of sludge with a strong odor, which requires additional treatment (Sravan et al., 2024).

Recent advancements in anaerobic digestion include:

- Combined Treatment Methods: These involve chemical, mechanical, and thermal pretreatment techniques to reduce volatile suspended solids and enhance methane yield (Yang et al., 2024).
- Microbial Ecology Optimization: This approach involves intentional microbial manipulation to improve biogas production, organic waste degradation, and process stability in anaerobic digesters.
- Hybrid Anaerobic Treatment Systems: These systems integrate anaerobic baffled reactors (ABRs) with expanded granular sludge bed (EGSB) reactors and micro-screens to enhance efficiency. The goal is to maximize methane production and energy recovery while maintaining operational simplicity.
- Full-Scale Implementation and Case Studies: Large-scale demonstrations of anaerobic digestion in wastewater treatment plants have provided valuable insights into its benefits and practical applications (Krohn et al., 2022).

Biological techniques are promising to recover valuable materials from diluted wastewater streams, even as emerging technologies make resource recovery more complex. Examples include using highly specialized metal-reducing and oxidizing microorganisms for metal recovery and applying rapidly growing heterotrophic, chemotrophic, phototrophic, and photosynthetic bacteria, as well as microalgae and terrestrial plants, for organic recovery. Additionally, biopolymers such as polyhydroxyalkanoates (PHAs) and alginates, produced by accumulating bacteria, can be recovered by organisms capable of absorbing complex organics (Puyol et al., 2017).

1.4.3 *Tertiary Water Treatment Technologies*

Tertiary treatment methods remove specific inorganic and organic pollutants that secondary treatment cannot, ensuring compliance with discharge regulations or reuse standards. As the final step in water purification, it is an advanced treatment process. Various methods are employed, including distillation, crystallization, evaporation, solvent extraction, oxidation, coagulation, precipitation, electrolysis, electrodialysis, ion exchange, reverse osmosis, and adsorption. The selection of an appropriate tertiary treatment system depends on the type of contamination present in the water (Gupta et al., 2012; Iyare et al., 2020; Sonune & Ghate, 2004).

Recent advancements in tertiary treatment include:

- Hybrid Electrochemical-Granular Activated Carbon Systems: These systems effectively eliminate persistent organic pollutants by combining electro-sorption and electrooxidation (Norra & Radjenovic, 2021).
- Nano-Photocatalysts in Advanced Oxidation Processes: These catalysts aid in the degradation of recalcitrant organic compounds by generating reactive oxygen species (ROS) (Katiyar et al., 2021).
- Integrated Wastewater Treatment Approaches: These approaches combine physical, chemical, and biological methods to enhance the efficiency and sustainability of wastewater treatment (Kumar et al., 2022).

Understanding the impact of various wastewater treatment infrastructure options on sustainability is essential. Decentralized systems have long been recognized as a more effective and sustainable alternative for wastewater treatment. One key advantage is the potential for resource separation within wastewater streams, allowing for the recovery of valuable resources such as metals, chemicals, fuels, nutrients, and energy (Vymazal, 2009).

Wastewater treatment processes not only eliminate contaminants from water but also provide opportunities for energy generation and the recovery of valuable byproducts, benefiting both the environment and the economy. Given the urgent environmental challenges being faced by the nation, the substantial potential for energy recovery from wastewater should be a key consideration when adopting new technologies (Ortiz-Sanchez & Cuevas-Rodriguez, 2023).

(a) **Energy**

There are several methods to harness solar energy, including biomass, photovoltaics, solar thermal, wind, ocean currents, and falling water (such as water wheels and hydroelectric turbines). Energy generation from wastewater falls under the category of biomass (Larsen et al., 2013). Although capturing energy from wastewater can provide significant benefits to municipalities and utilities—reducing greenhouse gas emissions and improving financial sustainability—the amount of energy contained in wastewater is insufficient to fully meet the demands of modern human society. Compared to fossil fuel energy generation and other alternative energy sources, utilizing wastewater as an energy source can help mitigate environmental problems and fulfill a portion of energy needs (Ortiz-Sanchez & Cuevas-Rodriguez, 2023).

During the electrolysis process, electrolyte temperature increases due to interfacial Joule heating. To evaluate the potential for heat recovery, an electrochemical reactor was designed and manufactured as part of an effluent remediation study (Pei et al., 2020). A study of a wastewater treatment plant (WWTP) in China found that at least 50% of the plant's energy consumption could be recovered through wastewater thermal energy, effectively compensating for energy sources such as hydroelectric, solar, and wind power. Additionally, hydroelectric power can be generated at WWTPs where elevation changes allow for it.

Solar energy can also be used in wastewater remediation through photocatalytic processes. For instance, ZnO has been employed as a photocatalyst to remove hazardous contaminants like methyl styrene from wastewater (Gu et al., 2016). In Mexico, several large-scale projects have successfully integrated wastewater treatment with renewable energy generation. Developing technologies for energy recovery from wastewater is crucial for countries facing both sanitation and energy supply challenges (Ortiz-Sanchez & Cuevas-Rodriguez, 2023).

Recent advancements in energy production from wastewater include microbial fuel cells (MFCs), bio-electrochemical treatment technology (BETT), and the generation of hydrogen gas and biogases, all of which present promising avenues for sustainable energy recovery.

(b) **Biofuels**

There has been a long history of converting wastewater streams rich in organic matter into bioenergy, particularly through anaerobic digestion (AD). Although various emerging technologies—such as biohydrogen, biodiesel, bioethanol, and microbial fuel cells—are being developed to convert organic matter into bioenergy, their current viability remains significantly lower than that of anaerobic systems (Puyol et al., 2017).

In addition to biofuel production, microalgae such as Chlorophyceae and *Chlorella vulgaris* are also utilized to produce vitamins, proteins, and carbohydrates. Cultivating algae in wastewater offers two key advantages: it effectively removes contaminants while serving as a feedstock for high-value bioproducts like biofuels. In recent years, the concept of an algal biorefinery has gained significant attention as a sustainable method for ethanol production.

An algal biorefinery integrates multiple processes, including wastewater treatment, biomass cultivation, biomass harvesting, bioproduct extraction, and biofuel conversion. High-rate algal ponds (HRAPs) are widely used for cultivating microalgae, which play a dual role in wastewater treatment and biofuel production. These systems have high energy input requirements but can yield non-food bioproducts with significant commercial value (Arashiro et al., 2018).

A study by Lee et al. (2023) demonstrated that combining industrial and domestic wastewater treatment processes led to enhanced biomass productivity and nutrient removal efficiency, resulting in higher biodiesel production.

(c) **Biopolymers**

Researchers have developed various methods to isolate and remove natural polymers from biomass generated in wastewater treatment facilities. These facilities have the potential to recover valuable materials using microorganisms, particularly extracellular polymeric substances (EPS), polyhydroxyalkanoates (PHAs), and polysaccharides. EPS is a high-molecular-weight biopolymer that plays a crucial role in biofilm formation, contaminant removal through the production of phycobiliproteins, soil water retention enhancement, and functioning as a biofertilizer (Cydzik-Kwiatkowska et al., 2024). PHAs, biodegradable polyesters stored in microalgal cells, offer a promising, environmentally friendly alternative to conventional plastics (Arashiro et al., 2020).

Polyhydroxyalkanoates (PHAs) are primarily synthesized from organic substrates, serving as biodegradable alternatives to conventional plastics like polyethylene and polypropylene. The most commonly used setup for PHA production follows a three-step technique involving various reactor designs and microorganisms. In the first step, the waste stream undergoes pre-fermentation, where hydrolytic and fermentative bacteria break down complex organic matter into easily biodegradable substances, such as volatile fatty acids (VFAs). The second step, enrichment, typically utilizes activated sludge as the predominant seed biomass. The third and final step involves PHA accumulation in batch systems, where the biopolymer content of the enriched microbial community is further enhanced. The composition of the biopolymer is determined by the distribution of volatile fatty acids (VFAs) during the accumulation phase (Laycock et al., 2013).

(d) **Metals**

Wastewater containing valuable metals is continuously released into the environment by various industries. Due to their remarkable physicochemical properties, these metals have a wide range of industrial applications. However, the scarcity and high market value of these metals have significantly impacted industrial operations, making their recovery from wastewater an attractive alternative to conventional extraction from natural resources (Gebreslassie et al., 2024). Depending on the chemical properties of the metals and the specific application requirements, recovery methods include electrochemical technology, adsorption, electrocoagulation, solvent extraction, leaching, ionic liquids, ion exchange resin, and precipitation. However, these techniques often require large amounts of chemicals and generate additional waste byproducts, necessitating further treatment. To address these challenges, metal recovery technologies that are highly selective and chemical-free are needed.

Membrane technology has emerged as the most efficient method for recovering metals from wastewater streams due to its numerous advantages, including high separation efficiency, low carbon footprint, ease of operation, minimal phase change, low energy consumption, compact space requirements, reduced chemical usage, and minimal solid waste generation (Gebreslassie et al., 2024). Specific metal recovery methods include oscillating electrodes for extracting heavy metals such as copper, neodymium, uranium, and tellurium, while microbial fuel cells (MFCs) and bio electrochemical technologies are used to recover nickel, zinc, ammonium, and cadmium. Additionally, sulphate-reducing bacteria facilitate the recovery of metal sulphides, while electrodialysis is used to recover salts such as potassium nitrate, sodium carbonate, and sodium sulphate (Puyol et al., 2017). Despite advancements in metal recovery techniques, high-value metals such as silver, gold, rhenium, platinum, ruthenium, palladium, rhodium, osmium, and iridium remain underexplored in wastewater recovery research, with only a few studies addressing their extraction (Gebreslassie et al., 2024).

(e) **Nutrients**

One of the main challenges in wastewater treatment is developing methods that efficiently recover essential nutrients while minimizing environmental harm. Enhancing nutrient recovery can be achieved by developing energy-efficient, cost-effective, portable, and multifunctional integrated systems (Babcock-Jackson et al., 2023). Phosphorus and nitrogen are crucial fertilizers for agriculture. However, nitrogenous fertilizer production requires high energy input, and global phosphorus reserves are depleting rapidly, with some estimates suggesting they may become critically scarce within the next decade. As a result, the recovery of nitrogen and phosphorus from wastewater is gaining significant attention.

Commercially, phosphorus recovery is conducted through processes such as AirPrex, PEARL, AshDec, and RecoPhos, primarily in the form of struvite. Studies have demonstrated that more than 94% of nitrogen can be recovered using a combination of microfiltration, ultrafiltration, and nanofiltration membranes (Babcock-Jackson et al., 2023). Additional recovery methods include electrodialysis, struvite precipitation, chemical precipitation, bio-electrochemical approaches, ion exchange, and gas-permeable membranes (Perera et al., 2019). Furthermore, the High-Rate Algal Pond (HRAP) system utilizes microalgae to extract nutrients from wastewater, offering a sustainable fertilizer alternative (Arashiro et al., 2018). With growing interest in membrane-based nutrient recovery technologies—such as forward osmosis, reverse osmosis, membrane distillation, and electrodialysis—these approaches have proven effective in extracting and concentrating essential nutrients, which can then be reused in industrial and agricultural applications (Saliu et al., 2024).

(f) **Hydrogen gas**

Hydrogen is increasingly being utilized as a clean fuel in fuel cells for both stationary and mobile applications. A 130L microbial electrolysis cell (MEC) pilot plant has been designed, constructed, and operated to produce hydrogen from urban wastewater, demonstrating the potential of this technology for sustainable energy generation. Microbial fuel cells (MFCs) and anaerobic hydrogen production processes optimize energy extraction from wastewater, enabling the simultaneous generation of both electricity and hydrogen gas (Govindarajan, 2018). Among the most environmentally friendly methods for hydrogen production are biological processes, which can be classified as autotrophic (e.g., bio-photolysis) and heterotrophic (e.g., photo-fermentation and dark fermentation) (Dincer & Acar, 2015). A comprehensive study conducted at the AI Ansab Sewage Treatment Plant (STP) in Muscat, Oman, thoroughly assessed the technical and financial viability of hydrogen production using treated wastewater for defined applications (Barghash et al., 2022).

Recent advancements include the Falling-Film Oscillating Water Splitting Assisted by Wastewater Electrolysis (FOWSAWE) system, which integrates forward osmosis (for water extraction from wastewater) with alkaline water electrolysis (for splitting water to produce H_2 and O_2). This method has demonstrated a 14-fold increase in hydrogen production compared to previous techniques, achieving 99% purity and significantly improving energy efficiency, with an energy consumption of less than 3.96 kWh Nm^{-3} (Cassol et al., 2024).

1.5 Challenges and Future Trends

The lifespan of existing wastewater treatment plants (WWTPs) presents a significant obstacle to adopting new technologies. Since most WWTPs have been operational for decades and continue to function effectively, replacing or upgrading them requires a complex business case. As a result, many facilities opt for minor modifications rather than implementing major technological changes. To achieve net-zero emissions and improve carbon neutrality, recent technological advancements focus on process intensification, resource recovery, and improved engineering systems to enhance effluent quality. Innovative wastewater treatment methods, such as the Cellvation and Nereda processes, aim to reduce environmental impacts while improving treatment efficiency (Soares, 2020).

- The Cellvation process extracts cellulose from wastewater using 350 μm Rotating Belt Filters (RBFs). This process has been tested at Geestmerambacht WWTP (Netherlands), Breivika WWTP (Norway), and California WWTP (USA). However, while cellulose is successfully recovered from sewage, the challenge lies in managing the extracted cellulose fibers (Marcelis & Wessels, 2019; Wiśniowska & Kowalczyk, 2022).
- The Nereda process utilizes aerobic granular sludge to purify wastewater, offering an advanced alternative to conventional treatment methods.

Developing energy-neutral WWTPs faces technological and environmental challenges e.g. Anammox technology, introduced in 1977, is still undergoing large-scale testing, as challenges vary based on the size and capacity of WWTPs. Low-load WWTPs struggle to achieve full energy independence due to limited organic matter available for energy recovery. High initial costs hinder the widespread adoption of energy-self-sufficient WWTPs, particularly when integrating combined heat and power (CHP) systems and photovoltaics, which require significant investment.

Furthermore, WWTPs must also address environmental risks associated with wastewater treatment, such as:

- Methane (CH_4) and nitrous oxide (N_2O) emissions, which contribute to global warming.
- Incomplete anaerobic treatment, which can negatively impact surrounding ecosystems.

The realization that conventional nitrogen removal methods (nitrification, denitrification, nitritation, de-ammonification, and biological nutrient removal) may not be sustainable due to N_2O emissions and the energy-intensive aeration process. Future advancements must focus on optimizing nitrogen management to ensure sustainability (Soares, 2020).

Besides, a more sustainable wastewater management system requires:

- Policy support to drive innovation and regulatory compliance.
- Stakeholder engagement from governments, industries, and local communities.
- Technological advancements that balance environmental benefits with economic feasibility.
- Sustainable practices that minimize energy consumption and waste production (Singh et al., 2023).

In rapidly urbanizing regions such as India, the construction of large-scale sewage networks and pumping stations presents additional challenges. Conventional wastewater systems require high capital, operational, and maintenance costs while offering minimal potential for energy recovery. Economic and energy-based analyses suggest that sustaining such systems is becoming increasingly difficult for fast-growing Indian cities (Masi et al., 2016).

Thus, the future of wastewater treatment lies in integrated, resource-efficient, and low-energy systems that align with the principles of circular economy and environmental sustainability.

1.6 Conclusion

Providing expanding cities worldwide with an adequate supply of high-quality water has always been a challenging task. The situation in India is no different, where rapid urbanization and population growth have placed immense pressure on natural water bodies. As a result, many municipal authorities have enacted legislation mandating the recycling and reuse of wastewater (Sushmitha et al., 2019; WHO Guidelines for Drinking-Water Quality, 2024). The concepts of wastewater treatment and reuse have become crucial strategies for ensuring a sustainable water supply. Additionally, new wastewater discharge regulations and the emergence of new pollutants have driven significant transformations in the wastewater industry. With the increasing demand for water in domestic and industrial sectors, the volume of wastewater discharged into sewage systems has also risen (Mojiri & Bashir, 2022).

To address water contamination, several physicochemical treatment techniques have been developed, including activated sludge, phytoremediation, bioremediation, anammox and hybrid methods. Despite these advancements, a universally effective treatment approach capable of eliminating all types of pollutants remains a distant goal. Continued research and technological innovations are necessary to develop more efficient, sustainable, and cost-effective wastewater treatment solutions.

References

Abdelsalam, E., Samer, M., Attia, Y. A., Abdel-Hadi, M. A., Hassan, H. E., & Badr, Y. (2017). Influence of zero valent iron nanoparticles and magnetic iron oxide nanoparticles on biogas and methane production from anaerobic digestion of manure. *Energy, 120*, 842–853. Available at: https://doi.org/10.1016/j.energy.2016.11.137

Al-Asheh, S., Bagheri, M., & Aidan, A. (2021). Membrane bioreactor for wastewater treatment: A review. *Case Studies in Chemical and Environmental Engineering, 4*, 100109. https://doi.org/10.1016/j.cscee.2021.100109

Anusuyadevi, P. R., Kumar, D. J. P., Jyothi, A. O., Patwardhan, N. S., Janani, V., & Mol, A. (2023). Towards viable eco-friendly local treatment of blackwater in sparsely populated regions. *Water, 15*(3), 542. Available at: https://doi.org/10.3390/w15030542

Arashiro, L. T., Boto-Ordóñez, M., Van Hulle, S. W., Ferrer, I., Garfí, M., & Rousseau, D. P. (2020). Natural pigments from microalgae grown in industrial wastewater. *Bioresource Technology, 303*, 122894. Available at: https://doi.org/10.1016/j.biortech.2020.122894

Arashiro, L. T., Montero, N., Ferrer, I., Acién, F. G., Gómez, C., & Garfí, M. (2018). Life cycle assessment of high rate algal ponds for wastewater treatment and resource recovery. *Science of the Total Environment, 622*, 1118–1130. Available at: https://doi.org/10.1016/j.scitotenv.2017.12.051

Babcock-Jackson, L., Konovalova, T., Krogman, J. P., Bird, R., & Díaz, L. L. (2023). Sustainable fertilizers: Publication landscape on wastes as nutrient sources, wastewater treatment processes for nutrient recovery, biorefineries, and green ammonia synthesis. *Journal of Agricultural and Food Chemistry, 71*(22), 8265–8296. https://doi.org/10.1021/acs.jafc.3c00454

Batool, M., & Shahzad, L. (2021). An analytical study on municipal wastewater to energy generation, current trends, and prospects in South Asian developing countries (an update on Pakistan scenario). *Environmental Science and Pollution Research, 28*(25), 32075–32094. Available at: https://doi.org/10.1007/s11356-021-14029-8

Burton, G. A. (2018). Sediment collection and processing: Factors affecting realism. In *Sediment toxicity assessment* (pp. 37–66). CRC Press.

Choudri, B. S., Al-Awadhi, T., Charabi, Y., & Al-Nasiri, N. (2020). Wastewater treatment, reuse, and disposal-associated effects on environment and health. *Water Environment Research, 92*(10), 1595–1602. Available at: https://doi.org/10.1002/wer.1406

Chun, J., Kim, S. M., Ko, G., Shin, H. J., Kim, M., & Cho, H. U. (2024). Thermophilic aerobic digestion using aquaculture sludge from rainbow trout aquaculture facilities: Effect of salinity. *Frontiers in Microbiology, 15*, 1488041. Available at: https://doi.org/10.3389/fmicb.2024.1488041

Crini, G., & Lichtfouse, E. (2019). Advantages and disadvantages of techniques used for wastewater treatment. *Environmental Chemistry Letters, 17*(1), 145–155. Available at: https://doi.org/10.1007/s10311-018-0785-9

Cydzik-Kwiatkowska, A., Ciesielski, S., Florczyk, M., Pasieczna-Patkowska, S., Komorowska-Kaufman, M., Pomian, W., Jóźwiak, K., & Oleskowicz-Popiel, P. (2024). Biopolymer production in a full-scale activated sludge wastewater treatment plant: Seasonal changes and promising bacterial producers. *Energies, 17*(24), 6231.

Devda, V., et al. (2021). Recovery of resources from industrial wastewater employing electrochemical technologies: Status, advancements and perspectives. *Bioengineered, 12*(1), 4697–4718. Available at: https://doi.org/10.1080/21655979.2021.1946631

Dincer, I., & Acar, C. (2015). Review and evaluation of hydrogen production methods for better sustainability. *International Journal of Hydrogen Energy, 40*(34), 11094–11111. https://doi.org/10.1016/j.ijhydene.2014.12.035

Dong, G., Chen, B., Liu, B., Hounjet, L. J., Cao, Y., Stoyanov, S. R., Yang, M., & Zhang, B. (2022). Advanced oxidation processes in microreactors for water and wastewater treatment: Development, challenges, and opportunities. *Water Research, 211*, 118047. https://doi.org/10.1016/j.watres.2022.118047

Gebreslassie, G., Desta, H. G., Dong, Y., Zheng, X., Zhao, M., & Lin, B. (2024). Advanced membrane-based high-value metal recovery from wastewater. *Water Research, 122122.* Available at: https://doi.org/10.1016/j.watres.2024.122122

Govindarajan, V. (2018). Recovery of different types of resources from wastewater–A structured review. *Vatten, 74*(1–2), 65–82.

Gu, Y., Dong, Y. N., Wang, H., Keller, A., Xu, J., Chiramba, T., & Li, F. (2016). Quantification of the water, energy, and carbon footprints of wastewater treatment plants in China considering a water–energy nexus perspective. *Ecological Indicators, 60*, 402–409. Available at: https://doi.org/10.1016/j.ecolind.2015.07.012

Gupta, V. K., Ali, I., Saleh, T. A., Nayak, A., & Agarwal, S. (2012). Chemical treatment technologies for waste-water recycling—an overview. *Rsc Advances, 2*(16), 6380–6388. Available at: https://doi.org/10.1039/C2RA20340E

http://www.indiaenvironmentportal.org.in/content/476091/water-purification-system-regulation-of-use-rules-2023. Accessed December 29, 2024.

https://www.niti.gov.in/sites/default/files/2022-09/Waste-Water-A4_20092022.pdf. Accessed January 6, 2025.

Iyare, P. U., Ouki, S. K., & Bond, T. (2020). Microplastics removal in wastewater treatment plants: A critical review. *Environmental Science: Water Research & Technology, 6*(10), 2664–2675. Available at: https://doi.org/10.1039/D0EW00397B

Jiang, J., Hou, R., Cui, H., Liu, D., Yan, G., Fan, Y., & Cao, Z. (2023). Occurrences of typical PPCPs during wastewater treatment and the composting of sewage sludge with micron-sized and nano-sized Fe_3O_4. *Environmental Pollution, 336*, 122386. Available at: https://doi.org/10.1016/j.envpol.2023.122386

Kamali, M., Persson, K. M., Costa, M. E., & Capela, I. (2019). Sustainability criteria for assessing nanotechnology applicability in industrial wastewater treatment: current status and future outlook. *Environment International, 125*, 261–276. Available at: https://doi.org/10.1016/j.envint.2019.01.055

Katiyar, J., Bargole, S., George, S., Bhoi, R., & Saharan, V. K. (2021). Advanced technologies for wastewater treatment: New trends. In *Handbook of nanomaterials for wastewater treatment* (pp. 85–133). Elsevier. Available at. https://doi.org/10.1016/B978-0-12-821496-1.00011-8

Kaur, R., Wani, S. P., Singh, A. K., & Lal, K. (2012). Wastewater production, treatment and use in India. In *National Report presented at the 2nd regional workshop on safe use of wastewater in agriculture*, 1–13.

Krohn, C., Khudur, L., Dias, D. A., Van den Akker, B., Rees, C. A., Crosbie, N. D., Surapaneni, A., O'Carroll, D. M., Stuetz, R. M., Batstone, D. J., & Ball, A. S. (2022). The role of microbial ecology in improving the performance of anaerobic digestion of sewage sludge. *Frontiers in Microbiology, 13*, 1079136. Available at: https://doi.org/10.3389/fmicb.2022.1079136

Kumar, M., Ambika, S., Hassani, A., & Nidheesh, P. V. (2023). Waste to catalyst: Role of agricultural waste in water and wastewater treatment. *Science of the Total Environment, 858*, 159762. Available at: https://doi.org/10.1016/j.scitotenv.2022.159762

Kumar, V., Bilal, M., & Ferreira, L. F. R. (2022). Recent Trends in integrated wastewater treatment for sustainable development. *Frontiers in Microbiology, 13*, 846503. Available at: https://doi.org/10.3389/fmicb.2022.846503

Larsen, T., Udert, K., & Lienert, J. (2013). *Source separation and decentralization for wastewater management*. Iwa Publishing. at: https://library.oapen.org/handle/20.500.12657/24364

Laycock, B., Halley, P., Pratt, S., Werker, A., & Lant, P. (2013). The chemomechanical properties of microbial polyhydroxyalkanoates. *Progress in Polymer Science, 38*(3–4), 536–583. Available at: https://doi.org/10.1016/j.progpolymsci.2012.06.003

Lee, S. Y., Lee, J. S., & Sim, S. J. (2023). Enhancement of microalgal biomass productivity through mixotrophic culture process utilizing waste soy sauce and industrial flue gas. *Bioresource Technology, 373*, 128719.

León, V. M., García-Agüera, I., Moltó, V., Fernández-González, V., Llorca-Pérez, L., Andrade, J. M., Muniategui-Lorenzo, S., & Campillo, J. A. (2019). PAHs, pesticides, personal care products and plastic additives in plastic debris from Spanish Mediterranean beaches. *Science of the Total Environment, 670*, 672–684. Available at: https://doi.org/10.1016/j.scitotenv.2019.03.216

Lin, L., Yang, H., & Xu, X. (2022). Effects of water pollution on human health and disease heterogeneity: A review. *Frontiers in Environmental Science, 10*, 880246. Available at: https://doi.org/10.3389/fenvs.2022.880246

Marcelis, P., & Wessels, C. (2019). Recovery and valorization of cellulose from waste water. Available at: https://www.smart-plant.eu/images/publications/circular-economy/30_2019-04-19_Recovery%20and%20valorisation%20of%20cellulose.pdf. Accessed January 6, 2025.

Masi, F., Bresciani, R., Rizzo, A., Edathoot, A., Patwardhan, N., Panse, D., & Langergraber, G. (2016). Green walls for greywater treatment and recycling in dense urban areas: A case-study in Pune. *Journal of Water, Sanitation and Hygiene for Development, 6*(2), 342–347. Available at: https://doi.org/10.2166/washdev.2016.019

McNamara, P., Sabba, F., Redmond, E., Dunlap, P., Worley-Morse, T., Marks, C., & Downing, L. (2022). Post-aerobic digestion (PAD) is a solids side-stream nutrient removal process that utilizes native carbon: Performance and key operational parameters from two full-scale PAD reactors. *Environmental Science Advances, 1*(2), 216–228. Available at: https://doi.org/10.1039/D2VA00045H

Mojiri, A., & Bashir, M. J. (2022). Wastewater treatment: Current and future techniques. *Water, 14*(3), 448.

Naveen, B. P., Sumalatha, J., & Malik, R. K. (2018). A study on contamination of ground and surface water bodies by leachate leakage from a landfill in Bangalore, India. *International Journal of Geo-Engineering, 9*(1), 27. https://doi.org/10.1186/s40703-018-0095-x

Norra, G. F., & Radjenovic, J. (2021). Removal of persistent organic contaminants from wastewater using a hybrid electrochemical-granular activated carbon (GAC) system. *Journal of Hazardous Materials, 415*, 125557. Available at: https://doi.org/10.1016/j.jhazmat.2021.125557

Ortiz-Sanchez, M. F., & Cuevas-Rodriguez, G. (2023). Energy recovery from wastewater in Mexico: A systematic review. *Frontiers in Environmental Science, 11*, 1116053. Available at: https://doi.org/10.3389/fenvs.2023.1116053

Pang, Y. L., & Abdullah, A. Z. (2013). Current status of textile industry wastewater management and research progress in Malaysia: A review. *CLEAN—Soil, Air Water, 41*(8), 751–764. Available at: https://doi.org/10.1002/clen.201000318

Pei, S., Teng, J., Ren, N., & You, S. (2020). Low-temperature removal of refractory organic pollutants by electrochemical oxidation: Role of interfacial joule heating effect. *Environmental Science & Technology, 54*(7), 4573–4582. Available at: https://doi.org/10.1021/acs.est.9b05929

Perera, M. K., Englehardt, J. D., & Dvorak, A. C. (2019). Technologies for recovering nutrients from wastewater: A critical review. *Environmental Engineering Science, 36*(5), 511–529. Available at: https://doi.org/10.1089/ees.2018.0436

Puyol, D., Batstone, D. J., Hülsen, T., Astals, S., Peces, M., & Krömer, J. O. (2017). Resource recovery from wastewater by biological technologies: Opportunities, challenges, and prospects. *Frontiers in Microbiology, 7*, 2106. Available at: https://doi.org/10.3389/fmicb.2016.02106

Rana, V., Satavan, R., & Vidyarthi, A. K. (2025) Industrial wastewater treatment by common effluent treatment plants located on the ganga river and its tributaries. In *Emerging trends and technologies in water management and conservation*, pp. 397–418. IGI Global Scientific Publishing.

Available at: https://www.igi-global.com/chapter/industrial-wastewater-treatment-by-common-effluent-treatment-plants-located-on-the-ganga-river-and-its-tributaries/364487.

Rodak, C. M., Jayakaran, A. D., Moore, T. L., David, R., Rhodes, E. R., & Vogel, J. R. (2020). Urban stormwater characterization, control, and treatment. *Water Environment Research, 92*(10), 1552–1586. https://doi.org/10.1002/wer.1403

Saliu, T. D., Oladoja, N. A., & Sauvé, S. (2024). Eco-friendly and sustainability assessment of technologies for nutrient recovery from human urine—a review. *Frontiers in Sustainability, 4*, 1338380.

Sanz, J. L., & Köchling, T. (2007). Molecular biology techniques used in wastewater treatment: an overview. *Process Biochemistry, 42*(2), 119–133. Available at: https://doi.org/10.1016/j.procbio.2006.10.003

Shen, C., Zhao, Y. Q., Liu, R. B., Morgan, D., & Wei, T. (2019). Enhancing wastewater remediation by drinking water treatment residual-augmented floating treatment wetlands. *Science of the Total Environment, 673*, 230–236. Available at: https://doi.org/10.1016/j.scitotenv.2019.04.035

Singh, B. J., Chakraborty, A., & Sehgal, R. (2023). A systematic review of industrial wastewater management: Evaluating challenges and enablers. *Journal of Environmental Management, 348*, 119230. https://doi.org/10.1016/j.jenvman.2023.119230

Soares, A. (2020). Wastewater treatment in 2050: Challenges ahead and future vision in a European context. *Environmental Science and Ecotechnology, 2*, 100030. Available at: https://doi.org/10.1016/j.ese.2020.100030

Sonune, A., & Ghate, R. (2004). Developments in wastewater treatment methods. *Desalination, 167*, 55–63. Available at: https://doi.org/10.1016/j.desal.2004.06.113

Sravan, J. S., Matsakas, L., & Sarkar, O. (2024). Advances in biological wastewater treatment processes: Focus on low-carbon energy and resource recovery in biorefinery context. *Bioengineering, 11*(3), 281. Available at: https://doi.org/10.3390/bioengineering11030281

Sushmitha, M. B., Chanakya, H. N., & Khuntia, H. K. (2019). Efficient grey water treatment and reuse options for India—A review. In S. K. Ghosh (Ed.), *Waste water recycling and management* (pp. 143–149. Available at:). Springer Singapore. https://doi.org/10.1007/978-981-13-2619-6_12

Ugrina, M., & Milojković, J. (2024). Advances in wastewater treatment. *Energies, 17*(6), 1400. Available at: https://doi.org/10.3390/en17061400.

Vymazal, J. (2009). The use of constructed wetlands with horizontal sub-surface flow for various types of wastewater. *Ecological Engineering, 35*(1), 1–17. Available at: https://doi.org/10.1016/j.ecoleng.2008.08.016

Wiśniowska, E., & Kowalczyk, M. (2022). Recovery of cellulose, extracellular polymeric substances and microplastics from sewage sludge: A review. *Energies, 15*(20), 7744. Available at: https://doi.org/10.3390/en15207744

Yang, J., Shi, L., Chen, X., Huang, X., Wang, X., Chen, S., Hua, Y., Gong, H., Dong, B., Liu, H., & Dai, X. (2024). Systematic review on the residual chemicals in wastewater treatment sludge: Specifically focusing on the occurrence state and anaerobic bioprocess. *Chemical Engineering Journal, 151563.* Available at. https://doi.org/10.1016/j.cej.2024.151563

Zhou, Y., Zhu, Y., Zhu, J., Li, C., & Chen, G. (2023). A comprehensive review on wastewater nitrogen removal and its recovery processes. *International Journal of Environmental Research and Public Health, 20*(4), 3429. https://doi.org/10.3390/ijerph20043429

Available at https://www.[illegible] [illegible] wastewater-treatment [illegible] [illegible] treatment plants located [illegible] [illegible]

[illegible] C. M., [illegible] A. [illegible], Moore, T. [illegible], David, R., Rhodes, [illegible] [illegible] [illegible] [illegible] [illegible]

[illegible] J. D. [illegible] [illegible] technologies [illegible] [illegible] [illegible]

[illegible] [illegible] 2010 [illegible] [illegible] [illegible] Available [illegible] [illegible] [illegible]

[illegible] (2020) [illegible] [illegible] [illegible] [illegible] [illegible]

[illegible] A. (2020). [illegible] [illegible] [illegible] [illegible] [illegible]

[illegible] [illegible] [illegible] [illegible] [illegible]

[illegible] [illegible] [illegible] (2018) [illegible] [illegible] [illegible] [illegible]

Chapter 2
Advanced Membrane Technologies: Membrane Bioreactors (MBRs) in Resource Recovery from Wastewater Streams

S. S. Kirthiga, R. Dhinesh, Anu Gopinath, and Mohammad Taherzadeh

Abstract Wastewater streams contain both contaminants and valuable resources, making advanced treatment technologies essential for sustainable resource recovery. Without effective treatment, wastewater discharge leads to eutrophication, hypoxia, and the accumulation of toxic pollutants, severely impacting aquatic ecosystems and public health. However, wastewater is also a reservoir of recoverable nutrients (nitrogen, phosphorus), organic matter, and high-quality water, which can be efficiently reclaimed through advanced membrane technologies. Membrane bioreactors (MBRs) represent a significant advancement in wastewater treatment, integrating biological processes with membrane filtration to achieve high-efficiency pollutant removal and resource recovery. Utilising microfiltration (MF) or ultrafiltration (UF), MBRs enhance biomass retention, improve effluent quality, and enable the recovery of nutrients and energy carriers. Despite these advantages, challenges such as membrane fouling, high energy demands, and limited membrane longevity necessitate continuous technological improvements. Recent advancements in MBR technology focus on hybrid configurations incorporating forward osmosis (FO), reverse osmosis (RO), and electrochemical processes to enhance resource recovery efficiency. Innovations in nanocomposite membranes, antifouling surface modifications, and biofilm-resistant coatings aim to extend membrane lifespan and reduce operational costs. Artificial intelligence (AI)-driven real-time monitoring is further optimising process efficiency and energy consumption. Additionally, the integration of bioelectrochemical systems (BES) with MBRs is being explored for simultaneous wastewater treatment and energy recovery. This chapter critically examines the latest advancements in MBR technology, highlighting their role in circular economy

S. S. Kirthiga (✉) · R. Dhinesh · A. Gopinath
Department of Aquatic Environment Management, Faculty of Fisheries Science, Kerala University of Fisheries and Ocean Studies, Kochi, Kerala, India
e-mail: sskkirthiga@gmail.com

M. Taherzadeh
Bioprocess Technology, Swedish Centre for Resource Recovery, University of Borås, Borås, Sweden
e-mail: mohammad.taherzadeh@hb.se

S. Dhiman and G. Mukherjee (eds.), *Resource Resurgence*, Springer Water,
https://doi.org/10.1007/978-3-032-02750-4_2

frameworks and their potential for large-scale application in sustainable wastewater resource recovery.

Keywords Membrane bioreactors · Wastewater · Contaminants · Resources · Technologies

2.1 Introduction

The increasing global water scarcity, along with the rising demand for clean water, has intensified the need for sustainable and integrated wastewater treatment strategies. Conventional treatment technologies, although effective for pollutant removal, are often inefficient in terms of recovering valuable resources embedded in wastewater streams, such as nutrients, energy, and reusable water (Diaz-Elsayed et al., 2019). Advanced membrane technologies, particularly membrane bioreactors (MBRs), have emerged as a transformative solution in wastewater treatment systems, integrating biological processes with membrane separation to achieve high-efficiency treatment and enable resource recovery (Namdeti et al., 2025). MBRs are designed to treat municipal, industrial, and agricultural wastewater, and their ability to produce high-quality effluent has positioned them at the forefront of modern wastewater management and circular economy practices. MBRs combine the principles of activated sludge processes with membrane filtration, typically ultrafiltration or microfiltration, offering superior solid–liquid separation. This integration eliminates the need for secondary clarifiers and allows operation at higher mixed liquor suspended solids (MLSS) concentrations, thereby enhancing treatment efficiency and reducing the footprint of the treatment plant (James, 2024). The compactness of the system and the consistent effluent quality make MBRs particularly advantageous for decentralised treatment facilities and water reuse applications. Notably, the permeate from MBRs often meets or exceeds the standards for non-potable reuse, and in some configurations, can be further treated for potable reuse (Yang et al., 2020). In addition to water reuse, one of the most compelling aspects of MBR technology is its potential for resource recovery.

Wastewater is increasingly being recognised not as a waste product, but as a reservoir of resources. The organic matter in wastewater can be harvested for biogas production through anaerobic digestion, while nutrients such as nitrogen and phosphorus can be recovered as fertilisers (Campos et al., 2019). MBRs, by maintaining high biomass concentrations and longer sludge retention times, can promote the growth of specific microbial consortia that facilitate nutrient accumulation and energy generation. The decoupling of hydraulic retention time (HRT) and sludge retention time (SRT) in MBRs also provides operational flexibility, optimising conditions for both pollutant degradation and metabolite production (Cheng et al., 2018). One technical advantage of MBRs lies in their ability to retain high-molecular-weight compounds and pathogens, resulting in effluents with significantly reduced bacterial and viral loads. This property makes MBRs particularly relevant for applications requiring stringent disinfection standards, such as irrigation, industrial process water, and aquifer recharge (Asif & Hai, 2019). Additionally, the integration of advanced oxidation processes (AOPs) or nanomaterials with MBR systems is being explored to further enhance the removal of micropollutants, including pharmaceuticals, personal care products, and endocrine-disrupting compounds, which are often resistant to conventional biological treatments (Silva et al., 2017).

From an engineering perspective, the configuration of MBRs, either submerged or side-stream, affects system design and energy requirements. Submerged MBRs, where membranes are immersed directly in the bioreactor, are more energy-efficient in terms of pumping but require intensive air scouring to control membrane fouling. Side-stream MBRs, conversely, have external membrane modules and allow better control over hydrodynamic conditions but are typically more energy-intensive due to higher recirculation rates (Judd, 2010). Recent developments in membrane materials, including the use of hydrophilic coatings, nanocomposites, and fouling-resistant surfaces, have significantly improved membrane longevity and performance, although membrane fouling remains a critical operational challenge (El Batouti et al., 2021). Moreover, advances in membrane surface modification and dynamic membrane systems are being actively researched to address the challenges in MBRs, like membrane fouling and enhance the viability of long-term operation. However, economically, while the capital and operational costs of MBR systems are higher than conventional treatment methods, the benefits of higher effluent quality, reduced sludge production, and potential for resource recovery can offset these costs, particularly in water-stressed regions or where stringent discharge standards apply. Life cycle assessments (LCA) have demonstrated that MBRs, especially when integrated with energy recovery units or nutrient harvesting systems, exhibit favourable environmental footprints over the long term.

The role of MBRs is expanding beyond mere wastewater treatment to encompass the broader objectives of resource recovery and environmental protection. Emerging trends include hybrid systems combining MBRs with anaerobic reactors, electrochemical cells, or forward osmosis modules to enhance the recovery of water, energy, and nutrients. For instance, anaerobic MBRs (AnMBRs) offer energy-positive treatment solutions by producing methane-rich biogas while minimising sludge production and nutrient loss (Aslam et al., 2022). These hybrid configurations are being tailored for specific wastewater types and recovery goals, representing the next frontier in resource-oriented wastewater treatment.

2.2 Potential of Wastewater Streams and the Need for Resource Recovery

Wastewater, once primarily viewed as an environmental burden, is increasingly being recognised as a valuable resource reservoir containing not only water but also critical energy, nutrients, and high-value materials. Both municipal and industrial effluents are rich in organic carbon, nitrogen, phosphorus, and other trace elements, including heavy metals and rare earth elements, all of which can be recovered and utilised through advanced treatment technologies. The recovery of essential nutrients like nitrogen and phosphorus, which are indispensable for agricultural productivity yet finite, has garnered significant attention. These nutrients can be recovered from wastewater in stable and usable forms, such as struvite (magnesium ammonium phosphate), offering several environmental benefits. By capturing these nutrients, wastewater treatment not only prevents nutrient overload and subsequent eutrophication of water bodies but also reduces reliance on mineral fertilisers, which are often associated with environmental degradation and unsustainable mining practices (Adam & Krüger, 2018; Muhmood et al., 2019). As the global challenge of freshwater scarcity becomes more pressing, treated wastewater is emerging as a viable alternative source for both non-potable and, in some cases, portable applications. Advances in membrane filtration, advanced oxidation, and biological treatment technologies have made it possible to treat wastewater to the extent that it meets stringent public health and safety standards, facilitating its reuse for irrigation, industrial processes, and even drinking water in water-stressed regions. This transformation of wastewater from a waste disposal problem to a resource recovery opportunity aligns closely with the principles of the circular economy, which emphasises the closing of resource loops and minimising waste through recycling and reuse. By shifting to a resource recovery model, wastewater management can be viewed not just as a cost centre but as a resource-generating process that recovers valuable materials while simultaneously protecting the environment and conserving natural resources.

The full potential of resource recovery from wastewater cannot be realised without a combination of technological innovation, supportive policy frameworks, and public acceptance. Advanced technologies such as anaerobic digestion for biogas production, membrane bioreactors (MBRs) for efficient filtration, and nutrient recovery systems are critical to the recovery of valuable resources like energy and nutrients. A Membrane Bioreactor (MBR) is an advanced wastewater treatment technology that combines biological treatment with membrane filtration, which uses a suspended growth bioreactor along with microfiltration or ultrafiltration membranes to achieve high-efficiency solid–liquid separation and the concept behind the membrane bioreactor is depicted in Fig. 2.1. This system produces high-quality effluent, suitable for water reuse, by effectively removing suspended solids, organic matter, and pathogens.

MBRs offer advantages such as compact design and enhanced treatment performance but are challenged by membrane fouling and higher operational costs. However, for these technologies to be effectively deployed, favourable policies that incentivise investment, set clear guidelines for the safe use of reclaimed products,

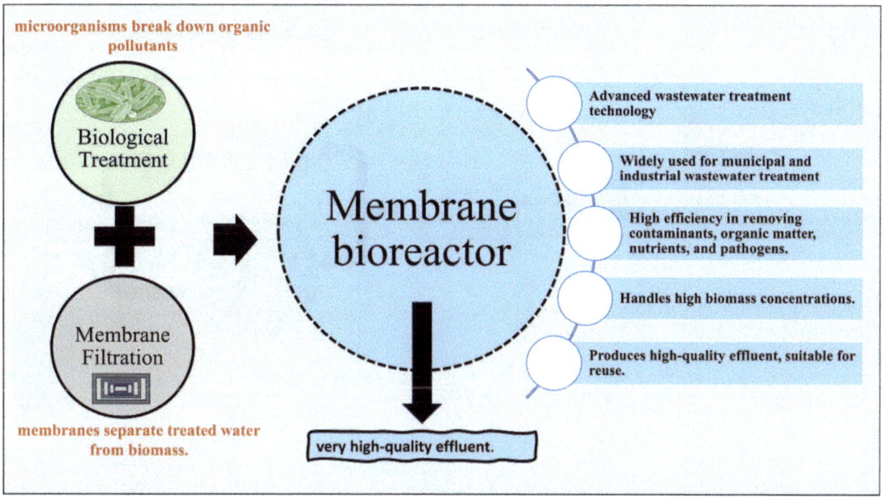

Fig. 2.1 Concept of membrane bioreactors and the advantages involved in MBR technologies

and ensure regulatory compliance are essential. Overcoming public concerns and fostering a positive perception of reclaimed water will be crucial for the widespread adoption of these technologies (Diaz-Elsayed et al., 2019). Ultimately, integrating resource recovery into wastewater treatment represents a paradigm shift in how we manage water, energy, and nutrients. It provides a sustainable pathway to address the global water crisis, reduce environmental pollution, and support the bioeconomy by transforming wastewater from a liability into a strategic asset. As the demand for natural resources intensifies globally, wastewater recovery presents an opportunity to create a more resilient and circular urban water infrastructure that contributes to achieving several Sustainable Development Goals (SDGs), including SDG 6 (Clean Water and Sanitation), SDG 12 (Responsible Consumption and Production), and SDG 13 (Climate Action). With continued research, technological advancement, and the establishment of supportive regulatory and policy frameworks, wastewater can serve as a cornerstone of sustainable resource management in the future.

2.3 Fundamentals and Mechanisms of Membrane Bioreactors (MBRs)

Membrane Bioreactors (MBRs) are an advanced wastewater treatment technology that integrates biological treatment processes with membrane filtration, combining the benefits of activated sludge systems with microfiltration (MF) or ultrafiltration (UF) membranes. The core function of MBRs is to replace the conventional secondary clarifier found in activated sludge systems, effectively separating biomass

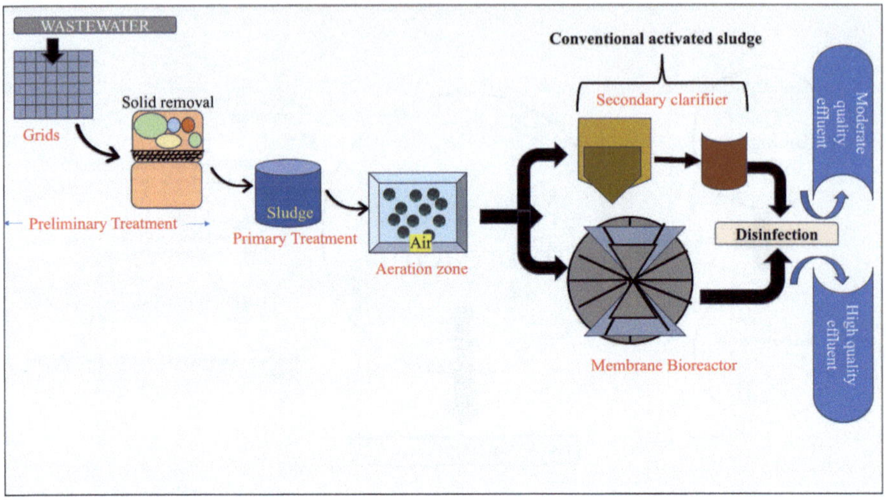

Fig. 2.2 Schematic representations of membrane bioreactors and conventional activated sludge systems for wastewater treatment

and improving the quality of the effluent which is represented in Fig. 2.2. This integration allows MBRs to achieve a more efficient treatment process by retaining microbial biomass and contaminants while allowing treated water to permeate through the membrane barrier. Typically, MBR systems consist of two key components: the aeration tank, where microorganisms break down organic pollutants, and the membrane tank, where filtration occurs. The membranes physically block suspended solids, pathogens, and colloids, ensuring the removal of contaminants and yielding high-quality effluent. There are two common MBR configurations: submerged MBRs, where the membrane modules are submerged directly in the bioreactor, and sidestream MBRs, where the membrane modules are externally placed with cross-flow filtration (Judd, 2010; Martínez et al., 2020).

The separation mechanism in MBRs is driven by size exclusion. Membranes with pore sizes ranging from 0.01 to 0.1 μm retain suspended solids, bacteria, and colloids, while allowing water to pass through. This mechanism results in a high Mixed Liquor Suspended Solids (MLSS) concentration, which typically ranges between 8000 and 15,000 mg/L, contributing to improved biodegradation efficiency and a reduced reactor volume (Judd, 2010). A notable advantage of MBRs over conventional systems is the decoupling of Hydraulic Retention Time (HRT) and Sludge Retention Time (SRT). With MBRs, longer SRTs—up to 60 days—can be achieved, promoting the breakdown of more complex pollutants and reducing the production of excess sludge. This decoupling allows for a more efficient treatment process, especially for recalcitrant compounds that are harder to degrade in traditional systems (Mannina et al., 2018).

In submerged MBRs, aeration serves a dual purpose: it provides necessary oxygen for microbial metabolism and helps to mitigate fouling by generating shear forces at the membrane surface. Fouling is a significant challenge in MBR systems, as the accumulation of particles, microorganisms, and precipitates can reduce membrane flux and overall system performance (Bera et al., 2022). These measures help to extend the lifespan of the membrane and improve the overall efficiency of the system (Rajendran et al., 2024; Turan, 2025).

MBRs are particularly suitable for decentralised wastewater treatment applications, especially in areas where space is limited or where consistent effluent quality is required for water reuse (Hube & Wu, 2021). Their compact design makes them well-suited for both urban and rural areas, where they can treat wastewater for non-potable uses such as irrigation, industrial processes, or even potable applications in water-scarce regions. Moreover, the removal of pathogens and contaminants makes MBRs highly effective for producing effluent that meets strict discharge standards (Bandh & Mushtaq, 2025). The modular design of MBRs also provides scalability, allowing the system to be adapted to different treatment capacities, from small communities to large urban centres (Capodaglio, 2017; Hube & Wu, 2021).

2.4 Membrane Technologies in Wastewater Treatment

Membrane technologies have emerged as key tools in advanced wastewater treatment systems due to their ability to selectively separate contaminants based on size, charge, and chemical affinity. They offer compactness, modular design, high separation efficiency, and compatibility with various treatment objectives, including water reuse, nutrient recovery, and contaminant removal (Judd, 2010). Table 2.1 shows a comparison of membrane bioreactors with conventional systems. Unlike conventional treatment systems that rely on sedimentation and biological degradation alone, membrane processes provide physical barriers that retain suspended solids, microorganisms, and even dissolved contaminants depending on membrane pore size and surface properties (Gkotsis et al., 2014). This enhanced separation efficiency makes membrane-based systems highly adaptable for decentralised treatment units, zero-liquid discharge systems, and integrated resource recovery platforms (Yadav et al., 2022).

2.4.1 Key Components of MBR and Process Parameters

The key components of membrane bioreactor (MBR) systems include the bioreactor, membrane unit, aeration system, and sludge recirculation loop, each playing a critical role in treatment efficiency and operational stability. The bioreactor facilitates biological degradation of organic matter through aerobic or anaerobic microbial processes, and its performance is heavily influenced by parameters such as sludge retention time (SRT), typically maintained between 20–60 days to ensure effective biomass acclimatisation and degradation of complex organics. The membrane unit, consisting

Table 2.1 Comparison of different criteria between membrane bioreactors and conventional activated sludge systems

Criteria	Membrane bioreactors (MBRS)	Conventional activated sludge (CAS)
Process integration	Combines biological treatment and membrane filtration	Separate biological treatment and secondary clarification
Footprint	Compact, requires less space	A large area is needed for sedimentation tanks
Effluent quality	Superior with low BOD, TSS, pathogens, and micropollutants	Moderate with higher BOD and TSS in effluent
Solid–liquid separation	Via membrane filtration (usually MF/UF)	Via gravity in secondary clarifiers
Sludge production	Reduced due to higher sludge retention time (SRT)	More sludge production
Flexibility and scalability	Modular design allows easy expansion	Less flexible, expansion requires a large infrastructure
Fouling issues	Membrane fouling is a key operational challenge	Not applicable
Operation and maintenance	Requires skilled operators and frequent membrane cleaning	Simpler operation, less technical expertise needed
Energy requirement	Higher due to membrane aeration and filtration pressure	Lower overall energy consumption
Cost	High capital and O&M costs	Lower capital and operational costs
Reuse potential	High–ideal for water reuse and recycling applications	Limited reuse potential
Applications	Municipal & industrial wastewater treatment, water reuse, space-limited areas	Municipal sewage treatment, basic industrial wastewater

of either hollow fibre or flat-sheet membranes with pore sizes in the range of 0.01–0.1 μm, performs solid–liquid separation based on size exclusion, ensuring complete removal of suspended solids, bacteria, and some viruses (Akinyemi et al., 2023). Aeration not only supplies dissolved oxygen to sustain microbial metabolism but also generates shear stress to limit membrane fouling; thus, the airflow rate must be carefully regulated to balance oxygen demand and energy consumption. Furthermore, pH, temperature, and nutrient loading also significantly affect microbial activity and membrane performance, necessitating integrated monitoring systems for effective control (Nabi et al., 2023). The interplay among these components and parameters directly determines the efficiency, stability, and fouling behaviour of the MBR system, thereby dictating its operational lifespan and overall treatment cost (Hamedi et al., 2019).

2.4.2 Classification of Membranes and Membrane Materials

Membranes can be classified based on several parameters, including pore size, driving force, configuration, and the nature of the material. Based on pore size, membranes are categorised into microfiltration (0.1–10 μm), ultrafiltration (0.01–0.1 μm), nanofiltration (1–10 nm), and reverse osmosis membranes (pore sizes < 1 nm) (Castro & Abejón, 2024). Each class serves distinct roles in wastewater treatment, from suspended solids removal in microfiltration to desalination and contaminant rejection in reverse osmosis. Membrane materials are primarily organic polymers or inorganic substances, each with specific advantages and limitations. Common polymeric membranes include polyvinylidene fluoride (PVDF), polysulfone (PS), polyethersulfone (PES), cellulose acetate (CA), and polyamide (PA). These materials offer flexibility, cost-effectiveness, and tunable properties such as hydrophilicity, charge density, and pore structure (Sahu et al., 2024). However, polymeric membranes are susceptible to fouling and chemical degradation, especially under harsh wastewater conditions (Naz et al., 2019). In contrast, inorganic membranes, made from materials like ceramic, titanium oxide, or zeolites, exhibit superior thermal and chemical resistance, longer operational life, and higher mechanical strength (Asif & Zhang, 2021). Hybrid membranes, combining organic and inorganic components (e.g., thin-film nanocomposites), represent an evolving class designed to synergise the permeability of polymers with the robustness of ceramics or nanomaterials like graphene oxide, carbon nanotubes, or metal–organic frameworks (MOFs) (Sahu et al., 2023). These materials are being developed to enhance antifouling behaviour, selectivity, and permeability while addressing sustainability and energy-efficiency goals (Osman et al., 2024). The choice of membrane material and configuration is thus pivotal to achieving tailored performance in diverse wastewater treatment applications.

2.4.3 Principles of Membrane Separation Processes

Membrane separation processes function based on selective permeability, where a semi-permeable membrane allows certain substances to pass while retaining others. The driving forces that govern separation vary with the membrane type pressure for microfiltration (MF), ultrafiltration (UF), nanofiltration (NF), and reverse osmosis (RO); concentration gradients for forward osmosis (FO); and electric potential for electrodialysis (ED) (Castro and Abejón, 2024) (Fig. 2.3). In pressure-driven processes, the separation mechanism is typically size exclusion, especially for MF and UF, where membranes retain particles, colloids, and bacteria while allowing water and solutes to pass (Nath, 2017). For NF and RO, separation is more complex and involves not just size but also electrostatic interactions and diffusion across the membrane matrix (Zhang et al., 2022). The performance of these membranes is influenced by factors such as transmembrane pressure, membrane porosity, surface

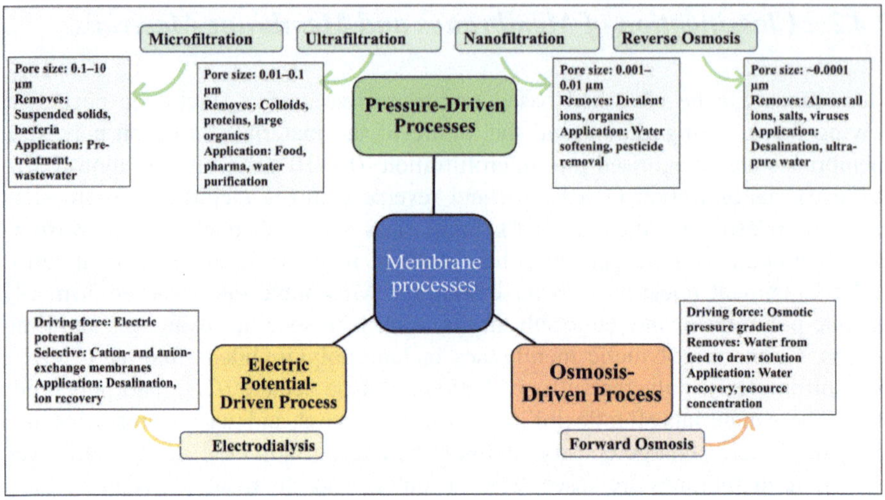

Fig. 2.3 Flowchart representing the processes and principles involved in the membrane bioreactors

charge, fouling resistance, and hydrophilicity (Choudhury et al., 2018). Membrane separation processes offer highly efficient solutions for filtering out contaminants, recovering valuable resources, and producing high-quality water, making them indispensable in both municipal and industrial applications.

2.4.3.1 Pressure-Driven Membrane Processes

Pressure-driven processes are the most widely used membrane filtration technologies. In these systems, pressure is applied to the feed solution to push water through a semi-permeable membrane. The nature of the separation in these systems is mainly based on size exclusion, although electrostatic interactions may also play a role in some cases. The main pressure-driven membrane processes include Microfiltration (MF), Ultrafiltration (UF), Nanofiltration (NF), and Reverse Osmosis (RO). Each of these processes varies in terms of pore size and the types of substances they can remove from the water.

- **Microfiltration** (MF): Microfiltration is a low-pressure membrane process that typically has a pore size between 0.1 to 10 microns. This allows it to effectively filter out larger particles such as suspended solids, bacteria, and other microorganisms. However, it does not remove dissolved salts or smaller molecules. MF is commonly used for pre-treatment in water treatment systems, including as a first stage before reverse osmosis (RO), to reduce the fouling load on finer membranes (Nath, 2017). In wastewater treatment, MF is effective for removing solids, colloids, and even some bacteria and pathogens.

- **Ultrafiltration (UF)**: Ultrafiltration membranes have smaller pores, ranging from 0.01 to 0.1 microns, which allows them to remove colloids, proteins, large organic molecules, and some bacteria. This makes UF a more selective filtration process than MF, but it still allows smaller molecules such as salts and low-molecular-weight organic compounds to pass through. UF is used extensively in the food and beverage industry, pharmaceutical applications, and water treatment for further purification (Zhang et al., 2022).
- **Nanofiltration (NF)**: Nanofiltration membranes have even smaller pores, typically between 0.001 to 0.01 microns. These membranes can reject divalent ions, such as calcium and magnesium (which contribute to water hardness), and remove organic compounds with molecular weights higher than 200–300 Daltons. NF can be used for water softening and partial desalination, as well as in applications that require the removal of pesticides, heavy metals, and other contaminants from water (Choudhury et al., 2018). Unlike reverse osmosis, NF membranes allow monovalent ions such as sodium and chloride to pass through to some extent.
- **Reverse Osmosis (RO)**: Reverse osmosis is the most stringent of the pressure-driven processes, with membranes that have pore sizes in the range of 0.0001 microns. RO membranes can remove almost all dissolved ions, salts, organic compounds, microorganisms, and even viruses from water. This process is commonly used in desalination plants to convert seawater into potable water. RO is also used in applications requiring ultra-pure water, such as in pharmaceutical manufacturing, electronics production, and medical fields (Castro & Abejón, 2024).

2.4.3.2 Osmosis-Driven Membrane Processes

Osmosis-driven processes, such as **Forward Osmosis (FO)**, rely on the osmotic pressure difference between two solutions to drive the movement of water through a membrane. Osmosis occurs naturally when water flows from an area of low solute concentration to an area of high solute concentration. By utilising this natural process, FO offers a more energy-efficient solution compared to pressure-driven processes.

- **Forward Osmosis (FO)**: In forward osmosis, water is drawn from a feed solution (such as wastewater) through a semi-permeable membrane and into a concentrated draw solution. The osmotic pressure difference between the feed and draw solutions drives the flow of water. FO has emerged as an attractive technology for water recovery due to its energy efficiency, particularly in applications like desalination, wastewater treatment, and food processing (Rajendran et al., 2024). It is also used in resource recovery, as FO can concentrate valuable nutrients or chemicals from wastewater streams before further treatment or reuse.

2.4.3.3 Electric-Driven Membrane Processes

Electrodialysis is an electric-driven membrane process that uses an electric potential to move ions through ion-exchange membranes. In electrodialysis, cation-exchange membranes allow positive ions (cations) to pass, while anion-exchange membranes allow negative ions (anions) to pass. This selective ion transport is driven by an electric field applied across the membrane stack. Electrodialysis is most commonly used for desalination of brackish water, selective ion recovery, and water softening. It is particularly useful for removing specific ions, such as heavy metals, from water while leaving other ions in place (Zhang et al., 2022). The process is energy-efficient and is increasingly being applied in industries that require selective ion separation.

2.4.3.4 Factors Influencing Membrane Performance

Several factors influence the performance of membrane separation processes. These include membrane characteristics, operating conditions, and the propensity for fouling. Understanding these factors is crucial to optimising membrane systems for specific applications and maintaining their performance over time.

- **Transmembrane Pressure (TMP)**: The transmembrane pressure is the difference in pressure between the feed side and permeate side of the membrane. TMP is a key factor in determining the flux or rate of filtration. Higher TMPs lead to higher flux rates but also increase the risk of fouling, which can decrease the efficiency of the membrane. Optimising TMP is critical to balancing high performance and minimising energy consumption (Nath, 2017).
- **Membrane Properties**: The physical properties of the membrane such as pore size, surface charge, and hydrophilicity which directly affect the separation efficiency and fouling resistance. Membranes with smaller pores provide higher selectivity, but they may be more prone to fouling because smaller pores are more likely to become blocked by suspended particles. Membranes with hydrophilic surfaces tend to resist fouling more effectively than hydrophobic membranes (Choudhury et al., 2018).
- **Fouling Resistance**: Fouling is the accumulation of particles, microorganisms, or chemical precipitates on the membrane surface, which decreases the water flux and membrane performance. Fouling is one of the main challenges in membrane filtration systems, particularly in applications involving wastewater treatment. To mitigate fouling, cleaning strategies such as chemical backwashing, physical cleaning, and surface modifications (e.g., applying hydrophilic coatings or zwitterionic polymers) are used (Rajendran et al., 2024).
- **Hydrodynamic Conditions**: The design and configuration of the membrane system, including whether it is flat-sheet, hollow-fibre, or spiral-wound, can significantly influence fouling rates and cleaning efficiency. The flow distribution across the membrane surface affects the formation of the boundary layer, which in turn influences the fouling process. Optimizing hydrodynamic conditions can help

reduce fouling and increase the operational life of the membrane (Kürzl et al., 2023).

2.5 Types of Membrane Bioreactor (MBR) Systems

Membrane Bioreactor (MBR) systems signify a significant innovation in wastewater treatment, integrating the conventional activated sludge process with membrane filtration to achieve superior effluent quality while maintaining a compact footprint (Judd, 2010). The classification of MBR systems is primarily based on the membrane placement and the hydraulic configuration of the system (Fig. 2.4). The two main types are Submerged MBRs (sMBRs) and Sidestream MBRs (ssMBRs) (Tran et al., 2024). In recent years, advances in materials and process integration have led to the development of Hybrid MBRs and Anaerobic MBRs (AnMBRs), which offer enhanced performance in terms of energy recovery, removal of emerging micropollutants, and adaptability to complex wastewater matrices (Bera et al., 2022).

2.5.1 *Submerged MBR Systems (sMBRs)*

Submerged MBR systems are designed with the membrane modules directly installed within the biological reactor or a separate membrane tank. Filtration occurs under low

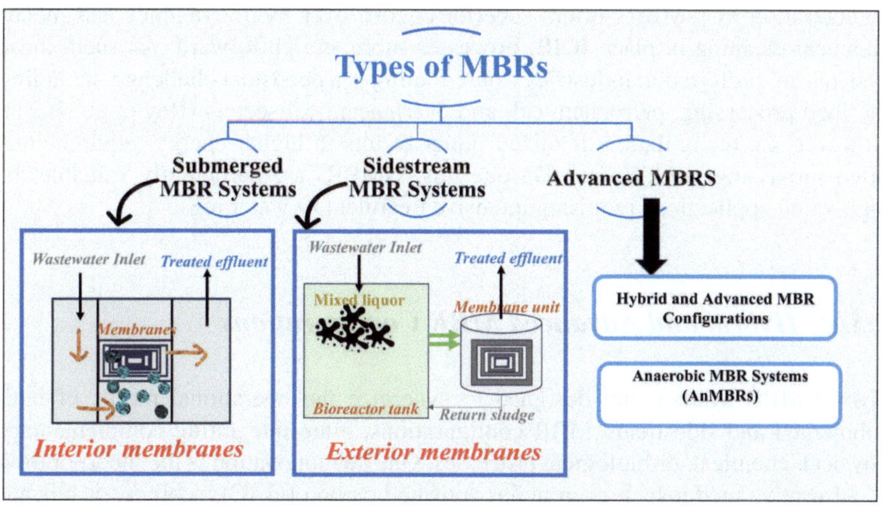

Fig. 2.4 Types of MBRs with the diagrammatic representions of the MBR systems

transmembrane pressure (TMP), usually ranging between 0.05 and 0.15 bar, facilitated by either gravity or vacuum suction (Turan, 2025). A key feature of sMBRs is air scouring, which provides both the necessary aerobic conditions and the shear force at the membrane surface to reduce fouling, particularly the formation of cake layers and biofilms (Namdeti et al., 2025). This fouling control mechanism contributes significantly to the system's operational reliability. A distinct advantage of sMBRs is the decoupling of the hydraulic retention time (HRT) and sludge retention time (SRT), allowing independent optimisation of biological processes. This flexibility promotes enhanced biological nutrient removal, especially for nitrification–denitrification and phosphorus uptake, critical for stringent effluent quality standards (Zhang et al., 2021). Additionally, sMBRs are energy-efficient, with specific energy consumption typically below 0.5 kWh/m^3 for filtration. Their compact design and ease of operation make them ideal for municipal and decentralised applications (Rahman et al., 2023).

2.5.2 *Sidestream MBR Systems (ssMBRs)*

Sidestream MBRs differ fundamentally from submerged systems in that their membrane modules are located outside the bioreactor. The mixed liquor is pumped at high velocities (2–4 m/s) through the membrane modules and recirculated back to the bioreactor. This crossflow operation induces high shear rates, often exceeding 1000 $s-^1$, which effectively minimizes concentration polarisation and reduces membrane fouling—an important consideration for treating industrial wastewaters with high organic loads or salinity (Giwa et al., 2022; Lim et al., 2021). The external configuration of ssMBRs offers superior control over hydrodynamics and makes chemical cleaning-in-place (CIP) processes more straightforward. As such, these systems are preferred in industries where fouling is a persistent challenge, including the food processing, petrochemical, and pharmaceutical sectors (Ray et al., 2021). However, the recirculation of mixed liquor results in higher energy requirements, often surpassing 1.0 kWh/m^3. Despite this, ssMBRs are particularly valuable for high-value applications requiring intensive treatment or water reuse.

2.5.3 *Hybrid and Advanced MBR Configurations*

Hybrid MBR systems are designed to synergize the operational merits of both submerged and sidestream MBR configurations, often integrating complementary physical, chemical, or biological processes. One key innovation is the incorporation of adsorptive media such as granular activated carbon (GAC), zeolites, or biochar into the bioreactor or membrane tank. These materials enhance the adsorption of recalcitrant organic micropollutants, including pharmaceuticals, personal care products, and endocrine-disrupting compounds (EDCs), which are otherwise resistant to conventional biological degradation pathways (Gutiérrez et al., 2021). The addition of adsorptive media not only provides a secondary removal mechanism via physical sorption but also acts as a microbial support medium, fostering biofilm formation and enabling simultaneous adsorption and biodegradation. Studies incorporating powdered activated carbon (PAC) and biochar into hybrid MBRs have reported significant improvements in the removal of EDCs and persistent organic pollutants (POPs), with reductions often exceeding 90% for compounds such as bisphenol A, triclosan, and sulfamethoxazole (Nguyen et al., 2022). Furthermore, these integrated systems exhibit increased membrane longevity due to reduced membrane fouling rates, attributed to lower concentrations of dissolved organics and soluble microbial products (SMPs) in the mixed liquor. Hybrid MBR systems may also utilize pre-treatment processes like ozonation or UV-based advanced oxidation processes (AOPs) to break down complex contaminants prior to membrane filtration. These pre-treatments enhance the biodegradability of influent organics and improve downstream membrane performance by reducing fouling potential. In addition, membrane surface modification techniques such as grafting hydrophilic polymers or incorporating antimicrobial nanoparticles are employed to mitigate biofouling and enhance flux stability over extended operational periods (Gutiérrez et al., 2021). As a result, hybrid MBRs represent a robust and versatile solution for advanced wastewater treatment, especially in contexts demanding high-quality effluent and micropollutant abatement.

2.5.4 Anaerobic MBR Systems (AnMBRs)

Anaerobic Membrane Bioreactor (AnMBR) systems represent a transformative shift in biological wastewater treatment by integrating membrane filtration with anaerobic digestion. Unlike aerobic MBRs, AnMBRs eliminate the need for oxygen supply, thereby significantly reducing energy consumption and enabling energy recovery in the form of methane-rich biogas. These systems are particularly suited for the treatment of high-strength industrial effluents such as those generated by food and beverage industries, distilleries, slaughterhouses, and dairy processing plants, where influent Chemical Oxygen Demand (COD) can range from 5000 to over 20,000 mg/L. AnMBRs operate under mesophilic (35–37 °C) or thermophilic (50–55 °C) conditions and are capable of sustaining high Organic Loading Rates (OLRs), often between 5–15 kg COD/m^3/day (Tran et al., 2024). The membrane component retains slow-growing anaerobic microorganisms, including methanogens, within the reactor, decoupling the hydraulic retention time (HRT) from the solids retention time (SRT). This facilitates the complete degradation of organic matter and results in minimal sludge production due to low microbial growth yields. Moreover, the absence of aeration and the production of methane enable AnMBRs to achieve a net positive energy balance, positioning them as a sustainable option for wastewater treatment plants (WWTPs) aiming for energy neutrality. However, several operational challenges remain in the widespread adoption of AnMBRs. Chief among these is membrane fouling, which is exacerbated under anaerobic conditions due to the accumulation of extracellular polymeric substances (EPS) and the lack of scouring air bubbles that aid in fouling mitigation in aerobic MBRs (Gao et al., 2018). Additionally, the solubilization and subsequent loss of dissolved methane in the effluent pose both environmental concerns and energy recovery inefficiencies. Sulfide generation from sulfate-reducing bacteria can also lead to membrane damage and toxicity to methanogens. Addressing these issues requires advanced membrane materials, such as antifouling coatings or ceramic membranes, and optimized operational strategies like periodic relaxation, backflushing, and biogas sparging (Gao et al., 2018). Despite these limitations, the AnMBR configuration continues to gain traction in both academic and industrial settings, particularly for decentralized wastewater treatment and zero-energy discharge systems. When integrated with post-treatment units such as constructed wetlands or polishing filters, AnMBRs can meet stringent discharge norms and enable water reuse while simultaneously valorizing waste streams.

2.6 Resource Recovery Potential of MBRs

2.6.1 Water Reuse and Quality Standards Achieved by MBRs

Membrane bioreactors (MBRs) provide a highly effective solution for water reuse due to their ability to consistently produce effluent with superior quality, meeting or exceeding international reuse standards. The integration of microfiltration or ultrafiltration membranes ensures the physical removal of suspended solids, bacteria, and many viruses, achieving turbidity levels < 0.2 NTU and total suspended solids (TSS) < 1 mg/L (Labhasetwar & Yadav, 2023). In addition, the biological treatment stage facilitates the degradation of biodegradable organic matter and partial nutrient removal. MBR-treated effluent typically meets stringent guidelines set by regulatory frameworks such as the U.S. EPA Class A + or WHO reuse standards, which require low biochemical oxygen demand (BOD) < 10 mg/L, E. coli < 1 CFU/100 mL, and total nitrogen and phosphorus within acceptable thresholds (Asif & Hai, 2019). These characteristics make MBRs suitable for non-potable reuse applications such as agricultural irrigation, industrial processes, groundwater recharge, and even potable reuse when followed by advanced post-treatment steps like reverse osmosis or advanced oxidation (Yang et al., 2019). Their robustness in handling variable influent loads further reinforces MBRs as reliable systems for ensuring consistent reclaimed water quality in decentralised and urban systems alike.

2.6.2 Nutrient Recovery (Nitrogen and Phosphorus Extraction)

MBRs offer unique opportunities for nitrogen and phosphorus recovery due to their decoupled hydraulic and sludge retention times, which allow for the integration of tailored nutrient removal and recovery processes (Nadeem et al., 2022). For nitrogen, MBR systems can be optimised to promote nitrification–denitrification cycles by controlling aeration patterns and anoxic zones, converting ammonium (NH_4^+) to nitrogen gas (N_2) through microbial pathways (Xie, 2021). Alternatively, ammonia stripping or recovery as ammonium sulfate can be integrated using ion exchange or gas-permeable membrane systems (Munasinghe-Arachchige et al., 2021). Phosphorus recovery is often achieved through chemical precipitation as struvite (magnesium ammonium phosphate) by dosing with magnesium salts, or biologically via enhanced biological phosphorus removal (EBPR), where phosphorus-accumulating organisms (PAOs) concentrate phosphorus intracellularly (Izadi et al., 2020). These strategies not only prevent eutrophication but also enable circular economy approaches by recovering essential nutrients for fertilizer production. Moreover, coupling MBRs with nutrient recovery technologies allows for the retention of high-value macronutrients from both domestic and industrial wastewaters, enhancing sustainability in wastewater treatment systems.

2.6.3 Organic Matter and Biogas Recovery from MBR Systems

The recovery of organic matter and its subsequent valorization into biogas is a significant advantage of MBR systems, particularly when configured as anaerobic membrane bioreactors (AnMBRs). These systems operate under oxygen-free conditions, allowing fermentative and methanogenic microbial consortia to convert organic matter into biogas, primarily methane (CH_4), with energy recovery potential (Musa et al., 2018). Unlike conventional anaerobic digestion, AnMBRs ensure complete biomass retention and separation through membranes, even for slow-growing microbes, enhancing process stability and organic conversion efficiency (An et al., 2023). The closed-loop nature of AnMBRs prevents biomass washout, enables operation at low hydraulic retention times, and allows for higher organic loading rates, particularly useful in high-strength industrial wastewater treatment. Additionally, residual solids from MBRs contain organic-rich biosolids that can be further digested anaerobically or composted, contributing to waste-to-energy pathways (Aryanfar et al., 2025). As energy neutrality becomes a global priority, MBRs, especially anaerobic configurations, offer promising solutions for carbon capture and renewable energy generation from wastewater streams.

2.6.4 Recovery of High-Value Byproducts (Bioactive Compounds, Proteins, and Acids)

Beyond water and energy recovery, MBRs have emerged as potential platforms for the extraction of high-value biochemical compounds such as proteins, extracellular polymeric substances (EPS), volatile fatty acids (VFAs), and bioactive metabolites from both domestic and industrial wastewater streams. High SRTs and concentrated biomass in MBRs enhance the production of microbial metabolites like EPS, which possess industrial relevance in bioplastics and bioflocculants (Mannina et al., 2023). Protein recovery, especially from agro-industrial effluents and food-processing wastewater, can be facilitated through ultrafiltration membranes that concentrate and isolate proteins with high selectivity and purity (Türker et al., 2022). Moreover, MBRs configured with fermentative pretreatment steps promote acidogenesis, enabling the recovery of VFAs, which serve as precursors for biopolymers or platform chemicals in biorefineries (Sukphun et al., 2021). The ability of MBRs to maintain steady-state microbial consortia and high biomass concentrations allows for targeted metabolite production and recovery, especially when integrated with downstream separation technologies like membrane distillation or chromatography. This value-added approach transforms wastewater treatment from a cost-intensive process to a resource-generating system, aligning with sustainable development goals and the principles of circular bioeconomy.

2.7 Advances and Enhancements in MBR Technology

2.7.1 Hybrid MBR Systems (MBR-RO, MBR-FO, MBR-Bioelectrochemical Integration)

Hybrid membrane bioreactor systems represent an important leap in wastewater treatment technology by integrating MBRs with complementary advanced treatment units to enhance performance and enable targeted resource recovery. The combination of MBR with reverse osmosis (MBR-RO) facilitates the production of high-purity water by eliminating micro-contaminants, dissolved solids, and pathogens beyond the capacity of conventional MBRs, making it suitable for potable reuse applications (Li et al., 2018). Similarly, forward osmosis (MBR-FO) integration leverages osmotic gradients for water extraction, reducing energy demands and offering improved resistance to membrane fouling due to low hydraulic pressure operation (Jha & Mishra, 2024). In MBR-bioelectrochemical systems, microbial fuel cells or microbial electrolysis cells are coupled with MBRs to simultaneously achieve wastewater treatment and energy generation, exploiting electroactive bacteria for bioelectricity or hydrogen production (Li et al., 2018). These hybrid configurations offer synergistic benefits, such as improved contaminant removal, reduced membrane fouling, and integrated recovery of clean water, energy, and nutrients, making them attractive for decentralised and high-strength wastewater applications.

2.7.2 *Integration of MBRs with Anaerobic Processes for Biogas Production*

The coupling of MBRs with anaerobic digestion technologies, particularly in the form of anaerobic MBRs (AnMBRs), is gaining prominence due to their ability to treat high-strength wastewaters while simultaneously producing biogas. Anaerobic conditions facilitate the microbial degradation of organic matter into methane-rich biogas, offering a pathway for renewable energy recovery (Kabeyi & Olanrewaju, 2022). In AnMBRs, the membrane ensures complete biomass retention, enhancing process stability and allowing higher organic loading rates than conventional anaerobic digesters (An et al., 2023). This integration minimizes sludge production and operational energy demand, as aeration is not required. AnMBRs are particularly beneficial for treating industrial effluents such as those from food, dairy, and pharmaceutical industries, which are rich in biodegradable organic matter (Mojiri et al., 2023). Moreover, the closed environment protects methanogens from inhibitory shocks, and the recovered biogas can offset energy requirements, contributing to the goal of energy-neutral wastewater treatment facilities.

2.7.3 *Nanotechnology and Advanced Membranes for Selective Recovery*

Nanotechnology has significantly influenced the advancement of membrane performance in MBRs, particularly for selective separation and recovery of valuable compounds (Pervez et al., 2020). Incorporation of nanoparticles such as TiO_2, ZnO, Ag, and carbon nanotubes into membrane matrices improves properties like hydrophilicity, mechanical strength, and antimicrobial resistance, which in turn enhance permeability and reduce fouling tendencies (Bolto & Xie, 2018). Nanostructured membranes can be engineered with molecular-level precision to allow selective recovery of nutrients like ammonium and phosphate or to retain micropollutants and pharmaceuticals. Functionalized nanomaterials provide reactive sites for adsorption or catalytic degradation of contaminants, offering both separation and transformation functionalities within a single membrane module. Furthermore, emerging developments in nano-enabled affinity membranes, such as metal–organic frameworks (MOFs) and graphene oxide composites, have shown promise for targeted recovery of rare earth elements and bioactive substances from industrial effluents, opening new avenues in wastewater valorisation (Faheem et al., 2024).

2.7.4 Development of Advanced Membrane Materials (Nanocomposites, Antifouling Coatings)

The development of novel membrane materials is central to enhancing MBR efficiency, particularly in mitigating membrane fouling, which is a persistent operational challenge. Nanocomposite membranes, fabricated by embedding functional nanomaterials like silica, zeolites, or carbon nanostructures into polymer matrices, offer improved porosity, hydrophilicity, and thermal stability (Saraswathi et al., 2019). These modifications enable higher flux and longer operational life. Concurrently, antifouling strategies such as surface grafting with hydrophilic polymers (e.g., polyethylene glycol) or zwitterionic coatings disrupt foulant adhesion and biofilm formation, reducing chemical cleaning frequency and extending membrane longevity (Maan et al., 2020). Photocatalytic coatings, especially those based on TiO_2, further assist in degrading organic foulants upon UV activation, providing a self-cleaning effect (Nguyen et al., 2022). The synthesis of asymmetric thin-film composite membranes with dense selective layers also allows for the fine-tuned separation of specific solutes while minimizing energy input. These material innovations collectively contribute to the economic and operational feasibility of MBR technology at larger scales.

2.7.5 AI-Driven Process Control and Performance Optimization

Artificial intelligence (AI) is revolutionizing MBR operation through real-time monitoring, predictive analytics, and autonomous process control. Machine learning algorithms can analyse complex datasets from sensors measuring parameters like transmembrane pressure, sludge retention time, and organic loading rate, thereby enabling early detection of membrane fouling and operational anomalies (Frontistis et al., 2023). Neural networks and decision tree models are employed to optimise aeration control, chemical dosing, and backwash cycles, reducing energy consumption and chemical use without compromising treatment performance (Ibrahim et al., 2024). Digital twins of MBR systems, built on AI frameworks, simulate process scenarios to support design, troubleshooting, and performance forecasting in real-time environments. Additionally, AI facilitates the integration of MBRs into smart water grids, allowing adaptive control based on fluctuating influent quality or environmental conditions (Chang et al., 2024). The incorporation of AI transforms MBRs into intelligent systems that can self-regulate, minimize downtime, and support predictive maintenance, thereby enhancing reliability and sustainability in modern wastewater treatment infrastructures.

2.8 Sustainability and Economic Aspects of MBR-Based Resource Recovery

Membrane bioreactor (MBR) systems are increasingly recognized not only for their efficiency in wastewater treatment but also for their potential to support sustainable resource recovery. A key method for assessing the environmental implications of these systems is Life Cycle Assessment (LCA), which evaluates impacts across the full system life from membrane production to the final disposal of sludge. Research consistently shows that energy use and chemical consumption, especially for membrane aeration and fouling control, are among the most significant contributors to environmental burden (Krzeminski et al., 2017). Although MBRs may require higher energy and material inputs at the outset, especially compared to conventional activated sludge systems, their long-term environmental advantages are notable. These include improved effluent quality, reduced chemical dosing, and enhanced water reuse potential. Technological developments such as low-energy membranes, automated control systems, and modular decentralised treatment plants have significantly decreased the global warming and eutrophication potential of MBR operations (Razman et al., 2022). Moreover, incorporating nutrient and energy recovery into LCA models shows that MBRs can potentially achieve net-zero or even net-positive resource performance (Diaz-Elsayed et al., 2021). From an economic perspective, a comprehensive cost–benefit analysis (CBA) of MBR systems must consider both direct and indirect values. These include the monetary value of reclaimed water, recovered nutrients, biogas production, and external environmental benefits like emission reduction and decreased sludge handling. While capital costs remain higher for MBRs due to membrane module expenses and system complexity, these are often balanced by lower operating costs (OPEX), particularly when treated water and energy are reused on-site (Krzeminski et al., 2017). MBRs also offer greater economic viability in regions with strict effluent discharge regulations, as they can remove the need for additional tertiary treatment (Judd, 2010). Additional economic gains are realized through the integration of anaerobic digestion, which enables energy recovery in the form of biogas, and nutrient recovery systems that reclaim phosphorus and nitrogen for reuse. Studies have shown that these combined approaches can deliver significant long-term returns and contribute to the overall financial sustainability of MBR operations (Zielińska & Bułkowska, 2024). Moreover, when considering avoided costs such as reduced potable water demand, minimized environmental fines, and decreased sludge disposal the economic case for MBRs becomes even more compelling. MBRs are also pivotal in advancing the circular economy within the water sector. Their ability to produce high-quality effluent suitable for industrial, agricultural, and even potable reuse extends the value chain of water resources (Jijingi et al., 2024). Anaerobic MBRs (AnMBRs) take this a step further by combining wastewater treatment with energy recovery, transforming waste into valuable products like methane and biosolids (Zielińska & Ojo, 2023). Additionally, nutrient recovery technologies embedded in MBR systems enable the extraction of phosphorus as struvite and nitrogen as ammonium sulfate, contributing to a sustainable nutrient management framework (Rey-Martínez et al., 2024). The adoption of MBRs has grown rapidly in both municipal and industrial sectors, particularly in high-density or water-scarce regions where water reuse is a policy priority. Countries such as Singapore have adopted MBRs to meet stringent reuse standards, with programs like NEWater serving as global benchmarks (Gu et al., 2019). In industry, sectors including pharmaceuticals, petrochemicals, and food processing rely on MBRs due to their compact footprint, operational robustness, and ability to handle high-strength or variable influent loads (Jijingi et al., 2024). Furthermore, the scalability and remote-operability of MBR units make them ideal for decentralized treatment in rural and peri-urban areas (Capodaglio, 2017). Case studies from across Europe, China, and the Middle East illustrate successful applications of MBRs that integrate goals such as energy self-sufficiency and nutrient recycling. To fully leverage the benefits of MBRs, there is a pressing need for standardized performance metrics and regulatory frameworks. Currently, the absence of unified global standards for membrane durability, fouling resistance, and recovery efficiency poses a barrier to broader deployment (Shehata et al., 2023). Organizations such as the International Water Association (IWA) and ISO technical committees are actively working on setting performance benchmarks and certification protocols. Simultaneously, national regulations must evolve to classify MBRs not only as treatment systems but also as resource recovery platforms. This includes establishing clear guidelines for the safe use and commercialization of recovered products, streamlining approval processes, and adopting performance-based regulatory approaches that support innovation and ensure product safety (Jijingi et al., 2024). Thus, MBR technology offers a robust pathway toward sustainable, circular water management. With the right policy support, technological innovation, and standardization, MBRs have the potential to transform wastewater treatment into a platform for resource recovery, environmental protection, and long-term economic resilience.

2.9 Challenges and Limitations of MBRs in Resource Recovery

Membrane bioreactors (MBRs), while representing a significant advancement in wastewater treatment and resource recovery, face several critical challenges and limitations that hinder their widespread implementation at larger scales (Aslam et al., 2022). Chief among these is membrane fouling, which arises from the accumulation of organic matter, microbial cells, extracellular polymeric substances (EPS), and inorganic scalants on membrane surfaces or within pores, resulting in decreased permeability, elevated transmembrane pressure (TMP), and increased operational costs. Fouling mechanisms can be biological, organic, inorganic, or particulate in nature, and their impacts are exacerbated under high-loading conditions and extended hydraulic retention times (Hanvajanawong et al., 2022). Mitigation strategies such as periodic backwashing, relaxation cycles, and advanced surface modifications like the incorporation of hydrophilic coatings, nanoparticles (e.g., TiO_2, Ag), and antifouling polymers have shown potential in reducing fouling but often fall short under real-world conditions (Bolto and Xie, 2018). Alongside fouling, the energy consumption of MBR systems remains a primary concern, particularly due to the high aeration requirements for biological treatment and membrane scouring, which can account for over 50% of the total energy demand (Hamedi et al., 2019). Innovative aeration control strategies, such as intermittent aeration and the use of fine bubble diffusers, along with the integration of anaerobic MBRs (AnMBRs), offer promising solutions to reduce energy intensity while enabling methane recovery for energy reuse. The durability and longevity of membranes are equally critical, as they are subjected to mechanical stress, chemical exposure from cleaning agents, and biological degradation, which cumulatively reduce service life and increase maintenance frequency (Rajendran et al., 2024). Scaling, particularly from calcium carbonate and metal oxides, further contributes to performance deterioration, especially in high-salinity or industrial effluents. Advanced materials such as ceramic membranes, polymeric nanocomposites, and zwitterionic surface coatings have improved resistance to these stressors, but high material costs remain a barrier (Li et al., 2022). Maintenance practices rely heavily on routine chemical cleaning, which, over time, can degrade membrane integrity and lead to operational downtime. Moreover, real-time membrane health monitoring and predictive maintenance, though promising with the advent of smart sensors and AI algorithms, are still in developmental stages (Villegas-Ch et al., 2024). Cost-effectiveness and scalability of MBRs are additional barriers; while MBRs provide high-quality effluent suitable for reuse, their capital and operational expenditures driven by membrane costs, energy use, and skilled labor requirements are significantly higher than conventional systems. Economic feasibility improves in scenarios prioritizing water reuse, nutrient recovery, and resource circularity, yet such benefits are often location-dependent and contingent on supporting infrastructure and regulatory frameworks. Life cycle assessments (LCA) suggest that with energy recovery integration and renewable energy use, the environmental and economic footprint of MBRs can be significantly reduced (Diaz-Elsayed et al., 2021; Krzeminski et al., 2017).

2.10 Conclusion

The advancement of membrane bioreactor (MBR) technologies has redefined wastewater treatment from a traditional pollutant removal operation to a sustainable platform for resource recovery, aligned with the principles of circular economy and environmental safety. MBRs effectively combine biological degradation and membrane filtration, providing high-quality effluent suitable for reuse and enabling the recovery of valuable resources such as nitrogen, phosphorus, organic matter, and energy in the form of biogas. The precision in contaminant separation achieved through ultrafiltration and microfiltration membranes integrated with biological systems allows for compliance with stringent water reuse regulations, particularly in water-scarce regions and industrial applications. Innovations in hybrid systems, including MBR-RO and MBR-AnMBR integrations, have expanded the functional potential of these systems, facilitating simultaneous wastewater reclamation and energy generation. Additionally, the development of advanced membrane materials such as nanocomposites, antifouling coatings, and selective separation layers has addressed some limitations associated with fouling, membrane degradation, and selective nutrient recovery. The integration of artificial intelligence and machine learning has further enhanced operational control, optimizing energy usage, membrane cleaning cycles, and predictive maintenance strategies. Despite these advancements, challenges such as membrane fouling, energy intensity, capital costs, and regulatory complexities continue to impact the feasibility of large-scale implementation. However, life cycle assessment studies underscore the potential of MBRs to reduce greenhouse gas emissions and environmental burdens when resource recovery and renewable energy are integrated. Moreover, supportive policy frameworks, industrial collaboration, and targeted investments in research and infrastructure are critical to accelerating the transition of MBRs from niche applications to mainstream sustainable wastewater management strategies. As the global demand for clean water and sustainable waste processing intensifies, MBRs stand as a cornerstone technology with immense potential to transform wastewater into a reservoir of recoverable resources, promoting resilience, efficiency, and environmental integrity in future water management systems. In conclusion, while membrane fouling, energy consumption, membrane longevity, and cost challenges currently limit the widespread adoption of MBRs, advancements in material science, process optimisation, and intelligent control are progressively bridging these gaps, paving the way for more sustainable and efficient deployment in the future of wastewater resource recovery.

References

Adam, C., & Krüger, O. (2018). Wastewater as a resource: From rare earth metals to phosphorus. *Phosphorus: Polluter and Resource*, 241. https://doi.org/10.2166/9781780408361_241

Akinyemi, B. T., Ogundele, O. D., & Afolabi, A. B. (2023). Advancements in sustainable membrane technologies for enhanced remediation and wastewater treatment: A comprehensive review. *Acadlore Transactions on Geosciences, 2*(4), 196–207. https://doi.org/10.56578/atg020402

Alayande, A. B., Qi, W., Karthikeyan, R., Popat, S., Ladner, D. A., & Amy, G. (2024). Use of reclaimed municipal wastewater in agriculture: Comparison of present practice versus an emerging paradigm of anaerobic membrane bioreactor treatment coupled with hydroponic controlled environment agriculture. *Water Research*, 122197. https://doi.org/10.1016/j.watres.2024.122197

An, Z., Zhu, J., Zhang, M., Zhou, Y., Su, X., Lin, H., & Sun, F. (2023). Anaerobic membrane bioreactor for the treatment of high-strength waste/wastewater: A critical review and update. *Chemical Engineering Journal, 470*, Article 144322. https://doi.org/10.1016/j.cej.2023.144322

Aryanfar, Y., García Alcaraz, J. L., Keçebaş, A., Fernandez, J. B., Arslan, B., Ilbas, M., & Irshad, K. (2025). Transforming food industrial sludge into sustainable resources: Innovations in waste management and renewable energy recovery. *Business Strategy and the Environment*.https://doi.org/10.1002/bse.4170

Asif, M. B., & Hai, F. I. (2019). Wastewater reuse applications and MBR. *Membr Biol React*, 183–205. https://doi.org/10.2166/9781780409177_0183

Asif, M. B., & Zhang, Z. (2021). Ceramic membrane technology for water and wastewater treatment: A critical review of performance, full-scale applications, membrane fouling and prospects. *Chemical Engineering Journal, 418*, Article 129481. https://doi.org/10.1016/j.cej.2021.129481

Aslam, A., Khan, S. J., & Shahzad, H. M. A. (2022). Anaerobic membrane bioreactors (AnMBRs) for municipal wastewater treatment-potential benefits, constraints, and future perspectives: An updated review. *Science of the Total Environment, 802*, Article 149612. https://doi.org/10.1016/j.scitotenv.2021.149612

Azis, K., Ntougias, S., & Melidis, P. (2019). Fouling control, using various cleaning methods, applied on an MBR system through continuous TMP monitoring. *Desalination and Water Treatment, 167*, 343–350. https://doi.org/10.5004/dwt.2019.24609

Bandh, S. A., & Mushtaq, B. (2025). Water and wastewater treatment technologies. In *Wastewater treatment technology: Technical progressions and advances* (pp. 67–163). Springer Nature.

Bera, S. P., Godhaniya, M., & Kothari, C. (2022). Emerging and advanced membrane technology for wastewater treatment: A review. *Journal of Basic Microbiology, 62*(3–4), 245–259. https://doi.org/10.1002/jobm.202100259

Bolto, B. A., & Xie, Z. (2018). Recent developments in fouling minimization of membranes modifed with silver nanoparticles. *Journal of Membrane Science and Research, 4*(3), 111–120. https://doi.org/10.22079/jmsr.2018.79056.1168

Campos, J. L., Crutchik, D., Franchi, Ó., Pavissich, J. P., Belmonte, M., Pedrouso, A., & Val del Río, Á. (2019). Nitrogen and phosphorus recovery from anaerobically pretreated agro-food wastes: A review. *Frontiers in Sustainable Food Systems, 2*, 91. https://doi.org/10.3389/fsufs.2018.00091

Capodaglio, A. G. (2017). Integrated, decentralized wastewater management for resource recovery in rural and peri-urban areas. *Resources, 6*(2), 22. https://doi.org/10.3390/resources6020022

Castro, K., & Abejón, R. (2024). Removal of heavy metals from wastewaters and other aqueous streams by pressure-driven membrane technologies: An outlook on reverse osmosis, nanofiltration, ultrafiltration and microfiltration potential from a bibliometric analysis. *Membranes, 14*(8), 180. https://doi.org/10.3390/membranes14080180

Chang, H. L., Liu, Y. L., Keng, C. J., Jiang, H. L., & Hu, J. (2024). Challenges of industrial wastewater treatment: Utilizing membrane bioreactors (MBRs) in conjunction with artificial intelligence (AI) technology. *Journal of Industrial and Production Engineering, 41*(5), 422–427. https://doi.org/10.1080/21681015.2024.2330401

Cheng, C., Zhou, Z., Pang, H., Zheng, Y., Chen, L., Jiang, L. M., & Zhao, X. (2018). Correlation of microbial community structure with pollutants removal, sludge reduction and sludge characteristics in micro-aerobic side-stream reactor coupled membrane bioreactors under different hydraulic retention times. *Bioresource Technology, 260*, 177–185. https://doi.org/10.1016/j.biortech.2018.03.088

Choudhury, R. R., Gohil, J. M., Mohanty, S., & Nayak, S. K. (2018). Antifouling, fouling release and antimicrobial materials for surface modification of reverse osmosis and nanofiltration membranes. *Journal of Materials Chemistry A, 6*(2), 313–333. https://doi.org/10.1039/C7TA08627J

Diaz-Elsayed, N., Rezaei, N., Guo, T., Mohebbi, S., & Zhang, Q. (2019). Wastewater-based resource recovery technologies across scale: A review. *Resources, Conservation and Recycling, 145*, 94–112. https://doi.org/10.1016/j.resconrec.2018.12.035

Diaz-Elsayed, N., Mo, W., & Zhang, Q. (2021). The sustainability dimensions of resource recovery from "Wastewater". In *Resource recovery from wastewater* (pp. 1–43). Apple Academic Press. https://doi.org/10.1007/s43615-023-00276-z

El Batouti, M., Alharby, N. F., & Elewa, M. M. (2021). Review of new approaches for fouling mitigation in membrane separation processes in water treatment applications. *Separations, 9*(1), 1. https://doi.org/10.3390/separations9010001

Faheem, M., Hassan, M. A., Du, J., & Wang, B. (2024). Harnessing potential of smart and conventional spent adsorbents: Global practices and sustainable approaches through regeneration and tailored recycling. *Separation and Purification Technology*, 128907. https://doi.org/10.1016/j.seppur.2024.128907

Frontistis, Z., Lykogiannis, G., & Sarmpanis, A. (2023). Machine learning implementation in membrane bioreactor systems: Progress, challenges, and future perspectives: A review. *Environments, 10*(7), 127. https://doi.org/10.3390/environments10070127

Gao, Y., Fang, Z., Liang, P., & Huang, X. (2018). Direct concentration of municipal sewage by forward osmosis and membrane fouling behavior. *Bioresource Technology, 247*, 730–735.

Giwa, A., Shehu, H., Ramalan, M., Orakwe, I., Abunomah, O., Ogunlude, P., Williamwest, T., Igbagara, W., Ogoun, E., Hashim, I., & Aisueni, F. (2022). Microporous alumina–silica composite membrane with very low N2 permeability but high CO2 selectivity considered for direct air capture. In A. A. Khan, M. L. Ciddi, & M. Unal (Eds.), *Proceedings of ICSEST* (pp. 182–210).

Gkotsis, P. K., Banti, D. C., Peleka, E. N., Zouboulis, A. I., & Samaras, P. E. (2014). Fouling issues in membrane bioreactors (MBRs) for wastewater treatment: Major mechanisms, prevention and control strategies. *Processes, 2*(4), 795–866. https://doi.org/10.3390/pr2040795

Gu, J., Liu, H., Wang, S., Zhang, M., & Liu, Y. (2019). An innovative anaerobic MBR-reverse osmosis-ion exchange process for energy-efficient reclamation of municipal wastewater to NEWater-like product water. *Journal of Cleaner Production, 230*, 1287–1293. https://doi.org/10.1016/j.jclepro.2019.05.198

Gutiérrez, M., Grillini, V., Pavlović, D. M., & Verlicchi, P. (2021). Activated carbon coupled with advanced biological wastewater treatment: A review of the enhancement in micropollutant removal. *Science of the Total Environment, 790*, Article 148050. https://doi.org/10.1016/j.scitotenv.2021.148050

Hamedi, H., Ehteshami, M., Mirbagheri, S. A., Rasouli, S. A., & Zendehboudi, S. (2019). Current status and future prospects of membrane bioreactors (MBRs) and fouling phenomena: A systematic review. *The Canadian Journal of Chemical Engineering, 97*(1), 32–58. https://doi.org/10.1002/cjce.23345

Hanvajanawong, K., Suyamud, B., Suwannasilp, B. B., Lohwacharin, J., & Visvanathan, C. (2022). Unravelling capability of two-stage thermophilic anaerobic membrane bioreactors for high organic loading wastewater: Effect of support media addition and irreversible fouling. *Bioresource Technology, 348*, Article 126725. https://doi.org/10.1016/j.biortech.2022.126725

Helmi, A., & Gallucci, F. (2020). Latest developments in membrane (bio) reactors. *Processes, 8*(10), 1239. https://doi.org/10.3390/pr8101239

Hube, S., & Wu, B. (2021). Mitigation of emerging pollutants and pathogens in decentralized wastewater treatment processes: A review. *Science of the Total Environment, 779*, Article 146545. https://doi.org/10.1016/j.scitotenv.2021.146545

Ibrahim, M., Haider, A., Lim, J. W., Mainali, B., Aslam, M., Kumar, M., & Shahid, M. K. (2024). Artificial neural network modeling for the prediction, estimation, and treatment of diverse wastewaters: A comprehensive review and future perspective. *Chemosphere*, 142860. https://doi.org/10.1016/j.chemosphere.2024.142860

Izadi, P., Izadi, P., & Eldyasti, A. (2020). Design, operation and technology configurations for enhanced biological phosphorus removal (EBPR) process: A review. *Reviews in Environmental Science and Bio/technology, 19*, 561–593. https://doi.org/10.1007/s11157-020-09538-w

James, L. (2024). *Development of innovative ultra-filtration fibre technologies for waste-water treatment in membrane bio-reactors (MBR)* (Doctoral dissertation, UNSW Sydney).

Jha, S., & Mishra, B. K. (2024). An overview of deploying different treatment processes with membrane bioreactor for enhanced treatment of wastewaters: synergistic performances and reduced fouling of membrane. *Environmental Science and Pollution Research*, 1–32.https://doi.org/10.1007/s11356-024-35459-0

Jijingi, H. E., Yazdia, S. K., Abakr, Y. A., & Etim, E. (2024). Evaluation of membrane bioreactor (MBR) technology for industrial wastewater treatment and its application in developing countries: A review. *Case Studies in Chemical and Environmental Engineering*, 100886.https://doi.org/10.2139/ssrn.4836303

Judd, S. (2010). In: *The MBR book: Principles and applications of membrane bioreactors for water and wastewater treatment*. Elsevier. ISBN: 978-0-08-096682-3

Kabeyi, M. J. B., & Olanrewaju, O. A. (2022). Biogas production and applications in the sustainable energy transition. *Journal of Energy, 2022*(1), 8750221. https://doi.org/10.1155/2022/8750221

Keyvan Hosseini, M., Liu, L., Keyvan Hosseini, P., Bhattacharyya, A., Lee, K., Miao, J., & Chen, B. (2022). Review of hollow fiber (HF) membrane filtration technology for the treatment of oily wastewater: Applications and challenges. *Journal of Marine Science and Engineering, 10*(9), 1313. https://doi.org/10.3390/jmse10091313

Krzeminski, P., Leverette, L., Malamis, S., & Katsou, E. (2017). Membrane bioreactors–A review on recent developments in energy reduction, fouling control, novel configurations, LCA and market prospects. *Journal of Membrane Science, 527*, 207–227. https://doi.org/10.1016/j.memsci.2016.12.010

Kürzl, C., & Kulozik, U. (2023). Comparison of the efficiency of pulsed flow membrane cleaning in hollow fibre (HFM) and spiral-wound microfiltration membranes (SWM). *Food and Bioproducts Processing, 139*, 166–177. https://doi.org/10.1016/j.fbp.2023.03.012

Labhasetwar, P. K., & Yadav, A. (2023). In *Membrane based point-of-use drinking water treatment systems*. IWA Publishing.https://doi.org/10.2166/9781789062724

Li, Y., Zhang, B., Li, G., & Luo, W. (2018). Osmotic membrane bioreactor and its hybrid systems for wastewater reuse and resource recovery: Advances, challenges, and future directions. *Current Pollution Reports, 4*(1), 23–34. https://doi.org/10.1007/s40726-018-0080-1

Li, Q., Wen, C., Yang, J., Zhou, X., Zhu, Y., Zheng, J., & Zhang, P. (2022). Zwitterionic biomaterials. *Chemical Reviews, 122*(23), 17073–17154. https://doi.org/10.1021/acs.chemrev.2c00344

Lim, S., Choi, P. J., An, A. K., Han, D. S., Phuntsho, S., & Shon, H. (2021). Submerged versus side-stream osmotic membrane bioreactors using an outer-selective hollow fiber osmotic membrane for desalination. *Desalination, 515*, Article 115196. https://doi.org/10.1016/j.desal.2021.115196

Maan, A. M., Hofman, A. H., de Vos, W. M., & Kamperman, M. (2020). Recent developments and practical feasibility of polymer-based antifouling coatings. *Advanced Functional Materials, 30*(32), 2000936. https://doi.org/10.1002/adfm.202000936

Maletskyi, Z. (2020). Advances in membrane materials and processes for water and wastewater treatment. In *Multidisciplinary advances in efficient separation processes* (pp. 3–35). American Chemical Society. https://doi.org/10.1021/bk 2020 1348.ch001

Mannina, G., Capodici, M., Cosenza, A., Di Trapani, D., & Ekama, G. A. (2018). The effect of the solids and hydraulic retention time on moving bed membrane bioreactor performance. *Journal of Cleaner Production, 170*, 1305–1315. https://doi.org/10.1016/j.jclepro.2017.09.200

Mannina, G., Ni, B. J., Makinia, J., Harmand, J., Alliet, M., Brepols, C., & Comas, J. (2023). Biological processes modelling for MBR systems: A review of the state-of-the-art focusing on SMP and EPS. *Water Research, 242*, Article 120275. https://doi.org/10.1016/j.watres.2023.120275

Martínez, R., Ruiz, M. O., Ramos, C., Cámara, J. M., & Diez, V. (2020). Comparison of external and submerged membranes used in anaerobic membrane bioreactors: Fouling related issues and biological activity. *Biochemical Engineering Journal, 159*, Article 107558. https://doi.org/10.1016/j.bej.2020.107558

Mojiri, A., Zhou, J. L., KarimiDermani, B., Razmi, E., & Kasmuri, N. (2023). Anaerobic membrane bioreactor (AnMBR) for the removal of dyes from water and wastewater: Progress, challenges, and future perspectives. *Processes, 11*(3), 855. https://doi.org/10.3390/pr11030855

Muhmood, A., Lu, J., Dong, R., & Wu, S. (2019). Formation of struvite from agricultural wastewaters and its reuse on farmlands: Status and hindrances to closing the nutrient loop. *Journal of Environmental Management, 230*, 1–13. https://doi.org/10.1016/j.jenvman.2018.09.030

Munasinghe-Arachchige, S. P., Abeysiriwardana-Arachchige, I. S. A., Delanka-Pedige, H. M. K., Cooke, P., & Nirmalakhandan, N. (2021). Nitrogen-fertilizer recovery from urban sewage via gas permeable membrane: Process analysis, modeling, and intensification. *Chemical Engineering Journal, 411*, Article 128443. https://doi.org/10.1016/j.cej.2021.128443

Musa, M. A., Idrus, S., Che Man, H., & Nik Daud, N. N. (2018). Wastewater treatment and biogas recovery using anaerobic membrane bioreactors (AnMBRs): Strategies and achievements. *Energies, 11*(7), 1675. https://doi.org/10.3390/en11071675

Nabi, M., Liang, H., Zhou, Q., Cao, J., & Gao, D. (2023). In-situ membrane fouling control and performance improvement by adding materials in anaerobic membrane bioreactor: A review. *Science of the Total Environment, 865*, Article 161262. https://doi.org/10.1016/j.scitotenv.2022.161262

Nadeem, K., Alliet, M., Plana, Q., Bernier, J., Azimi, S., Rocher, V., & Albasi, C. (2022). Modeling, simulation and control of biological and chemical P-removal processes for membrane bioreactors (MBRs) from lab to full-scale applications: State of the art. *Science of the Total Environment, 809*, Article 151109. https://doi.org/10.1016/j.scitotenv.2021.151109

Namdeti, R., Lakkimsetty, N. R., Rao, G. B., Thandlam, A. K., Karu, C. V., Doddamani, D., & Prema, P. M. (2025). Membrane bioreactors: Integration of biological processes with membrane technology for wastewater treatment. *Journal of Membrane Science and Research, 11*(2). https://doi.org/10.22079/jmsr.2025.2040886.1677

Nath, K. (2017). In *Membrane separation processes*. PHI Learning Pvt. Ltd.

Naz, A., Sattar, R., & Siddiq, M. (2019). Polymer membranes for biofouling mitigation: A review. *Polymer-Plastics Technology and Materials, 58*(17), 1829–1854. https://doi.org/10.1080/25740881.2019.1576200

Nguyen, H. T., Pham, M. T., Nguyen, T. M. T., Bui, H. M., Wang, Y. F., & You, S. J. (2022). Modifications of conventional organic membranes with photocatalysts for antifouling and self-cleaning properties applied in wastewater filtration and separation processes: A review. *Separation Science and Technology, 57*(9), 1471–1500. https://doi.org/10.1080/01496395.2021.1982981

Osman, A. I., Chen, Z., Elgarahy, A. M., Farghali, M., Mohamed, I. M., Priya, A. K., & Yap, P. S. (2024). Membrane technology for energy saving: Principles, techniques, applications, challenges, and prospects. *Advanced Energy and Sustainability Research, 5*(5), 2400011. https://doi.org/10.1002/aesr.202400011

Pervez, M. N., Balakrishnan, M., Hasan, S. W., Choo, K. H., Zhao, Y., Cai, Y., & Naddeo, V. (2020). A critical review on nanomaterials membrane bioreactor (NMs-MBR) for wastewater treatment. *NPJ Clean Water, 3*(1), 43. https://doi.org/10.1038/s41545-020-00090-2

Rahman, T. U., Roy, H., Islam, M. R., Tahmid, M., Fariha, A., Mazumder, A., & Islam, M. S. (2023). The advancement in membrane bioreactor (MBR) technology toward sustainable industrial wastewater management. *Membranes, 13*(2), 181. https://doi.org/10.3390/membranes13020181

Rajendran, D. S., Devi, E. G., Subikshaa, V. S., Sethi, P., Patil, A., Chakraborty, A., ... & Kumar, V. V. (2024). Recent advances in various cleaning strategies to control membrane fouling: A comprehensive review. *Clean Technologies and Environmental Policy*, 1–16.https://doi.org/10.1007/s10098-024-03000-z

Ray, P., Polisetti, V., & Singh, P. S. (2021). Membrane bioreactors: an advanced technology to treat industrial waste water.https://doi.org/10.1039/9781839165399-00174

Razman, K. K., Hanafiah, M. M., & Mohammad, A. W. (2022). An overview of LCA applied to various membrane technologies: Progress, challenges, and harmonization. *Environmental Technology and Innovation, 27*, Article 102803. https://doi.org/10.1016/j.eti.2022.102803

Rey-Martínez, N., Torres-Sallan, G., Morales, N., Serra, E., Bisschops, I., van Eekert, M. H., ... & Sanchis, S. (2024). Combination of technologies for nutrient recovery from wastewater: A review. *Cleaner Waste Systems*, 100139.https://doi.org/10.1016/j.clwas.2024.100139

Sahu, A., Dosi, R., Kwiatkowski, C., Schmal, S., & Poler, J. C. (2023). Advanced polymeric nanocomposite membranes for water and wastewater treatment: A comprehensive review. *Polymers, 15*(3), 540. https://doi.org/10.3390/polym15030540

Sahu, L. R., Yadav, D., Borah, D., Gogoi, A., Goswami, S., Hazarika, G., & Ingole, P. G. (2024). Polymeric membranes for liquid separation: Innovations in materials, fabrication, and industrial applications. *Polymers, 16*(23), 3240.https://doi.org/10.3390/polym16233240

Saraswathi, M. S. S. A., Nagendran, A., & Rana, D. (2019). Tailored polymer nanocomposite membranes based on carbon, metal oxide and silicon nanomaterials: A review. *Journal of Materials Chemistry A, 7*(15), 8723–8745. https://doi.org/10.1039/C8TA11460A

Shehata, N., Egirani, D., Olabi, A. G., Inayat, A., Abdelkareem, M. A., Chae, K. J., & Sayed, E. T. (2023). Membrane-based water and wastewater treatment technologies: Issues, current trends, challenges, and role in achieving sustainable development goals, and circular economy. *Chemosphere, 320*, Article 137993. https://doi.org/10.1016/j.chemosphere.2023.137993

Silva, L. L., Moreira, C. G., Curzio, B. A., & da Fonseca, F. V. (2017). Micropollutant removal from water by membrane and advanced oxidation processes—A review. *Journal of Water Resource and Protection, 9*(5), 411–431. https://doi.org/10.4236/jwarp.2017.95027

Sukphun, P., Sittijunda, S., & Reungsang, A. (2021). Volatile fatty acid production from organic waste with the emphasis on membrane-based recovery. *Fermentation, 7*(3), 159. https://doi.org/10.3390/fermentation7030159

Tran, D. P. H., You, S. J., Bui, X. T., Wang, Y. F., & Ramos, A. (2024). Anaerobic membrane bioreactors for municipal wastewater: Progress in resource and energy recovery improvement approaches. *Journal of Environmental Management, 366*, Article 121855. https://doi.org/10.1016/j.jenvman.2024.121855

Turan, M. (2025). Membrane fouling and control approaches in membrane bioreactor systems: A review. *Water Air Soil Pollution, 236*(3), 1–40. https://doi.org/10.1007/s11270-025-07794-6

Türker, M., Selimoğlu, S. M., & Taşpınar-Demir, H. (2022). Waste (water) to feed protein: Effluent characteristics, protein recovery, and single-cell protein production from food industry waste streams. In *Clean energy and resource recovery* (pp. 201–244).https://doi.org/10.1016/B978-0-323-90178-9.00017-2

Villegas-Ch, W., García-Ortiz, J., & Sánchez-Viteri, S. (2024). Towards intelligent monitoring in iot: Ai applications for real-time analysis and prediction. *IEEE Access*.https://doi.org/10.1109/ACCESS.2024.3376707

Xie, Y. (2021). *Bioreactor strategies for sustainable nitrogen cycling based on mineralization/nitrification, partial nitritation/anammox or sulfur-based denitratation* (Doctoral dissertation, University of Antwerp).

Yadav, A., Labhasetwar, P. K., & Shahi, V. K. (2022). Membrane distillation crystallization technology for zero liquid discharge and resource recovery: Opportunities, challenges and futuristic perspectives. *Science of the Total Environment, 806*, Article 150692. https://doi.org/10.1016/j.scitotenv.2021.150692

Yang, J., Monnot, M., Ercolei, L., & Moulin, P. (2020). Membrane-based processes used in municipal wastewater treatment for water reuse: State-of-the-art and performance analysis. *Membranes, 10*(6), 131. https://doi.org/10.3390/membranes10060131

Yang, Y., Qiao, S., Jin, R., Zhou, J., & Quan, X. (2019). A novel aerobic electrochemical membrane bioreactor with CNTs hollow fiber membrane by electrochemical oxidation to improve water quality and mitigate membrane fouling. *Water Research, 151*, 54–63.

Zhang, J., Xiao, K., Liu, Z., Gao, T., Liang, S., & Huang, X. (2021). Large-scale membrane bioreactors for industrial wastewater treatment in China: Technical and economic features, driving forces, and perspectives. *Engineering, 7*(6), 868–880. https://doi.org/10.1016/j.eng.2020.09.012

Zhang, K., Wu, H. H., Huo, H. Q., Ji, Y. L., Zhou, Y., & Gao, C. J. (2022). Recent advances in nanofiltration, reverse osmosis membranes and their applications in biomedical separation field. *Chinese Journal of Chemical Engineering, 49*, 76–99. https://doi.org/10.1016/j.cjche.2022.06.017

Zielińska, M., & Ojo, A. (2023). Anaerobic membrane bioreactors (ANMBRs) for wastewater treatment: Recovery of nutrients and energy, and management of fouling. *Energies, 16*(6), 2829. https://doi.org/10.3390/en16062829

Zielińska, M., & Bułkowska, K. (2024). Sustainable management and advanced nutrient recovery from biogas energy sector effluents. *Energies, 17*(15), 3705.

Chapter 3
Nanofiltration: Selective Separation of Nutrients from Wastewater Streams

Shilpi R. Sindhu, Anita Verma, Reena Rani, Nisha Boora, Bharat Garg, Mandeep Redhu, and Shikha Yashveer

Abstract Wastewater streams have drawn more attention recently as a potential alternate source of nutrients including phosphorous and nitrogen. It is possible to isolate and recover valuable forms of essential elements, such as Nitrogen and Phosphorous, from wastewater streams using appropriate treatment procedures. However, the effectiveness of water purification and nutrients recovery depends on the selective exclusion or recovery of selected solutes, such as micropollutants, precious elements, and minerals from wastewater. However, this process is both difficult and essential. An efficient treatment procedure for the selective exclusion of nutrients from wastewater streams is necessary to mitigate this problem. Firstly, in this chapter we discuss the basic principal concepts, functioning, and uses of nanofiltration in the recovery and selective removal of important nutrients from various wastewater streams, including phosphorus, nitrogen, and trace elements. Next, we go into the material characteristics of nanofiltration membranes, the function of pressure-driven procedures, and how operating parameters affect separation effectiveness. This chapter explores the challenges and opportunities of using highly selective nanofiltration membranes in realistic environmental applications. It highlights the existing information gap that needs to be addressed. Bridging this gap will guide future research aimed at enhancing environmental sustainability through more advanced and adaptable nanofiltration membrane separation. Finally, this chapter concludes with a discussion of the difficulties and potential paths ahead for improving the

S. R. Sindhu · A. Verma · R. Rani · S. Yashveer (✉)
Department of Molecular Biology and Biotechnology, College of Biotechnology, CCS HAU, Hisar, India
e-mail: shikhayashveer@gmail.com

N. Boora
Department of Bioinformatics and Computational Biology, College of Biotechnology, CCS HAU, Hisar, India

B. Garg
Centre for Crop Health, University of Southern Queensland, Toowoomba, QLD, Australia

M. Redhu
Department of Plant, Soil and Agricultural Systems, College of Agricultural, Life and Physical Sciences, Southern Illinois University, Carbondale, IL, USA

S. Dhiman and G. Mukherjee (eds.), *Resource Resurgence*, Springer Water,
https://doi.org/10.1007/978-3-032-02750-4_3

sustainability and effectiveness of nanofiltration in the selective removal of nutrients from wastewater.

Keywords Nanofiltration · Wastewater streams · Nutrients · Environmental sustainability

3.1 Introduction

Domestic wastewater streams represent a majorresource of essential nutrients such as nitrogen (N) and phosphorus (P). The key challenges of the twenty-first century is sustainably meeting global food demands while minimizing agriculture's environmental impact (Foley et al., 2011; West et al., 2014; Xie et al., 2016). This challenge is underscored by the anticipated 4% annual increase in fertilizer demand required to feed a further 2.3 billion people by 2050, which highlights the ongoing need for a stable fertilizer supply (Elser & Bennett, 2011). In 2023, global fertilizer demand reached 2.1×10^8 tons, with projections (FAO, 2023) indicating a 2% annual growth. The world fertilizer market was valued at over US$191 billion in 2022. Modern agriculture relies heavily on nitrogen (N) and phosphorus (P) fertilizers, with global demand for N and P_2O_5 fertilizers estimated at 1.12×10^{11} kg/year and 0.50×10^{11} kg/year, respectively, by 2022 (FAO, 2023). The Haber–Bosch process is mainly used for nitrogen fertilizers production but it is energy intensive process which account for 1% of the world's total energy consumption (Capdevila-Cortada, 2019). Furthermore, phosphate rock, the key raw material for phosphorus fertilizers, is a finite resource, with estimates suggesting it may be depleted in 50–100 years (Cordell, 2009).

The dependence on agriculture fertilizers to meet food demand for growing population also imposes a significant strain on wastewater treatment processes. The production and use of fertilizers lead to the release of essential nutrients up to 30% of nitrogen and 16% of phosphorus in wastewater (Verstraete et al., 2009; Rahman et al., 2014). In United States 4% of the country's electricity consumptions takes place in wastewater treatment plants, out of which 77% of this energy being dedicated to activated sludge aeration during the nitrification process (Tsalaset al., 2024; Zhang & Ma, 2020). The process of removing and release of nitrogen gas from wastewater to atmosphere require about 45 MJ of energy per kg of nitrogen, despite this, it also contributes to greenhouse gas emissions, releasing 0.9 kg of CO_2 percubic meter of treated wastewater (Kurt, 2024). The considerable energy consumption and environmental impact associated with nutrient removal exacerbate the sustainability challenges tied to fertilizer production for food security. Consequently, nutrient recovery from wastewater streams is recognized as a efficient solution which promote the sustainability of both fertilizer production and food security, while offering efficient benefits to wastewater treatment process (Grant et al., 2012; Saliu et al., 2021).

Domestic wastewater (DWW) streams has gained growing attention as a valuable alternative source for recovery of nutrients like nitrogen and phosphorous and also helping to reduce energy consumption. If these nutrients are effectively recovered then it could be estimated that these nutrients could potentially meet more than half of the world fertilizers demand (Batstone et al., 2015; Zhao et al., 2021). Therefore, there should be significant interest in developing technologies that can efficiently recover these nutrients from wastewater streams. As compared to traditional methods, membrane separation technologies are mainly used in water purification and desalination for water and wastewater treatment. Membrane separation process helps in minimizing the risk associated with generating toxic byproducts which is crucial for wastewater reuse. Recently, synthetic membranes with varying pore sizes are used in water purification processes such as microfiltration (MF), ultrafilteration (UF), nanofiltration (NF), and reverse osmosis (RO) (Werber et al., 2016).

3.2 What is Nanofiltration?

In 1984, Dr. Peter Eriksson introduced nanofiltration (NF) membranes, a new class of membranes that have since gained significant market application. The term "NF" refers to the membrane's pore size, typically defined by its capacity to remove solutes based on their molecular weight. Nanofiltration (NF) membranes belong to the fourth category of pressure-driven membrane technologies, alongside reverse osmosis (RO), ultrafiltration (UF), and microfiltration (MF). These membranes are unique in their ability to separate small solutes through two main mechanisms. The first, widely acknowledged in scientific literature, is ionic separation, which divides molecules based on their charge. The second mechanism is sieving, which separates uncharged solutes according to their molecular size. Nanofiltration is primarily used for separating and purifying solutes in liquid mixtures. NF membranes have pore sizes ranging from 0.001 to 0.01 microns, allowing them to retain divalent ions, larger monovalent ions, organic compounds, and certain salts, while permitting water and smaller solutes to pass through (Khulbe & Matsuura, 2021). This makes NF a flexible and efficient technology for a range of applications, including water treatment and resource recovery (Shen et al., 2020).

In the nanofiltration (NF) process, a portion of the feed stream passes through the semi-permeable membrane, separating the inlet stream into two parts: the permeate, which is the filtered portion, and the retentate or concentrate, which is the portion that is rejected and not filtered (Zheng et al., 2024). NF has been shown to be highly effective in removing organic compounds. However, chlorine disinfection plays a key role in controlling microbial growth, which has been observed in NF distribution systems. To minimize microbial growth, NF membranes that have low retention of inorganic materials and high efficiency in removing organic substances can produce water of the highest quality (Zheng et al., 2024).

3.2.1 Basic Principle of Nanofiltration

Nanofiltration (NF) operates on the principles of size exclusion, charge exclusion, and hydrophobic-hydrophilic interactions, which work in concert to achieve selective separation of solutes from a feed solution. The basic operating mechanisms of NF can be summarized as follows:

Pressure-Driven Separation:

- A hydraulic pressure gradient is applied to push the feed solution against the semi-permeable NF membrane. This pressure needs to be greater than the osmotic pressure of the solution to drive the solvent, usually water, through the membrane. The pressure-driven process facilitates the separation of solutes based on factors such as size, charge, and their interactions with the membrane.

Selective Permeability:

- **Size Exclusion:** NF membranes allow molecules smaller than their pore size—such as water and some monovalent salts—to pass through, while rejecting larger solutes, including divalent or multivalent ions, as well as organic molecules.
- **Charge Exclusion:** NF membranes typically possess a slight negative charge, which enhances their ability to reject ions based on their charge. In particular, divalent and multivalent ions are rejected more effectively than monovalent ions, owing to electrostatic interactions between the ions and the membrane surface.
- **Solubility and Diffusion:** Nonpolar or hydrophobic molecules may be partially excluded based on their solubility and affinity for the membrane material. This interaction influences the permeation rate of certain organic compounds and solutes.

Cross-Flow Filtration:

- In the NF process, the feed solution flows tangentially to the membrane surface, creating a sweeping action that helps reduce fouling and allows for continuous separation. This cross-flow configuration ensures that the rejected solutes are continuously removed from the membrane surface, preventing accumulation and promoting higher membrane performance over time.

Overall, nanofiltration provides a highly selective and energy-efficient method for separating solutes from water and wastewater, making it suitable for a broad range of applications including water softening, desalination, and the exclusion of organic and inorganic contaminants. The combination of size and charge exclusion, along with the cross-flow filtration process, allows NF membranes to effectively balance solute retention and water recovery.

3.3 Nanofiltration Membrane—Properties and Structure

Nanofiltration (NF) membranes are semi-permeable membranes that function between ultra-filtration (UF) and reverse osmosis (RO) in terms of separation properties. They are extensively employed in water and wastewater treatment due to their selective ability to reject divalent and multivalent ions, organic molecules, and nutrients, while allowing the passage of mono-valent ions and water molecules. A thorough understanding of the properties and structure of NF membranes is essential to optimize their use in various applications, such as water purification, nutrient recovery, and contaminant removal (Kim et al., 2018).

3.3.1 *Properties of Nanofiltration Membranes*

Molecular Weight—NF membranes typically have a molecular weight ranges between 200 and 1000 Dalton, which defines their ability to reject organic molecules, macromolecules, and divalent ions (Wei et al., 2018). This molecular weight cutoff (MWCO) range makes NF membranes suitable for applications requiring the selective exclusion of larger solutes and organic contaminants, while allowing smaller mono-valent ions, such as sodium and chloride, to pass through.

Charge Characteristics—Many NF membranes display a negative surface charge because of the functional groups, like carboxyl or sulfonate groups, present in their polymer structure. This negative charge plays a crucial role in the electrostatic interactions between the membrane and solutes, particularly in rejecting divalent and multivalent anions more efficiently than monovalent ions (López et al., 2020). The charge characteristics of NF membranes contribute significantly to their selectivity, enabling enhanced performance for water treatment processes, especially in removing hardness-causing ions like calcium and magnesium.

Water Permeability—NF membranes typically demonstrate higher water permeability compared to RO membranes, making them more energy-efficient for applications that require moderate rejection rates. The water permeability of nanofilter membranes is primarily influenced by their pore size distribution, surface hydrophilicity, and the nature of the polymer materials used in their fabrication (Lee et al., 2019). This enhanced permeability enables NF membranes to maintain higher flux rates, making them an attractive option for large-scale water treatment processes.

Salt Rejection—NF membranes are capable of rejecting 50–80% of monovalent salts and more than 99% of divalent and multivalent salts (Tang et al., 2021). This selective salt rejection is primarily attributed to the combination of **size exclusion** and **charge exclusion** effects. Larger ions or those with higher charges, such as divalent and multivalent ions, experience stronger rejection, while smaller monovalent ions pass through more readily (Tang et al., 2021). The ability to selectively reject salts is especially beneficial for applications such as water softening, nutrient recovery, and wastewater treatment.

Chemical and Thermal Stability—NF membranes exhibit moderate resistance to chemical oxidants and can generally withstand a temperature range of up to 45 °C (Wang et al., 2017). However, prolonged exposure to harsh chemicals or extreme temperatures can degrade membrane performance and reduce their lifespan (Wang et al., 2017). The stability of NF membranes is affected by the material composition, with polyamide-based membranes offering good resistance to chemical degradation but potentially being susceptible to fouling and chlorine degradation.

3.3.2 Structure of Nanofiltration Membranes

Thin-Film Composite (TFC) composition—The mainstreams of NF membranes are constructed as thin-film composite (TFC) membranes, which typically consist of three distinct layers:

Polyamide Active Layer: The active layer, usually made of a thin polyamide film, is responsible for the selective separation of solutes. It provides high selectivity due to its dense polymer network, which offers the primary resistance to solute passage (Kim et al., 2018).

Microporous Support Layer: Beneath the polyamide layer, a microporous support layer, typically composed of materials such as polysulfone or polyester, provides mechanical strength to the membrane, ensuring its structural integrity and support (Basile et al., 2021).

Nonwoven Fabric Base Layer: The base layer, often made of nonwoven fabric, serves as the structural substrate for the membrane. This layer enhances the durability, flexibility, and mechanical stability of the membrane during use.

Pore Size Distribution—The pore sizes of NF membranes usuallyvary from 0.5 to 2 nm, which allows for selective separation based on both size exclusion and electrostatic interactions. The small pore size enables NF membranes to reject larger solutes and ions, while permitting smaller molecules such as monovalent ions and water to exceed through the membrane (Shen et al., 2020).

Hydrophilicity—The hydrophilic nature of NF membranes is an important property that enhances their water permeability and reduces fouling caused by hydrophobic contaminants. Surface modifications or the incorporation of hydrophilic polymers can improve the membrane's hydrophilicity, further enhancing its performance in water treatment applications by minimizing fouling and increasing longevity (Liu et al., 2016).

Surface Morphology—The surface morphology of NF membranes typically features a ridge-and-valley structure in the active layer, which increases the effective surface area available for filtration. However, this increased surface area can also promote fouling, particularly if not adequately managed. The design of membrane surface morphology is crucial in minimizing fouling and ensuring long-term performance (Yang et al., 2019).

3.3.3 Advances in Nanofiltration Membrane Design

Surface Modifications—Recent advancements in NF membrane design focus on surface modifications to enhance membrane performance, particularly in provisions of hydrophilicity, fouling resistance, and permeability. Functional nanomaterials, such as graphene oxide and carbon nanotubes, are being incorporated into the membrane surface to enhance these properties (Zhao et al., 2022). These modifications also improve the membrane's resistance to fouling and extend its operational lifespan.

Hybrid Membranes—Incorporating inorganic nanoparticles, such as silica or titanium dioxide, into the polymer matrix of NF membranes has led to improvements in their thermal stability, mechanical strength, and rejection efficiency. Hybrid membranes have shown enhanced performance, especially in challenging applications involving high temperature or chemically aggressive environments (Chen et al., 2020).

Layer-by-Layer Assembly—Layer-by-layer (LbL) assembly techniques have been explored as a way to precisely control the membrane thickness and functionality at the nanoscale. This approach enables the development of membranes with tailored properties, including improved selectivity, enhanced fouling resistance, and greater durability. By adjusting the number of layers and the type of materials used, LbL assembly provides a method to create highly customizable NF membranes for specific applications (Wang et al., 2021).

3.4 Selective Separation and Recovery of Nutrients in Wastewater Streams

Traditional methods of nitrogen (N)and phosphorus (P)removal from wastewater have primarily focused on eliminating these nutrients rather than recovering them. Nitrogen removal is mostly achieved through biologically mediated nitrification/denitrification, a process that converts nitrogen compounds into nitrogen gas (N_2). However, this process is energy-rigorous and results in the loss of valuable nitrogen, as ammonia, a critical nutrient, is converted into N_2, which cannot be utilized in agricultural systems, leading to wasted energy (Foley et al., 2010; Ledezma et al., 2015).

Similarly, conventional phosphorus recovery methods, such as simultaneous precipitation in aeration tanks, have faced several limitations. These methods are often criticized for losing phosphate to biosolid waste and causing precipitative fouling (scaling) of equipment, which decreases the efficiency and scalability of phosphorus recovery (Gurreri et al., 2020). Consequently, there has been an increasing focus on developing more sustainable and efficient approaches for nutrient recovery from wastewater.

3.4.1 Shift Towards Membrane-Based Nutrient Recovery

Recent trends in nutrient recovery from wastewater are increasingly centered on the use of membrane technologies and hybrid multi-stage processes. A review of recent advancements in membrane technology indicates that processes driven by external pressures, electric fields, vapor pressure, and chemical potential are being explored for effective nutrient recovery (Yu et al., 2021). Among these, pressure-driven membrane processes like reverse osmosis (RO) and nanofiltration (NF) have become increasingly popular in water treatment, wastewater recycling, and reuse, especially when high-quality water is needed (Osman et al., 2024).

Membrane technologies are being recognized for their ability to efficiently separate and recover nutrients like nitrogen and phosphorus. Methods such as microfiltration (MF), ultrafiltration (UF), nanofiltration (NF), and reverse osmosis (RO) are already widely used in various industries, including water and wastewater treatment, desalination, and pharmaceuticals (Beckinghausen et al., 2020; Vineyard et al., 2021). These techniques offer a promising alternative to traditional methods by selectively removing contaminants based on their size and charge, making them particularly suitable for nutrient recovery.

3.4.2 Pressure-Driven Membrane Technologies

Pressure-driven membranes are a proven technology with significant potential for nutrient recovery. These membranes function by applying a hydraulic pressure gradient to force the feed solution through a semi-permeable membrane, separating contaminants based on size or charge. Different membrane processes have distinct pore sizes: microfiltration (MF) has a pore size of 0.1 μm, ultrafiltration (UF) operates at 0.01 μm, nanofiltration (NF) at 0.001 μm, and reverse osmosis (RO) is essentially non-porous (Crittenden et al., 2012). The pressure required for each process depends on the quality of the feed stream. For example, wastewater streams, such as those from animal waste, require transmembrane pressures ranging from 100–180 kPa for

MF, $\leq$ 800 kPa for UF, 350–3000 kPa for NF, and 3.5–6.5 MPa for RO (Hjorth et al., 2011).

RO membranes can achieve exceptionally high operating pressures, reaching up to 150 MPa in specially designed systems. However, the economic feasibility of using such high-pressure processes for nitrogen recovery remains a challenge. Membrane technologies differ in their ability to reject contaminants. MF removes particles larger than its pore size, including algae and bacteria, while UF is effective at removing small colloids and viruses. NF is particularly effective for nutrient recovery, as it removes dissolved organic matter and multivalent ions. RO, on the other hand, can also reject monovalent species, making it ideal for desalination and producing ultrapure water.

3.4.2.1 Nitrogen Recovery

Concerns about reactive nitrogen levels surpassing planetary boundaries are well-established and frequently addressed in the literature. A significant amount of anthropogenic nitrogen enters wastewater systems, making its removal a major challenge. Traditional nitrogen removal methods in wastewater treatment are energy-intensive, typically relying on biologically mediated nitrification/denitrification processes that convert nitrogen compounds into nitrogen gas (N2). Although these conventional methods effectively reduce nitrogen levels, they consume a lot of energy and result in the loss of valuable nitrogen compounds that could be recovered and reused (Foley et al., 2010; Ledezma et al., 2015). This has led to increased interest in nitrogen recovery technologies that can save energy, recover valuable nutrients, and help meet stringent discharge regulations. Consequently, nitrogen recovery has become a key area of focus in the development of more sustainable wastewater treatment technologies.

Membrane technologies, particularly reverse osmosis (RO) and nanofiltration (NF), have gained significant attention for their role in nutrient recovery. These technologies are highly effective in selectively removing contaminants, including nitrogen compounds, from wastewater streams. Their application is seen as an efficient way to address the energy consumption associated with traditional nitrogen removal techniques while facilitating the recovery of nitrogen for reuse in agricultural or industrial applications. Membrane processes offer promising alternatives due to their capability to selectively reject nitrogen compounds while allowing water to exceed through, thus reducing energy consumption (Al-Juboori et al., 2022).

The separation mechanisms involved in nitrogen removal using NF and RO membranes are intricate and involve factors such as Donnan potential, steric effects, membrane potential, and size exclusion (Singh et al., 2020). These mechanisms are affected by the molecular charge, size, and structure of the nitrogen compounds, which influence their rejection efficiency. Common nitrogen forms in wastewater include urea, ammonium ions, ammonia, nitrate, and nitrite. The predominance of one nitrogen species over another is mainly determined by the pH and the biological processes occurring in the wastewater. For example, urea is enzymatically converted into ammonium, which is then further transformed into ammonia, nitrate, and nitrite through nitrification and denitrification, especially in alkaline conditions.

Studies, including those by Lee and Lueptow (2001), have examined the rejection of nitrogen compounds through NF and RO membranes both theoretically and experimentally. Their results indicate that the primary rejection mechanisms for nitrogen compounds in these membrane processes are mainly influenced by the Donnan potential and steric effects, which depend on the ratio of solute radius to pore radius. They found that nitrogen compound transport through NF and RO membranes is mainly driven by diffusion, with convection and electromigration having a smaller impact, particularly for charged molecules. For uncharged nitrogen compounds like urea, convection accounted for about 20% of transport through RO membranes and 30% through NF membranes. However, due to urea's small size and charge neutrality, achieving high rejection levels with RO or NF membranes is difficult, as it is less effectively rejected compared to other nitrogen compounds (Lee & Lueptow, 2001).

Nitrate and nitrite, in contrast, are additionally rejected by RO membranes, with rejection rates similar to ammonium. However, when using NF membranes, the rejection of nitrate and nitrite is affected by the nature of the counterions present in the solution. Specifically, NF membranes exhibit higher rejection efficiencies for nitrate and nitrite when the counterions are multivalent, as opposed to mono-valent ions. This highlights the importance of understanding the ion exchange and electrostatic interactions that govern the rejection of various nitrogen species during membrane filtration processes (Lee & Lueptow, 2001; Singh et al., 2020).

The shift in focus from nitrogen concentration to nitrogen recovery is reflective of broader trends in the field of wastewater treatment. Membrane technologies have become central to this transition, offering a more energy-efficient and sustainable means of nutrient recovery. The advancements in membrane materials and design are paving the way for more effective systems that not only meet regulatory standards but also contribute to resource recovery in wastewater treatment systems. The growing body of research on membrane technologies for nitrogen recovery indicates that these methods are becoming an increasingly viable alternative to traditional nitrogen removal processes, offering both environmental and economic benefits (Al-Juboori et al., 2022).

In conclusion, the use of membrane technologies like NF and RO for nitrogen recovery represents a promising field of research and development. These technologies utilize advanced separation mechanisms, such as size exclusion, charge exclusion, and steric effects, to selectively remove nitrogen compounds from wastewater. By focusing on recovering nitrogen instead of just removing it, membrane processes can help save energy, minimize nutrient loss, and provide valuable nitrogen resources for reuse. Ongoing advancements in membrane design and material innovation are expected to enhance the efficiency and performance of these systems, making them an essential part of sustainable wastewater treatment practices.

3.4.2.2 Phosphorous Recovery

Phosphorus (P) is an essential nutrient for all life forms and a vital component of agricultural fertilizers, playing a crucial role in global food production. However, the depletion of commercially available phosphate resources is expected within the next 50–100 years, posing significant risks to global food security (Cordell et al., 2009). This looming shortage highlights the urgent need for sustainable technologies to recover and reuse phosphorus. Urban wastewater offers a promising source of phosphorus, with sewage sludge serving as a particularly rich reservoir, containing more than 90% of the phosphorus in the influent (Petzet et al., 2012). As a result, phosphorus recovery has become a key area of focus, with recent research concentrated on developing efficient and cost-effective recovery technologies to address this growing challenge.

The advanced method for phosphorus recovery in wastewater treatment is struvite precipitation, a process that leads to the formation of magnesium ammonium phosphate (struvite) crystals. This technique is widely used because it efficiently recovers phosphorus from wastewater and produces a valuable byproduct that can be utilized as a slow-release fertilizer (Amann et al., 2018). However, the process also presents challenges, particularly with the potential co-precipitation of harmful substances, such as emerging micropollutants (EMPs) and heavy metals like lead (Pb), nickel (Ni), and zinc (Zn) (Ebbers et al., 2015; Thong et al., 2016). Even in trace amounts, these pollutants can accumulate in soil and enter the food chain, leading to unintended environmental and health risks. Therefore, the development of technologies that can selectively recover phosphorus while minimizing the retention of these harmful substances is crucial.

Nanofiltration (NF) offers a promising alternative for phosphorus recovery, with selective separation capabilities that could address some of the issues associated with struvite precipitation. NF membranes have the potential to recover phosphorus while effectively rejecting harmful substances like heavy metals and other contaminants, provided the membrane is sufficiently selective. A key challenge in using NF for phosphorus recovery is converting organic phosphorus into soluble inorganic forms, which can be facilitated through processes like anaerobic digestion (Song et al., 2018). Once converted, the phosphorus-enriched permeate from NF membranes can be used directly for irrigation or further concentrated to produce fertilizers through additional processes, such as reverse osmosis (RO) followed by precipitation.

NF membranes can also serve as a post-treatment stage following other phosphorus recovery techniques, such as acidic hydrolysis or wet oxidation, which transform sewage sludge into an inorganic phosphorus-rich liquid phase. Research has shown that lowering the pH of the feed solution (e.g., below 3.0) can enhance phosphorus permeation while increasing the retention of metal cations when using commercial thin-film composite polyamide (TFC-PA) NF membranes (Niewersch et al., 2010, 2014). The acidic conditions reduce the electrostatic attraction between the membrane surface and metal cations, improving phosphorus selectivity. However, the dense structure of conventional NF membranes leads to phosphorus retention rates of 30% to 50%, which is considered too high for economically feasible phosphorus recovery (Niewersch et al., 2010, 2014; Schütte et al., 2015). Therefore, membranes that allows phosphorus, along with essential minerals and nutrients, to pass freely while efficiently rejecting coexisting contaminants are required for energy- and cost-effective phosphorus recovery. These membranes should ideally produce high-quality permeates suitable for agricultural irrigation without the need for chemical additives.

However, the development of NF membranes for phosphorus recovery comes with its own set of challenges. Achieving the required selectivity for phosphorus recovery, while simultaneously rejecting pollutants, remains a significant hurdle. The precision required to design membranes capable of selectively permeating phosphorus while rejecting trace contaminants is still an area of active research, with progress limited by the inherent complexities of membrane selectivity.

An alternative method for phosphorus recovery from wastewater involves the direct treatment of biologically treated effluent, where phosphorus mainly exists as inorganic ions. McCarty et al. (2011) proposed a low-strength anaerobic treatment process that removes solids and soluble organic materials while preserving essential nutrients like nitrogen and phosphorus. This approach enables efficient nutrient recovery while minimizing the loss of valuable resources. Building on this concept, Mrayed et al. (2011) showed how NF membranes can be used to treat effluent from conventional wastewater treatment plants. The phosphorus-enriched NF retentate was then diluted with RO permeate for use in irrigation. While this method holds potential, it faces limitations related to water product quality, as undesirable heavy metals and emerging micropollutants (EMPs) are often concentrated in the NF retentate, rendering it unsuitable for direct agricultural use. To address this challenge, membrane technologies need to be developed that can retain high phosphorus levels in the permeate while effectively removing pollutants from the wastewater.

The recovered product water, now enriched with valuable nutrients, could be directly utilized for agricultural irrigation or in controlled environment agriculture (CEA) systems, such as urban hydroponics, where nutrient-rich water is essential for plant growth (Benke et al., 2017). However, this approach requires the development of suitable membranes that can efficiently recover phosphorus while ensuring the removal of harmful contaminants, ultimately producing high-quality effluent for agricultural and industrial applications.

In conclusion, phosphorus recovery from wastewater streams is a critical area of research aimed at addressing the impending global phosphorus shortage. Membrane technologies, particularly nanofiltration, offer a promising alternative to traditional recovery methods, providing a sustainable solution for nutrient recycling. While significant progress has been made, challenges remain in achieving the required membrane selectivity to recover phosphorus efficiently while rejecting harmful pollutants. Ongoing research and technological advancements in membrane materials and design will be essential to overcoming these challenges and making phosphorus recovery from wastewater a viable and economically feasible option for the future.

3.5 Various Factors Affecting Nanofiltration Efficiency

For long-term, sustainable membrane filtration aimed at nutrient separation from various wastewater streams, it is crucial to align several key factors related to the membrane, medium, and operating conditions. The use of nanofiltration (NF) membranes for nutrient recovery, particularly nitrogen and phosphorus, has proven to be both effective and sustainable. However, the success of nutrient recovery is highly influenced by various parameters. These can be categorized into two main groups: membrane properties and medium characteristics. Membrane properties include factors like material/composition, pore size, charge, hydrophilicity/phobicity, and permeability, while medium characteristics involve feed concentrations, pH, temperature, pressure, and ionic strength.

Membrane Material/Composition

The material composition of the membrane is crucial in determining its filtration efficiency. Various materials, such as metals, ceramics, hollow fibers, and polymers, have been used for nutrient recovery in pressure-driven membrane filtration systems (Choi & Ng, 2008; Sadeghi et al., 2018). Metallic membranes, for example, have shown promising performance due to their resistance to high-temperature oxidation and enhanced durability, which helps minimize fouling during filtration (Zhang et al., 2005). Ceramic membranes are widely adopted in commercial applications, offering several benefits, including high flux, corrosion resistance, low fouling tendencies, and the ability to undergo backwashing for cleaning (Ersu & Ong, 2008). Ceramic membranes are also advantageous because they have fewer interactions between the membrane surface and foulants, leading to higher flux rates (Baker, 2012).

Despite their advantages, the use of ceramic NF membranes remains less common than polymeric membranes, even though significant advancements have been made in ceramic membrane technology in recent years. Ceramic membranes are recognized for their exceptional chemical, structural, and thermal stability. They do not deform under pressure, swell, or require complex cleaning processes. However, their brittleness and higher cost compared to polymeric membranes limit their broader adoption (Lu et al., 2007; Verweij, 2012).

Membrane Pore Size and Molecular Weight

The pore size of a membrane is a key factor in determining its separation properties, especially in nanofiltration processes. It is well-established that pore size directly influences the permeability and retention of solutes during filtration (Dizge et al., 2011). The selectivity of microfiltration (MF) membranes is primarily governed by their pore size, whereas nanofiltration membranes are typically characterized by their molecular weight (MW), measured in Dalton (Da) or kilo-dalton (kDa). NF membranes generally have smaller pores than MF membranes, enabling them to more efficiently retain or remove low molecular weight compounds and ions.

The MW of NF membranes typically ranges between 200–1000 Da, which makes them well-suited for nutrient recovery, particularly for wastewater treatment (Sewerin et al., 2021). These membranes are able to selectively eliminate contaminants based on their molecular size, making them effective for removing unwanted solutes while allowing nutrients like phosphate and nitrogen to pass through for recovery.

Effects of pH on Filtration Efficiency

The performance of NF membranes is highly influenced by variations in pH and salt conditions, which can have a considerable effect on membrane flux, solute rejection, and fouling behavior. Changes in pH and salt concentration alter the interactions between solutes and the membrane surface, thereby affecting the overall efficiency of the filtration process (Luo & Wan, 2013).NF membranes possess dissociable functional groups that are sensitive to pH, and they operate through a combination of size exclusion and electrostatic effects.

For example, when salt or high pH is introduced to the system, the efficient pore size of the membrane can increase, thereby enhancing membrane permeability. However, the addition of salt may not necessarily improve permeate flux, as the increased viscosity of the solution in the pores can counteract this effect. The impact of pH on NF performance involves several mechanisms, such as membrane swelling, changes in charge distribution, electrostatic effects, competition between ions, and the Donnan effect (charge balance). Understanding these interactions is crucial for optimizing NF processes to maximize nutrient recovery while minimizing fouling.

Fouling

Membrane fouling is a significant issue in membrane filtration processes, including NF, as it leads to a decline in membrane performance over time. Fouling happens when inorganic and organic substances from the feedwater accumulate on the membrane surface and within its pores, which reduces membrane flux and raises costs due to increased energy consumption and the need for membrane replacement (Rajendran et al., 2024). The negative impacts of fouling are substantial, requiring frequent cleaning and maintenance, reducing recovery rates, and causing losses in feedwater.

Fouling can be mitigated through various preventive measures, such as pretreating the feedwater, early prediction of fouling, selecting the right membrane material, and optimizing module design and operational parameters. Regular cleaning of the membranes is also essential for maintaining their performance. The effectiveness of chemical cleaning methods relies on various factors, such as the type of foulant, the selection of cleaning agents, the temperature and pH of the cleaning solution, the concentration of chemicals, the duration of contact between the solution and the membrane, and operational parameters like cross-flow velocity and pressure.

Effective fouling control is essential for maintaining the effectiveness of the NF system. Ideally, controlling fouling can reduce the need for frequent cleaning, which, in turn, minimizes membrane downtime and prolongs the lifespan of the membranes, thus improving the overall economic feasibility of nutrient recovery systems.

In conclusion, the successful application of nanofiltration membranes for nutrient recovery from wastewater is dependent on a delicate balance of various membrane, medium, and operational factors. Membrane material, pore size, pH, fouling control, and other operational parameters all play critical roles in determining the efficiency and sustainability of the process. Ongoing research and development of new membrane materials and technologies will continue to enhance the efficiency of NF systems, making them a key component of future nutrient recovery strategies.

3.6 Challenges in Nanofiltration for Wastewater Treatment

Nanofiltration (NF) has emerged as a promising technology for wastewater treatment and nutrient recovery. However, its widespread application faces several challenges that limit its effectiveness and scalability. These challenges can be categorized into operational, material, environmental, and economic factors.

Fouling and Scaling

Membrane fouling is a significant obstacle in NF systems, leading to performance deterioration and operational inefficiency. Fouling occurs in various forms, including organic fouling, biofouling, and scaling. Organic fouling is caused by the accumulation of organic substances such as humic acids and oils on the membrane surface, which impedes water permeability and reduces the efficiency of nutrient recovery (Li et al., 2021). Biofouling, driven by microbial growth, further exacerbates the issue by clogging the membrane pores and degrading its performance (Matin et al., 2019). Scaling, which results from the precipitation of inorganic salts like calcium carbonate and magnesium sulfate, adds to the operational challenges of NF systems (Yu et al., 2021). While chemical cleaning strategies can help mitigate fouling, repeated cleaning can degrade the membrane material, reducing its lifespan and efficiency over time (Lau et al., 2022).

Selectivity versus Permeability Trade-Off

Achieving a balance between membrane selectivity and permeability remains a key challenge for NF applications. High selectivity is crucial for efficient nutrient recovery, but it often comes at the expense of reduced water flux, leading to increased energy consumption (Al-Rowaili et al., 2023). This trade-off poses a barrier to the optimal performance of NF systems, especially when treating wastewater streams with diverse and complex compositions. The need for high selectivity without compromising permeability is essential to enhance the efficiency of NF-based nutrient recovery systems.

Membrane Degradation

Membrane degradation is another significant issue faced by NF systems. Membranes are vulnerable to various forms of degradation, including chemical, mechanical, and thermal damage. Exposure to chemicals such as oxidants (e.g., chlorine) can accelerate the degradation of polymeric membranes, shortening their operational lifespan (Wang et al., 2017). Physical damage can also occur due to high-pressure operations and the presence of abrasive particles in wastewater. Additionally, temperature fluctuations can induce thermal instability, further compromising membrane performance (Liu et al., 2021). The combined impact of these factors can lead to increased membrane replacement costs and operational downtime.

Energy Requirements

The energy demands of NF systems represent a significant challenge, particularly when treating wastewater streams with high salinity or suspended solids. To achieve the required separation efficiency, high pressure must be applied, which results in elevated energy consumption and operational costs (Yusuf et al., 2024). This high energy requirement can make large-scale adoption of NF systems economically unfeasible, particularly in resource-limited settings. Efforts to reduce energy consumption, such as optimizing system design or utilizing energy recovery technologies, are essential to making NF more economically viable.

Pretreatment Requirements

Effective pretreatment of wastewater is often necessary to prevent fouling and scaling in NF systems. Common pretreatment methods, such as sedimentation and filtration, add complexity and increase operational costs (Zhang et al., 2023). These additional steps make the overall process less economically attractive, especially when considering the long-term costs of system maintenance and the need for frequent cleaning. Streamlining pretreatment methods or developing more effective alternatives is crucial to improving the feasibility of NF systems for nutrient recovery.

Concentrate Disposal

NF processes produce a concentrated waste stream with retained contaminants, creating difficulties for its proper disposal. Proper disposal of this concentrate is critical, especially when it contains hazardous or toxic substances. The disposal process is subject to stringent environmental regulations, which can increase operational costs and complicate the implementation of NF systems (Yaacoubi & Dumée, 2023). Addressing the environmental impact of concentrate disposal and developing sustainable methods for handling waste streams is an important area for future research.

Material and Cost Challenges

The high cost of manufacturing and replacing NF membranes is a significant barrier to their widespread use. Ongoing research is focused on creating cost-effective membranes with improved durability and performance (Zhang et al., 2023). Advances in materials science, particularly the development of stronger and more affordable membrane materials, could help reduce the high operational costs of NF systems and make them more viable for large-scale applications.

Variability in Wastewater Composition

The composition of wastewater streams varies widely, presenting a challenge for the standardization of NF systems. Variability in nutrient concentrations, suspended solids, and other contaminants requires the customization of membrane selection and system design to suit specific wastewater characteristics (Wang et al., 2022). This variability adds complexity to the process and maintenance of NF systems and complicates the process of scaling up the technology for broader use.

Environmental and Regulatory Concerns

The potential release of microplastics or nanomaterials from NF membranes into treated water raises serious environmental concerns (Poerio et al., 2019). These materials can accumulate in the environment and enter the food chain, leading to potential ecological and health risks. Moreover, strict regulatory requirements regarding treated water quality and waste disposal further complicate the implementation of NF systems. Ensuring that NF systems meet environmental standards while minimizing their ecological footprint is essential for their long-term sustainability.

Limited Recovery Efficiency for Specific Nutrients

NF membranes may struggle to recover certain nutrients, such as ammonia or nitrate, due to their small size and charge characteristics. Despite advances in membrane modification techniques aimed at enhancing selectivity for specific nutrients, these adjustments often come with trade-offs, such as reduced overall performance (Xie et al., 2016). Achieving high selectivity for nutrients like ammonia and nitrate without compromising the overall efficiency of the NF system remains a critical challenge in nutrient recovery research.

In conclusion, while NF technology offers significant potential for wastewater treatment and nutrient recovery, various challenges must be addressed to optimize its performance and scalability. Operational issues like fouling and scaling, the selectivity-permeability trade-off, and membrane degradation are major factors that limit the effectiveness of NF systems. Additionally, high energy requirements, pretreatment complexities, and waste disposal concerns further complicate its widespread adoption. Overcoming these barriers through continued research and development will be essential to unlocking the complete potential of NF technology in sustainable nutrient recovery from wastewater.

3.7 Cost Effectiveness of NFs

The profitableacceptance of nanofiltration (NF) in the water treatment sector has been hindered primarily by its high operating costs, despite its potential for efficient nutrient removal and wastewater treatment. According to the Water Policy and Regulatory Reform Project, NF systems have yet to see widespread implementation in large-scale water treatment due to financial constraints. As operational NF plants are still limited, cost data is often derived from real-world pilot-scale results, alongside theoretical calculations based on experimental data (Chellam et al., 1998). A strong correlation between the flux rate and operational cost has been observed, though the recovery effectiveness of the system has a relatively smaller impact on the overall costs in industrial applications. Despite this, other investment factors also contribute to the total cost of NF implementation, including plant size and investment in membrane technology.

Investment Costs and Scale Effects

The investment costs associated with NF plants depend significantly on the scale of the facility. Studies have found that plants with smaller membrane capacities typically incur membrane-related investment costs amounting to 20–30% of the total project costs, whereas larger plants (with capacities varying from 53,000 to 125,000 m^3/d) experience membrane investment costs approaching 50% of total capital costs (Chellam et al., 1998). The high cost of membrane modules, particularly for larger-scale plants, highlights the importance of economies of scale in reducing the per-unit cost of membrane investment. As the scale of the NF plant increases, the cost per membrane unit becomes less critical, but operating costs such as maintenance, labor, and energy consumption continue to play a crucial role in determining the overall cost-effectiveness of NF systems.

Operational and Maintenance Costs

Operational and maintenance (O&M) costs account for a significant portion of the total expenses in NF systems. These costs include both fixed expenditures (e.g., labor, general maintenance) and variable costs (e.g., replacement costs, chemicals, and energy consumption). Higher operational efficiency and membrane lifespan are crucial in reducing O&M costs. Studies indicate that the total O&M cost can be significantly impacted by the energy consumption and the replacement frequency of the membranes (Moslehyani et al., 2019). For example, fluctuations in material costs, membrane lifespan, and energy prices can significantly impact the economic feasibility of an NF system (Abdel-Fatah, 2018). Additionally, operating NF systems at higher flow rates is essential for minimizing the cost per unit of treated water and improving overall efficiency.

Impact of Energy and Material Costs

Energy and material costs, particularly those associated with membrane replacement, constitute approximately 42–61% of the total cost of NF systems (Moslehyani et al., 2019). This emphasizes the substantial impact that fluctuations in energy prices and materials cost can have on the viability of NF-based treatment systems. For example, if electricity prices increase by 29%, the overall cost of NF systems could rise by 4–6% (Chellam et al., 1998). The energy consumption required for pressurizing wastewater to pass through NF membranes—especially in systems dealing with high salinity or high suspended solids—represents one of the largest ongoing expenses for NF systems. Consequently, minimizing energy usage through optimized system design or incorporating energy recovery technologies could offer significant cost savings, improving the economic competitiveness of NF.

Membrane Lifespan and Cost Efficiency

The longevity and durability of NF membranes play a pivotal task in reducing the operational costs of the system. While membranes tend to degrade over time due to factors such as chemical exposure, fouling, and physical wear, advancements in membrane materials aim to enhance their resistance to these issues, thereby extending their service life and reducing the need for frequent replacements. Longer-lasting membranes could substantially lower the total cost of NF systems by reducing material and replacement costs, which are among the largest contributors to the overall expense (Wang et al., 2022). This underscores the importance of ongoing research into the development of more durable and cost-effective membrane materials to make NF systems more commercially viable.

Comparison to Other Treatment Technologies

While NF systems face challenges in terms of high initial investment and operational costs, they remain competitive with other water treatment technologies, particularly for small- to medium-scale units. The economic feasibility of NF systems can improve when compared to alternative treatment methods that may require higher energy consumption, chemical additives, or maintenance costs (Zhang et al., 2023). In particular, NF is often preferred for applications requiring selective separation of ions, organic compounds, and nutrients, where traditional methods may fall short in terms of efficiency or effectiveness.

Economic Potential and Future Outlook

Given the increasing global demand for clean water, the application of NF technology is expected to expand significantly in both industrial and potable water treatment sectors. As the technology continues to mature, economies of scale and technological advancements in membrane materials, energy recovery systems, and overall process optimization will likely reduce the cost per unit of treated water, making NF more commercially attractive. Furthermore, the growing demand for efficient wastewater treatment and nutrient recovery, combined with improvements in membrane technology and energy efficiency, suggests that NF systems could become a key player in the water treatment sector in the coming decades (Yusuf et al., 2024).

In conclusion, the cost-effectiveness of NF systems is influenced by various factors, including membrane material costs, energy consumption, system size, and operational efficiency. While the high initial costs of NF membranes and energy consumption currently limit its widespread commercial adoption, ongoing advancements in membrane technology, energy recovery, and process optimization are expected to enhance the economic viability of NF systems. As the demand for clean water continues to rise, NF technology holds significant potential for large-scale applications, provided that economic and technical challenges are addressed through continued innovation and development.

3.8 Conclusion and Future Prospective

Nanofiltration (NF) has become a promising and effective membrane technology for selectively separating nutrients from wastewater, providing potential solutions to significant environmental and sustainability issues. NF membranes' ability to selectively remove ions, organic compounds, and nutrients like nitrogen and phosphorus makes them ideal for both wastewater treatment and nutrient recovery, which are essential aspects of the circular economy in water management.

From the perspective of membrane characteristics, NF systems benefit from a well-established principle of separation, combining size exclusion with electrostatic interactions. This enables the selective retention of contaminants while allowing valuable nutrients to pass through. The recovery of nitrogen and phosphorus from urban wastewater has garnered considerable interest, particularly because it offers opportunities for resource recovery while reducing the environmental impacts linked to the excessive discharge of nutrients. Despite its advantages, the widespread adoption of NF for wastewater treatment and nutrient recovery is hindered by several operational, material, and economic challenges, including fouling and scaling, high energy consumption, membrane degradation, and high initial capital and operational costs.

In terms of cost-effectiveness, the high operating costs of NF systems—driven by factors such as membrane material, energy consumption, and maintenance requirements—remain a major limitation for their large-scale deployment. However, ongoing research and technological advancements are progressively addressing these challenges. Innovations in membrane materials, such as the development of more durable and cost-effective membranes, are expected to improve the economic feasibility of NF systems. Furthermore, the integration of energy recovery technologies and optimization of system design for enhanced operational efficiency will likely reduce energy consumption and overall costs. The adoption of advanced pretreatment processes and better fouling management strategies will also contribute to improving the long-term sustainability and cost-effectiveness of NF systems.

Looking toward the future, the role of NF in wastewater treatment and nutrient recovery is expected to expand as the demand for clean water increases globally, along with the need for sustainable resource management. Continued research into membrane selectivity, novel membrane materials, and energy-efficient processes will be critical in unlocking the full potential of NF technology. Additionally, future developments in membrane recycling and the management of concentrate waste will be crucial for ensuring the environmental sustainability of NF systems. It is anticipated that the application of NF will grow exponentially, particularly in small- to medium-scale water treatment facilities, where its efficiency and selective nutrient recovery capabilities offer substantial benefits.

In the context of regulatory and environmental concerns, it is imperative that future NF systems are designed to not only meet the growing demand for clean water but also ensure compliance with stringent water quality standards and environmental regulations. The potential environmental impact of NF processes, such as the release of microplastics and other by-products, requires further investigation and the development of strategies to mitigate such risks.

Overall, while challenges remain, the future of NF technology is promising. As technological advancements continue to address the economic and operational constraints, NF systems are poised to become an integral part of sustainable water treatment strategies, contributing to the circular economy by recovering valuable nutrients, improving wastewater quality, and reducing the environmental impact of industrial and municipal effluents.

3.9 Future Directions

Future research should focus on the following key areas to improve the performance and cost-effectiveness of NF systems:

Membrane Development and Innovation: Research into new materials and membrane structures that offer improved resistance to fouling, enhanced selectivity, and increased durability will be crucial. The development of more cost-effective, high-performance membranes that can function efficiently in diverse wastewater streams will facilitate wider adoption.

Energy Efficiency and Recovery: The integration of energy recovery systems in NF processes is critical to reducing the high energy demand associated with membrane filtration. Future research should explore novel energy-efficient designs and strategies to lower the operational costs and make NF more economically viable.

Fouling Control and Management: Investigating more effective fouling control mechanisms, such as pre-treatment methods or membrane coatings, will be essential for improving the long-term performance of NF systems. The development of predictive models for fouling and real-time monitoring systems will also help mitigate fouling issues before they significantly impact the operation.

Concentrate Management: As NF systems produce concentrated waste streams that may contain hazardous materials, innovative approaches to managing and disposing of these concentrates will be necessary. Research into the recycling or safe disposal of concentrates could greatly enhance the sustainability of NF systems.

Pilot-scale and Real-world Applications: The transition from laboratory-scale studies to large-scale, real-world applications is vital for demonstrating the practicality and scalability of NF technology. Further pilot-scale studies and field trials are needed to refine NF processes for specific wastewater streams and optimize their performance for industrial and municipal applications.

Policy and Regulatory Development: As NF technology progresses, it is essential to develop clear regulations and guidelines to ensure that it meets environmental standards, particularly with respect to the release of treated effluent and waste by-products. Collaborative efforts between researchers, policymakers, and industry stakeholders will be key to fostering the adoption of NF in water treatment.

In conclusion, NF technology holds considerable promise as an efficient and sustainable solution for nutrient recovery and wastewater treatment. With continued advancements in membrane technology, process optimization, and cost reduction strategies, NF is poised to play a crucial role in addressing global water scarcity and environmental challenges. Through collaborative efforts and ongoing innovation, NF systems could contribute significantly to the development of a sustainable and circular water economy.

Acknowledgements The authors are highly grateful to the Department of Molecular Biology and Biotechnology, CCS Haryana Agricultural University, Hisar, India.

References

Abdel-Fatah, M. A. (2018). Nanofiltration systems and applications in wastewater treatment. *Ain Shams Engineering Journal, 9*(4), 3077–3092.

Al-Juboori, R. A., Al-Shaeli, M., Aani, S. A., Johnson, D., & Hilal, N. (2022). Membrane technologies for nitrogen recovery from waste streams: Scientometrics and technical analysis. *Membranes, 13*(1), 15.

Al-Rowaili, F. N., Khaled, M., Jamal, A., & Zahid, U. (2023). Mixed matrix membranes for H_2/CO_2 gas separation-a critical review. *Fuel, 333*, Article 126285.

Amann, A., Zoboli, O., Krampe, J., Rechberger, H., Zessner, M., & Egle, L. (2018). Environmental impacts of phosphorus recovery from municipal wastewater. *Resources, Conservation and Recycling, 130*, 127–139.

Baker, R. W. (2012). In *Membrane technology and applications* (pp. 1). Wiley & Sons.

Basile, A., Ghasemzadeh, K., & Piemonte, V. (2021). Recent advances in nanofiltration membrane technology. *Membrane Science and Technology, 12*(4), 567–589.

Batstone, D. J., Hülsen, T., Mehta, C. M., & Keller, J. (2015). Platforms for energy and nutrient recovery from domestic wastewater: A review. *Chemosphere, 140*, 2–11.

Beckinghausen, A., Odlare, M., Thorin, E., & Schwede, S. (2020). From removal to recovery: An evaluation of nitrogen recovery techniques from wastewater. *Applied Energy, 263*, Article 114616.

Benke, K., & Tomkins, B. (2017). Future food-production systems: Vertical farming and controlled-environment agriculture. *Sustainability: Science, Practice and Policy, 13*(1), 13–26.

Capdevila-Cortada, M. (2019). Electrifying the Haber-Bosch. *Nature Catalysis, 2*(12), 1055–1055.

Chellam, S., Serra, C. A., & Wiesner, M. R. (1998). Estimating costs for integrated membrane systems. *Journal-American Water Works Association, 90*(11), 96–104.

Chen, R., Zhang, T., & Huang, L. (2020). Hybrid nanofiltration membranes for wastewater treatment: Properties and applications. *Journal of Membrane Science, 601*, Article 117840.

Choi, J. H., & Ng, H. Y. (2008). Effect of membrane type and material on performance of a submerged membrane bioreactor. *Chemosphere, 71*(5), 853–859.

Cordell, D., Drangert, J. O., & White, S. (2009). The story of phosphorus: Global food security and food for thought. *Global Environmental Change, 19*(2), 292–305.

Crittenden, J. C., Trussell, R. R., Hand, D. W., Howe, K. J., & Tchobanoglous, G. (2012). *MWH's water treatment: Principles and design.* Wiley & Sons.

Dizge, N., Soydemir, G., Karagunduz, A., & Keskinler, B. (2011). Influence of type and pore size of membranes on cross flow microfiltration of biological suspension. *Journal of Membrane Science, 366*(1–2), 278–285.

Ebbers, B., Ottosen, L. M., & Jensen, P. E. (2015). Electrodialytic treatment of municipal wastewater and sludge for the removal of heavy metals and recovery of phosphorus. *Electrochimica Acta, 181*, 90–99.

Elser, J., & Bennett, E. (2011). A broken biogeochemical cycle. *Nature, 478*(7367), 29–31.

Ersu, C. B., & Ong, S. K. (2008). Treatment of wastewater containing phenol using a tubular ceramic membrane bioreactor. *Environmental Technology, 29*(2), 225–234.

Foley, J., De Haas, D., Hartley, K., & Lant, P. (2010). Comprehensive life cycle inventories of alternative wastewater treatment systems. *Water Research, 44*(5), 1654–1666.

Foley, J. A., Ramankutty, N., Brauman, K. A., Cassidy, E. S., Gerber, J. S., Johnston, M., & Zaks, D. P. (2011). Solutions for a cultivated planet. *Nature, 478*(7369), 337–342.

Food and Agriculture Organization (FAO). (2023). *Fertilizer use by crop in 2023*. FAO Statistical Yearbook.

Grant, S. B., Saphores, J. D., Feldman, D. L., Hamilton, A. J., Fletcher, T. D., Cook, P. L., ... & Marusic, I. (2012). Taking the "waste" out of "wastewater" for human water security and ecosystem sustainability. *Science, 337*(6095), 681–686.

Gurreri, L., Tamburini, A., Cipollina, A., & Micale, G. (2020). Electrodialysis applications in wastewater treatment for environmental protection and resources recovery: A systematic review on progress and perspectives. *Membranes, 10*(7), 146.

Hjorth, M., Christensen, K. V., Christensen, M. L., & Sommer, S. G. (2011). Solid–liquid separation of animal slurry in theory and practice. *Sustainable Agriculture, 2*, 953–986.

Khulbe, K. C., & Matsuura, T. (2021). Nanotechnology in membrane processes.

Kim, J. H., Moon, S. J., Park, S. H., Cook, M., Livingston, A. G., & Lee, Y. M. (2018). A robust thin film composite membrane incorporating thermally rearranged polymer support for organic solvent nanofiltration and pressure retarded osmosis. *Journal of Membrane Science, 550*, 322–331.

Kurt, S. (2024). Energetic Optimisation of Wastewater Treatment Plant and Evaluation of Green House Gases (Doctoral dissertation, Politecnico di Torino).

Lau, H. S., Lau, S. K., Soh, L. S., Hong, S. U., Gok, X. Y., Yi, S., & Yong, W. F. (2022). State-of-the-art organic-and inorganic-based hollow fiber membranes in liquid and gas applications: Looking back and beyond. *Membranes, 12*(5), 539.

Ledezma, P., Kuntke, P., Buisman, C. J., Keller, J., & Freguia, S. (2015). Source-separated urine opens golden opportunities for microbial electrochemical technologies. *Trends in Biotechnology, 33*(4), 214–220.

Lee, S., & Lueptow, R. M. (2001). Membrane rejection of nitrogen compounds. *Environmental Science and Technology, 35*(14), 3008–3018.

Lee, S., Tang, C. Y., & Aravind, P. (2019). Energy-efficient nanofiltration membranes: Advances and challenges. *Water Research, 163*, Article 114871.

Li, X., Shen, S., Xu, Y., Guo, T., Dai, H., & Lu, X. (2021). Application of membrane separation processes in phosphorus recovery: A review. *Science of the Total Environment, 767*, Article 144346.

Liu, Z., Zhao, X., & Feng, Y. (2016). Hydrophilic modifications of nanofiltration membranes: A review. *Journal of Membrane Science, 504*, 1–10.

Liu, Y., Deng, Y. Y., Zhang, Q., & Liu, H. (2021). Overview of recent developments of resource recovery from wastewater via electrochemistry-based technologies. *Science of the Total Environment, 757*, Article 143901.

López, A., Torres, J., & Martinez, J. (2020). Charge-driven separation mechanisms in nanofiltration membranes. *Environmental Science and Technology, 54*(3), 1221–1232.

Lu, G. Q., Da Costa, J. D., Duke, M., Giessler, S., Socolow, R., Williams, R. H., & Kreutz, T. (2007). Inorganic membranes for hydrogen production and purification: A critical review and perspective. *Journal of Colloid and Interface Science, 314*(2), 589–603.

Luo, J., & Wan, Y. (2013). Effects of pH and salt on nanofiltration—A critical review. *Journal of Membrane Science, 438*, 18–28.

Matin, A., Khan, Z., & Zaidi, S. (2019). Biofouling in nanofiltration membranes: Mechanisms and mitigation. *Desalination, 468*, Article 114084.

Moslehyani, A., Ismail, A. F., Matsuura, T., Rahman, M. A., & Goh, P. S. (2019). Recent progresses of ultrafiltration (UF) membranes and processes in water treatment. In *Membrane separation principles and applications* (pp. 85–110). Elsevier.

Mrayed, S. M., Sanciolo, P., Zou, L., & Leslie, G. (2011). An alternative membrane treatment process to produce low-salt and high-nutrient recycled water suitable for irrigation purposes. *Desalination, 274*(1–3), 144–149.

Niewersch, C., Meier, K., Wintgens, T., & Melin, T. (2010). Selectivity of polyamide nanofiltration membranes for cations and phosphoric acid. *Desalination, 250*(3), 1021–1024.

Niewersch, C., Bloch, A. B., Yüce, S., Melin, T., & Wessling, M. (2014). Nanofiltration for the recovery of phosphorus—Development of a mass transport model. *Desalination, 346*, 70–78.

Osman, A. I., Chen, Z., Elgarahy, A. M., Farghali, M., Mohamed, I. M., Priya, A. K., & Yap, P. S. (2024). Membrane technology for energy saving: Principles, techniques, applications, challenges, and prospects. *Advanced Energy and Sustainability Research, 5*(5), 2400011.

Petzet, S., Peplinski, B., & Cornel, P. (2012). On wet chemical phosphorus recovery from sewage sludge ash by acidic or alkaline leaching and an optimized combination of both. *Water Research, 46*(12), 3769–3780.

Poerio, T., Piacentini, E., & Mazzei, R. (2019). Membrane processes for microplastic removal. *Molecules, 24*(22), 4148.

Rahman, M. M., Salleh, M. A. M., Rashid, U., Ahsan, A., Hossain, M. M., & Ra, C. S. (2014). Production of slow release crystal fertilizer from wastewaters through struvite crystallization–A review. *Arabian Journal of Chemistry, 7*(1), 139–155.

Rajendran, D. S., Devi, E. G., Subikshaa, V. S., Sethi, P., Patil, A., Chakraborty, A., ... & Kumar, V. V. (2024). Recent advances in various cleaning strategies to control membrane fouling: a comprehensive review. *Clean Technologies and Environmental Policy*, 1–16.

Sadeghi, I., Kaner, P., & Asatekin, A. (2018). Controlling and expanding the selectivity of filtration membranes. *Chemistry of Materials, 30*(21), 7328–7354.

Saliu, T. D., & Oladoja, N. A. (2021). Nutrient recovery from wastewater and reuse in agriculture: A review. *Environmental Chemistry Letters, 19*(3), 2299–2316.

Schütte, T., Niewersch, C., Wintgens, T., & Yüce, S. (2015). Phosphorus recovery from sewage sludge by nanofiltration in diafiltration mode. *Journal of Membrane Science, 480*, 74–82.

Sewerin, T., Elshof, M. G., Matencio, S., Boerrigter, M., Yu, J., & de Grooth, J. (2021). Advances and applications of hollow fiber nanofiltration membranes: A review. *Membranes, 11*(11), 890.

Shen, Y., Xu, G., & Wang, H. (2020). Pore size characterization of nanofiltration membranes: Advances and challenges. *Separation and Purification Technology, 239*, Article 116477.

Singh, R., Bhadouria, R., Singh, P., Kumar, A., Pandey, S., & Singh, V. K. (2020). Nanofiltration technology for removal of pathogens present in drinking water. In *Waterborne pathogens* (pp. 463–489). Butterworth-Heinemann.

Song, X., Luo, W., Hai, F. I., Price, W. E., Guo, W., Ngo, H. H., & Nghiem, L. D. (2018). Resource recovery from wastewater by anaerobic membrane bioreactors: Opportunities and challenges. *Bioresource Technology, 270*, 669–677.

Tang, W., Zhang, X., & Wang, R. (2021). Salt rejection mechanisms in nanofiltration: A comprehensive review. *Membrane Science Reviews, 6*(1), 45–60.

Thong, Z., Cui, Y., Ong, Y. K., & Chung, T. S. (2016). Molecular design of nanofiltration membranes for the recovery of phosphorus from sewage sludge. *ACS Sustainable Chemistry and Engineering, 4*(10), 5570–5577.

Tsalas, N., Golfinopoulos, S. K., Samios, S., Katsouras, G., & Peroulis, K. (2024). Optimization of energy consumption in a wastewater treatment plant: An overview. *Energies, 17*(12), 2808.

Verweij, H. (2012). Inorganic membranes. *Current Opinion in Chemical Engineering, 1*(2), 156–162.

Vineyard, D., Hicks, A., Karthikeyan, K. G., Davidson, C., & Barak, P. (2021). Life cycle assessment of electrodialysis for sidestream nitrogen recovery in municipal wastewater treatment. *Cleaner Environmental Systems, 2*, Article 100026.

Wang, Z., Zhu, J., & Liu, P. (2017). Chemical stability of polyamide nanofiltration membranes. *Water Research, 123*, 205–213.

Wang, Q., Lin, X., & Gao, F. (2021). Layer-by-layer assembly of nanofiltration membranes: Recent progress and applications. *Chemical Engineering Journal, 415*, Article 129224.

Wang, K., Wang, X., Januszewski, B., Liu, Y., Li, D., Fu, R., & Huang, X. (2022). Tailored design of nanofiltration membranes for water treatment based on synthesis–property–performance relationships. *Chemical Society Reviews, 51*(2), 672–719.

Wei, C., He, Z., Lin, L., Cheng, Q., Huang, K., Ma, S., & Chen, L. (2018). Negatively charged polyimide nanofiltration membranes with high selectivity and performance stability by optimization of synergistic imidization. *Journal of Membrane Science, 563*, 752–761.

Werber, J. R., Osuji, C. O., & Elimelech, M. (2016). Materials for next-generation desalination and water purification membranes. *Nature Reviews Materials, 1*(5), 1–15.

West, P. C., Gerber, J. S., Engstrom, P. M., Mueller, N. D., Brauman, K. A., Carlson, K. M., & Siebert, S. (2014). Leverage points for improving global food security and the environment. *Science, 345*(6194), 325–328.

Williams, A. T., Zitomer, D. H., & Mayer, B. K. (2015). Ion exchange-precipitation for nutrient recovery from dilute wastewater. *Environmental Science: Water Research and Technology, 1*(6), 832–838.

Xie, M., Shon, H. K., Gray, S. R., & Elimelech, M. (2016). Membrane-based processes for wastewater nutrient recovery: Technology, challenges, and future direction. *Water Research, 89*, 210–221.

Yaacoubi, H., & Dumée, L. F. (2023). Sustainable organic solvent nanofiltration membranes. In *Green membrane technologies towards environmental sustainability* (pp. 515–554). Elsevier.

Yang, X., Li, J., & Zhao, H. (2019). Surface morphology and fouling behavior of nanofiltration membranes. *Journal of Environmental Management, 246*, 794–802.

Yu, C., Yin, W., Yu, Z., Chen, J., Huang, R., & Zhou, X. (2021). Membrane technologies in toilet urine treatment for toilet urine resource utilization: A review. *RSC Advances, 11*(56), 35525–35535.

Yusuf, M., Kurzina, I., Voronova, G., Mohamed, M., Mohammed, S. D., & Oladoja, N. A. (2024). Trends in the energy and environmental application of metal-organic framework-based materials. *Energy Advances.*

Zhang, S., Qu, Y., Liu, Y., Yang, F., Zhang, X., Furukawa, K., & Yamada, Y. (2005). Experimental study of domestic sewage treatment with a metal membrane bioreactor. *Desalination, 177*(1–3), 83–93.

Zhang, J., Li, G., Yuan, X., Li, P., Yu, Y., Yang, W., & Zhao, S. (2023). Reduction of ultrafiltration membrane fouling by the pre-treatment removal of emerging pollutants: A review. *Membranes, 13*(1), 77.

Zhang, M., & Ma, Y. (2020). Energy use and challenges in current wastewater treatment plants. In *AB processes: Towards energy self-sufficient municipal wastewater treatment* (pp. 1–28).

Zhao, Y., Tong, T., Wang, X., Lin, S., Reid, E. M., & Chen, Y. (2021). Differentiating solutes with precise nanofiltration for next generation environmental separations: A review. *Environmental Science and Technology, 55*(3), 1359–1376.

Zhao, Y., Liu, X., & Sun, W. (2022). Nanomaterial-modified nanofiltration membranes for advanced water treatment. *Advances in Water Research, 34*(2), 233–246.

Zheng, S., Lin, T., Chen, H., Zhang, X., & Jiang, F. (2024). Impact of changes in biofilm composition response following chlorine and chloramine disinfection on nitrogenous disinfection byproduct formation and toxicity risk in drinking water distribution systems. *Water Research, 253*, Article 121331.

Chapter 4
Microbial Fuel Cells: Harnessing Power from Wastewater

Sukhvinder Dhiman, Sapna Devi, Satish Koundal, and Manoj Kumar

Abstract The energy demand has been dramatically increasing due to rise in global population and rapid industrial growth depleting non-renewable resources such as fossil fuels, minerals and nuclear energy. In addition to this, large amount of wastewater is generated from domestic and industrial activities which needs to be treated wisely to meet the demand of future generations. Recently, microbial fuel cell (MFC) emerged as a promising novel technology to generate electricity using wastewater sources, based on inherent property of microbial electron transport chain system. MFCs uses catabolic potential to break down complex organic compounds present in wastewater to release source of electrons that can be further used for electricity generation. The present chapter delves into the fundamental working principle, key features, types and various components used for making an efficient MFC system. In addition, it discusses various types of microbes and factors influencing the performance of MFCs. Besides this, various advantages and disadvantages of using MFCs including the economic and scientific challenges for greater aims and better efficiencies for its possible practical applications and its alternative as a source of renewable energy for its future use.

Keywords Microbial fuel cell · Wastewater treatment · Electricity generation from waste · Bioenergy · Waste-to-fuel · Microbial fuel cell reactor

---

S. Dhiman
Institute of Nano Science and Technology, Mohali, Punjab, India

Institute of Organic Chemistry, Polish Academy of Sciences, Warsaw, Poland

S. Devi
Department of Microbiology, DAV University, Jalandhar, Punjab, India

S. Koundal
Department of Medical Lab Technology, DAV University, Jalandhar, Punjab, India

M. Kumar (✉)
Department of Microbiology, Guru Nanak Dev University, Amritsar, Punjab, India

S. Dhiman and G. Mukherjee (eds.), *Resource Resurgence*, Springer Water,
https://doi.org/10.1007/978-3-032-02750-4_4

4.1 Introduction

Wastewater treatment is an energy-intensive and costly process that addresses both the growing global energy demands and the removal of large volumes of wastewater due to its environmental concerns (Malik et al., 2023; Siddiqui et al., 2023). As a result, wastewater treatment powered by generation of renewable energy sources has emerged as a promising solution to tackle both power generation challenges and environmental pollution. To meet the demand of future generations, microbial fuel cells (MFC) have emerged as a highly efficient and sustainable solution for wastewater treatment, enabling the concurrent recovery of energy and nutrients (Sarangi et al., 2023). This process generates bioelectricity due to the electrogenic potential of microbes which can oxidize organic contaminants present in the wastewater (Bazina et al., 2023; Kusmayadi et al., 2020; Mahto & Das, 2025).

Due to the limitations of conventional wastewater treatment methods, the sustainable applications of MFCs offer a promising alternative for wastewater treatment, carbon sequestration, biohydrogen production, green electricity generation, and friendly sewage treatment (Aelterman et al., 2008; Chaudhary et al., 2025; Ye et al., 2019). MFCs use anodic oxidation and cathodic reduction for electrochemical treatment-a mechanism refined over decades of research. However, generating electricity in MFCs continues to be a significant hurdle for practical applications, mainly because of the challenge in achieving a balance between energy yield and system scalability (Molognoni et al., 2016; Ortiz-Martínez et al., 2015; Sleutels et al., 2011).

MFC functions through electrochemical energy conversion, where microorganisms generate electrochemical compounds. These compounds interact at the anode, forming a biofilm that functions as a biocatalyst to generate electricity. The structure of MFCs comprises two chambers constructed from constituents such plexiglass, polycarbonate with electrodes typically made from graphite and carbon paper and cloth. First chamber, known as anodic chamber, is anaerobic, with nitrogen gas used to maintain the required conditions. In this chamber, wastewater containing various microorganisms undergoes treatment through electrochemical processes. The other chamber, the cathodic chamber, is aerobic and contains highly potential electron acceptor electrode, typically O_2. The proton exchange membrane (PEM) typically constructed from materials like Nafion or Ultrex, separates the two chambers in a microbial fuel cell. These membranes are essential in PEM fuel cells, as they allow cations to pass through, enabling the electrochemical process (Aelterman et al., 2008; Molognoni et al., 2016; Ortiz-Martínez et al., 2015; Sleutels et al., 2011).

MFCs offer various features that make them more suitable for wastewater treatment (i) Direct alteration of substrate energy into electricity, (ii) Reduced additional activated sludge than anaerobic digesters and conventional aerobic activated sludge processes, (iii) Resilience to various operating environments at low temp, (iv) No need for gas treatment, (v) No energy required for conversion, and (vi) Potential for widespread application in areas with limited electrical infrastructure. MFCs have emerged as a promising yet challenging technology for tackling energy and environmental issues, particularly in remote areas. MFCs are prepared with biosensors, enable biohydrogen production, and serve as in-situ power sources for bioremediation and wastewater treatment. This chapter aims to explore the challenges and applications of MFCs in wastewater treatment (Mohan et al., 2014; Kumar et al., 2016; Mercuri et al., 2016).

4.2 Working Principle, General Features and Types of Microbial Fuel Cell (MFC)

4.2.1 Working Principle

Microbial Fuel Cells consist of electrodes, microorganisms, a substrate/analyte, and an external circuit. The main reaction involves anaerobic bacteria oxidizing acetate at anode, generating electrons and protons. The electrons flow through the external circuit to cathode, while protons move through electrolyte to complete the circuit (Li et al., 2010; Malik et al., 2023).

$$CH_3COO^- + 4H_2O \rightarrow 2HCO_3 + 9H^+ + 8e^- \tag{4.1}$$

In microbial fuel cells (MFCs), electroactive bacteria (EAB) act as biocatalysts, oxidizing the substrate and transferring electrons to anode via direct electron transfer or exoelectrogens. The electrons flow over the external circuit to cathode, while protons pass through the PEM toward maintain electroneutrality, enabling power generation in MFCs (Fig. 4.1).

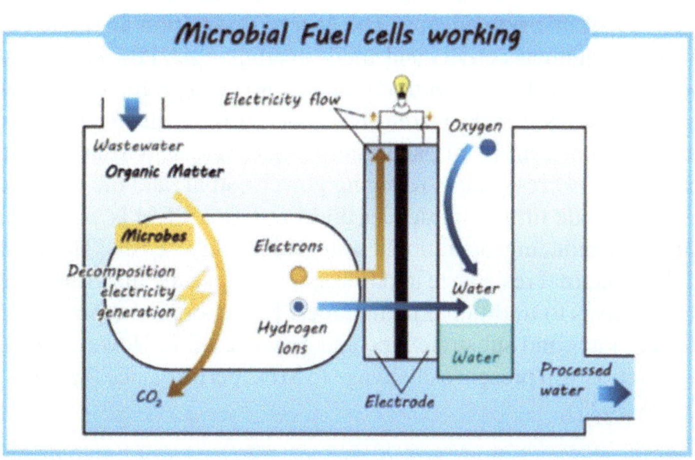

Fig. 4.1 Working principle of MFCs (Adapted from reference source Nawaz et al., 2022)

In MFCs, indirect electron transfer occurs through soluble electron shuttles, which aid in moving electrons to the anode. These shuttles can be natural or synthetic, with synthetic mediators like phenothiazine, phenazine, indophenol, and thionine commonly used. Electrons generate at anode travel through the external circuit to cathode, generating an electrical current, which is fundamental to electricity production in MFCs. The cathode chamber comprises electron acceptors, such as oxygen, which facilitate reduction reactions as given below (Heijne et al., 2010; Malik et al., 2023).

$$2H_2O + O_2 + 4e^- \rightarrow 4OH^- \quad (4.2)$$

As electrons travel through the external circuit, they drive protons over the PEM. Protons combine with oxygen in cathode to produce water molecules. Equation given below (Aelterman et al., 2008; Ahmad et al., 2023; Akram et al., 2024):

$$2H^+ + O_2 + 2e^- \rightarrow 2H_2O \quad (4.3)$$

Wastewater from industrial, domestic, and agricultural sources includes organic constituents that act as serve fuel for MFCs. MFCs can be developed as self-sustaining wastewater treatment technology, operating without external power while generating energy. He present approach offers economic advantages, such as improved energy efficiency, reduced sludge production, minimal toxic byproducts, and recovery of valuable products like nutrients and electricity. It also ensures operational flexibility under diverse conditions, providing reliable and feasible solutions (Malik et al., 2023; Nawaz et al., 2022).

The efficiency of MFCs depends on electrochemical, biological, physicochemical and operational factors. Biological factors include prosperity, diversity, and catalytic activity of microorganisms. The energy dissipation at anode results from the loss of electrochemical activity and anode overpotential. Physicochemical factors include electrolytic resistance, electrode surface area and proton transfer rate, reduction reactions at cathode and external resistance. Electrochemical factors include charge transfer, ohmic, diffusion, and internal resistances. A smaller PEM surface area increases internal resistance, reducing power output, and the distance between the anode and cathode further influences this resistance. PEM is vital for improving performance and reducing operational and maintenance costs. The carbon paper can also reduce internal resistance and lower costs. Ceramic materials provide cost-effective alternatives to traditional membranes. Key operational parameters include organic loading rates and substrate type and concentration. However, substrate and anolyte evaporation still present a challenge to MFC performance (Malik et al., 2023; Siddiqui et al., 2023).

4.2.2 General Features of MFC

In MFC, microorganisms metabolize substrates and transfer electrons to an electrode. Anaerobic-respiring bacteria at the anode reduce organic matter into CO_2, generating electrons and protons. The electrons transfer through an external circuit and protons travel between electrodes to maintain electroneutrality. MFCs are often "half biological" with microorganisms at the anode and an abiotic cathode. Hydrogen peroxide, ferricyanide and Oxygen and act as electron acceptors at cathode, with oxygen being the most suitable. The air cathode has a catalyst layer, electrode, and separator. The centrifuge increases internal resistance, reducing MFC performance and limiting practical applications (Malik et al., 2023; Sarangi et al., 2023; Siddiqui et al., 2023).

Previous efforts have concentrated on developing advanced abiotic catalysts to enhance oxygen reduction reactions. However, these catalysts remain costly for MFCs due to their lower power and current densities compared to conventional fuel cells. Biocathodes, which use bacteria as biocatalysts on cost-effective carbon cathodes, offer an attractive alternative. Despite their advantages, research on the metabolic mechanisms of biocathodes is still limited, requiring more fundamental studies for optimization. The strategy and architecture of MFCs significantly affect their application in the treatment of wastewater (Siddiqui et al., 2023). Logan and Regan (2006) emphasize that system architecture, rather than the bacterial community composition, is the key to achieving higher power density in MFCs. Studies show that MFCs inoculated with mixed cultures outperform those with pure cultures, offering higher power densities and better contaminant removal efficiency.

The complex composition of various waste biomass sources necessitates a diverse microbial community for effective degradation. MFC designs that are scalable, low-cost, high-performance, and easy to integrate with reported systems could facilitate commercialization. While MFCs are still in the laboratory stage, innovative strategies have been established to implant them in other wastewater treatment systems. However, the energy harvest from MFCs is currently else low to achieve energy-neutral method at practical scales, and costs of both capital and operational remain high. Long-term stability, power output, and cost remain significant challenges. Integrating MFCs with other processes is seen as a promising strategy to advance their role in wastewater treatment (Aelterman et al., 2008; Ye et al., 2019).

MFCs in wastewater treatment are typically characterized by energy savings, reduced sludge production, and lower energy consumption. Researchers have explored their potential to treat various types of waste water, and studies have shown that MFCs can remove a wide range of contaminants, including biological waste, heavy metals, polyalcohol, dyes, phenols, furan, quinolone and pyridine derivatives. However, to implement MFCs in practical applications, their performance needs to be evaluated using real wastewater (Nawaz et al., 2022).

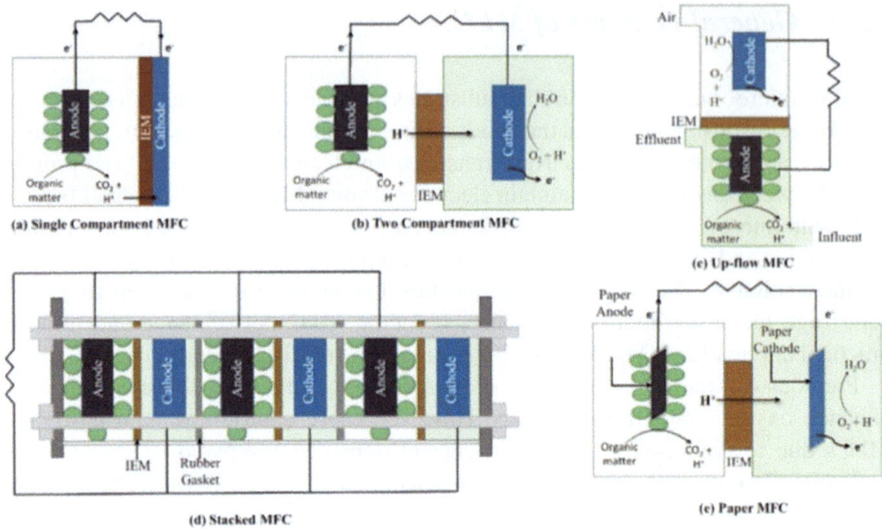

Fig. 4.2 Different types of MFCs (Adapted from reference source Malik et al., 2023)

4.2.3 Types of Microbial Fuel Cell (MFC)

MFCs are classified based on anode and cathode chamber configuration, electrode assembly, and the use of either a proton exchange membrane (PEM) or a salt bridge. MFCs can be categorized into different types based on their design and operating mechanisms, are illustrated in Fig. 4.2.

(i) **Single-Compartment MFCs**

Single-compartment MFCs have a simple design with anode chamber and a cathode exposed to the environment, removing the need for an oxygen supply. This makes them productive and easy to scale. Constructed with polycarbonate plates, the MFC features a porous carbon paper anode, a platinum-catalysed carbon cloth cathode, and a Nafion proton exchange membrane. Copper wire connects the electrodes to external circuits. Incorporating ceramics enhances performance, offering low-cost ion exchange, stability for bacteria, and efficient energy harvesting (Chaudhuri & Lovley, 2003; Malik et al., 2023).

(ii) **Two-Compartment MFCs**

This two-compartment MFC features separate anode and cathode chambers divided by PEM or salt bridge. The anode contains microbes, a medium, and an electrode, while cathode holds freshwater or buffer, an electrode, and oxygen. Common electrodes include stainless steel mesh, copper, graphite, and carbon materials. Anaerobic conditions in the anode may require a nitrogen source. The MFC can be built using borosilicate glass bottles with a Nafion PEM separator. Platinum-free catalysts like

Pd-Cu, manganese oxides, and activated carbon nickel serve as alternatives. The microbes are cultivated in a mineral salt medium and kept at 4 °C for later usage (Malik et al., 2023; Siddiqui et al., 2023; Winfield et al., 2016).

(iii) **Up-Flow MFCs**

Up-flow MFCs operate in persistent mode, where wastewater is injected from the bottom under high pressure and exits from the top. The system uses a tubular structure made of polyacrylic plastic without a proton exchange membrane. It describes a microbial fuel cell setup where graphite felt acts as both anode and cathode, separated by glass beads and glass wool. Artificial wastewater with glucose and glutamate fuels the system, while cathode is aerated. Platinum wires connect the electrodes to an external circuit. While the strategy reduces costs by omitting the PEM, its net energy efficiency is limited, as the energy essential for wastewater pumping exceeds the energy produced (Malik et al., 2023; Roy et al., 2023).

(iv) **Stacked MFCs**

This design features numerous MFCs organized in parallel to improve output efficiency and enhance COD elimination. It consists of six individuals continuous MFCs stacked systematized. The graphite granules serve as both the anode and cathode, with an Ultrex CMI7000 PEM acting as the separator. Parallel cell connections have been shown to outperform series connections, subsequent in higher efficiency and better COD removal (Lee et al., 2016; Malik et al., 2023).

(v) **Paper MFCs**

Paper-based MFCs provide advantages such as cost efficiency, chemical resistance, and simple removal. The strategy consists of anode and cathode, where graphite electrodes are placed onto paper using four pencil strokes. Parchment paper functions as the proton exchange membrane (PEM), with crayons applied to improve hydrophobicity. Microbes like *Shewanella oneidensis* are introduced into anode along with a suitable growth medium (Ieropoulos et al., 2008; Malik et al., 2023).

4.3 Microorganisms Used in MFC

The efficient working of a microbial fuel cell (MFC) depends upon the kind of microorganism employed for harvesting energy in the form of electricity. The microorganisms that donate electrons to extracellular electron acceptors are called as electrogens whereas, which consume electron from extracellular electron donors called as electrotrophs (Nawaz et al., 2022). In literature, various microorganisms are reported for their potential to transfer electrons either as electrogens and electrophs represented in Figs. 4.3 and 4.4. These microorganisms play a central role in the biochemical reactions of MFC working as either as pure culture or mixed cultures (Siddiqui et al., 2023). The working of MFCs depends on two types of microorganisms, one involved in the reduction of electrode and electron movement towards the anode. This kind of electron transfer is generally observed in *Geobacter sulfurreducens* and *Shewanella oneidensis* (Shi et al., 2019). The movement of electrons by microorganisms can be either carried out by direct transfer method using conductive nanowires or outer membrane proteins/cytochromes or pili present on the surface or by some mediator molecules present in the solution. Whereas, some the microorganisms involved in the oxidation of electrodes obtain electrons from organic components such as acetate and carbon dioxide called electricigens (Nawaz et al., 2022; Siddiqui et al., 2023). This kind of electron transfer is generally observed in *Clostridium butyricum* and *Bacteroides spp.* Hence, the selection of microorganisms must be done keeping in mind their capacity to transfer electrons from substrate to anode (Nosek et al., 2023). The microorganisms used in various MFC system can be either pure culture or mixed cultures. Among the most commonly used microbes in MFC includes *Geobacter* and *Shewanella* among the predominant ones, but more in-depth research in this area can unravel novel microbes having potential application in MFC large scale applications (Nawaz et al., 2022; Siddiqui et al., 2023).

In literature, various bacterial species were reported that can be used in MFC either as electrogens and electrotrophs represented in Figs. 4.3 and 4.4, respectively based on molecular phylogenetic analysis of 16S rRNA gene sequences (Saitou & Nei, 1987). Among the bacteria, Proteobacteria group constitutes the major portion which can be used in MFC. Some microorganisms carried out indirect transfer of electrons using electron mediators such as flavins, quinones and phenazines as observed in *Escherichia coli* and *Enterococcus faecalis* (Nosek et al., 2023). In addition, some microorganisms can utilize light energy for the generation of electrons such as observed in *Rhodopseudomonas palustris* and *Synechococcus spp.* (Li et al., 2022; Tanvir et al., 2021). Some bacteria form biofilms on the electrode surface enhancing electron transfer efficiency such as *Geobacter sulfurreducens* (Liu et al., 2020).

4.4 Factors Affecting the Performance of MFCs

The performance of microbial fuel cells (MFCs) can be based on multiple factors such as factors biological, chemical, and physical that can be used to determine their efficiency in electricity generation. All of these factors might influence electron transfer, microbial activity, and overall system performance (Mohan et al., 2014). The below-mentioned factors can affect MFC performance. One of the important factors affecting MFC performance is biological factors such as composition of microbes which may be categorized into electrogens or electrotrophs (Nawaz et al., 2022). Also, the inclusion of stable and enriched electrogenic community enhances power output of MFC. Some of these microbes help in the formation of biofilm on the anode, enhancing power output by improving electron transfer efficiency (Wang et al., 2019). The efficient working of microbes depends upon the growth conditions maintained in the MFC chamber such as optimal temperature in the range of 25–30°C, optimal pH conditions between 6.5–7.5 and availability of nutrients viz. carbon source & energy sources (Angelaalincy et al., 2018). In addition to this, the substrate

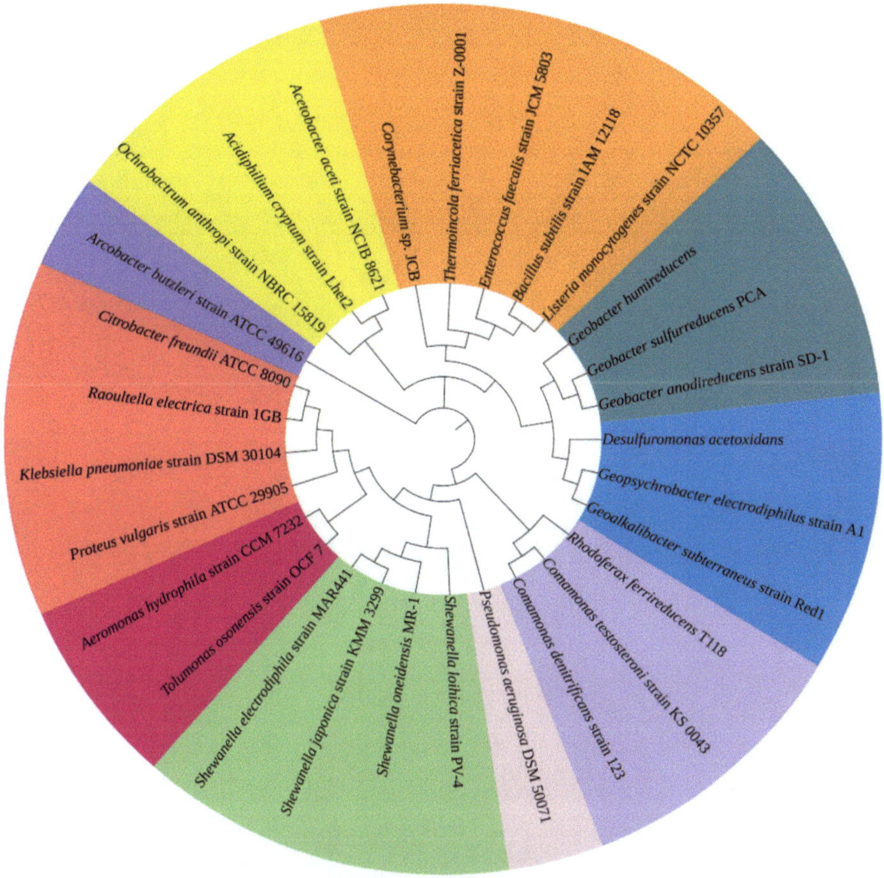

Fig. 4.3 Diversity of electrogens which can be used in MFCs. The figure represents molecular phylogenetic analysis based on 16S rRNA gene sequences. The evolutionary history was inferred using the Neighbor-Joining method (Saitou and Nei, 1987). The optimal tree with the sum of branch length = 1.69243903 is shown. The percentage of replicate trees in which the associated taxa clustered together in the bootstrap test (1000 replicates) are shown next to the branches (Felsenstein, 1985). The tree is drawn to scale, with branch lengths in the same units as those of the evolutionary distances used to infer the phylogenetic tree. The evolutionary distances were computed using the Tajima-Nei method (Tajima & Nei, 1984) as per best-fit model and are in the units of the number of base substitutions per site. The rate variation among sites was modeled with a gamma distribution (shape parameter = 1). The analysis involved 29 nucleotide sequences. Codon positions included were 1st + 2nd + 3rd + Noncoding. All positions containing gaps and missing data were eliminated manually. There were a total of 1112 positions in the final dataset. Evolutionary analyses were conducted in MEGA7 (Kumar et al., 2016). Tree was edited in iTol server (Letunic and Bork, 2016)

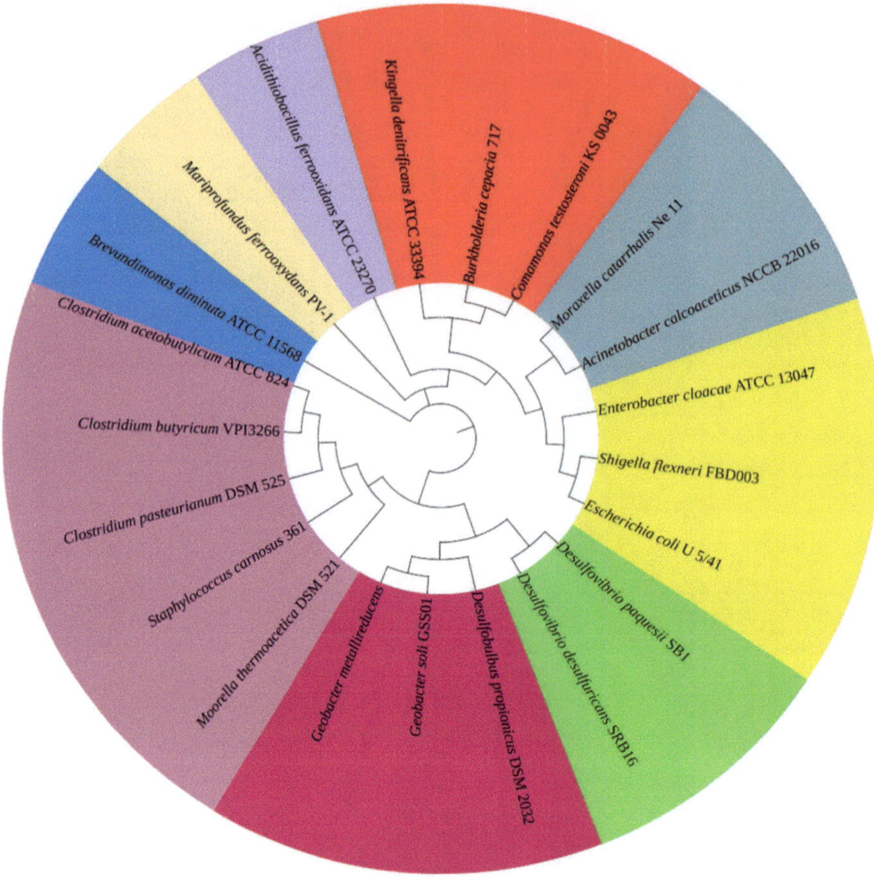

Fig. 4.4 Diversity of electrotrophs which can be used in MFCs. The figure represents molecular phylogenetic analysis based on 16S rRNA gene sequences. The evolutionary history was inferred using the Neighbor-Joining method (Saitou and Nei, 1987). The optimal tree with the sum of branch length = 1.56512242 is shown. The percentage of replicate trees in which the associated taxa clustered together in the bootstrap test (1000 replicates) are shown next to the branches (Felsenstein, 1985). The tree is drawn to scale, with branch lengths in the same units as those of the evolutionary distances used to infer the phylogenetic tree. The evolutionary distances were computed using the Kimura 2-parameter method (Kimura, 1980) as per best-fit model and are in the units of the number of base substitutions per site. The rate variation among sites was modeled with a gamma distribution (shape parameter = 1). The analysis involved 21 nucleotide sequences. Codon positions included were 1st + 2nd + 3rd + Noncoding. All positions containing gaps and missing data were eliminated manually. There were a total of 1247 positions in the final dataset. Evolutionary analyses were conducted in MEGA7 (Kumar et al., 2016). Tree was edited in iTol server (Letunic and Bork, 2016)

present in the wastewater may also hinder the working of MFC such as readily degradable substrates (e.g., acetate, glucose) enhance electricity generation whereas complex organic sources require additional breakdown by fermentative bacteria, affecting electron flow. Also, thc wastewater containing toxic compounds, presence of heavy metals such as Cr, Cd, Hg etc. (Kumar & Saini, 2024) and inhibitors can reduce microbial activity of MFC, hence optimal nutrient balance supports microbial growth (Nawaz et al., 2022; Pant et al., 2010).

The working of MFC can be affected by type of electron acceptors at the cathode such as O_2 is the most common electron acceptor, but alternatives like ferricyanide or permanganate can enhance cathode performance (Ucar et al., 2017). Sometimes, electrode material can affect the electron transfer capability of MFC such as carbon-based electrodes (carbon felt, graphite, graphene) offer good conductivity and biocompatibility; also, platinum-based cathodes have high efficiency but are costly. The closer spacing of the anode and cathode reduces internal resistance and improves power density (Borja-Maldonado & Zavala, 2022).

4.5 Benefits of MFC's for Electricity Generation

The use of microbial fuel cells (MFCs) to generate electricity from wastewater has emerged as a promising technology for sustainable and environmentally friendly power generation (Ali et al., 2025; Arun et al., 2024). The MFCs depends on biological activity instead of fossil fuels, which are finite and environmentally harmful (Hoang et al., 2022). Using the metabolic activity of microorganisms, bioelectrochemical systems generate electrical energy while simultaneously treating the organic waste (Pandya et al., 2024). Moreover, MFCs have a number of distinct advantages over conventional fuel cells as they are more efficient and produce less pollution. In a process known as electro-hydrogenesis, when the electrons produced by microbial metabolism are used to reduce protons (H^+) to form hydrogen (H_2) at the cathode, MFCs can even produce hydrogen along with electricity thus solving the hydrogen problem at the same time (Logan & Regan, 2006). The use of conventional batteries is unfriendly to the environment as they contain heavy metals, and are powered by electricity whereas MFCs need to be charged before usage (Rabaey et al., 2005). Decentralized and off-grid electricity generation through MFCs is an ideal solution for remote and rural areas without conventional power infrastructure (Mondal & Goswami, 2024). For small-scale uses like powering sensors, remote monitoring systems and environmental monitoring equipment, microbial fuel cells can be especially advantageous (Santoro et al., 2017). In addition to removing organics from wastewater, microbe-electrode interactions are being used to develop a variety of systems that produce and recover valuable products such as bio-flocculants, bio-plastics, bio-surfactants, chemicals, electricity, hydrogen, methane and a host of others (Akram et al., 2024).

MFCs offer the potential for two functions: electricity generation and wastewater treatment. Microorganisms can convert organic compounds into water, carbon dioxide, and energy (Li & Yu, 2011). In addition to providing an environment for microbial growth and metabolism, MFCs also serve as an efficient means of harvesting these microbially generated energies. Microorganisms can produce electrons by breaking down organic matter in wastewater, which improves water quality by reducing chemical oxygen demand (COD) and biochemical oxygen demand (BOD). The production of electricity by power plants utilize coal which lead to the production of large amounts of greenhouse gases (GHGs) that contribute to global warming. Microorganisms have gained attention for their ability to directly produce electricity from organic sources via extracellular electron transfer (Rabaey & Rozendal, 2010). In a microbial fuel cell (MFC), during anaerobic respiration of organic carbon, microorganisms are able to use an electrode (the anode) as an electron acceptor. The microbial decomposition of organic matter can be directly harvested as electrical current by connecting a bioanalyzed anode (bioanode) to a (bio) cathode in a bio-electrochemical system (BES) (Yang et al., 2014). Logan and Regan (2006) reported that H_2 can also be used in the anodic process, acetate is the most common substrate for the process. Microbial oxidation of organic matter is a low-carbon process, MFCs produce electricity with minimal greenhouse gas emissions (Wang et al., 2011). The development of low-cost materials and more efficient systems, with higher power output and durability, are important for the application of MFCs in industrial/large scale (Boas et al., 2022). Compared to other bioenergy technologies, MFCs require lower energy input and operational costs. Unlike conventional wastewater treatment plants that rely on aeration (which is energy-intensive), MFCs operate under anaerobic or low-oxygen conditions, reducing the overall energy demand of the system (Logan & Regan, 2006). In lab and pilot-scale experiments, MFC technology has demonstrated promising scalability. MFCs can also be integrated into existing wastewater treatment plants to improve their efficiency and to recover energy from organic waste streams (Li et al., 2014). Furthermore, MFCs can produce value-added byproducts, such as hydrogen, biohydrogen, and various bio-based chemicals, enhancing their economic feasibility and sustainability (Wilberforce et al., 2021).

4.6 Challenges and Limitations of MFCs

Microbial Fuel Cells (MFCs) are a promising approach to generate renewable energy with promoting environmental sustainability (Khandaker et al., 2021). MFCs harness the metabolic processes of microorganisms to transform the chemical energy stored in organic matter into electricity. Despite their potential, several challenges limit their large-scale adoption (Fig. 4.5). The biggest limitation of microbial fuel cells (MFCs) is their relatively low power output compared to conventional systems (Slate et al., 2019). MFCs produce few ranges from milliwatts per square meter, which is insufficient for practical applications. Different factors are responsible for this limitation, including inefficient electron transfer, high internal resistance, constraints in microbial metabolism, and material related challenges (Obileke et al., 2021). To address these challenges, research and technological advancements are essential.

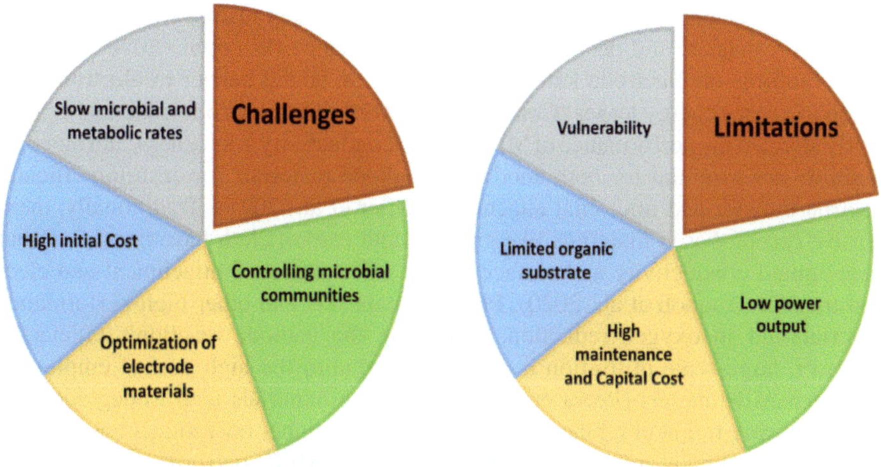

Fig. 4.5 Challenges and limitations of microbial fuel cells (MFCs)

Advances in microbial engineering, electrode materials, and system design could significantly improve MFC efficiency and cost-effectiveness (Do et al., 2018).

The primary causes of MFCs low power density are the inefficient transfer of electrons (Kumar et al., 2018). The inadequate optimization of microbial electron transport limits the overall efficiency of MFCs. In the anode chamber microorganisms metabolize organic substrates and produce electrons as part of their metabolism. These electrons need to be efficiently transferred from the microbes to the electrode for effective electricity generation (Thapa et al., 2022). In many microbial species, the processes behind extracellular electron transport (EET) are unknown. For MFCs to function well, robust microbial communities must be carefully chosen and effectively maintained. Some high-performing microbes, such as *Geobacter sulfurreducens* and *Shewanella oneidensis* are known for their superior electron transfer capabilities (Xie et al., 2021). The microorganisms greatly influence the efficiency of electron transfer and substrate metabolism. These bacteria have long doubling times, making it difficult to establish and maintain stable, high-performing microbial populations in MFCs (Dwivedi et al., 2022). Moreover, these communities are sensitive to environmental factors like pH, temperature, and substrate availability. Variations in these parameters can cause fluctuations in power output and efficiency (Guo et al., 2021). In mixed-culture systems, often used in practice, competition among microbial species for limited resources can lower overall efficiency, as some microbes may outcompete others without aiding electron transfer. Ensuring the stability and dominance of electrogenic bacteria in mixed cultures remains a significant challenge for large-scale MFCs applications (Kumar et al., 2019).

High internal resistance, especially at the electrode–electrolyte interface, decreases the efficiency of electron flow in the system. The resistance is caused by inefficient proton exchange, poor ionic conductivity, and suboptimal electrode materials (Ahmad et al., 2023). The proton exchange membrane (PEM) separates the anode and cathode compartments and is prone to degradation over time, which increases resistance and decreases system performance (Wu et al., 2020). Furthermore, biofilms on electrode surfaces can create an added barrier to electron flow, which increases energy losses (Conners et al., 2022). The electrode materials significantly affected the performance of MFCs. High conductivity, robustness, and biocompatibility are essential for both anode and cathode materials for ensuring efficient electron transfer and microbial attachment (Mier et al., 2021). Traditionally, inexpensive carbon-based materials like graphite and carbon cloth have been used, but their limited conductivity and surface area restrict microbial attachment and electron transfer (Yaqoob et al., 2020). Platinum catalysts and other high-performance materials aid in oxygen reduction, increasing the cathode reaction's efficiency. However, large-scale utilisation is not feasible due to the high cost of employing these expensive metals (Peera et al., 2021). These materials tend to degrade over time, requiring frequent replacement and leading to higher operational costs. Additionally, biofouling is another major issue that affects MFCs performance. Microbial communities form biofilms on electrode surfaces over time, which can lead to clogging and decreased conductivity (Jadhav et al., 2021), lowering power output and electron transfer efficiency. Various methods are used for the reduction of biofouling such as surface modifications and routine electrode cleaning, but these practices increase maintenance costs and operational complexity (Koók et al., 2019; Sharif et al., 2021). Various challenges arise when MFCs are scaled up from laboratory to industrial applications. For large-scale design, even substrate distribution, effective electron collection, and minimal energy losses are required (Mittal et al., 2024). The power output decreases with system size because substrate transport and metabolic byproduct elimination are less effective at larger systems (Seelam et al., 2018). To prevent localised nutricnt dcplction and the accumulation of hazardous metabolites, large-scale MFCs hydrodynamics must be carefully optimised.

The commercialisation of MFCs infrastructure is restricted by high initial investments (Bagheri & Mirbagheri, 2018). Capital expenses are raised by the high price of anodes, cathodes, and membranes in addition to the necessity of specialised materials and construction for large-scale reactors. The energy delivered per unit cost from MFCs continues to be not comparable with conventional energy systems, despite innovations in reactor design and materials research (Santoro et al., 2017). The lack of standardized designs and protocols complicates comparisons between MFCs systems, hindering commercialization and preventing the establishment of best practices. Developing standardized MFCs design and operation guidelines is essential for promoting widespread adoption.

4.7 Future Prospects and Conclusion

Microbial fuel cells (MFCs) have great potential as a sustainable energy technology, but in order to operate economically, a number of issues need to be resolved. Large-scale adoption of MFCs is hampered by several challenges, chief among them being ineffective electron transmission, high internal resistance, limited electrode material, and problems with microbial stability. Overcoming these barriers will be crucial for realizing the full potential of MFCs as a viable alternative energy source in the future. The novel microbes with enhanced electron transport pathways and faster growth rates can significantly boost MFC efficiency. Moreover, synthetic biology and genetic engineering technologies offer exciting possibilities for improving microbial electron transfer. Aside from the microbial perspective, material innovations are also crucial for addressing enhanced conductivity and durability for electrode-related limitations. The use of nanomaterials and highly conductive polymer-based electrodes could help enhance efficiency while minimizing biofouling. Combining MFCs with wastewater treatment systems not only generates bioelectricity but also helps with environmental remediation, making the technology more economically feasible. However, innovations in microbial engineering, material science, and system design are paving the way for significant improvements. With continued research and technological advancements, MFCs could become a reliable alternative energy source in the future.

References

Aelterman, P., Versichele, M., Marzorati, M., Boon, N., & Verstraete, W. (2008). Loading rate and external resistance control the electricity generation of microbial fuel cells with different three-dimensional anodes. *Bioresource Technology, 99*(18), 8895–8902. https://doi.org/10.1016/j.biortech.2008.04.061

Ahmad, D., Haroon, M., Haq, F., Zeb, A., Mahmood, S., & Kiran, M. (2023). Microbial fuel cell, their type, working principle and different factors affecting their performance: A review. *Sustainable Chemical Engineering*, 121–146. https://doi.org/10.37256/sce.4220233125

Akram, N., Zia, K. M., Usman, M., Anjum, F., & Pervaiz, S. (2024). Recovery of value-added products from sewage sludge using biological processes. In *Sewage and biomass from wastewater to energy* (pp. 327–345). https://doi.org/10.1002/9781394204502.ch12

Ali, H. E., Hemdan, B. A., El-Naggar, M. E., El-Liethy, M. A., Jadhav, D. A., El-Hendawy, H. H., ... & El-Taweel, G. E. (2025). Harnessing the power of microbial fuel cells as pioneering green technology: Advancing sustainable energy and wastewater treatment through innovative nanotechnology. *Bioprocess and Biosystems Engineering*, 1–24. https://doi.org/10.1007/s00449-024-03115-z

Angelaalincy, M. J., Navanietha Krishnaraj, R., Shakambari, G., Ashokkumar, B., Kathiresan, S., & Varalakshmi, P. (2018). Biofilm engineering approaches for improving the performance of microbial fuel cells and bioelectrochemical systems. *Frontiers in Energy Research, 6*, 63. https://doi.org/10.3389/fenrg.2018.00063

Arun, J., SundarRajan, P., Pavithra, K. G., Priyadharsini, P., Shyam, S., Goutham, R., Le, Q. H., & Pugazhendhi, A. (2024). New insights into microbial electrolysis cells (MEC) and microbial fuel cells (MFC) for simultaneous wastewater treatment and green fuel (hydrogen) generation. *Fuel, 355*, Article 129530. https://doi.org/10.1016/j.fuel.2023.129530

Bagheri, M., & Mirbagheri, S. A. (2018). Critical review of fouling mitigation strategies in membrane bioreactors treating water and wastewater. *Bioresource Technology, 258*, 318–334. https://doi.org/10.1016/j.biortech.2018.03.026

Bazina, N., Ahmed, T. G., Almdaaf, M., Jibia, S., & Sarker, M. (2023). Power generation from wastewater using microbial fuel cells: A review. *Journal of Biotechnology, 374*, 17–30. https://doi.org/10.1016/j.jbiotec.2023.07.006

Boas, J. V., Oliveira, V. B., Simões, M., & Pinto, A. M. (2022). Review on microbial fuel cells applications, developments and costs. *Journal of Environmental Management, 307*, Article 114525. https://doi.org/10.1016/j.jenvman.2022.114525

Borja-Maldonado, F., & Zavala, M. Á. L. (2022). Contribution of configurations, electrode and membrane materials, electron transfer mechanisms, and cost of components on the current and future development of microbial fuel cells. *Heliyon, 8*(7). https://doi.org/10.1016/j.heliyon.2022.e09849

Chaudhary, T. N., Ahmed, S., Usman, M., Maka, A. O., Rasool, S., Ghaleeh, M., & Chen, B. (2025). Utilizing different food cultures for wastewater treatment and enhanced power generation in microbial fuel cells. *Biomass and Bioenergy, 195*, Article 107706. https://doi.org/10.1016/j.biombioe.2025.107706

Chaudhuri, S. K., & Lovley, D. R. (2003). Electricity generation by direct oxidation of glucose in mediatorless microbial fuel cells. *Nature Biotechnology, 21*(10), 1229–1232. https://doi.org/10.1038/nbt867

Conners, E. M., Rengasamy, K., & Bose, A. (2022). Electroactive biofilms: How microbial electron transfer enables bioelectrochemical applications. *Journal of Industrial Microbiology and Biotechnology, 49*(4), kuac012. https://doi.org/10.1093/jimb/kuac012

Do, M. H., Ngo, H. H., Guo, W. S., Liu, Y., Chang, S. W., Nguyen, D. D., Nghiem, L. D., & Ni, B. J. (2018). Challenges in the application of microbial fuel cells to wastewater treatment and energy production: A mini review. *Science of the Total Environment, 639*, 910–920. https://doi.org/10.1016/j.scitotenv.2018.05.136

Dwivedi, K. A., Huang, S. J., & Wang, C. T. (2022). Integration of various technology-based approaches for enhancing the performance of microbial fuel cell technology: A review. *Chemosphere, 287*, Article 132248. https://doi.org/10.1016/j.chemosphere.2021.132248

Felsenstein, J. (1985). Confidence limits on phylogenies: An approach using the bootstrap. *Evolution, 39*(4), 783–791. https://doi.org/10.1111/j.1558-5646.1985.tb00420.x

Guo, F., Luo, H., Shi, Z., Wu, Y., & Liu, H. (2021). Substrate salinity: A critical factor regulating the performance of microbial fuel cells, a review. *Science of the Total Environment, 763*, Article 143021. https://doi.org/10.1016/j.scitotenv.2020.143021

Heijne, A. T., Liu, F., Weijden, R. V. D., Weijma, J., Buisman, C. J., & Hamelers, H. V. (2010). Copper recovery combined with electricity production in a microbial fuel cell. *Environmental Science and Technology, 44*(11), 4376–4381. https://doi.org/10.1021/es100526g

Hoang, A. T., Nižetić, S., Ng, K. H., Papadopoulos, A. M., Le, A. T., Kumar, S., & Hadiyanto, H. (2022). Microbial fuel cells for bioelectricity production from waste as sustainable prospect of future energy sector. *Chemosphere, 287*, Article 132285. https://doi.org/10.1016/j.chemosphere.2021.132285

Ieropoulos, I., Greenman, J., & Melhuish, C. (2008). Microbial fuel cells based on carbon veil electrodes: Stack configuration and scalability. *International Journal of Energy Research, 32*(13), 1228–1240. https://doi.org/10.1002/er.1419

Jadhav, D. A., Pandit, S., Sonawane, J. M., Gupta, P. K., Prasad, R., & Chendake, A. D. (2021). Effect of membrane biofouling on the performance of microbial electrochemical cells and mitigation strategies. *Bioresource Technology Reports, 15*, Article 100822. https://doi.org/10.1016/j.biteb.2021.100822

Khandaker, S., Das, S., Hossain, M. T., Islam, A., Miah, M. R., & Awual, M. R. (2021). Sustainable approach for wastewater treatment using microbial fuel cells and green energy generation–A comprehensive review. *Journal of Molecular Liquids, 344*, Article 117795. https://doi.org/10.1016/j.molliq.2021.117795

Kimura, M. (1980). A simple method for estimating evolutionary rates of base substitutions through comparative studies of nucleotide sequences. *Journal of Molecular Evolution, 16*, 111–120. https://doi.org/10.1007/BF01731581

Koók, L., Bakonyi, P., Harnisch, F., Kretzschmar, J., Chae, K. J., Zhen, G., Kumar, G., Rózsenberszki, T., Tóth, G., Nemestóthy, N., & Bélafi-Bakó, K. (2019). Biofouling of membranes in microbial electrochemical technologies: Causes, characterization methods and mitigation strategies. *Bioresource Technology, 279*, 327–338. https://doi.org/10.1016/j.biortech.2019.02.001

Kumar, R., Singh, L., & Zularisam, A. W. (2016a). Exoelectrogens: Recent advances in molecular drivers involved in extracellular electron transfer and strategies used to improve it for microbial fuel cell applications. *Renewable and Sustainable Energy Reviews, 56*, 1322–1336. https://doi.org/10.1016/j.rser.2015.12.029

Kumar, R., Singh, L., Zularisam, A. W., & Hai, F. I. (2018). Microbial fuel cell is emerging as a versatile technology: A review on its possible applications, challenges and strategies to improve the performances. *International Journal of Energy Research, 42*(2), 369–394. https://doi.org/10.1002/er.3780

Kumar, S. S., Kumar, V., Kumar, R., Malyan, S. K., & Pugazhendhi, A. (2019). Microbial fuel cells as a sustainable platform technology for bioenergy, biosensing, environmental monitoring, and other low power device applications. *Fuel, 255*, Article 115682. https://doi.org/10.1016/j.fuel.2019.115682

Kumar, M., & Saini, H. S. (2024). Novel approaches for sustainable management of chromium contaminated wastewater. IntechOpen. https://doi.org/10.5772/intechopen.1003944

Kumar, S., Stecher, G., & Tamura, K. (2016). MEGA7: Molecular evolutionary genetics analysis version 7.0 for bigger datasets. *Molecular Biology and Evolution, 33*(7), 1870–1874. https://doi.org/10.1093/molbev/msw054

Kusmayadi, A., Leong, Y. K., Yen, H. W., Huang, C. Y., Dong, C. D., & Chang, J. S. (2020). Microalgae-microbial fuel cell (mMFC): An integrated process for electricity generation, wastewater treatment, CO_2 sequestration and biomass production. *International Journal of Energy Research, 44*(12), 9254–9265. https://doi.org/10.1002/er.5531

Lee, S. H., Ban, J. Y., Oh, C. H., Park, H. K., & Choi, S. (2016). A solvent-free microbial-activated air cathode battery paper platform made with pencil-traced graphite electrodes. *Scientific Reports, 6*(1), 28588. https://doi.org/10.1038/srep28588

Letunic, I., & Bork, P. (2016). Interactive tree of life (iTOL) v3: An online tool for the display and annotation of phylogenetic and other trees. *Nucleic Acids Research, 44*(W1), W242–W245. https://doi.org/10.1093/nar/gkw290

Li, W. W., & Yu, H. Q. (2011). From wastewater to bioenergy and biochemicals via two-stage bioconversion processes: A future paradigm. *Biotechnology Advances, 29*(6), 972–982. https://doi.org/10.1016/j.biotechadv.2011.08.012

Li, F., Sharma, Y., Lei, Y., Li, B., & Zhou, Q. (2010). Microbial fuel cells: The effects of configurations, electrolyte solutions, and electrode materials on power generation. *Applied Biochemistry and Biotechnology, 160*, 168–181. https://doi.org/10.1007/s12010-008-8516-5

Li, W. W., Yu, H. Q., & He, Z. (2014). Towards sustainable wastewater treatment by using microbial fuel cells-centered technologies. *Energy and Environmental Science, 7*(3), 911–924. https://doi.org/10.1039/C3EE43106A

Li, M., Ning, P., Sun, Y., Luo, J., & Yang, J. (2022). Characteristics and application of *Rhodopseudomonas palustris* as a microbial cell factory. *Frontiers in Bioengineering and Biotechnology, 10*, Article 897003. https://doi.org/10.3389/fbioe.2022.897003

Liu, P., Mohamed, A., Liang, P., & Beyenal, H. (2020). Effect of electrode spacing on electron transfer and conductivity of *Geobacter sulfurreducens* biofilms. *Bioelectrochemistry, 131*, Article 107395. https://doi.org/10.1016/j.bioelechem.2019.107395

Logan, B. E., & Regan, J. M. (2006). Electricity-producing bacterial communities in microbial fuel cells. *Trends in Microbiology, 14*(12), 512–518. https://doi.org/10.1016/j.tim.2006.10.003

Mahto, K. U., & Das, S. (2025). Electroactive biofilm communities in microbial fuel cells for the synergistic treatment of wastewater and bioelectricity generation. *Critical Reviews in Biotechnology, 45*(2), 434–453. https://doi.org/10.1080/07388551.2024.2372070

Malik, S., Kishore, S., Dhasmana, A., Kumari, P., Mitra, T., Chaudhary, V., Kumari, R., Bora, J., Ranjan, A., Minkina, T., & Rajput, V. D. (2023). A perspective review on microbial fuel cells in treatment and product recovery from wastewater. *Water, 15*(2), 316. https://doi.org/10.3390/w15020316

Mercuri, E. G. F., Kumata, A. Y. J., Amaral, E. B., & Vitule, J. R. S. (2016). Energy by microbial fuel cells: Scientometric global synthesis and challenges. *Renewable and Sustainable Energy Reviews, 65*, 832–840. https://doi.org/10.1016/j.rser.2016.06.050

Mier, A. A., Olvera-Vargas, H., Mejía-López, M., Longoria, A., Verea, L., Sebastian, P. J., & Arias, D. M. (2021). A review of recent advances in electrode materials for emerging bioelectrochemical systems: From biofilm-bearing anodes to specialized cathodes. *Chemosphere, 283*, Article 131138. https://doi.org/10.1016/j.chemosphere.2021.131138

Mittal, Y., Ul, Z., Saquib, S., Gupta, S., Saeed, T., Imteaz, M. A., & Yadav, A. K. (2024). Challenges in up-scaling of microbial electrochemical technologies for practical environmental applications. In *Emerging trends and advances in microbial electrochemical technologies* (pp. 711–753). Elsevier. https://doi.org/10.1016/B978-0-443-15557-4.00021-6

Mohan, S. V., Velvizhi, G., Modestra, J. A., & Srikanth, S. (2014). Microbial fuel cell: Critical factors regulating bio-catalyzed electrochemical process and recent advancements. *Renewable and Sustainable Energy Reviews, 40*, 779–797. https://doi.org/10.1016/j.rser.2014.07.109

Molognoni, D., Puig, S., Balaguer, M. D., Capodaglio, A. G., Callegari, A., & Colprim, J. (2016). Multiparametric control for enhanced biofilm selection in microbial fuel cells. *Journal of Chemical Technology and Biotechnology, 91*(6), 1720–1727. https://doi.org/10.1002/jctb.4760

Mondal, S., & Goswami, S. S. (2024). Microbial fuel cells: Advancements, challenges, and applications in sustainable energy and environmental remediation. *Journal of Sustainable Energy, 3*(3), 198–221. https://doi.org/10.56578/jse030305

Nawaz, A., ul Haq, I., Qaisar, K., Gunes, B., Raja, S. I., Mohyuddin, K., & Amin, H. (2022). Microbial fuel cells: Insight into simultaneous wastewater treatment and bioelectricity generation. *Process Safety and Environmental Protection, 161*, 357–373. https://doi.org/10.1016/j.psep.2022.03.039

Nosek, D., Mikołajczyk, T., & Cydzik-Kwiatkowska, A. (2023). Anode modification with Fe_2O_3 affects the anode microbiome and improves energy generation in microbial fuel cells powered by wastewater. *International Journal of Environmental Research and Public Health, 20*(3), 2580. https://doi.org/10.3390/ijerph20032580

Obileke, K., Onyeaka, H., Meyer, E. L., & Nwokolo, N. (2021). Microbial fuel cells, a renewable energy technology for bio-electricity generation: A mini-review. *Electrochemistry Communications, 125*, Article 107003. https://doi.org/10.1016/j.elecom.2021.107003

Ortiz-Martínez, V. M., Salar-García, M. J., De Los Ríos, A. P., Hernández-Fernández, F. J., Egea, J. A., & Lozano, L. J. (2015). Developments in microbial fuel cell modeling. *Chemical Engineering Journal, 271*, 50–60. https://doi.org/10.1016/j.cej.2015.02.076

Pandya, R. S., Kaur, T., Bhattacharya, R., Bose, D., & Saraf, D. (2024). Harnessing microorganisms for bioenergy with microbial fuel cells: Powering the future. *Water-Energy Nexus, 7*, 1–12. https://doi.org/10.1016/j.wen.2023.11.004

Pant, D., Van Bogaert, G., Diels, L., & Vanbroekhoven, K. (2010). A review of the substrates used in microbial fuel cells (MFCs) for sustainable energy production. *Bioresource Technology, 101*(6), 1533–1543. https://doi.org/10.1016/j.biortech.2009.10.017

Peera, S. G., Maiyalagan, T., Liu, C., Ashmath, S., Lee, T. G., Jiang, Z., & Mao, S. (2021). A review on carbon and non-precious metal based cathode catalysts in microbial fuel cells. *International Journal of Hydrogen Energy, 46*(4), 3056–3089. https://doi.org/10.1016/j.ijhydene.2020.07.252

Rabaey, K., Lissens, G., & Verstraete, W. (2005). Microbial fuel cells: performances and perspectives. In *Biofuels for fuel cells: Renewable energy from biomass fermentation* (pp. 377–399). IWA Publishing. http://hdl.handle.net/1854/LU-328755

Rabaey, K., & Rozendal, R. A. (2010). Microbial electrosynthesis—Revisiting the electrical route for microbial production. *Nature Reviews Microbiology, 8*(10), 706–716. https://doi.org/10.1038/nrmicro2422

Roy, H., Rahman, T. U., Tasnim, N., Arju, J., Rafid, M. M., Islam, M. R., Pervez, M. N., Cai, Y., Naddeo, V., & Islam, M. S. (2023). Microbial fuel cell construction features and application for sustainable wastewater treatment. *Membranes, 13*(5), 490. https://doi.org/10.3390/membranes13050490

Saitou, N., & Nei, M. (1987). The neighbor-joining method: A new method for reconstructing phylogenetic trees. *Molecular Biology and Evolution, 4*(4), 406–425. https://doi.org/10.1093/oxfordjournals.molbev.a040454

Santoro, C., Arbizzani, C., Erable, B., & Ieropoulos, I. (2017a). Microbial fuel cells: From fundamentals to applications. A review. *Journal of Power Sources, 356*, 225–244. https://doi.org/10.1016/j.jpowsour.2017.03.109

Santoro, C., Serov, A., Gokhale, R., Rojas-Carbonell, S., Stariha, L., Gordon, J., Artyushkova, K., & Atanassov, P. (2017b). A family of Fe-NC oxygen reduction electrocatalysts for microbial fuel cell (MFC) application: Relationships between surface chemistry and performances. *Applied Catalysis B: Environmental, 205*, 24–33. https://doi.org/10.1016/j.apcatb.2016.12.013

Sarangi, P. K., Srivastava, R. K., Singh, A. K., Sahoo, U. K., Prus, P., & Sass, R. (2023). Municipal-based biowaste conversion for developing and promoting renewable energy in smart cities. *Sustainability, 15*(17), 12737. https://doi.org/10.3390/su151712737

Seelam, J. S., Rundel, C. T., Boghani, H. C., & Mohanakrishna, G. (2018). Scaling up of MFCs: Challenges and case studies. In: Microbial Fuel Cell: A Bioelectrochemical System that Converts Waste to Watts (pp. 459–481). https://doi.org/10.1007/978-3-319-66793-5_24

Sharif, H. M. A., Farooq, M., Hussain, I., Ali, M., Mujtaba, M. A., Sultan, M., & Yang, B. (2021). Recent innovations for scaling up microbial fuel cell systems: Significance of physicochemical factors for electrodes and membranes materials. *Journal of the Taiwan Institute of Chemical Engineers, 129*, 207–226. https://doi.org/10.1016/j.jtice.2021.09.001

Shi, M., Jiang, Y., & Shi, L. (2019). Electromicrobiology and biotechnological applications of the exoelectrogens *Geobacter* and *Shewanella* spp. *Science China Technological Sciences, 62*, 1670–1678. https://doi.org/10.1007/s11431-019-9509-8

Siddiqui, S., Bhatnagar, P., Dhingra, S., Upadhyay, U., & Sreedhar, I. (2023). Wastewater treatment and energy production by microbial fuel cells. *Biomass Conversion and Biorefinery, 13*(5), 3569–3592. https://doi.org/10.1007/s13399-021-01411-2

Slate, A. J., Whitehead, K. A., Brownson, D. A., & Banks, C. E. (2019). Microbial fuel cells: An overview of current technology. *Renewable and Sustainable Energy Reviews, 101*, 60–81. https://doi.org/10.1016/j.rser.2018.09.044

Sleutels, T. H., Darus, L., Hamelers, H. V., & Buisman, C. J. (2011). Effect of operational parameters on Coulombic efficiency in bioelectrochemical systems. *Bioresource Technology, 102*(24), 11172–11176. https://doi.org/10.1016/j.biortech.2011.09.078

Tajima, F., & Nei, M. (1984). Estimation of evolutionary distance between nucleotide sequences. *Molecular Biology and Evolution, 1*(3), 269–285. https://doi.org/10.1093/oxfordjournals.molbev.a040317

Tanvir, R. U., Zhang, J., Canter, T., Chen, D., Lu, J., & Hu, Z. (2021). Harnessing solar energy using phototrophic microorganisms: A sustainable pathway to bioenergy, biomaterials, and environmental solutions. *Renewable and Sustainable Energy Reviews, 146*, Article 111181. https://doi.org/10.1016/j.rser.2021.111181

Thapa, B. S., Kim, T., Pandit, S., Song, Y. E., Afsharian, Y. P., Rahimnejad, M., Kim, J. R., & Oh, S. E. (2022). Overview of electroactive microorganisms and electron transfer mechanisms in microbial electrochemistry. *Bioresource Technology, 347*, Article 126579. https://doi.org/10.1016/j.biortech.2021.126579

Ucar, D., Zhang, Y., & Angelidaki, I. (2017). An overview of electron acceptors in microbial fuel cells. *Frontiers in Microbiology, 8*, 643. https://doi.org/10.3389/fmicb.2017.00643

Wang, A., Sun, D., Cao, G., Wang, H., Ren, N., Wu, W. M., & Logan, B. E. (2011). Integrated hydrogen production process from cellulose by combining dark fermentation, microbial fuel cells, and a microbial electrolysis cell. *Bioresource Technology, 102*(5), 4137–4143. https://doi.org/10.1016/j.biortech.2010.10.137

Wang, Y., Chen, Y., Wen, Q., Zheng, H., Xu, H., & Qi, L. (2019). Electricity generation, energy storage, and microbial-community analysis in microbial fuel cells with multilayer capacitive anodes. *Energy, 189*, Article 116342. https://doi.org/10.1016/j.energy.2019.116342

Wilberforce, T., Sayed, E. T., Abdelkareem, M. A., Elsaid, K., & Olabi, A. G. (2021). Value added products from wastewater using bioelectrochemical systems: Current trends and perspectives. *Journal of Water Process Engineering, 39*, Article 101737. https://doi.org/10.1016/j.jwpe.2020.101737

Winfield, J., Gajda, I., Greenman, J., & Ieropoulos, I. (2016). A review into the use of ceramics in microbial fuel cells. *Bioresource Technology, 215*, 296–303. https://doi.org/10.1016/j.biortech.2016.03.135

Wu, D., Peng, C., Yin, C., & Tang, H. (2020). Review of system integration and control of proton exchange membrane fuel cells. *Electrochemical Energy Reviews, 3*, 466–505. https://doi.org/10.1007/s41918-020-00068-1

Xie, Q., Lu, Y., Tang, L., Zeng, G., Yang, Z., Fan, C., Wang, J., & Atashgahi, S. (2021). The mechanism and application of bidirectional extracellular electron transport in the field of energy and environment. *Critical Reviews in Environmental Science and Technology, 51*(17), 1924–1969. https://doi.org/10.1080/10643389.2020.1773728

Yang, Y., Xiang, Y., Xia, C., Wu, W. M., Sun, G., & Xu, M. (2014). Physiological and electrochemical effects of different electron acceptors on bacterial anode respiration in bioelectrochemical systems. *Bioresource Technology, 164*, 270–275. https://doi.org/10.1016/j.biortech.2014.04.098

Yaqoob, A. A., Ibrahim, M. N. M., & Rodríguez-Couto, S. (2020). Development and modification of materials to build cost-effective anodes for microbial fuel cells (MFCs): An overview. *Biochemical Engineering Journal, 164*, Article 107779. https://doi.org/10.1016/j.bej.2020.107779

Ye, Y., Ngo, H. H., Guo, W., Chang, S. W., Nguyen, D. D., Liu, Y., Nghiem, L. D., Zhang, X., & Wang, J. (2019). Effect of organic loading rate on the recovery of nutrients and energy in a dual-chamber microbial fuel cell. *Bioresource Technology, 281*, 367–373. https://doi.org/10.1016/j.biortech.2019.02.108

Chapter 5
Maximizing Biogas Yields: Technological Insights into Wastewater Anaerobic Digestion

Abdulhalim Musa Abubakar, Rashid Shamsuddin, Archana Kasinath, Muhammad Asif, and Kiran Batool

Abstract The present chapter explores the mechanisms and processes involved in the anaerobic digestion (AD) of organic waste present in wastewater, which seldom leads to the production of biogas—a renewable energy source rich in methane. By laying emphasis on the dual benefits of waste reduction and energy generation, the study provides a comprehensive overview of the latest advancements in AD technologies. It discusses the integration of diverse technologies into existing wastewater treatment facilities, also highlighting the economic and environmental advantages of biogas production. As well, the study examines case studies where AD has been successfully implemented, and demonstrated its feasibility and scalability. Again, we discussed the optimization techniques essential for maximizing biogas yield,

Senior Contributor: Abdulhalim Musa Abubakar.

A. M. Abubakar (✉)
Department of Chemical Engineering, Faculty of Engineering, Modibbo Adama University (MAU), P.M.B. 2076, Yola 640261, Adamawa State, Nigeria
e-mail: abdulhalim@mau.edu.ng

R. Shamsuddin
Department of Chemical Engineering, Faculty of Engineering, Islamic University of Madinah, Madinah 42311, Saudi Arabia
e-mail: r.shamsuddin@iu.edu.sa

A. Kasinath
Department of Water and Wastewater Technology, Faculty of Civil and Environmental Engineering, Gdansk University of Technology, 80-233 Gdańsk, Poland
e-mail: archana.kasinath@pg.edu.pl

M. Asif
Department of Energy and Environment, Faculty of Agricultural Engineering (FAE), Sindh Agriculture University, Tando Jam, Hyderabad 70060, Pakistan
e-mail: engr.asif86@yahoo.com

K. Batool
Beijing Key Laboratory of New Energy and Low Carbon Development, School of Economics and Management, North China Electric Power University, Changpin District, Beijing 102206, China
e-mail: kiranbatool@ncepu.edu.cn

S. Dhiman and G. Mukherjee (eds.), *Resource Resurgence*, Springer Water,
https://doi.org/10.1007/978-3-032-02750-4_5

including temperature control, pH regulation, and co-digestion strategies with other organic wastes. The environmental impact of biogas production is also analyzed, highlighting the reduction in greenhouse gas emissions compared to conventional waste treatment methods. Artificial intelligence (AI) integration for process control and the use of genetically engineered microorganisms to enhance efficiency are novel/smart technologies reviewed. Ample knowledge of scalability of bioenergy recovery from wastewater will trigger its recommendations for improvements to encourage broader implementation.

Keywords Wastewater · Bioenergy · Smart recycling · Anaerobic digestion · Biogas production

5.1 Introduction

Anaerobic digestion (AD) is no doubt a cornerstone technology in addressing the dual challenges of wastewater treatment and renewable energy generation. As global concerns over environmental sustainability and energy security grow (to wit, Germany) (Thrän et al., 2023), the valorization of organic waste in wastewater through AD is increasingly recognized as a transformative solution. Wastewater rich in organic matter, presents an untapped resource for biogas production, with methane as the primary energy component. Despite its potential, many wastewater treatment facilities underutilize AD, leaving significant opportunities for energy recovery and waste management unexplored. The pressing problem lies in the inefficiency of conventional wastewater treatment methods, which often result in high energy consumption and limited resource recovery. Traditional aerobic processes require significant operational energy and produce substantial volumes of sludge, which pose disposal challenges. AD, however, offers a sustainable alternative by converting organic waste into biogas, thereby reducing greenhouse gas emissions, and stabilizing sludge. Debowski and Zielinski (2022) found that the technology, when optimized, can transform wastewater treatment plants into energy producers rather than energy consumers, also addressing both environmental and economic concerns. The significance of AD is highlighted by its ability to contribute to circular economy principles and the capability to turn waste into resource (Kabeyi & Olanrewaju, 2022a). It aligns with global efforts to mitigate climate change by reducing methane emissions from untreated waste and replacing fossil fuels with renewable biogas. Studies have documented the successful integration of AD into wastewater facilities in various regions, showcasing its feasibility, scalability, and economic benefits. For example, case studies in Europe and North America demonstrate how biogas production from wastewater supports grid energy, reduces operational costs, and promotes sustainable practices.

Previous studies have also delved into optimizing AD processes for wastewater, by focusing on enhanced biogas yields through co-digestion, temperature control, and advanced reactor designs. Kesharwani and Bajpai (2024) also explained resource recovery potential from wastewater AD. Lansing et al. (2017) solved the energy, economic and sanitary issues of shared toilet AD systems in Haiti community. Energy recovery from wastewater was the main study direction of Hao et al. (2019). Innovations in smart technologies, such as AI-driven process monitoring and blockchain for traceability, further underscore the transformative potential of AD. An avalanche of those advancements have provided a robust foundation for exploring new frontiers in wastewater-based bioenergy recovery. The objective of this study is to explore the mechanisms, technological advancements, and optimization strategies involved in harvesting bioenergy from wastewater using AD. By addressing critical aspects such as the integration of smart technologies, scalability, and environmental impacts, this study seeks to provide a holistic view of AD's potential in transforming wastewater into a renewable energy resource. The methodological approach followed combines a comprehensive review of existing literature with case study analysis and a focus on recent technological innovations, such as AI-driven monitoring, blockchain for process transparency, and adaptive bioreactor systems. These methodologies not only ensure a thorough examination of the subject but also position the study to highlight emerging trends and address gaps in current knowledge. The unique contribution of this review lies in its focus on integrating novel smart technologies and advanced optimization techniques into the AD process. Challenges such as foam formation, analyzed by Moeller and Görsch (2015), represent a significant instance where this technology has not yet been utilized for a solution. While previous studies have primarily concentrated on traditional AD parameters, such as temperature and pH regulation, this chapter emphasizes the role of intelligent systems, such as AI, in process monitoring and decision-making, as well as blockchain for ensuring traceability and accountability. Additionally, the discussion on scalability, particularly in adapting AD systems for varying wastewater compositions in both municipal and industrial settings, offers new perspectives on implementation challenges and solutions.

5.2 AD of Wastewater: Mechanisms and Processes

5.2.1 Organic Matter in Wastewater: Composition and Characteristics

Organic matter in wastewater is very important in AD for bioenergy generation, as it acts as a substrate upon which microbial communities' depolymerization produces biogas, mainly methane (Abendroth et al., 2018). The composition and characteristics of organic matter vary greatly depending on the source of the wastewater, and such variation have significant impacts on efficiency and effectiveness in the AD process. The characteristics of the organic matter, such as biodegradability, presence of inhibiting substances, and nutrient balance (Zacharof et al., 2019), directly impact the AD process. Certain measurements, such as Biochemical Oxygen Demand

Fig. 5.1 Classification of wastewater

(BOD), Chemical Oxygen Demand (COD), and Total Organic Carbon (TOC), give important information about the organic load and the potential methane production in the system (Aguilar-Torrejón et al., 2023). Relatively biodegradable organic components, namely, carbohydrates and proteins, are associated with significant methane generation. In contrast, complex compounds or those with toxic properties inhibit the production of methane and decrease total biogas. Thus, the types of wastewaters in question are categorized into three groups: domestic, industrial, and agricultural, as shown in Fig. 5.1.

Domestic wastewater is mainly composed of biodegradable organic matter, including proteins, carbohydrates, fats, and oils (Adeniran et al., 2016; Lohani et al., 2015). Mahendra and Patil (2013) and Zeeman and Lettinga (1999) described this type of wastewater as being able to be broken down by anaerobic microorganisms; hence, domestic wastewater is a good candidate for AD systems to produce bioenergy. The presence of non-biodegradable organic compounds/substances (NBOC), such as detergents and some chemicals, may inhibit the activity of microorganisms, thereby reducing the efficiency of the anaerobic process (Khalil et al., 1988; Mensah & Forster, 2003). High concentrations of fats and oils also cause scum formation on the surface of the digester, which then impacts mixing and overall system efficiency (Pan et al., 2019; Thakur & Barjibhe, 2019; Thakur et al., 2019; Yangın-Gömeç, 2006). Lastly, industrial wastewater is more complex owing to the diverse organic compounds in it (Choi et al., 2017; van der Berg et al., 2024). Such wastes from food processing industries, including dairy (Pang et al., 2020), meat (Hamawand, 2015; Luste & Luostarinen, 2010), and vegetable processing (Agrawal et al., 2023; Dubrovskis & Plume, 2015; Ifa et al., 2023; Viswanath et al., 1992), contain large amounts of organic matter that can easily be converted to biogas (Logan et al., 2019), primarily because of the presence of carbohydrates and proteins (Yang et al., 2015). According to Elmekawy et al. (2015), industrial wastewater may also contain toxic compounds, chemicals, and high concentrations of inorganic materials, which are challenging for anaerobic microorganisms to degrade. The wastewaters generated by paper mills, textiles (Tharmarajah et al., 2024; Wei et al., 2019), and petrochemical industries are significantly rich in NBOC (Tian et al., 2020), which inhibits the activity of microbes (Yang, 2020). Kong et al. (2019) mentioned that it significantly diminishes the chances of bioenergy production via AD. On the contrary, agricultural wastewaters become a very crucial source of organic matter for AD (Li et al., 2021). This is because wastewater from agricultural runoff, livestock, and crop irrigation systems typically contains a high concentration of organic materials, including animal manure, plant residues, and organic fertilizers (Palatsi et al., 2011; Usman et al., 2021). Based on the research carried out by Parawira (2004), the materials are readily biodegradable. Therefore, agricultural wastewater is considered ideal for AD. However, according to Campos et al. (2018), high nitrogen and phosphorus levels in agro-food waste or agricultural runoff create difficulties during digestion. While the inherent nutrients promote bacterial growth, excessively high levels can cause imbalances or inhibit nutrient utilization, particularly in closed-loop systems, where nutrient recycling is essential, as described by Sforza et al. (2017) and Romero-Güiza et al. (2016). Table 5.1 presents the nutritional content and characteristics of the various types of wastewaters discussed.

NBOC are vital in industrial wastewater, especially from textile (Abubakar et al., 2023a, 2023b), petrochemical (Wei et al., 2019), and pharmaceutical industries (Rana et al., 2017), where toxic chemicals, dyes, and solvents are present (Pereira et al., 2021; Vysokomornaya et al., 2015), which are challenging for biodegradation (Pradt & Meidl, 1993). Table 5.1 clearly shows that carbohydrate content is generally high in food processing wastewater and agricultural runoff, which contain plant residues and sugars (Fuentes-Santiago et al., 2023); protein content is high in food processing wastewater (especially from meat, dairy, and beverage industries), and livestock manure and; lipids are high in food processing industries (dairy, meat, oils), petrochemical industries (Wei et al., 2019), and livestock wastewater. Wastewaters with high levels of BOC, such as domestic wastewater, food processing, and livestock wastewaters, typically have high BOD. Food processing, petrochemical reviewed by Aljuboury et al. (2017), and pharmaceutical wastewater (Mouele et al., 2021), tend to have high COD due to the presence of both BOC and NBOCs. Lastly, wastewaters with high organic matter (e.g., food processing and livestock wastewater) show moderate to high TOC. However, the biodegradability of wastewater varies significantly across different types and ranked from high/most, mildly and low degrading, according to Abubakar et al. (2023a, 2023b). First, domestic wastewater and agricultural wastewater are highly biodegradable due to their high content of biodegradable organic matter, such as proteins, carbohydrates, fats, and oils, making them ideal for biogas production (Liu et al., 2016; Makowska & Spychała, 2014; Saha et al., 2024). Secondly, food and beverage industry wastewater and food processing wastewater are moderately biodegradable, with high levels of organic matter, though the presence of oils and solids may require pretreatment for optimal digestion efficiency. In contrast,

Table 5.1 Composition and characteristics of organic matter in different wastewater types

Type	Example	Carbohydrate content	Protein content	Fat and oil content	BOD	COD	TOC	NBOCs
Domestic wastewater	General household wastewater	Medium	Medium	Medium	High	High	Medium	Low
	Municipal wastewater	Medium	Medium	Medium	High	High	Medium	Low
Industrial wastewater	Food and beverage industry	High	High	Medium	High	High	Medium	Low to medium
	Pulp and paper industry	Medium	Medium	Low	High	High	Medium	High (lignin)
	Textile industry	Low	Low	Medium	Medium	High	Medium	High (dyes, chemicals)
	Petrochemical industry	Low	Medium	High	High	High	High	High (solvents, toxins)
	Pharmaceutical industry	Low	Medium	Low	High	High	High	High (APIs, solvents)
Agricultural wastewater	Livestock wastewater (manure)	Medium	High	High	High	High	Medium	Low
	Food processing wastewater	High	Medium	Medium	High	High	Medium	Low to medium
	Runoff from agricultural fields	High	Medium	Low	Medium	High	High	Medium (fertilizers, pesticides)
	Dairy farm wastewater	Medium	High	High	High	High	Medium	Low
	Irrigation runoff	Low	Low	Low	Low	Medium	Medium	Medium (nutrients)

industrial wastewater from sectors such as textile (Sforza et al., 2020), petrochemical (Janajreh et al., 2020; Varjani et al., 2020), pharmaceutical (Álvarez-Torrellas et al., 2017), and pulp and paper industries (Bajpai, 2017; Ekstrand, 2019; Hassan et al., 2020; Meyer & Edwards, 2014), is the least biodegradable (Sánchez-Contreras et al., 2021).

5.2.2 *Stages of Anaerobic Digestion Specific to Wastewater*

AD is a complex biochemical process used to treat wastewater and other biowaste materials (Mohan, 2008). It occurs in a series of sequential stages: hydrolysis, acidogenesis, acetogenesis, and methanogenesis. Each stage is facilitated by specific microbial communities and environmental conditions, said Sahm (2005). When comparing AD in wastewater treatment to other biowaste materials, the stages remain fundamentally similar, as the microbial processes are consistent. However, the composition of the feedstock influence the rate and efficiency of each stage. For instance, wastewater typically contains high levels of dissolved organic matter, which undergo hydrolysis more rapidly than solid biowastes (including agricultural residues or food waste). Solid biowastes often require pre-treatment to enhance hydrolysis and overall digestion efficiency. The simulation of AD, including these stages, is feasible using advanced modeling software, e.g., Aspen Plus (Ravendran et al., 2019), MATLAB (I. C. Offie, 2023), and specialized bioprocess simulation tools such as BioWin or GPS-X (Callahan et al., 2022; Jasim, 2020). The platforms open the door for a detailed representation of the biochemical kinetics, mass transfer, and thermodynamics involved in AD. Aspen Plus, for example, can model the chemical and phase equilibrium processes in digestion (Menacho et al., 2022), while MATLAB can simulate microbial growth dynamics and reactor behavior (Ganeshan & Rajendran, 2022). Such tools are invaluable for optimizing process design, scaling up reactors, and performing sensitivity analyses to predict performance under varying operational conditions.

5.2.2.1 Hydrolysis of Wastewater Organics

Hydrolysis is the initial and often the rate-limiting step in the AD of wastewater organics (Menzel et al., 2020). During this stage, complex organic molecules such as proteins, carbohydrates, and lipids are enzymatically broken down into simpler amino acids, sugars, and fatty acids soluble compounds (Mulat et al., 2018). In wastewater treatment, the speed of hydrolysis vary widely depending on the nature of the organic matter. Soluble organic materials hydrolyze relatively quickly, whereas particulate or complex organics take longer due to the additional enzymatic and diffusion processes required (Guo et al., 2021). Typically, hydrolysis is slower when large or structurally complex molecules dominate the wastewater composition (Qu et al., 2021; Saha et al., 2021). Several factors influence the hydrolysis stage. Temperature plays a significant role, with higher temperatures generally enhancing enzymatic activity and reaction rates. pH levels also affect hydrolysis, with optimal activity often observed in a near-neutral to slightly acidic range (Ravi et al., 2018). Additionally, the particle size and bioavailability of organic matter are critical—smaller particles with larger surface areas hydrolyze more efficiently (Aldin, 2010). The presence and activity of hydrolytic enzymes, influenced by microbial population dynamics, are also key determinants of hydrolysis efficiency (Sanders, 2001).

The type of wastewater seriously impacts the hydrolysis stage (Coelmont et al., 2023; Zarina & Mezule, 2024). In trinary wastewater (mixtures of industrial, agricultural, and domestic wastes), the heterogeneity and varying proportions of soluble and particulate organics lead to variable hydrolysis rates. Slaughterhouse wastewater, rich in proteins and lipids, requires longer hydrolysis times due to the complexity of the compounds (Kothari et al., 2024; Saddoud & Sayadi, 2007). In contrast, treatment plant sludge, which contains partially degraded organics (Yang et al., 2015), may experience relatively faster hydrolysis if the sludge has undergone prior treatments via thickening or disintegration (Elsaesser et al., 2012; Radetic, 2022). As such, the differences highlighted, demonstrate that hydrolysis is not uniform across wastewater types; it is highly dependent on the composition (Benit & Roslin, 2015; Litvinov et al., 2022) and pre-treatment of the effluent (Prabakar et al., 2018).

5.2.2.2 Acidogenesis in Wastewater Treatment

Acidogenesis is the second stage of AD in wastewater treatment, where the simpler soluble compounds produced during hydrolysis—such as sugars, amino acids, and fatty acids—are metabolized by acidogenic bacteria into volatile fatty acids (VFAs), alcohols, hydrogen, and carbon dioxide (CO_2) (Saddoud & Sayadi, 2007). It is critical for preparing the substrates for subsequent acetogenesis and methanogenesis. Acidogenesis typically occurs rapidly under anaerobic conditions, as acidogenic bacteria are highly active and adaptable to diverse organic substrates (Li, 2011). It is possible to conduct acidogenesis in a separate reactor from hydrolysis (Yu et al., 2003). As observed in a study by Cui et al. (2010) and Cremonez et al. (2021), the separation (often part of a two-phase AD system), can enhance process control and efficiency by optimizing conditions such as pH and retention time for each stage. For instance, hydrolysis benefits from slightly acidic to neutral conditions, while acidogenesis thrives under slightly acidic pH (5.5–6.5) (Abubakar et al., 2022; Guerrero et al., 1999). By separating the stages in line with a study by Janesch et al. (2021), the growth environment for acidogenic bacteria is finely tuned and lead to potential increase in their activity, as well as an improved overall digestion performance (Carneiro et al., 2020; Mathur et al., 2024; Yu et al., 2003).

Several strategies accelerate acidogenesis, as evidenced during the treatment of wastewater containing copper and 2-nitroaniline by Jinadasa (2003). Increasing the temperature to mesophilic (30–40 °C) or thermophilic (50–60 °C) ranges stimulate bacterial metabolism. Maintaining optimal pH levels is crucial to avoid inhibiting acidogenic bacteria. Pretreatment of feedstock through mechanical, thermal, or chemical methods can increase the availability of hydrolysates, speeding up acidogenesis in the process (Domínguez-Maldonado et al., 2020; Lu et al., 2013; Pereira et al., 2022). Seeding the reactor with a high concentration of acidogenic bacteria or microbial consortia also promote faster VFA production. The efficiency of acidogenesis is influenced by multiple factors (Ghosh et al., 1975). The composition of the feedstock—whether the wastewater contains readily biodegradable organics or recalcitrant compounds—affects bacterial activity. Musolino (2016) mentioned that, nutrient availability (e.g., nitrogen and phosphorus) is essential for bacterial growth and function. In that cse, the presence of inhibitory substances, such as toxic heavy metals or excessive salts, hinders acidogenesis. Effective reactor design and operational strategies are therefore key to optimizing this vital stage of AD, especially a study by Liu et al. (2012) where FeO was added.

5.2.2.3 Acetogenesis and Methanogenesis for Biogas Production

In acetogenesis, acetogenic bacteria convert the VFAs produced during acidogenesis into acetic acid, hydrogen, and CO_2. Its products serve as direct substrates for methanogenesis, where methanogenic archaea generate methane and additional CO_2 (Wolfson et al., 2018). Hassan et al. (2020) pointed out that the stages are often interdependent, with hydrogen and acetic acid concentrations tightly regulated through syntrophic interactions between acetogens and methanogens. The speed of these stages depends on several factors, including temperature, pH, and microbial activity, but they are generally slower than acidogenesis due to the need for precise environmental conditions and the metabolic constraints of the microorganisms involved (Zeppilli et al., 2022). Several factors can inhibit acetogenesis and methanogenesis (Jones et al., 1982). High concentrations of VFAs or accumulated hydrogen sometimes suppress acetogenic bacteria by disrupting their metabolic pathways, as stated by (Magdalena et al., 2019). Similarly, methanogenesis is highly sensitive to pH fluctuations, with optimal activity occurring between 6.8 and 7.2. Toxic substances (ammonia, sulfides, heavy metals, or organic pollutants) can inhibit methanogenic archaea (Vítězová et al., 2020). In real time, temperature stress—whether from low temperatures in psychrophilic systems or excessive heat in mesophilic or thermophilic systems—can reduce the metabolic activity of both acetogens and methanogens (Fu et al., 2019).

Khandaker et al. (2021) discover that acetogenic pretreatment of textile wastewater is energy-efficient. Effective monitoring and management of the foregone parameters are crucial to ensure efficient biogas production. Jameel et al. (2024) show that the four stages of AD—hydrolysis, acidogenesis, acetogenesis, and methanogenesis—are sequential and interconnected. While they must all occur for complete digestion and biogas production, there is some overlap in timing. For example, hydrolysis and acidogenesis can begin simultaneously in a mixed reactor (Janesch et al., 2021). However, skipping any stage disrupts the process, as each provides essential substrates for the next. For instance, bypassing hydrolysis leaves complex organics unavailable for acidogenesis and subsequent stages, leading to poor biogas yields.

5.2.3 Key Factors Affecting AD in Wastewater Systems

AD in wastewater systems is influenced by numerous key factors that collectively determine its efficiency and bioenergy yields (Induchoodan et al., 2022). Among the most critical are temperature, which directly impacts microbial activity, with mesophilic (30–40 °C) and thermophilic (50–60 °C) ranges generally offering the best performance (Cioabla et al., 2012). pH levels are equally vital, as most stages of AD operate optimally within specific ranges; methanogenesis, for example, thrives at pH 6.8–7.2 (Yadav & Singh, 2018). Hydraulic retention time (HRT) and organic loading rate (OLR) must be carefully balanced to prevent system overload while allowing adequate time for microbial processes (Lohani & Havukainen, 2018). The composition of wastewater, including the ratio of soluble to particulate organics and the presence of inhibitory substances that has to do with ammonia, sulfides, or heavy metals, significantly influences digestion (Ahmad et al., 2016). Carbon-to-nitrogen (C/N) ratio is suitable for maintaining microbial growth and avoiding ammonia toxicity or nutrient limitation (Pang et al., 2020). Mixing ensures uniform substrate distribution and avoids the formation of inhibitory zones (Lindmark et al., 2014). Microbial population dynamics also play a major role, as diverse and active communities of hydrolytic, acidogenic, acetogenic, and methanogenic microorganisms are essential for efficient digestion (Shah et al., 2014; Vítězová et al., 2020).

Less influential but impactful factors include feedstock particle size (Sebola et al., 2015), which matters more for solid biowastes than for soluble-rich wastewater, and pre-treatment methods; namely, thermal or enzymatic treatments, which are more beneficial for lignocellulosic biowastes (Cabeza et al., 2023). As such, the factor have less pronounced effects in wastewater systems, where substrates are typically more bioavailable. Some factors, such as feedstock pre-treatment or specific microbial strains, may be less critical for wastewater-based AD and can sometimes be overlooked without significantly reducing bioenergy yields (Anjum et al., 2016). However, ignoring fundamental factors encompassing pH balance, temperature control, or proper HRT would drastically compromise efficiency. Effective prioritization and automation of the most influential parameters will ensure high biogas yields and system reliability. Designed biodigesters and bioreactors can control many of the factors explained. For example, temperature and pH regulation can be achieved through heating systems and buffer additions (Li et al., 2022a, 2022b; Wang et al., 2017). Automated systems equipped with real-time monitoring sensors can maintain optimal OLR, HRT, and mixing (Vanrolleghem, 1999), while feedback controls adjust for changes in wastewater composition (Bernard et al., 2001). Table 5.2 lists the systems that play an impactful role in ensuring process efficiency and maximizing bioenergy output (primarily, methane).

Table 5.2 Systems for optimizing bioenergy generation in wastewater anaerobic digestion

System	Type	Uses	Role in AD	Discussed by
Supervisory control and data acquisition (SCADA)	Control system	Monitoring and control of plant operations, real-time data collection	Ensures continuous monitoring, optimization of process parameters, and automation of control processes for efficient bioenergy generation	Dieu (2001), Gospodinov et al. (2014) and Newsom et al. (2022)
Programmable logic controller (PLC)	Control system	Automation of equipment and processes (e.g., valves, pumps)	Direct control of critical components in the AD process, ensuring accurate operation and safety	Kim et al. (1991), Kumar et al. (2014) and Song et al. (2016)
Distributed control system (DCS)	Control system	Process control for complex operations	Centralized monitoring of multiple process variables, enhancing system stability and performance	Ashok et al. (2017) and Selişteanu et al. (2018)
Real-time monitoring systems	Monitoring system	Continuous tracking of key parameters (e.g., temperature, pH, biogas production)	Provides data for controlling conditions within the digester to maintain optimal bioenergy production	Offie et al. (2022)
Gas flow meters and sensors	Instrumentation	Measures biogas production and flow	Ensures efficient gas production and quantification for bioenergy recovery	Pereda et al. (2010) and Vanrolleghem (1999)
Biogas analyzer	Instrumentation	Analysis of biogas composition (CH_4, CO_2, H_2S, etc.)	Optimizes methane recovery for bioenergy generation and monitors the digester's health	Abakaka et al. (2022)
Bioreactor control system	Control system	Regulates conditions within the anaerobic reactor	Ensures optimum operational conditions (e.g., temperature, pH, mixing) for maximum methane production	Wang et al. (2020)

(continued)

Table 5.2 (continued)

System	Type	Uses	Role in AD	Discussed by
Data acquisition system (DAQ)	Monitoring system	Collects data from various sensors throughout the system	Gathers real-time data, providing insight into the performance of the AD process for optimization	Echiegu and Ghaly (2014) and Teixeira et al. (2017)
Microbial fuel cell (MFC) systems	Bioelectrochemical system	Converts chemical energy from organic waste into electricity	Generates electricity from the organic matter in the wastewater, contributing to bioenergy recovery	Ardakani and Gholikandi (2020) and Makhtar et al. (2018)
Biomethanation systems	Biological system	Methanogenesis process for biogas production	Converts organic material into biogas, maximizing bioenergy generation potential in AD	Dev et al. (2019) and Nan et al. (2020)
Heat recovery systems	Energy recovery system	Captures heat from the digester and other system components	Enhances energy efficiency by reusing heat, supporting the AD process for sustained bioenergy generation	Mikkonen et al. (2013) and Nagpal et al. (2021)

SCADA systems remained part of automation technologies that are increasingly integrated into modern AD facilities (Abilash et al., 2016; Cairone et al., 2024; Gospodinov et al., 2014), of which an example is the Marselisborg WWTP in Denmark (UN, 2020). Even though SCADA, DCS, and MFC systems tend to be more expensive, while gas flow meters (Martínez-Sibaja et al., 2011), DAQ (Teixeira et al., 2017), and real-time monitoring systems are more affordable (Cruz et al., 2019). Also, SCADA, DCS, and MFC systems have complex installation requirements, while simpler systems, viz., gas flow meters and biogas analyzers are easier to install. SCADA and DCS systems, while providing comprehensive control, are more complex to operate. However, gas flow meters (Pereda et al., 2010), real-time monitoring (Offie et al., 2022), and DAQ systems are easier to use once set up (Echiegu & Ghaly, 2014).

5.3 Wastewater as a Resource: Dual Benefits of AD

5.3.1 Reduction of Organic Load in Wastewater

The reduction of organic load in wastewater is a principal process in wastewater treatment (WWT), aimed at mitigating pollution and enhancing environmental sustainability (Mahmoud et al., 2020; Pani et al., 2024). Organic load refers to the amount of organic matter, such as proteins, carbohydrates, and fats, present in wastewater, which can lead to environmental contamination if not properly managed (Kanat & Yeter, 2020). Various methods are employed to reduce organic load, including biological (Banadaki et al., 2024; Huang et al., 2008), chemical (Matheson, 2016), and physical treatments (Laaksonen et al., 2017), as shown in Fig. 5.2. Biological treatments described by Narayanan and Narayan (2019), particularly activated sludge processes, rely on microorganisms to break down organic materials into simpler compounds. Aerobic and anaerobic processes, where bacteria degrade organic matter in the presence or absence of oxygen, are commonly used (Bala et al., 2015). Constructed wetlands and biofilm reactors also provide effective biological treatment (Cocozza et al., 2023; Li et al., 2022a, 2022b). Simply because they utilize plant roots or attached microbial communities to remove organic pollutants. Chemical treatments, such as coagulation-flocculation, involve the addition of chemicals that bind to organic particles, forming flocs that are removable through sedimentation or filtration. Naguib et al. (2024) and Matheson (2016) signals the importance of advanced oxidation processes (AOPs), stating that it uses powerful oxidants to break down complex organic compounds into less harmful substances.

Lastly, physical methods, such as filtration (Laaksonen et al., 2017), membrane technologies (e.g., reverse osmosis) (Al-Sayed et al., 2023; Zacharof et al., 2019), and adsorption onto activated carbon materials, are also employed to remove suspended

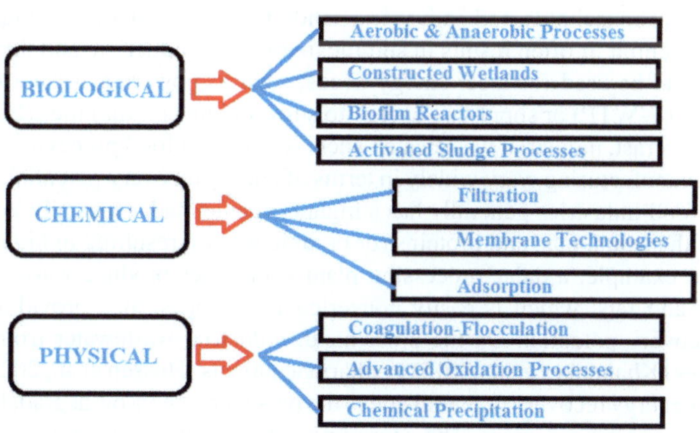

Fig. 5.2 Methods of reducing organic loads from wastewater

organic solids and dissolved organic compounds (Barno & Abbas, 2020). Reducing the organic load in wastewater is significant because it directly impacts the quality of water bodies into which treated water is released (Trilita et al., 2016). Lower organic loads decrease the BOD and COD, thereby reducing the strain on aquatic ecosystems and preventing eutrophication, which then leads to oxygen depletion and again, poses harm to aquatic life (Ghosh et al., 2010; Madukasi & Zhang, 2010; Naguib et al., 2024). Furthermore, effective reduction of organic load is essential for meeting environmental regulations (Madukasi & Zhang, 2010), promoting the reuse of treated wastewater, and ensuring the health of human populations by preventing the spread of waterborne diseases (Maruthi et al., 2011). Therefore, combination of several treatment methods is vital, in order to achieve a sustainable water management and safeguard natural water resources in the process (Tompkins, 2021).

5.3.2 Biogas Generation from Wastewater: Energy Recovery Potential

Biogas generation from wastewater is an effective method for energy recovery, where organic matter present in wastewater is anaerobically digested by microorganisms to produce methane-rich biogas (Ijoma et al., 2022). This biogas is most times harnessed as a renewable energy source, and used for electricity generation, heating, or even as vehicle fuel (Simwambi et al., 2020). The energy recovery potential from wastewater varies significantly depending on the type of wastewater being treated, as different sources of wastewater contain different concentrations of organic matter and nutrients (Hernandez & Martín, 2022). In Table 5.3, energy recovered through AD from various wastewater, is reported, given credit to Tsiakiri et al. (2021) who highlighted how it is estimated.

Municipal wastewater, as revealed by Gingerich and Mauter (2017), typically sourced from households and industries, tends to have moderate to high energy recovery potential. It often results in significant biogas production, and the methane generated can be used for local energy needs, such as powering wastewater treatment plants (WWTP) or supplying energy to surrounding communities (Bachmann, 2015). In contrast, industrial wastewater, such as that from food processing, dairy, or chemical manufacturing, vary widely in terms of energy recovery potential. Wastewater from food industries generally has a higher organic load, especially in facilities dealing with sugars, fats, and proteins (or organic waste), resulting in higher biogas yields. For example, a dairy processing plant sometimes produce wastewater rich in lactose and fats, which is easily converted into biogas; also providing a high energy recovery potential. On the other hand, industrial wastewater from sectors, viz., textiles (Khandaker et al., 2015) or pharmaceuticals (Hossen et al., 2024), might have lower energy recovery potential due to the presence of non-biodegradable chemicals and low organic content. The energy potential from wastewater also depends on temperature, pH, and retention time factors in the treatment process. The biogas

Table 5.3 Energy and biogas potential of various wastewater

Type	Biogas/methane recovered	Energy content	References
Textile wastewater + cow dung and wheat straw	415–590 mL/kg VS biogas	NR	Abubakar et al. (2023a, 2023b)
Textile processing plants	NR	185,538 KWh/day	Khandaker et al. (2015)
Sewer screenings	355 m^3/kg VS CH_4	NR	Wid and Horan (2016, 2018)
Organic waste from pharmaceutical operations	NR	100 GWh/year	Fick (2025)
Sewage sludge	3,276,497 Nm^3/y	75,347.06 GJ	Mukawa et al. (2022)
Municipal wastewater	600 g	2.6 and 0.1 kWh	Simwambi et al. (2020)
Beverage wastewater sludge	4.4 and 10.8 m^3	18.9 and 22.1 kWh	Admasu et al. (2022)

NR not reported

generated from WWT is basically converted into various applications, primarily through combustion to produce electricity and heat. In some cases, purified methane is upgraded to biomethane, a natural gas substitute that can be injected into the grid or used as fuel for vehicles. The use of biogas for energy recovery does not solely provides a sustainable energy source but also helps to mitigate greenhouse gas emissions—as methane, a potent greenhouse gas, is captured and utilized rather than being released into the atmosphere.

5.3.3 Enhancing Wastewater Treatment Efficiency Through AD

To achieve optimal WWT efficiency through AD, several strategies must be implemented (O'Dwyer et al., 2020). All the factors affecting its production must be ensured to be kept at the desired condition for maximum biogas yield (Abubakar et al., 2022). The use of co-digestion is another effective strategy to enhance AD efficiency. Pre-treatment of wastewater also improve AD efficiency (Anjum et al., 2016). Methods such as mechanical, thermal, or chemical pre-treatment break down complex organic compounds, and hence makes them more accessible to anaerobic microbes. Microbial management plays a critical role in optimizing AD (Zhang et al., 2016). Introducing specific microbial strains that are particularly effective at breaking down certain types of organic waste can improve the digestion process. Managing the sludge retention time (SRT) and volatile solids content within the digester is important for maximizing AD efficiency. Regular monitoring of biogas production, methane content, and gas flow rates proposed by Martínez-Sibaja et al. (2011), will aid operators to optimize operational parameters (e.g., feed rate and temperature). The integration of biogas upgrading technologies, which purify the biogas into high-quality methane, can increase the value of the energy recovered from the process. Thus, itenables its use as a renewable energy source for electricity or heat production (Nguyen et al., 2021).

5.4 Technological Advancements in Wastewater AD Systems

5.4.1 High-Rate Anaerobic Reactors for Wastewater Treatment

High-rate anaerobic reactors (HRARs) are advanced systems used in WWT that aim to enhance the efficiency of AD processes by increasing the rate at which organic matter is broken down (Parawira, 2004). They are designed to maximize biogas production and minimize the footprint and operational costs associated with traditional anaerobic systems at the same time (Biese, 2016). HRARs operate under conditions that allow for higher concentrations of organic material to be treated more rapidly (Jeihanipour et al., 2013), making them particularly suitable for high-strength wastewater with elevated levels of organic load, such as industrial effluents, food processing waste, or municipal wastewater with high BOD (Elliott et al., 2024). One of the primary features of HRARs is the ability to maintain a high concentration of active biomass in the reactor, which is crucial for enhancing the digestion process. This is typically achieved through the use of specialized reactor designs, such as the Upflow Anaerobic Sludge Blanket (UASB) (Mainardis et al., 2020), Expanded Granular Sludge Bed (EGSB) (Connelly et al., 2017), or Anaerobic Baffled Reactor (ABR) (Hassan & Dahlan, 2013; Olayanju et al., 2024), shown in Fig. 5.3. In those systems, wastewater flows upwards or through a series of compartments where solid particles and microorganisms form granules or aggregates that facilitate efficient biogas production. Nevertheless, granular sludge particles consisting of tightly packed microbial communities are highly efficient at breaking down organic material, and are more resistant to washout than traditional suspended biomass, thereby permitting higher hydraulic and OLR (Kazimierowicz & Dębowski, 2022; Tassew et al., 2019; Zhou et al., 2022).

The design of HRARs also allows for better contact between wastewater and microorganisms, leading to improved mass transfer and enhanced AD. The compact nature of such reactors means that they occupy less space, which is beneficial in situations where land availability is limited or expensive (van Lier et al., 1997). Moreover, HRARs can operate at higher OLRs, which reduces the retention time needed for digestion, increasing throughput and reducing the size and cost of the treatment system (Biese, 2016; Jeganathan et al., 2006). Another advantage of HRARs is their

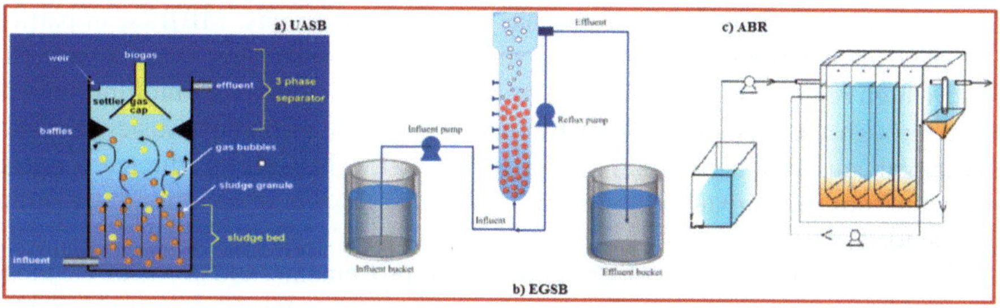

Fig. 5.3 Typical examples of HRARs

energy recovery potential, as mentioned in Joshi (1998). The biogas produced during AD, primarily composed of methane, can be captured and utilized as a renewable energy source for electricity generation, heating, or even vehicle fuel. This energy recovery capability not only makes HRARs more sustainable but also helps offset the operational costs of WWT (Li et al., 2023). To add to this, the treated effluent from HRARs typically has lower concentrations of suspended solids and organic pollutants, and is suitable for further treatment or discharge into the environment with minimal ecological impact (Lettinga et al., 1999). To achieve optimal performance; however, HRARs require careful operational management, such as maintaining stable HRT, controlling the pH, and monitoring microbial health (Batstone, 2000; Biese, 2016; Joshi, 1998). Proper start-up procedures and maintaining the granulation process for sludge are also essential to prevent washout or biofilm sloughing operational issues (Picavet & Alves, 2010). For sure, the performance of HRARs is occasionally influenced by the type and strength of wastewater being treated, as well as by the temperature and nutrient conditions within the reactor.

5.4.2 Innovations in Wastewater-Specific Reactor Design

Traditional WWT systems face challenges related to space limitations, energy consumption, and handling varying wastewater characteristics, such as high organic loads or complex chemical pollutants. In response, engineers and researchers have developed several innovative reactor designs tailored to meet specific WWT needs, in a move to address these challenges while enhancing performance. One notable innovation is the membrane bioreactor (MBR), in which Gienau et al. (2018) and Ardakani and Gholikandi (2020) revealed that it combines conventional activated sludge treatment with membrane filtration to achieve high-quality effluent with a smaller footprint. Another significant development is the biofilm reactor, including the moving bed biofilm reactor (MBBR) (Aygun et al., 2008) and the fixed film reactor (FFR) (Nandy & Kaul, 2001). Both systems support the growth of microorganisms on biofilm carriers that provide a large surface area for microbial colonization and ensure more efficient removal of organic matter and nutrients. MBBRs, in particular, have gained popularity in treating high-strength wastewater because they can achieve high removal efficiencies with minimal energy input, as the biofilm carriers move through the reactor, in order to boost the contact between wastewater and microorganisms (Baddour et al., 2016; Borkar, 2013). Majority of the reactors are well-suited for municipal and industrial WWT and flexible in handling fluctuating organic loads.

The UASB reactor has also seen significant improvements, and is one of the most efficient designs for treating high-strength organic wastewater (Kavan et al., 2024), especially from food processing (Schmidt & Ahring, 1997) and breweries industries (Basandorj, 2007; Chen et al., 2016; Parawira et al., 2005; Smetana & Grosser, 2024). Recent innovations in UASB reactors discussed in Mou et al. (2024) involve enhanced sludge granulation techniques that ensures a better settling of anaerobic biomass and more efficient biogas production. Daud et al. (2018) have shown that the improvements help increase the overall treatment capacity and reduce the need for additional downstream treatment processes. Consequently, Mainardis et al. (2020) describe UASBs as a cost-effective solution for energy recovery through biogas production. For nutrient removal, particularly nitrogen and phosphorus discussed in Macura et al. (2019), innovations in sequencing batch reactor (SBR) (Dutta & Sarkar, 2015) and anaerobic/aerobic membrane bioreactors (AnMBR) (Chen et al., 2016; Nguyen et al., 2020), have made it easier to meet stringent effluent quality standards (Alagha et al., 2020; Vigneswaran et al., 2004). SBRs, in particular, dish-out greater flexibility by operating in batch cycles (Kannan & Singaram, 2012; Kim, 2008). It then facilitates an optimized treatment phases for nitrification, denitrification, and biological phosphorus removal, which is ideal for treating municipal and industrial wastewater with varying pollutant loads (Chinh et al., 2018; Irvine et al., 1989; Zaiat et al., 2001). Moreover, advances in hybrid reactor designs, combining multiple treatment technologies in one system, have proven effective in tackling complex wastewater streams, as researched by Saini et al. (2019). For example, integrating electrocoagulation with biofilm or anaerobic processes enhances the removal of metals, organic contaminants, and suspended solids in a single treatment unit (Derakhshesh et al., 2022; Liu et al., 2020). Hybrid reactors offer high treatment efficiency, reduced chemical usage, and improved energy recovery (Sutthiprapa et al., 2024). They are becoming increasingly popular for treating industrial effluents, especially in sectors such as textiles (Jeihanipour et al., 2013; Sforza et al., 2020), pharmaceuticals (Abubakar et al., 2024), and petrochemicals (Derakhshan & Fazeli, 2018; Varjani et al., 2020). Citrus industry wastewater is an additional example where AD hybrid system was previously implemented (Rosas-Mendoza et al., 2018).

5.4.3 Materials and Technologies Optimized for Wastewater AD

Key aspect of optimizing AD for WWT is the use of high-performance materials for reactor construction and components. Stainless steel and fiberglass are the most commonly used materials for reactors due to their durability, resistance to corrosion, and ability to withstand harsh anaerobic conditions. The use of composite materials in reactor design is also gaining traction, as these materials offer both structural strength and thermal insulation. Thereby improving the efficiency of thermophilic AD processes, where higher temperatures enhance microbial activity. The development of specialized biofilm carriers is another technological advancement that reportedly optimizes AD. Biofilm-based reactors (i.e., MBBR and UASB reactors), rely on materials that provide a large surface area for the attachment of microorganisms. In view of that, polypropylene or polyethylene are commonly used to make biofilm carriers because they are lightweight, chemically resistant, and have high surface area for microbial colonization. The aforementioned materials promote the growth of anaerobic bacteria in the reactor, leading to more efficient organic matter degradation and increased biogas production. The development of biodegradable carriers is another innovation aimed at reducing environmental impact and improving long-term system sustainability.

Technologies designed to surge biogas recovery are also a key area of optimization. Membrane technologies that includes gas separation membranes, are being increasingly integrated into AD systems to purify and upgrade biogas, as well as during the removal of impurities such as CO_2, hydrogen sulfide (H_2S), and siloxanes (Aguilloso et al., 2024; Chen et al., 2015). Methane content of the biogas is thereby increased by the process, which instigate it to be more suitable for use as a renewable energy source for electricity generation, heating, or even as vehicle fuel. Gas storage and compression technologies are also optimized to capture biogas efficiently, and later ensure that it can be stored or transported for later use without loss of energy potential (Kapdi et al., 2005; Nallamothu et al., 2013; Ray et al., 2016). Additionally, the inclusion of advanced monitoring and control technologies helps optimize the performance of AD systems. The combined use of sensors to monitor pH, temperature, biogas production rate, and OLR lets operators to modify operational settings in real time, which leads to optimal microbial activity and biogas output (Lopez-Arenas et al., 2004; Singh & Kumar, 2021). Automation systems, which adjust reactor conditions based on sensor data, can help maintain stable operating conditions, improving both treatment efficiency and energy recovery (Abdallah & Kennedy, 2013). Real-time data analytics also assist in detecting operational issues associated with organic overload or microbial imbalance, so as to enable preventive measures to be taken before problems arise. Finally, pre-treatment technologies are increasingly being employed to improve the biodegradability of wastewater before it enters the AD process. Pre-treatment methods such as thermal hydrolysis (Westerholm et al., 2019), chemical dosing (Sarto et al., 2019), and mechanical shearing

(Tsapekos et al., 2018), break down complex organic compounds to be more accessible to anaerobic microbes. Implementing them aid the reduction of the retention time needed for digestion, leads to increase in biogas yield, and enhance the overall efficiency of the AD process.

5.5 Integration of AD into Existing Wastewater Treatment Infrastructure

5.5.1 Retrofitting Conventional Wastewater Treatment Plants

Traditional WWTPs, which often rely on aerobic processes for organic matter degradation are energy-intensive and require significant operational costs, especially when treating high-strength or large volumes of wastewater. By retrofitting treatment plants with AD technology, WWT becomes more energy-efficient and sustainable, as AD offers the potential to produce biogas—a renewable energy source—while reducing the environmental footprint of WWT operations. The integration of AD into existing WWTPs begins with the installation of anaerobic digesters for the breakdown of organic pollutants in the wastewater under anaerobic conditions into biogas. This biogas can then be captured and used as a renewable energy source for electricity generation, heating, or even vehicle fuel, in order to reduce the need for external energy sources and lower operational costs. One of the main benefits of this integration is that the biogas generated from AD offset the energy required for aerobic processes, potentially transforming WWTPs into net energy producers rather than consumers.

The retrofitting process typically involves the modification of existing infrastructure to accommodate AD, which may include adding digesters, upgrading sludge handling systems, or installing biogas storage and upgrading facilities. In some cases, existing anaerobic processes, such as the UASB reactor or ABR is enhanced or expanded to increase biogas production capacity. It is basically achieved through improvements in sludge management and granulation technologies that enhance microbial efficiency in digesters. Moreover, the integration of AD into existing WWTPs provides a way to address challenges related to excess sludge production, which is a byproduct of conventional WWT processes. Through AD, the excess sludge are stabilized and further reduced in volume, to make sludge disposal more manageable and less costly. In addition, the upgrading of biogas produced in AD systems through membrane filtration (Aguilloso et al., 2024; Rodero et al., 2024; Tomczak et al., 2024; Vrbová & Ciahotný, 2017), pressure swing adsorption (Barbera et al., 2018b), microbial electrolysis cell (MEC) (Sher et al., 2024) and/or biological/chemical/physical approaches (Ahmed et al., 2021; Muñoz et al., 2015), allows the methane content to be increased, so as to make the biogas more efficient for energy applications in accordance with Yuan et al. (2022). This retrofit not only improves the overall sustainability of WWTPs but also aligns with global objectives of reducing greenhouse gas emissions and advancing the circular economy. By harvesting biogas from wastewater, utilities reduce their reliance on fossil fuels, lower carbon footprints, and contribute to energy resilience. The integration of AD into existing infrastructure also improve the quality of treated effluent, as AD has been shown to remove specific organic matter, nitrogen, and phosphorus pollutants, depending on the configuration of the treatment system.

5.5.2 Case Studies: Successful AD Integration in Wastewater Facilities

Di Costanzo et al. (2021) pinpoints sewage sludge as containing significant amount of recoverable nutrients. In North America, several WWTPs have incorporated AD to optimize energy recovery while reducing operational costs (Boyd et al., 2019). For instance, in the United States, large facilities, the likes of Orange County Sanitation District in California have successfully integrated AD to process sewage sludge togenerate considerable amount of biogas that are used for electricity production; in order to offset a large portion of the facility's energy needs (Lippman et al., 2023; Shen et al., 2015; Wanke & Gallagher, 2024). Europe's AD has become a cornerstone of WWT strategies, driven by strict environmental regulations and a strong emphasis on sustainability (Raboni & Urbini, 2014; Silva et al., 2024). A notable example is the Kaiseraugst WWTP in Switzerland that uses advanced AD systems to treat wastewater and recover biogas. The plant produces enough biogas to meet its own energy needs and even provides excess electricity to the local grid. Similarly, Thames Water's WWTP in the UK have integrated AD to enhance sludge management, improve effluent quality, and recover biogas for energy production. This integration has not only helped reduce greenhouse gas emissions but has also led to operational savings. Raudkivi and Tenno (2021)'s study show that the Baltic Sea Region is not void of nutrients, such as phosphorus, that can be recovered. While in Asia, countries with rapidly growing populations and WWT needs (such as China and India), have increasingly turned to AD to manage wastewater effectively, in addition to addressing energy demands. The Shanghai Minhang WWTP in China, for example, implemented AD systems that produce biogas from the treatment of municipal sewage, by supporting the facility's power needs and contributing to the country's renewable energy goals (Jiang, 2011). Similarly, in India, the Kolkata Sewage Treatment Plant integrated AD for waste-to-energy recovery (Khan et al., 2024). That way, it led to a significant reduction in its reliance on grid power and contributed to the country's efforts to improve waste management (Sarkar et al., 2009). Energy or power equivalent of biogas recovered from notable WWTP are shown in Table 5.4.

Africa saw a successful AD integration in wastewater facilities, particularly in South Africa (Ngema et al., 2023) and Kenya (Owuor, 2020). The Balfour WWTP

Table 5.4 Case studies on the integration of AD for biogas production in WWT facilities

WWTP name	Biogas recovered	Energy equivalent	Location	References
East Bay Municipal Utility District (EBMUD)	18,000 m^3/day	90% of 5 MW (4.5 MW)	Oakland, California, USA	Hagey (2011)
Gresham WWTP	65.83 m^3/h	395 kW	Oregon, USA	CHP (2018)
Aarhus Vand Marselisborg WWTP	1,604,667 m^3	9,628 MWh	Aarhus, Denmark	UN (2020)
Boston Deer Island WWTP	12,000 m^3/day	3 MW	Massachusetts, USA	Rea (2014)
Lille Metropole Urban Community	7,400,000 m^3/y	44,400 MWh/y	Lille, France	Kovács (2016)

Biogas energy content = 6 kWh/m^3

in South Africa utilizes AD to reduce organic pollutants in wastewater and generate biogas, which is used for electricity generation, as well as meet local energy demands (Steenkamp, 2023). In Kenya, the Nairobi Industrial Area Sewerage Treatment Plant introduced AD systems to manage high-strength industrial wastewater while producing biogas for energy recovery, by simultaneously addressing both WWT and energy access challenges (Miruka et al., 2018). Others in the same country include the Ruai WWTP (Mwaniki et al., 2021), Dandora Estate Sewerage Treatment Works (Mabonga et al., 2020) and Kariobangi treatment plants (Kilingo et al., 2021). In South America, Brazil (Udaeta et al., 2019) and Argentina (Morero et al., 2017) made strides in integrating AD into their WWT infrastructure. The Sao Paulo WWTP in Brazil uses AD to treat wastewater and generate biogas to power the plant and reduce its carbon footprint (Oliveira et al., 2006; Udaeta et al., 2019). Likewise, in Argentina, the Sewage Treatment Plant in Buenos Aires adopted AD technology to reduce organic waste and produce biogas for energy production, thereby contributing to the country's sustainability objectives (Cirelli & Ojeda, 2008). In Australia, the integration of AD into wastewater facilities has helped address both environmental and energy-related challenges (Chapman & Cuthbertson, 1999; Radcliffe & Page, 2020). The Sydney Water Treatment Plant has been using an energy-efficient AD for over a decade to reduce sludge volume and produce biogas for energy generation. (Chapman & Cuthbertson, 1999; Chen et al., 2013). Majority of the shift to renewable energy from biogas recorded, had helped to minimize operational costs involved and reduces the facility's reliance on fossil fuels.

5.5.3 Challenges and Solutions in Wastewater AD Integration

5.5.3.1 Common Challenges Faced

AD integration into WWT systems faces numerous challenges. Of primary difficulty is maintaining an optimal operational environment, as the microorganisms responsible for AD are sensitive to fluctuations in temperature, pH, and other factors. Variability in wastewater composition, including fluctuating OLRs and the presence of toxic compounds affects the performance of AD systems. Additionally, high concentrations of nitrogen and sulfur compounds, common in industrial wastewater, sometimes lead to inhibition of microbial activity, and results in reduced biogas production and poor treatment efficiency (Vergara-Araya et al., 2022). Another challenge is the slow start-up time of anaerobic digesters, as the microbial communities take time to establish and reach stable operation. The physical infrastructure, including the need for proper mixing and effective sludge handling are expensive to install and maintain, adding to the overall operational costs. The requirement for a long retention time to achieve optimal treatment further limits the practicality of integrating AD into fast-flowing or high-throughput wastewater systems. Moreover, Fang et al. (2019) mentioned that the need for pretreatment of wastewater to enhance the biodegradability of organic matter and increase the overall efficiency of AD present additional costs and complexity. In some cases, such as with high-strength or high-concentration industrial wastewaters, even pretreatment may not be sufficient to ensure effective digestion. The integration of AD also faces regulatory hurdles, as effluent quality standards may not align with the treated output from AD systems, and so limits their widespread adoption. Finally, the management of the produced biogas containing H_2S, poses operational and environmental risks if not properly handled.

5.5.3.2 Solution to AD Integration Challenges

To overcome the challenges associated with integrating AD into WWT systems, several solutions must be implemented. One of the key strategies to address the sensitivity of anaerobic microorganisms to fluctuating operational conditions is the precise control of environmental parameters. This includes maintaining optimal temperature and pH levels through advanced monitoring and control systems, with the aim of stabilizing the microbial activity within digesters. To manage variations in wastewater composition, pre-treatment methods such as coagulation, flocculation (Qian et al., 2023), or membrane filtration (Gienau et al., 2018) should be employed to remove or neutralize contaminants that may inhibit microbial activity or disrupt the digestion process. Pre-treatment steps function by adjusting the OLRs and ensuring a more stable and consistent feed to the digester. To mitigate the slow start-up time, the introduction of microbial inoculants or seed cultures with established microbial communities speed up the acclimatization of the digester and enhance its operational efficiency. For quicker stabilization of the AD system, this approach reduce the need for extended retention times. To add to this, integrating systems that allow for better mixing and sludge circulation, namely advanced pumping and agitation technologies, improves the mass transfer within the digester and prevent scum formation and uneven digestion that hinders performance.

Managing high-strength or toxic wastewaters presents a significant challenge; however, implementing multi-stage or hybrid systems combining AD with other treatment technologies improve overall treatment efficiency. For instance, combining AD with aerobic processes or advanced oxidation enhance the degradation of complex organic compounds, to make possible the effective handling of industrial wastewaters. For wastewater streams containing high concentrations of nitrogen and sulfur, specific adjustments in controlled ammonia stripping discussed by Qian et al. (2023) for piggery effluent, nitrification–denitrification processes, or sulfur-reducing bacteria can prevent the inhibition of microbial activity and improve biogas production. To ensure regulatory compliance and improve the reliability of the effluent quality, continuous monitoring of BOD, COD, and residual contaminants parameters should be implemented. This would let operators make real-time adjustments to the system to meet effluent standards. Furthermore, to address the management of biogas, biogas desulfurization systems are employed to remove contaminants such as H_2S, thus preventing corrosion in pipelines and equipment and making the biogas suitable for energy generation or other applications.

5.6 Economic and Environmental Impact of AD in Wastewater Treatment

The cost analysis of implementing AD in WWTPs involves assessing both the initial capital investment and the ongoing operational costs (Kechrist, 2022). The initial investment primarily consist of the cost of constructing the anaerobic digester, which involves substantial expenditure on equipment such as reactors, mixers, pumps, and gas collection systems. There are other costs associated with land acquisition, civil works, and infrastructure development, which vary based on the size of the treatment plant and the complexity of the installation. For large-scale WWT facilities, the capital cost of AD implementation is significant, especially when considering the need for ancillary systems (such as biogas handling, sludge dewatering, and pre-treatment processes) (Amigun & von Blottnitz, 2010). On the operational side, the costs are influenced by energy consumption, maintenance, and labor. Energy costs for heating the digester to maintain the optimal temperature for microbial activity are a major ongoing expense; although biogas produced during the digestion process offset some of these costs if the gas is used for energy generation (Vasco-Correa et al., 2018). Maintenance costs, including repairs, replacement of parts, and monitoring systems, are also crucial factors in the long-term sustainability of AD systems. The labor costs involved in operating and maintaining the system must be considered, as skilled personnel are required to ensure the digester functions very well and to handle any operational issues that arise.

Another vital consideration is the cost of pre-treatment and post-treatment processes, collectively studied by Amoo et al. (2023). For example, high-strength wastewaters does require more advanced pre-treatment technologies, to wit, membrane filtration or chemical dosing that adds to the overall costs. Similarly, the final effluent from the AD system require further treatment to meet regulatory standards, and contribute additional costs for post-treatment. However, these costs are offset by benefits reaped from biogas production, which are later used for energy recovery or sold as potential revenue stream (Thrän et al., 2023). The biogas is sometimes upgraded to natural gas quality in order to open opportunities for further revenue generation (Jameel et al., 2024). A key factor in the cost-effectiveness of AD is the scale of implementation. Larger facilities typically benefit from economies of scale, as the capital investment per unit of treated wastewater decreases with size. Smaller plants, on the other hand, face higher per-unit costs, making AD less economically viable unless supported by incentives, subsidies, or integration with other waste management systems. To add to it, cost savings are mostly realized through the reduction of waste disposal fees (Domini et al., 2022; Hau & Sculli, 1991), as AD reduces the volume of residual sludge that would otherwise need to be disposed of in landfills (Kumari et al., 2023; Siddique et al., 2023).

Revenue from biogas produced through AD in WWTPs presents a significant opportunity for offsetting operational costs and creating a sustainable source of income. Biogas is a valuable renewable energy resource that can be harnessed for electricity generation (Amponsem et al., 2023), heating, or even upgraded to natural gas quality for sale (Kovács, 2016). The potential revenue depends on the volume and methane content of the biogas produced, which in turn is influenced by the organic load of the wastewater, the efficiency of the AD process, and the scale of the treatment facility. In wastewater plants where AD is integrated, biogas recovery systems generate electricity to power the plant's operations, thereby reducing energy costs and making the facility more self-sufficient. In addition to on-site energy generation, surplus biogas are sold to local utilities or industries to pave way for a steady revenue stream. Biogas can also be processed into biomethane through upgrading technologies associated with CO_2 removal, which is then injected into the natural gas grid or sold as a renewable energy product. Upgraded biogas command a higher market price due to its higher energy content and suitability for use in existing gas infrastructure. Furthermore, biogas are utilized for transportation fuel by converting it into compressed natural gas (CNG) or liquefied natural gas (LNG), opening up additional market opportunities (Yang et al., 2014).

Another avenue for revenue generation is through the sale of carbon credits or renewable energy certificates (RECs) (Awonuga et al., 2023; Vaz & Mattos, 2020). WWTPs that utilize AD to produce biogas and reduce greenhouse gas emissions qualify for carbon credit programs. Doing so, it lets them sell those credits to organizations seeking to offset their emissions. Later on, a significant financial return is provided, especially in regions where carbon pricing mechanisms are in place. Moreover, the revenue from biogas cover the costs of WWT operations by reducing dependency on external energy sources and lowering waste disposal costs. The act of reducing the volume of residual sludge, which would otherwise need to be treated and disposed of, consequentially lowers the operational expenses of biogas production further. When biogas is used in combined heat and power (CHP) systems, the revenue generated from both electricity and thermal energy production contribute to the overall financial sustainability of the WWTP (Kozak & Majchrzycka, 2009; Yin et al., 2021).

In addition to controlling methane emissions, AD plays a crucial role in reducing other harmful pollutants in wastewater. The process efficiently removes organic matter, including BOD (Khotimah et al., 2021) and COD (Ghosh et al., 2010), which are key indicators of water pollution (Zahmatkesh et al., 2023). By breaking down organic contaminants in wastewater, AD reduces the potential for eutrophication and the degradation of water quality in receiving bodies, which results in the loss of biodiversity and the disruption of aquatic ecosystems. Furthermore, AD decreases the volume of residual sludge, typically difficult and costly to dispose of, thereby reducing the need for landfilling or incineration and thus preventing the release of harmful substances into the environment. AD quietly offers environmental benefits by promoting nutrient recovery (Hernandez & Martín, 2022; Romero-Güiza et al., 2016; Zacharof et al., 2019). For example, the nitrogen and phosphorus present in wastewater are partially removed during the digestion process, to reduce nutrient pollution in waterways (Kilingo et al., 2021). This is particularly important in regions where excessive nutrient runoff from wastewater and agricultural activities contributes to harmful algal blooms (Barbera et al., 2018a; Musolino, 2016), water toxicity, and the degradation of aquatic ecosystems (Benbow et al., 2020). On top of that, the treated effluent from AD systems is often less harmful than untreated wastewater. By implications, it becomes safer for discharge into water bodies or for reuse in irrigation and other non-potable applications.

5.7 Optimization Techniques for Biogas Yield from Wastewater

Optimization techniques for biogas yield from wastewater are instrumental in boosting the efficiency and sustainability of AD systems (Shahidul et al., 2018). Mathematical models, computational algorithms, and experimental designs techniques are employed to identify the ideal conditions for maximum biogas production (Thirugnanasambandham et al., 2014). As well, optimization minimizes resource wastage by determining the most effective pre-treatment methods for feedstocks, thereby reducing energy inputs and operational costs. The result is not only an increase in energy recovery but also a reduced environmental impact through efficient organic load degradation and lower greenhouse gas emissions. Through the optimization of substrate concentration, temperature, pH, HRT, and C/N ratio, these methods ensure that microbial activity is maintained at its peak, leading to improved methane/biogas yields (Djimtoingar et al., 2022). Tools, videlicet, Response Surface Methodology (RSM) and Artificial Neural Networks (ANNs) highlighted by Thomaz et al. (2023), allow researchers to model complex interactions between variables, predict outcomes, and fine-tune operations with precision. Table 5.5 presents previous researches in this aspect.

Optimization also supports the handling of diverse waste streams, enabling the AD system to adapt to variations in wastewater quality through co-digestion, as shown in Table 5.6.

Some researchers study the co-digestion technique without optimization while others looked at an optimized co-digestion process. Tables 5.5 and 5.6 both show that the optimization of biogas from AD of diverse wastewaters mentioned in this review, is still not captured.

5.8 Smart Technologies in Wastewater AD

5.8.1 Smart Wastewater Source Segregation

Several benefits of segregating waste have been documented by Sanathkumar et al. (2021). This method protects digesters from being contaminated with unsuitable materials while also boosting feedstock biodegradability, resulting in increased biogas yield and nutrient recovery (Elliott et al., 2024; Lansing et al., 2017). At the forefront of this innovation is AI-guided Waste Characterization (AIWC), where advanced AI systems analyze wastewater streams in real-time (D. Qu, 2022). Sensors and data-driven algorithms in such systems are used to analyze the composition of organic and inorganic matter, forecast biodegradability, and detect inhibitors such as heavy metals and sulfides (Vanrolleghem, 1999; Yan et al., 2021). According to Carrere et al. (2016), it enables operators to tailor pre-treatment processes and prioritize high-quality feedstocks for digestion. Complementing this is Automated Source Tracking (AST), which employs IoT devices as studied by Sowndharya et al. (2019), Lingaraju et al. (2023), Martínez et al. (2020), Logan et al. (2019) and Chitale et al. (2023), as well as cloud-based systems, can perhaps be used to monitor and trace the origin of wastewater streams (Uwamungu et al., 2022). Smart waste segregation technology will always ensure that high-value waste (such as food processing residues or agricultural runoff) is directed to anaerobic digesters while diverting non-degradable or hazardous materials for alternative treatment. At the moment, there exist an intelligent app-based systems for waste segregation and collection described in Sinduja and Kumar (2024).

Table 5.5 Previous RSM optimization studies on AD of wastewater

References	Wastewater type	Design type	Process parameter	Response(s)
Ngema et al. (2023)	Municipal	Box-Behnken Design (BBD)	pH (5–9); HRT (15–30 days) and magnetite load (0.125–0.875 g/L)	Biogas produced and color and COD removal
Chong et al. (2023)	Palm oil mill effluent (POME)	Completely randomised design (CRD)	Digester temp. = 38.9 °C; pH = 7.03 and recirculation ratio = 1.89	Biogas and methane yield
Pratap et al. (2024)	Pretreatment sludge	Specified RSM design	Dosage = 80 mg/g TS and temp. = 50 °C	Biogas and methane yield
Zwain et al. (2022)	Recycled paper mill wastewater (RPMW)	Factorial design	COD_{in} = 4,000 mg/L and HRT = 2 days	COD, BOD and lignin removal; CH_4 content and CH_4 production
Thirugnanasambandham et al. (2014)	Rice mill effluent	BBD	Temp. = 40 °C; alkalinity dose = 1.5 g $CaCO_3$/L and flow rate = 6 L/day	COD removal and biogas production
Choo (2015)	Poultry manure wastewater	Central composite design (CCD)	Agitation speed = 120 rpm and retention time = 3 days	Biogas yield and COD removal
Srivichai and Chavalparit (2020)	Waste activated sludge and modified tapioca starch sludge	CCD	TVS added = 2% and C/N ratio = 23.56	Biogas yield
Mazza et al. (2014)	Sewage sludge	CCD	COD = 1061.06 mg/L; OLR = 0.90 kg/m^3 day and pH = 7.74	Methane content
Tsegaye and Leta (2022)	Slaughterhouse wastewater	NR	Total VFA = 520 mg/L; total alkalinity = 1424 mg/L; ratio = 0.36 and salinity = 1172 mg/L	Volatile solids, methane yield, biogas produced and COD consumed

NR not reported

Table 5.6 Literature record of wastewater anaerobic co-digestion RSM optimization

Feedstock blend	Bioreactor type	Process parameter	Response(s)	References
Sewage sludge + municipal solid waste (MSW)	Biogas plant	Glycerol additive	Biomethane and CO_2 removal	Morero et al. (2017)
Banana peels/ magnetite + WWTP sample	African WWTP	Nanoparticle and biochar load	pH, biogas production and COD removal	Adedeji et al. (2022)
Oily-biological sludge + sugarcane bagasse	Batch reactor (around 5 L gallon)	C/N ratio and co-substrate/ inoculum ratio	Methane yield	Ghaleb et al. (2020)
Sweet corncob waste + corn cob milling wastewater	1 L batch reactor	Initial pH, total solids and ratio of milling and production wastewater	Biogas production	Chanathaworn et al. (2018)
Distillery wastewater sludge + disposable spent yeast	Bioreactor	Temp., pH and volume ratio	Biogas yield	Debele et al. (2024)
Sugarcane bagasse + South African industrial wastewater	1000 mL Duran Schott Bottles	Temp. and OLR	Biogas yield	Armah et al. (2020)
Sewage sludge + Cladophora green algae	Batch reactor	Substrate composition ratios	Biogas production	Rasouli et al. (2023)

5.8.2 Adaptive and Self-healing Systems

Adaptive (Yoshida et al., 2020) and self-healing systems highlighted in Sarathkumar et al. (2021), represent a cutting-edge approach to enhancing the reliability and efficiency of AD processes. Petre et al. (2013) pointed out that both systems are designed to respond dynamically to changes in operational conditions, such as fluctuations in feedstock quality, environmental variables, or system disturbances. In view of that, there would be a consistent performance without the need for constant manual intervention. A key feature of such systems is the use of Self-regulating Bioreactors (SRB), which are equipped with advanced sensors and control algorithms to monitor critical parameters—to wit, pH, temperature, and VFA levels in real-time (Lamb et al., 2019; Yan et al., 2021). When deviations from optimal conditions are detected, bioreactors automatically adjust operational settings. Possible adjustments are agitation speed, substrate flow rates, or heating, to restore balance and maintain microbial activity. Another vital component is Dynamic Load Balancing (DLB) (Thakor & Patel, 2019). It is capable of optimizing the distribution of organic load across multiple digesters in response to variations in feedstock availability or composition (Jendželovský & Uhlířová, 2021). By dynamically allocating waste streams to different units based on their capacity and current load, this technology will prevent overloading (Sundh et al., 2003), under-utilization (Mertins & Wawer, 2022), and system failures (Hegazy et al., 2024; Hewitt et al., 2022), thereby maximizing biogas production and process stability.

5.8.3 Blockchain Technology

Blockchain technology is now a transformative tool in the AD and wastewater management sectors, offering unparalleled transparency, traceability, and efficiency (Mould et al., 2022). The utilization of decentralized and immutable ledgers with-blockchain ensures that every stage of the AD process, from feedstock sourcing to biogas production and byproduct utilization, is securely recorded and verifiable (Onu et al., 2023). A notable application is in traceability and accountability, where blockchain tracks the origins and composition of waste streams entering the AD system (Lv, 2023). A transparency of such, will instigate immediate compliance with regulations, reduces the risk of contamination, and provides a reliable audit trail for stakeholders. It will also allow consumers or industries using the biogas or biofertilizers produced to verify their environmental credentials, in order to foster trust and market credibility (Alzoubi & Mishra, 2023; Erwin & Yang, 2023). Another significant application is Carbon Credit Monitoring (CCM) (Tolman et al., 2024; Vaz & Mattos, 2020), where blockchain automates the calculation and tracking of emissions reductions achieved through AD (Baashar et al., 2021; Taherdoost, 2024). Through the recording data such as methane capture and biogas utilization, blockchain ensures accurate and tamper-proof documentation of carbon offsets, facilitating participation in carbon trading markets (Baashar et al., 2021; Santos et al., 2024). Hence, it provides financial incentives for adopting sustainable practices and also supports global efforts to combat climate change (Alzoubi & Mishra, 2023; Krajnakova et al., 2019). Presently, Power (2023) mention the possibility of converting methane gas from animal waste into energy for mining Bitcoin, pinpointing the ability of CHAINERGY company in the United Kingdom to do so.

5.8.4 Remote Monitoring and Cloud Computing

Bernard et al. (2005), Steyer et al. (2001) and Dhayalan et al. (2024) describe remote monitoring and cloud computing as having the ability to transform the management and optimization of AD systems through the provision of real-time access to key data and enabled predictive maintenance. Cloud-based platforms play a pivotal role by storing and processing large volumes of operational data from sensors embedded in the AD system, according to O'Donovan et al. (2015). Both devices offer operators continuous, remote access to critical pH, temperature, gas composition, and organic load parameters to facilitate seamless monitoring across various locations (Reardon, 2021; Salem et al., 2022; Soehartanto et al., 2023; Udomsom et al., 2022). Thus, advanced analytics and visualization features built into these platforms allow operators to spot inefficiencies, optimize efficiency, and make educated decisions regardless of location (Steyer et al., 1999). Complementing this capability is remote diagnostics (Mandhadi & Iyer, 2012; Visali et al., 2017), which leverages cloud computing and machine learning (ML) algorithms to predict and identify potential issues before they escalate into critical failures (Alavijeh et al., 2024; BPC, 2022; Rahimzadeh et al., 2024). Through real-time analysis of system data, remote diagnostics pinpoint temperature imbalances, equipment wear, or gas leakage anomalies (Jagadeesh & Samuel, 2010; Visali et al., 2017). Thus, volume expansion issues due to gas hold up reported by Bartek et al. (2015) and Chapman and Krugel (2011), can be solved. López-Estrada et al. (2015) said that early detection minimizes downtime, reduces maintenance costs, and ensures consistent biogas production.

5.8.5 *Smart Waste Input Management*

Smart waste input management is a transformative approach in AD operations, designed to optimize the utilization of diverse feedstocks by leveraging advanced technologies (Cheema et al., 2022; Nwazor et al., 2019; Olawade et al., 2024). At the heart of this strategy is Feedstock Optimization Software (FOS) (Dolat et al., 2024; Topić et al., 2020), such as AFOS-feed formation and BioSOR, which analyze the chemical and biological characteristics of incoming waste streams (Buelna et al., 1999; Carter, 2024; H2O-Innovation, 2005). In real time, those applications calculate essential parameters e.g., C/N ratio, moisture content, and potential biogas yield in real-time. The provision of data-driven insights is known to help operators blend feedstocks effectively, balancing nutrients and organic loads to ensure consistent digestion performance and prevent system overload. If an automated wastewater collection is actualized as reported by Abdullahi et al. (2024), it will boost the volume of wastewater collection and optimize biogas production as well as the lifespan of the digester. Equally important are Pre-treatment Monitoring Tools (PMT), such as NIR for ammonia inhibition tracking (Raju et al., 2012) and ThermoScan for thermal pre-treatment efficiency (Saha et al., 2021). The aforementioned tools monitor and fine-tune enzymatic hydrolysis, chemical pretreatment or heat application, and in the process, ensure that feedstocks are adequately prepared for microbial breakdown. From Donatien et al. (2024) who analyzed the metallic corrosive inhibitors, it can be said that, the tool help detect them to avoid disruptions and maximize substrate availability (and/or including suboptimal conditions).

5.8.6 Renewable Energy Integration

Renewable energy integration in AD systems has foundationally advanced their role as sustainable energy solutions during seamless incorporation of biogas into energy networks (Panchenko et al., 2023). Smart Grid Systems (SGS) synchronizes biogas energy production with the electricity grid (Khalid, 2024). Grid systems utilize real-time data from advanced sensors and communication networks to monitor energy demand and supply, to facilitate efficient distribution and minimal energy loss (Kabeyi & Olanrewaju, 2022b). Smart grids enable biogas plants to respond dynamically to fluctuations in grid requirements, and in the long run, facilitate the export of surplus energy during peak demand and storage during low demand periods (Chaikovskaya, 2024). Consequently, it leads to a two-fold benefit of stabilizing energy supply and maximizing the economic returns from biogas production (Davoli et al., 2012). On a smaller scale, microgrid systems are currently transforming energy management for decentralized and off-grid AD installations (Davoli et al., 2012). It is observed in Chaikovskaya (2024) that those systems integrate biogas with other renewable sources (solar and wind) to create resilient, localized energy networks. As such, intelligent controllers and energy storage technologies will ensure that reliable power supply from microgrid to remote areas or during grid outages, is obtained (Loboichenko et al., 2024). They also enable energy self-sufficiency for communities or industries, whose goal is to reduce the dependency on fossil fuels.

5.8.7 Artificial Intelligence and Data Analytics

Artificial intelligence (AI) as per Thomaz et al. (2023)'s study and data analytics in Kim et al. (2022) revealed that they are transforming AD systems via the provision of the tools to enhance efficiency, optimize operations, and predict outcomes with high accuracy (Cinar et al., 2021; Offie, 2023; Offie et al., 2023). At the core of this transformation are ML algorithms explained by Rajakumaran et al. (2023) and De Clercq et al. (2019), which analyze historical and real-time data from AD systems to predict biogas yields, detect anomalies, and optimize pH, temperature, and organic load. As such, the two algorithms instigate adaptive decision-making, so that operators respond proactively to fluctuations in feedstock quality or system performance, thereby preventing downtime and maximizing productivity. Complementing this capability is Big Data Analytics (BDA), which processes vast amounts of information generated by sensors and operational logs to identify trends and correlations that may not be immediately evident (Chiang et al., 2017; Sadiku et al., 2019). BDA, for example, reveal insights about substrate mixing and resource usage by studying trends in feedstock variability and biogas output (Zhang et al., 2018). In this case, long-term planning and operational efficiency are supported by this comprehensive analysis. Adding to this suite are expert systems practically implemented on biogas plants, based on studies by Djatkov et al. (2014), Qiu et al. (2005) and Kharpude et al.

(2017). It serves as decision-support tools that emulate human expertise in managing AD systems (Alexandra, 1998). Systems of such use rule-based logic and knowledge bases to make recommendations on the best system settings, troubleshooting tactics, and maintenance plans (Calzada et al., 1993; Erdmann et al., 1988; Hanratty, 1991).

5.9 Scalability and Feasibility of AD in Wastewater Management

In New Zealand, Canada, Australia, United States and France (Dennis et al., 2020), municipal wastewater scalability encompasses the evaluation of the ability of AD systems to handle variable waste loads produced by the rising populations and urbanization (Arias et al., 2020; Lu et al., 2022), as also exemplified in the case of Egypt, and as reported by Reymond et al. (2018). Influential energy demands (Morán et al., 2023), land availability (Ciliberti et al., 2016; Levavasseur et al., 2023), and integration with prevailing infrastructure factors, play substantial roles (Gamble et al., 2015). Accordingly, modular AD systems, which can be expanded as desired, propose a scalable solution for municipalities (Maleki-Ghelichi & Sharifi, 2017), letting them to acclimatize to shifting waste volumes, negative reactor conditions and comply with stricter environmental regulations (Bartek et al., 2017; Bartek et al., 2015; Chapman & Krugel, 2011)—sometimes implemented in Europe (Alberici & Dekker, 2021). Connelly et al. (2017), Elmekawy et al. (2015) and Nguyen et al. (2020) has it that; in industrial settings, scalability valuations concentrate on the compatibility of AD systems with high-strength effluents that may comprise of inhibitory substances. Thus, industries profit from customized AD systems designed to manage definite food processing (Pennington, 2019; Pilarska et al., 2023; Rahman et al., 2018) or chemical manufacturing waste streams (Bochmann et al., 2020; Verbeeck et al., 2018). It is worthy of note that advanced technologies, namely, multi-stage digesters (Biey et al., 2003; Moedinger et al., 2012; Stampfli, 2020; Stewart, 2014; Suhan et al., 2019; USEPA, 2006) and pre-treatment processes highlighted in Bochmann et al. (2020), are frequently required to enhance scalability while maintaining efficiency. Another important factor to evaluate is economic feasibility, previously conducted for dairy farm in Florida (Giesy et al., 2005), which includes the potential for nutrient recycling (Lam et al., 2020; Pang et al., 2020; Romero-Güiza et al., 2016), energy recovery (Fang et al., 2019; Rachbauer et al., 2024), and money creation from biogas or biofertilizer (Hernandez & Martín, 2022; Kataki et al., 2018; Macura et al., 2019).

Barth et al. (2014) highlight a robust framework incentivize investment in AD technologies by providing financial support, as in United States (Sam et al., 2017), tax credits (WSDR, 2018), and subsidies for renewable energy production (Nijman, 2022; Pfau et al., 2017; Thrän et al., 2023) and waste management initiatives (Vasco-Correa et al., 2018). As discovered, feed-in tariffs (FITs) and renewable portfolio standards policies encourage energy recovery from biogas (Duale & Alex-Adedipe, 2023; Hoo et al., 2018; Pastor, 2020; Romero et al., 2017), makes AD a viable economic option for WWT facilities or agro-wastes (Bangalore et al., 2016). Basically, regulatory mandates such as waste diversion goals or greenhouse gas (GHG) reduction targets, further propel the implementation of AD by prioritizing sustainable waste treatment and carbon–neutral energy solutions (EPA, 2024; Jameel et al., 2024). Research show that environmental regulations enable AD systems meet stringent standards for effluent quality (McDonnell et al., 2018; MPCA, 2023; USDA, 2017), odor control (Levey, 2016; Luo et al., 2022), and emissions (Gingerich & Mauter, 2017), thereby fostering public acceptance (Bharadwaj et al., 2023a, 2023b; Lisiak-Zielińska et al., 2023; Mazzanti et al., 2021; Meidiana et al., 2017, 2019) and environmental compliance (McDonnell et al., 2018). Moreover, policies that support public–private partnerships (Fantozzi et al., 2014; Taron et al., 2023) and research funding accelerate technological innovation (KENPRO, 2023; Yamaji et al., 2024), also boosting the efficiency and scalability of AD systems. Luckily, international agreements, the likes of the Paris Agreement, deliver a global framework to integrate AD into climate change mitigation strategies by recognizing its role in reducing methane emissions and promoting renewable energy (Obaideen et al., 2022).

5.10 Conclusions

This chapter has demonstrated the potential of AD to convert organic matter in wastewater into biogas, a renewable energy source, while simultaneously reducing greenhouse gas emissions and stabilizing sludge. By leveraging advanced technological innovations, such as AI-driven monitoring systems, blockchain for transparency, and adaptive bioreactors, AD systems can achieve unprecedented efficiency and scalability, catering to both municipal and industrial wastewater streams. The integration of smart technologies has been identified as a key driver in optimizing AD processes. AI and big data analytics allow for real-time monitoring and adaptive control, ensuring consistent performance and higher biogas yields. Blockchain technology enhances traceability and accountability, thereby promoting environmental compliance and enabling participation in carbon credit markets. Furthermore, advancements in reactor design and co-digestion strategies have expanded the applicability of AD, enabling it to handle complex and high-strength wastewater streams effectively. Notwithstanding its advantages, the finding acknowledges challenges such as high capital costs, variability in wastewater composition, and the need for pre-treatment in certain cases. However, solutions concerning modular AD systems, renewable energy integration, and supportive policy frameworks seriously boost the feasibility and adoption of this technology. Therefore, the insights provided in this study emphasize that AD is more than a WWT solution—it is a pathway toward a sustainable circular economy.

References

Abakaka, I. M., Koda, J. T., Tsuanyo, D., Aziz, A., Fotso, J. M. K., & Nanfack, P. F. T. (2022). Design, implementation and testing of a biogas analyzer. *International Energy 2021-Conference on "Renewable Energy and Digital Technologies for the Development of Africa," 354*(03001), 1–5. https://doi.org/10.1051/e3sconf/202235403001

Abdallah, M., & Kennedy, K. (2013). Advanced monitoring and control of anaerobic digestion in bioreactor landfills. In R. Chamy & F. Rosenkranz (Eds.), *Biodegradation—Engineering and technology* (pp. 1–30). InTech Open. https://doi.org/10.5772/55715

Abdullahi, A., Mohammed, A., Bonet, M. U., El-Suleiman, A., Ahmad, R. B., & Teng, D. (2024). Development of a Smart Waste Management System with automatic Bin Lid Control for smart city environment. *EAI Endorsed Transactions on Smart Cities, 7*(3), 1–8. https://doi.org/10.4108/eetsc.4385

Abendroth, C., Hahnke, S., Simeonov, C., Klocke, M., Casani-Miravalls, S., Ramm, P., Bürger, C., Luschnig, O., & Porcar, M. (2018). Microbial communities involved in biogas production exhibit high resilience to heat shocks. *Bioresource Technology, 249*, 1074–1079. https://doi.org/10.1016/j.biortech.2017.10.093

Abilash, D. A., Kayalvizhi, P., Rakesh, R., & Balamurugan, S. (2016). Automation in biomethanation plant using PLC and SCADA. *International Journal of Bio-Science and Bio-Technology, 8*(2), 171–180. https://doi.org/10.14257/ijbsbt.2016.8.2.16

Abubakar, A. M., Silas, K., & Aji, M. M. (2022). An elaborate breakdown of the essentials of biogas production. *Journal of Engineering Research and Sciences (JENRS), 1*(4), 93–118. https://doi.org/10.55708/js0104013

Abubakar, A. M., Abdo, A., Mengstie, M. A., Nayem, Z., Selele, M. I., Abubakar, M. A., & Rohaeni, E. S. (2023). Fecal sludge recycling to useful products: Environmental concerns, viability and potential. *Applied Research and Innovation (ARI), 1*(1), 1–17. https://doi.org/10.54536/ari.v1i1.1526

Abubakar, M. S., Yazid, A. B., Sabo, A. A., Ibrahim, A., Aliyu, A. U., & Saleh, M. (2023). Effects of textile wastewater pre-treatment on lignocellulosic biomass for solid-state anaerobic digestion. *FUDMA Journal of Sciences, 7*(1), 156–164. https://doi.org/10.33003/fjs-2023-0701-1267

Abubakar, A. M., Haruna, N. M., Ahmed, Z. D., Abdul, A. Z., Abubakar, M. A., Ali, A. M., Sabo, M. J. U., & Thiagarajan, B. (2024). Unlocking biogas production potential: Evaluating the environmental impact and biodegradability of pharmaceutical and medical wastes. *Sigma Journal of Engineering and Natural Sciences, 42*(4), 1261–1291. https://doi.org/10.14744/sigma.2024.000

Adedeji, J. A., Tetteh, E. K., Amo-Duodu, G., Armah, E. K., Rathilal, S., & Chetty, M. (2022). Central composite design optimisation of banana peels/magnetite for anaerobic biogas production from wastewater. *Applied Sciences, 12*(12037). https://doi.org/10.3390/app122312037

Adeniran, A. E., Adetinuke, A., & Omolaraeni, O. (2016). An evaluation of biogas production from anaerobic digester of a constructed wetland domestic wastewater treatment plant. *African Journal of Environmental Science and Technology, 10*(10), 329–337. https://doi.org/10.5897/AJEST2016.2153

Admasu, A., Bogale, W., & Mekonnen, Y. S. (2022). Experimental and simulation analysis of biogas production from beverage wastewater sludge for electricity generation. *Scientific Reports, 12*(9107), 1–15. https://doi.org/10.1038/s41598-022-12811-3

Agrawal, A., Chaudhari, P. K., & Ghosh, P. (2023). Anaerobic digestion of fruit and vegetable waste: A critical review of associated challenges. *Environmental Science and Pollution Research, 30*, 24987–25012. https://doi.org/10.1007/s11356-022-21643-7

Aguilar-Torrejón, J. A., Balderas-Hernández, P., Roa-Morales, G., Barrera-Díaz, C. E., Rodríguez-Torres, I., & Torres-Blancas, T. (2023). Relationship, importance, and development of analytical techniques: COD, BOD, and TOC in water—An overview through time. *SN Applied Sciences, 5*(118), 1–13. https://doi.org/10.1007/s42452-023-05318-7

Aguilloso, G., Arpia, K., Khan, M., Sapico, Z. A., & Lopez, E. C. R. (2024). Recent advances in membrane technologies for biogas upgrading †. *Energy Proceedings, 67*(57), 1–21. https://doi.org/10.3390/engproc2024067057

Ahmad, H. R., Aziz, T., Zia-ur-Rehman, M., Sabir, M., & Khalid, H. (2016). Sources and composition of waste water: Threats to plants and soil health. In K. R. Hakeem, J. Akhtar, & M. Sabir (Eds.), *Soil science: Agricultural and environmental prospectives* (pp. 349–370). Springer, Cham. https://doi.org/10.1007/978-3-319-34451-5_16

Ahmed, S. F., Rahman, M., Tarannum, K., Chowdhury, A. T., Rafa, N., Nuzhat, S., Kumar, P. S., Vo, D.-V. N., Lichtfouse, E., & Mahlia, T. M. I. (2021). Biogas upgrading, economy and utilization: A review. *Environmental Chemistry Letters*, 1–65. https://doi.org/10.1007/s10311-021-01292-x

Alagha, O., Allazem, A., Bukhari, A. A., Anil, I., & Mu'azu, N. D. (2020). Suitability of SBR for wastewater treatment and reuse: Pilot-scale reactor operated in different anoxic conditions. *International Journal of Environmental Research and Public Health, 17*(1617), 1–13. https://doi.org/10.3390/ijerph17051617

Alavijeh, M. K., Lee, Y. Y., & Gras, S. L. (2024). A perspective-driven and technical evaluation of machine learning in bioreactor scale-up: A case-study for potential model developments. *Engineering, 24*(7). https://doi.org/10.1002/elsc.202400023

Alberici, S., & Dekker, H. (2021). Scaling up biomethane in the European Union. In S. Twinning, D. Mockly, & K. Simson (Eds.), *The biomethane declaration, organized by European biogas Association [EBA, Brussels 7th December 2021]* (pp. 1–41). https://www.europeanbiogas.eu/wp-content/uploads/2021/12/

Aldin, S. (2010). In G. Nakhla & M. B. Ray (Eds.), *The effect of particle size on hydrolysis and modeling of anaerobic digestion.* Electronic Thesis and Dissertation Repository. https://ir.lib.uwo.ca/etd/60/

Alexandra, A. A. (1998). *Using expert systems for fault detection and diagnosis in industrial applications* (pp. 1–9). https://apps.dtic.mil/sti/tr/pdf/ADA347351.pdf

Aljuboury, D. A. D. A., Palaniandy, P., Abdul Aziz, H. B., & Feroz, S. (2017). Treatment of petroleum wastewater by conventional and new technologies—A review. *Global NEST Journal, 19*(3), 439–452. https://journal.gnest.org/sites/default/files/

Al-Sayed, A., Hassan, G. K., Al-Shemy, M. T., & El-gohary, F. A. (2023). Effect of organic loading rates on the performance of membrane bioreactor for wastewater treatment behaviours, fouling, and economic cost. *Scientific Reports, 13*(15601). https://doi.org/10.1038/s41598-023-42876-7

Álvarez-Torrellas, S., Peres, J. A., Gil-Álvarez, V., Ovejero, G., & García, J. (2017). Effective adsorption of non-biodegradable pharmaceuticals from hospital wastewater with different carbon materials. *Chemical Engineering Journal, 320*, 319–329. https://doi.org/10.1016/j.cej.2017.03.077

Alzoubi, Y. I., & Mishra, A. (2023). Green blockchain—A move towards sustainability. *Journal of Cleaner Production, 430*(139541), 1–13. https://doi.org/10.1016/j.jclepro.2023.139541

Amigun, B., & von Blottnitz, H. (2010). Capacity-cost and location-cost analyses for biogas plants in Africa. *Resources, Conservation and Recycling, 55*(1), 63–73. https://doi.org/10.1016/j.resconrec.2010.07.004

Amoo, A. O., Ijanu, E. M., Haruna, A., Adeleye, A. O., & Sabo, A. (2023). Review on the pretreatment advancements of biogas production barriers. *UMYU Journal of Microbiology Research (UJMR), 8*(1), 6–30. https://doi.org/10.47430/ujmr.2381.002

Amponsem, B., Bensah, E., Antwi, E., Ahiekpor, J., Boahen, B., Mensah, I., & Narra, S. (2023). Electricity generation from biogas as resource recovery potential from solid waste composition in a mixed-income municipality. *Cleaner Waste Systems, 4*(100067), 1–12. https://doi.org/10.1016/j.clwas.2022.100067

Anjum, M., Al-Makishah, N. H., & Barakat, M. A. (2016). Wastewater sludge stabilization using pre-treatment methods. *Process Safety and Environmental Protection, 102*, 615–632. https://doi.org/10.1016/j.psep.2016.05.022

Ardakani, M. N., & Gholikandi, G. B. (2020). Microbial fuel cells (MFCs) in integration with anaerobic treatment processes (AnTPs) and membrane bioreactors (MBRs) for simultaneous efficient wastewater/sludge treatment and energy recovery—A state-of-the-art review. *Biomass and Bioenergy, 141*(105726). https://doi.org/10.1016/j.biombioe.2020.105726

Arias, A., Feijoo, G., & Moreira, M. T. (2020). What is the best scale for implementing anaerobic digestion according to environmental and economic indicators? *Journal of Water Process Engineering, 35*(101235). https://doi.org/10.1016/j.jwpe.2020.101235

Armah, E. K., Chetty, M., & Deenadayalu, N. (2020). Biogas production from sugarcane bagasse with South African industrial wastewater and novel kinetic study using response surface methodology. *Scientific African, 10*(e00556), 1–9. https://doi.org/10.1016/j.sciaf.2020.e00556

Ashok, A., T, A. S., Jithesh, K. S., Sukanya, T. R., & Rajani, S. H. (2017). Waste water treatment plant using distributed control system. In *National conference on "emerging research trends in electrical, electronics & instrumentation" (ERTEEI'17)* (pp. 5–13). https://iosrjournals.org/iosr-jeee/Papers/Conf.17017/

Awonuga, Y., Ogba, D., & Gaduya, N. (2023). *Exploring renewable energy certificates for Nigeria's decarbonisation strategy in the power sector* (pp. 1–5). TEMPLARS. https://www.templars-law.com/app/uploads/2023/01/RECs-in-Nigeria.pdf

Aygun, A., Nas, B., & Berktay, A. (2008). Influence of high organic loading rates on COD removal and sludge production in moving bed biofilm reactor. *Environmental Engineering Science, 25*(9), 1311–1316. https://doi.org/10.1089/ees.2007.0071

Baashar, Y., Alkawsi, G., Alkahtani, A. A., Hashim, W., Razali, R. A., & Tiong, S. K. (2021). Toward blockchain technology in the energy environment. *Sustainability, 13*(1008), 1–20. https://doi.org/10.3390/su13169008

Bachmann, N. (2015). In J. C. Jansen, D. Baxter, G. Bochmann, & N. Montpart (Eds.), *Sustainable biogas production in municipal wastewater treatment plants*. IEA Bioenergy. https://task37.iea bioenergy.com/wp-content/uploads/sites/32/2022/02/Wastewater_biogas_grey_web-1.pdf

Baddour, E. M., Farhoud, N., Sharholy, M., & Abdel-Magid, I. M. (2016). Biological treatment of poultry slaughterhouses wastewater by using aerobic moving bed biofilm reactor. *International Research Journal of Public and Environmental Health, 3*(5), 96–106. https://doi.org/10.15739/irjpeh.16.013

Bajpai, P. (2017). Anaerobic technology in pulp and paper industry. In J. Kacprzyk (Ed.), *SpringerBriefs in applied sciences and technology* (pp. 1–106). Springer Nature. https://doi.org/10.1007/978-981-10-4130-3

Bala, J. D., Lalung, J., & Ismail, N. (2015). Studies on the reduction of organic load from palm oil mill effluent (POME) by bacterial strains. *International Journal of Recycling of Organic Waste in Agriculture, 4*(1), 1–10. https://doi.org/10.1007/s40093-014-0079-6

Banadaki, F. D., Nematollahi, M. A., Jamali, H. A., & Hamidi, Z. (2024). The use of Chlorella vulgaris in reducing the organic load of poultry slaughterhouse wastewater: Modeling and optimization of influential factors in the process. *Environmental Health Engineering and Management Journal, 11*(2), 147–159. https://doi.org/10.34172/EHEM.2024.15

Bangalore, M., Hochman, G., & Zilberman, D. (2016). Policy incentives and adoption of agricultural anaerobic digestion: A survey of Europe and the United States. *Renewable Energy, 97*, 559–571. https://doi.org/10.1016/j.renene.2016.05.062

Barbera, E., Bertucco, A., & Kumar, S. (2018a). Nutrients recovery and recycling in algae processing for biofuels production. *Renewable and Sustainable Energy Reviews, 90*, 28–42. https://doi.org/10.1016/j.rser.2018.03.004

Barbera, E., Menegon, S., Banzato, D., D'Alpaos, C., & Bertucco, A. (2018b). From biogas to biomethane: A process simulation-based techno-economic comparison of different upgrading technologies in the Italian context. *Renewable Energy*, 1–35. https://doi.org/10.1016/j.renene.2018.12.052

Barno, S. K. A., & Abbas, A. S. (2020). Reduction of organics in dairy wastewater by adsorption on a prepared charcoal from Iraqi sugarcane. *Energy Security and Chemical Engineering Congress, 736*, 1–8. https://doi.org/10.1088/1757-899X/736/2/022096

Bartek, N. W., Higgins, M. J., Murthy, S., Beightol, S., & Peaslee, T. (2015). Causes and cures of rapid volume expansion in anaerobic digesters due to gas holdup. In *WEF IWA residuals and biosolids conference 2015: At Washington, D.C* (pp. 1–5). https://www.researchgate.net/publication/317450006

Bartek, N., Higgins, M. J., Murthy, S. N., Beightol, S., & Al-Omari, A. (2017). Understanding the role of mixing and viscosity in rapid volume expansion due to gas holdup in anaerobic digesters. In *WEF residuals and biosolids conference 2017* (pp. 1–11). https://www.academia.edu/66239712

Barth, C. A., Bellaver, C., Bush, C., Chaiyakul, A., Costa, A., Dhussa, A., Dimapilis, M., Duke, C., Elias, H. T., Fikreyesus, D., Freig, E., Green, J., Grullon, F., Hilbert, J., Jain, S., Kinh, L., Martin, T., Massé, D., Mateen, A., … Vainio-Mattila, B. (2014). A global perspective of anaerobic digestion policies and incentives. In *Global methane initiative agricultural subcommittee* (pp. 1–39). GMI. https://www.globalmethane.org/documents/tools/a-global-perspective-of-ad-policies-incentives.pdf

Basandorj, D. (2007). Anaerobic treatment of brewery wastewater using UASB (up flow anaerobic sludge blanket) reactors seeded with activated sludge. *International Forum on Strategic Technology*. https://doi.org/10.1109/IFOST.2007.4798679

Batstone, D. J. (2000). *High rate anaerobic treatment of complex wastewater* [Bachelor of Chemical Engineering (Hons) Degree of Doctor of Philosophy Thesis in Chemical Engineering, The University of Queensland]. https://www.researchgate.net/publication/37616650

Benbow, M. E., Receveur, J. P., & Lamberti, G. A. (2020). Death and decomposition in aquatic ecosystems. *Frontiers in Ecology and Evolution, 8*. https://doi.org/10.3389/fevo.2020.00017

Benit, N., & Roslin, A. S. (2015). Physicochemical properties of wastewater collected from different sewage sources. *International Journal of Innovative Science, Engineering & Technology (IJISET), 2*(11), 691–696. https://ijiset.com/vol2/v2s11/IJISET_V2_I11_89.pdf

Bernard, O., Polit, M., Hadj-Sadok, Z., Pengov, M., Dochain, D., Estaben, M., & Labat, P. (2001). Advanced monitoring and control of anaerobic wastewater treatment plants: software sensors and controllers for an anaerobic digester. *Water Science & Technology, 43*(7), 175–182. https://pubmed.ncbi.nlm.nih.gov/11385844/

Bernard, O., Chachuat, B., Hélias, A., Dantec, B. Le, Sialve, B., Steyer, J.-P., Lardon, L., Neveu, P., Lambert, S., Gallop, J., Dixon, M., Ratini, P., Quintaba, A., Frattesi, S., Lema, J. M., Bordello, E. R., Ruiz-Filippi, G., Rodriguez, J., Franco, A., … Lavigne, J.-F. (2005). An integrated system to remote monitor and control anaerobic wastewater treatment plant through the internet. *Water Science & Technology, 52*(1–2), 457–464. https://doi.org/10.2166/wst.2005.0553

Bharadwaj, B., Kambo, A., Arratia-Solar, A., & Ashworth, P. (2023b). Social acceptance of biogas in Australia. *Proceedings of the Australian Hydrogen Research Conference 2023 (AHRC 2023) 8–10 February 2023*, 1–1. https://doi.org/10.2139/ssrn.4479079

Bharadwaj, B., Kambo, A., Arratia-Solar, A., & Ashworth, P. (2023a). Awareness of biogas increases its social acceptance. *Journal of Cleaner Production, 421*(138432), 1–9. https://doi.org/10.1016/j.jclepro.2023.138432

Biese, T. J. (2016). In T. G. Ellis, K. Ikuma, & T. E. Loynachan (Eds.), *High rate anaerobic digestion of primary and secondary sludge using the static granular bed reactor (SGBR)* [Degree of Master of Science Thesis, Civil Engineering (Environmental Engineering) Major, Iowa State University]. https://dr.lib.iastate.edu/server/api/core/bitstreams

Biey, E. M., Musibono, E. D., & Verstraete, W. (2003). Start-up of a multi-stage system for biogas production and solid waste treatment in low-tech countries. *Water Science & Technology, 48*(4), 239–243. https://doi.org/10.2166/wst.2003.0262

Bochmann, G., Pesta, G., Rachbauer, L., & Gabauer, W. (2020). Anaerobic digestion of pretreated industrial residues and their energetic process integration. *Frontiers in Bioengineering and Biotechnology, 8*(487), 1–11. https://doi.org/10.3389/fbioe.2020.00487

Borkar, R. P. (2013). Moving bed biofilm reactor—A new perspective in wastewater treatment. *IOSR Journal of Environmental Science Toxicology and Food Technology, 6*(6), 15–21. https://doi.org/10.9790/2402-0661521

Boyd, G., Na, D., Li, Z., Snowling, S., Zhang, Q., & Zhou, P. (2019). Influent forecasting for wastewater treatment plants in North America. *Sustainability, 11*(1764), 1–14. https://doi.org/10.3390/su11061764

BPC. (2022). BioReactor simulator—A universal simulation platform. In *BPC Instruments AB* (pp. 1–8). BPC Instruments AB. https://bpcinstruments.com/

Buelna, G., Dube, R., Michel, M.-C., Turgeon, N., Bernard, Y., & Lessard, P. (1999). *Comprehensive pig manure treatment using the BIOSOR bio-filtration process* (pp. 389–404). https://ramiran.uvlf.sk/doc98/FIN-ORAL/BUELNA.pdf

Cabeza, C., van Lier, J. B., & van der Steen, P. (2023). Effects of thermal and enzymatic pretreatments on the solubilisation of extracellular polymeric substances (EPS) and subsequent anaerobic digestion of microalgae-bacterial biomass. *Algal Research, 42*(103130), 1–10. https://doi.org/10.1016/j.algal.2023.103130

Cairone, S., Hasan, S. W., Choo, K.-H., Lekkas, D. F., Fortunato, L., Zorpas, A. A., Korshin, G., Zarra, T., Belgiorno, V., & Naddeo, V. (2024). Revolutionizing wastewater treatment toward circular economy and carbon neutrality goals: Pioneering sustainable and efficient solutions for automation and advanced process control with smart and cutting-edge technologies. *Journal of Water Process Engineering, 63*(105486), 1–15. https://doi.org/10.1016/j.jwpe.2024.105486

Callahan, J. L., Pfluger, A. R., Figueroa, L. A., & Munakata-Marr, J. (2022). BioWin® modeling of anaerobic sludge blanket treatment of domestic wastewater. *Bioresource Technology Reports, 20*(101231). https://doi.org/10.1016/j.biteb.2022.101231

Calzada, M., Espasa, M. L., & Barrera, A. (1993). Expert systems applied to safety in chemical engineering plants. *Transactions on Information and Communications Technologie, 1*, 683–696. https://www.witpress.com/Secure/elibrary/papers/AIENG93/AIENG93048FU.pdf

Campos, J. L., Crutchik, D., Franchi, Ó., Pavissich, J. P., Belmonte, M., Pedrouso, A., Mosquera-Corral, A., & Del Rio, Á. V. (2018). Nitrogen and phosphorus recovery from anaerobically pretreated agro-food wastes: A review. *Frontiers in Sustainable Food Systems, 2*(91), 1–11. https://doi.org/10.3389/fsufs.2018.00091

Carneiro, R. B., Gonzalez-Gil, L., Londoño, Y. A., Zaiat, M., Carballa, M., & Lema, J. M. (2020). Acidogenesis is a key step in the anaerobic biotransformation of organic micropollutants. *Journal of Hazardous Materials, 389*(121888). https://doi.org/10.1016/j.jhazmat.2019.121888

Carrere, H., Antonopoulou, G., Affes, R., Passos, F., Battimelli, A., Lyberatos, G., & Ferrer, I. (2016). Review of feedstock pretreatment strategies for improved anaerobic digestion: From lab-scale research to full-scale application. *Bioresource Technology, 199*, 386–397. https://doi.org/10.1016/j.biortech.2015.09.007

Carter, J. (2024). *Innovative feed formulation made simple.* AgriPro Feed Co. https://www.animalfeedsoftware.com/

Chaikovskaya, E. (2024). Biogas heat pump power supply based on smart grid functioning solar electric system network. *Journal of Solar Energy Research Updates, 11*, 55–67. https://doi.org/10.31875/2410-2199.2024.11.07

Chanathaworn, J., Thararux, C., & Paoaiang, J. (2018). Determining optimization of factors affecting biogas production by co-digestion of sweet corncob waste with wastewater using response surface methodology. *Journal of Renewable Energy and Smart Grid Technology, 13*(1), 1–11. https://ph01.tci-thaijo.org/index.php/RAST/article/view/65162

Chapman, R., & Cuthbertson, S. (1999). Sydney's water—A suitable case for private treatment? In *Public policy for the private sector* (Vol. 80, pp. 1–7). The World Bank Group. https://cdi.mecon.gob.ar/bases/docelec/bm/ppps/n80.pdf

Chapman, T., & Krugel, S. (2011). Rapid volume expansion—An investigation into digester overflows and safety. *Proceedings of the Water Environment Federation, 2011*(4), 1016–1040. https://doi.org/10.2175/193864711802863229

Cheema, S. M., Hannan, A., & Pires, I. M. (2022). Smart waste management and classification systems using cutting edge approach. *Sustainability, 14*(10226), 1–21. https://doi.org/10.3390/su141610226

Chen, X. Y., Vinh, H. duc, Ramirez, A. A., Rodrigue, D., & Kaliaguine, S. (2015). Membrane gas separation technologies for biogas upgrading. *RSC Advances*, *5*(31), 24399–24448. https://doi.org/10.1039/C5RA00666J

Chen, H., Chang, S., Guo, Q., Hong, Y., & Wu, P. (2016). Brewery wastewater treatment using an anaerobic membrane bioreactor. *Biochemical Engineering Journal*, *105*(Part B), 321–331. https://doi.org/10.1016/j.bej.2015.10.006

Chen, Z., Ngo, H. H., Guo, W., & Wang, X. (2013). Analysis of Sydney's recycled water schemes. *Frontiers of Environmental Science & Engineering, 7*, 608–615. https://doi.org/10.1007/s11783-012-0468-x

Chiang, L., Lu, B., & Castillo, I. (2017). Big data analytics in chemical engineering. *Annual Review of Chemical and Biomolecular Engineering, 8*, 63–85. https://doi.org/10.1146/annurev-chembioeng-060816-101555

Chinh, T. T., Hieu, P. D., Cuong, B. V., Linh, N. N., Lan, N. N., Nguyen, N. S., Hung, N. Q., & Hien, L. T. T. (2018). Sequencing Batch Reactor and bacterial community in aerobic granular sludge for wastewater treatment of noodle-manufacturing sector. *Applied Sciences, 8*(509), 1–14. https://doi.org/10.3390/app8040509

Chitale, M. R., Chitpur, S. J., Chivate, A. B., Chopade, P. D., Deshmukh, S. M., & Marathe, A. A. (2023). Automated smart waste segregation system using IoT technology. In *3rd international conference on intelligent robotics, mechatronics and automation systems (IRMAS 2023), 2601*, 1–9. https://doi.org/10.1088/1742-6596/2601/1/012015

Choi, Y.-Y., Baek, S.-R., Kim, J.-I., Choi, J.-W., Hur, J., Lee, T.-U., Park, C.-J., & Lee, B. J. (2017). Characteristics and biodegradability of wastewater organic matter in municipal wastewater treatment plants collecting domestic wastewater and industrial discharge. *Water*, *9*(409). https://doi.org/10.3390/w9060409

Chong, D. J. S., Chan, Y. J., Arumugasamy, S. K., Yazdi, S. K., & Lim, J. W. (2023). Optimisation and performance evaluation of response surface methodology (RSM), artificial neural network (ANN) and adaptive neuro-fuzzy inference system (ANFIS) in the prediction of biogas production from palm oil mill effluent (POME). *Energy, 266*(126449), 1–18. https://doi.org/10.1016/j.energy.2022.126449

Choo, W. C. (2015). In S. M. F. S. Abdul Aziz & N. Zainol (Eds.), *Production of biogas from poultry manure wastewater: Optimization* [Bachelor of Chemical Engineering Undergraduate Project, Faculty of Chemical & Natural Resources Engineering, Universiti Malaysia Pahang]. http://umpir.ump.edu.my/id/eprint/11012/

CHP. (2018). *Gresham wastewater treatment plant net zero energy use with CHP* (pp. 1–2). CHP Technical Assistant Partnership. https://chptap.ornl.gov/profile/90/Gresham_WWTF-Project_Profile.pdf

Ciliberti, C., Jordaan, S. M., Smith, S. V., & Spatari, S. (2016). A life cycle perspective on land use and project economics of electricity from wind and anaerobic digestion. *Energy Policy, 89*, 52–63. https://doi.org/10.1016/j.enpol.2015.11.015

Cinar, S., Cinar, S. O., Wieczorek, N., Sohoo, I., & Kuchta, K. (2021). Integration of artificial intelligence into biogas plant pperation. *Processes, 9*(85), 1–18. https://doi.org/10.3390/pr9010085

Cioabla, A. E., Ionel, I., Dumitrel, G.-A., & Popescu, F. (2012). Comparative study on factors affecting anaerobic digestion of agricultural vegetal residues. *Biotechnology for Biofuels and Bioproducts, 5*(39), 1–9. https://doi.org/10.1186/1754-6834-5-39

Cirelli, A. F., & Ojeda, C. (2008). Wastewater management in Greater Buenos Aires, Argentina. *Desalination, 218*(1–3), 52–61. https://doi.org/10.1016/j.desal.2006.10.040

Cocozza, C., Iaconi, C. Di, Murgolo, S., Traversa, A., De Mastro, F., De Sanctis, M., Altieri, V. G., Cacace, C., Brunetti, G., & Mascolo, G. (2023). Use of constructed wetlands to prevent overloading of wastewater treatment plants. *Chemosphere, 311*(Part 2). https://doi.org/10.1016/j.chemosphere.2022.137126

Coelmont, T., Gaelen, P. V., & Smets, I. (2023). Quantification of hydrolysis activity in a biological wastewater treatment context. *Applied Microbiology and Biotechnology, 107*(7–8), 2143–2153. https://doi.org/10.1007/s00253-023-12465-9

Connelly, S., Shin, S. G., Dillon, R. J., Ijaz, U. Z., Quince, C., Sloan, W. T., & Collins, G. (2017). Bioreactor scalability: Laboratory-scale bioreactor design influences performance, ecology, and community physiology in expanded granular sludge bed bioreactors. *Frontiers in Microbiology, 8*(664), 1–15. https://doi.org/10.3389/fmicb.2017.00664

Cremonez, P. A., Teleken, J. G., Meier, T. R. W., & Alves, H. J. (2021). Two-stage anaerobic digestion in agroindustrial waste treatment: A review. *Journal of Environmental Management, 281*(111854). https://doi.org/10.1016/j.jenvman.2020.111854

Cruz, I. A., De Melo, L., Leite, A. N., Sátiro, J. V. M., Andrade, L. R. S., Torres, N. H., Padilla, R. Y. C., Bharagava, R. N., Tavares, R. F., & Ferreira, L. F. R. (2019). A new approach using an open-source low cost system for monitoring and controlling biogas production from dairy wastewater. *Journal of Cleaner Production, 241*(118284). https://doi.org/10.1016/j.jclepro.2019.118284

Cui, Y., Liu, C., & Wang, L. (2010). Feasibility of acidogenesis reactor with shell carrier for domestic wastewater treatment. In *4th international conference on bioinformatics and biomedical engineering, Chengdu, China, on 18–20 June 2010* (pp. 1–3). https://doi.org/10.1109/ICBBE.2010.5517515

Daud, M. K., Rizvi, H., Akram, M. F., Ali, S., Rizwan, M., Nafees, M., & Jin, Z. S. (2018). Review of upflow anaerobic sludge blanket reactor technology: Effect of different parameters and developments for domestic wastewater treatment. *Journal of Chemistry*. https://doi.org/10.1155/2018/1596319

Davoli, F., Repetto, M., Tornelli, C., Proserpio, G., & Cucchietti, F. (2012). In C. Buet, H. Ota, & E. Campilongo (Eds.), *Boosting energy efficiency through smart grids*. International Telecommunication Union (ITU). https://www.itu.int/dms_pub/itu-t/oth/4B/01/T4B010000050001PDFE.pdf

De Clercq, D., Jalota, D., Shang, R., Ni, K., Zhang, Z., Khan, A., Wen, Z., Caicedo, L. M., & Yuan, K. (2019). Machine learning powered software for accurate prediction of biogas production: A case study on industrial-scale Chinese production data. *Journal of Cleaner Production, 218*, 1–2. https://doi.org/10.1016/j.jclepro.2019.01.031

Debele, A. D., Bekele, T., & Geda, A. (2024). Biogas production through anaerobic codigestion of distillery wastewater sludge and disposable spent yeast. *International Journal of Chemical Engineering*. https://doi.org/10.1155/2024/5510471

Debowski, M., & Zielinski, M. (2022). Wastewater treatment and biogas production: Innovative technologies, research and development directions. *Energies, 15*(2122), 1–4. https://doi.org/10.3390/en15062122

Dennis, S., Whitehead, J., Patterson, R., & Goonetilleke, A. (2020). *Scalability of advanced passive wastewater systems* (V8; Whitehead & Associates). https://www.academia.edu/44530528/Scalability_of_Advanced_Passive_Wastewater_Systems

Derakhshan, M., & Fazeli, M. (2018). Improved biodegradability of hardly-decomposable wastewaters from petrochemical industry through photo-Fenton method and determination of optimum operational conditions by response surface methodology. *Journal of Biological Engineering, 12*(10), 1–10. https://doi.org/10.1186/s13036-018-0104-9

Derakhshesh, S., Sharghi, E. A., Bonakdarpour, B., & Khoshnevisan, B. (2022). Integrating electrocoagulation process with up-flow anaerobic sludge blanket for in-situ biomethanation and performance improvement. *Bioresource Technology, 360*(127536). https://doi.org/10.1016/j.biortech.2022.127536

Dev, S., Saha, S., Kurade, M., Salama, E.-S., El-Dalatony, M. M., & Jeon, B.-H. (2019). Perspective on anaerobic digestion for biomethanation in cold environments. *Renewable and Sustainable Energy Reviews, 103*, 85–95. https://doi.org/10.1016/j.rser.2018.12.034

Dhayalan, V., Raman, R., Kalaivani, N., Shrirvastava, A., Rajasekar, G., & Thamizhamuthu, R. (2024). Cloud-based smart sewage management system for efficient wastewater treatment. In

4th international conference on innovative practices in technology and management (ICIPTM), 21–23 February 2024. https://doi.org/10.1109/ICIPTM59628.2024.10563306

Di Costanzo, N., Cesaro, A., Di Capua, F., & Esposito, G. (2021). Exploiting the nutrient potential of anaerobically digested sewage sludge: A review. *Energies, 14*(8149), 1–25. https://doi.org/10.3390/en14238149

Dieu, B. (2001). Application of the SCADA system in wastewater treatment plants. *ISA Transactions, 40*(3), 267–281. https://doi.org/10.1016/S0019-0578(00)00053-7

Djatkov, D., Effenberger, M., & Martinov, M. (2014). Method for assessing and improving the efficiency of agricultural biogas plants based on fuzzy logic and expert systems. *Applied Energy, 134*, 163–175. https://doi.org/10.1016/j.apenergy.2014.08.021

Djimtoingar, S. S., Sarfo, N., Kuranchie, F. A., & Kusi, J. Y. (2022). A review of response surface methodology for biogas process optimization. *Cogent Engineering, 9*(1). https://doi.org/10.1080/23311916.2022.2115283

Dolat, M., Murali, R., Zarei, M., Zhang, R., Pincam, T., Liu, Y.-Q., Sadhukhan, J., Bywater, A., & Short, M. (2024). Dynamic feed scheduling for optimised anaerobic digestion: An optimisation approach for better decision-making to enhance revenue and environmental benefits. *Digital Chemical Engineering, 13*(100191), 1–14. https://doi.org/10.1016/j.dche.2024.100191

Domínguez-Maldonado, J. A., Alzate-Gaviria, L., Milquez-Sanabria, H. A., Tapia-Tussell, R., Leal-Bautista, R. M., & España-Gamboa, E. I. (2020). Chemical pretreatments to enrich the acidogenic phase in a system coupled packed bed reactor with a UASB reactor using peels and rotten onion waste. *Waste and Biomass Valorization, 11*, 4181–4194. https://doi.org/10.1007/s12649-019-00751-5

Domini, M., Abbà, A., & Bertanza, G. (2022). Analysis of the variation of costs for sewage sludge transport, recovery and disposal in Northern Italy: A recent survey (2015–2021). *Water Science & Technology, 85*(3), 1167–1175. https://doi.org/10.2166/wst.2022.040

Donatien, E. A., Rokia, T. H., Daouda, O., Souleymane, S., Remis, K. G., Rodrigue, K. A., Amara, B., Marius, N. P., & Benjamin, Y. (2024). Study of aluminum corrosion in contact with biogas before and after purification on different carbons. *Applied Chemical Engineering (ACE), 7*(3), 1–12. https://doi.org/10.59429/ace.v7i3.5524

Duale, O., & Alex-Adedipe. (2023). *Renewables Portfolio Standards as a tool for the adoption of renewable energy in Nigeria*. DOA. https://www.doa-law.com/

Dubrovskis, V., & Plume, I. (2015). Anaerobic digestion of vegetables processing wastes with catalyst metaferm. *Agronomy Research, 13*(2), 294–302. https://doi.org/10.5555/20153188924

Dutta, A., & Sarkar, S. (2015). Sequencing batch reactor for wastewater treatment: Recent advance. *Current Pollution Reports, 1*, 177–190. https://doi.org/10.1007/s40726-015-0016-y

Echiegu, E. A., & Ghaly, A. (2014). An in-house computer-based data acquisition and control system for anaerobic fermentation processes. *American Journal of Engineering and Applied Sciences, 7*(3), 313–323. https://doi.org/10.3844/ajeassp.2014.313.323

Ekstrand, E.-M. (2019). In C. Perhammar & L. LiU-Tryck (Eds.), *Anaerobic digestion in the kraft pulp and paper industry—Challenges and possibilities for implementation* (Vol. 1:1, Issue 769) [Doctorak Thesis from Department of Thematic Studies—Environmental Change, Faculty of Arts and Sciences, Linköping University]. https://www.diva-portal.org/smash/get/diva2:1313936/FULLTEXT01.pdf

Elliott, J. A. K., Krohn, C., & Ball, A. S. (2024). Source-separated industrial wastewater is a candidate for biogas production through anaerobic digestion. *Fermentation, 10*(165), 1–23. https://doi.org/10.3390/fermentation10030165

Elmekawy, A., Mohanakrishna, G., Sandipam, S., & Pant, D. (2015). The role of bioreactors in industrial wastewater treatment. In R. Chandra (Ed.), *Environmental waste management* (1st ed., pp. 1–28). CRC Press. https://doi.org/10.1201/b19243-8

Elsaesser, T., Houdkova, L., Belohradsky, P., & Stehlik, P. (2012). Disintegration of sewage sludge for improved dewaterability. *Chemical Engineering Transactions (CEt), 29*, 1273–1278. https://doi.org/10.3303/CET1229213

EPA. (2024). *An overview of renewable natural gas from biogas*. https://www.epa.gov/system/files/documents/2024-01/lmop_rng_document.pdf

Erdmann, H.-H., Engelmann, H.-D., Burgard, W., & Simmrock, K. H. (1988). The use of expert systems in chemical engineering. *Chemical Engineering and Processing: Process Intensification, 23*(2), 125–133. https://doi.org/10.1016/0255-2701(88)80006-0

Erwin, T., & Yang, B. (2023). Green energy, emissions, and blockchain technology. *SSRN*, 1–24. https://doi.org/10.2139/ssrn.4475737

Fang, C., Huang, R., Dykstra, C. M., Jiang, R., Pavlostathis, S. G., & Tang, Y. (2019). Energy and nutrient recovery from sewage sludge and manure via anaerobic digestion with hydrothermal pretreatment. *Environmental Science & Technology, 54*(2), 1147–1156. https://doi.org/10.1021/acs.est.9b03269

Fantozzi, F., Bartocci, P., D'Alessandro, B., Arabatzis, S., & Manos, B. (2014). Public–private partnerships value in bioenergy projects: Economic feasibility analysis based on two case studies. *Biomass and Bioenergy, 66*, 387–397. https://doi.org/10.1016/j.biombioe.2014.04.006

Fick, M. (2025). AstraZeneca opens biomethane plant in UK to cut emissions. *Reuters*. https://www.reuters.com/sustainability/climate-energy/astrazeneca-opens

Fu, B., Jin, X., Conrad, R., Liu, H., & Liu, H. (2019). Competition between chemolithotrophic acetogenesis and hydrogenotrophic methanogenesis for exogenous H_2/CO_2 in anaerobically digested sludge: Impact of temperature. *Frontiers in Microbiology, 10*(2418), 1–9. https://doi.org/10.3389/fmicb.2019.02418

Fuentes-Santiago, V., Valdez-Vazquez, I., Vital-Jácome, M., Zavala-Méndez, M., Razo-Flores, E., & Carrillo-Reyes, J. (2023). Carbohydrates/acid ratios drives microbial communities and metabolic pathways during biohydrogen production from fermented agro-industrial wastewater. *Journal of Environmental Chemical Engineering, 11*(3). https://doi.org/10.1016/j.jece.2023.110302

Gamble, K. J., Houser, J. B., Hambourger, M. S., & Hoepfl, M. C. (2015). Anaerobic digestion from the laboratory to the field: An experimental study into the scalability of anaerobic digestion. In *2015 ASEE southeast section conference*, pp 1–15. http://se.asee.org/proceedings/ASEE2015/papers2015/46.pdf

Ganeshan, P., & Rajendran, K. (2022). Dynamic simulation and optimization of anaerobic digestion processes using MATLAB. *Bioresource Technology, 351*(126970). https://doi.org/10.1016/j.biortech.2022.126970

Ghaleb, A. A. S., Kutty, S. R. M., Ho, Y.-C., Jagaba, A. H., Noor, A., Al-Sabaeei, A. M., & Almahbashi, N. M. Y. (2020). Response Surface Methodology to optimize methane production from mesophilic anaerobic co-digestion of oily-biological sludge and sugarcane bagasse. *Sustainability, 12*(2116), 1–11. https://doi.org/10.3390/su12052116

Ghosh, S., Conrad, J. R., & Klass, D. L. (1975). Anaerobic acidogenesis of wastewater sludge. *Water Pollution Control Federation, 47*(1), 30–45. https://www.jstor.org/stable/25038594

Ghosh, P., Samanta, A. N., & Ray, S. (2010). COD reduction of petrochemical industry wastewater using Fenton's oxidation. *The Canadian Journal of Chemical Engineering, 88*(6), 1021–1026. https://doi.org/10.1002/cjce.20353

Gienau, T., Brüß, U., Kraume, M., & Rosenberger, S. (2018). Nutrient recovery from anaerobic sludge by membrane filtration: Pilot tests at a 2.5 MWe biogas plant. *International Journal of Recycling of Organic Waste in Agriculture, 7*, 325–334. https://doi.org/10.1007/s40093-018-0218-6

Giesy, R., Wilkie, A. C., De Vries, A., & Nordstedt, R. A. (2005). Economic feasibility of anaerobic digestion to produce electricity on Florida Dairy Farm. In *The Institute of Food and Agricultural Sciences (IFAS)* (pp. 1–8). Department of Animal Sciences, Florida Cooperative Extension Service, Institute of Food and Agricultural Sciences, University of Florida. http://edis.ifas.ufl.edu/

Gingerich, D. B., & Mauter, M. S. (2017). Air emission reduction benefits of biogas electricity generation at municipal wastewater treatment plants. *Environmental Science & Technology, 52*(3), 1633–1643. https://doi.org/10.1021/acs.est.7b04649

Gospodinov, M., Gospodinova, E., & Cheshmedjiev, K. (2014). SCADA system for management and visualization in generation of renewable energy from biomass. *International Journal of Engineering Research & Technology (IJERT), 3*(5), 882–887. https://www.ijert.org/research/

Guerrero, L., Omil, F., Méndez, R., & Lema, J. M. (1999). Anaerobic hydrolysis and acidogenesis of wastewaters from food industries with high content of organic solids and protein. *Water Research, 33*(15), 3281–3290. https://doi.org/10.1016/S0043-1354(99)00041-X

Guo, H., Oosterkamp, M. J., Tonin, F., Hendriks, A., Nair, R., van Lier, J. B., & De Kreuk, M. (2021). Reconsidering hydrolysis kinetics for anaerobic digestion of waste activated sludge applying cascade reactors with ultra-short residence times. *Water Research, 202*(117398). https://doi.org/10.1016/j.watres.2021.117398

H_2O-Innovation. (2005). H_2O innovation to acquire biosor technologies. *Membrane Technology, 2005*(7), 5–6. https://doi.org/10.1016/S0958-2118(05)70454-7

Hagey, P. (2011). Utility district ramps up food waste to energy program. *BioCycle, 52*(11), 1–39. https://www.biocycle.net/utility-district-ramps-up-food-waste-to-energy-program/

Hamawand, I. (2015). Anaerobic digestion process and bio-energy in meat industry: A review and a potential. *Renewable and Sustainable Energy Reviews, 44*, 37–51. https://doi.org/10.1016/j.rser.2014.12.009

Hanratty, P. J. (1991). In B. Joseph & M. P. Dudukovic (Eds.), *Expert system applications in chemical engineering* [Degree of Doctor of Science Thesis, Sever Institute of Technology, Washington University]. https://www.proquest.com/openview/50b5261

Hao, X., Li, J., van Loosdrecht, M. C. M., Jiang, H., & Liu, R. (2019). Energy recovery from wastewater: Heat over organics. *Water Research, 161*, 74–77. https://doi.org/10.1016/j.watres.2019.05.106

Hassan, S. R. B., Abu, M. J., & Dahlan, I. (2020). Industrial wastewater treatment: Recycled paper mill effluent treatment using Modified Anaerobic Hybrid Baffled (MAHB) bioreactor. In *What is acetogenesis* (pp. 1–21). IGI Global. https://doi.org/10.4018/978-1-7998-0369-0.ch014

Hassan, S. R., & Dahlan, I. (2013). Anaerobic wastewater treatment using anaerobic baffled bioreactor: A review. *Open Engineering, 3*(3), 389–399. https://doi.org/10.2478/s13531-013-0107-8

Hau, K. C., & Sculli, D. (1991). Costs of disposal of sewage sludge: A case study. *Engineering Costs and Production Economics, 21*(2), 133–141. https://doi.org/10.1016/0167-188X(91)90027-Y

Hegazy, H., Saady, N. M. C., Khan, F., Zendehboudi, S., & Albayati, T. M. (2024). Biogas plants accidents: Analyzing occurrence, severity, and associations between 1990 and 2023. *Safety Science, 177*(106597), 1–18. https://doi.org/10.1016/j.ssci.2024.106597

Hernandez, E. M., & Martín, M. (2022). Anaerobic digestion and nutrient recovery. In M. Martín (Ed.), *Sustainable design for renewable processes* (pp. 239–281). Elsevier-Principles and Case Studies. https://doi.org/10.1016/B978-0-12-824324-4.00017-2

Hewitt, J., Holden, M., Robinson, B. L., Jewitt, S., & Clifford, M. J. (2022). Not quite cooking on gas: Understanding biogas plant failure and abandonment in Northern Tanzania. *Renewable and Sustainable Energy Reviews, 165*(112600), 1–9. https://doi.org/10.1016/j.rser.2022.112600

Hoo, P. Y., Hashim, H., & Ho, W. S. (2018). Economic feasibility of feed-in tariff (FiT) for biomethane injection into natural gas distribution grid. *Chemical Engineering Transactions (CEt), 70*, 631–636. https://doi.org/10.3303/CET1870106

Hossen, A., Sattar, G. S., & Mostafa, M. G. (2024). Factors affecting the performance of a pharmaceutical wastewater treatment plant: Characterization of effluent and environmental risk. *Heliyon, 10*, 1–13. https://doi.org/10.1016/j.heliyon.2024.e29165

Huang, L., Cheng, S., Rezaei, F., & Logan, B. E. (2008). Reducing organic loads in wastewater effluents from paper recycling plants using microbial fuel cells. *Environmental Technology, 30*(5), 499–504. https://doi.org/10.1080/09593330902788244

Ifa, L., Nurjannah, Yani, S., & Jaya, F. (2023). Anaerobic digestion of vegetable wastes to produce biogas as a substitute for LPG. *Young Scholar Symposium on Science and Mathematics Education, and Environment, 15–16 January 2022, Bandar Lampung, Indonesia, 2595*(1). https://doi.org/10.1063/5.0145996

Ijoma, G. N., Mutungwazi, A., Mannie, T., Nurmahomed, W., Matambo, T. S., & Hildebrandt, D. (2022). Addressing the water-energy nexus: A focus on the barriers and potentials of harnessing wastewater treatment processes for biogas production in Sub Saharan Africa. *Heliyon, 8*(5). https://doi.org/10.1016/j.heliyon.2022.e09385

Induchoodan, T. G., Haq, I., & Kalamdhad, A. S. (2022). Factors affecting anaerobic digestion for biogas production: A review. In C. Hussain & S. Hait (Eds.), *Advanced organic waste management* (pp. 223–233). Elsevier-Sustainable Practices and Approaches. https://doi.org/10.1016/B978-0-323-85792-5.00020-4

Irvine, R. L., Ketchum, L. H., & Asano, T. (1989). Sequencing Batch Reactor for biological wastewater treatment. *Critical Reviews in Environmental Control, 18*(4), 255–294. https://doi.org/10.1080/10643388909388350

Jagadeesh, C. A. P., & Samuel, R. D. S. (2010). Modeling, simulation and control of bioreactors process parameters – Remote experimentation approach. *International Journal of Computer Applications, 1*(10), 81–88. https://citeseerx.ist.psu.edu/

Jameel, M. K., Mustafa, M. A., Ahmed, H. S., Mohammed, A. J., Ghazy, H., Shakir, M. N., Lawas, A. M., Mohammed, S. khudhur, Idan, A. H., Mahmoud, Z. H., Sayadi, H., & Kianfar, E. (2024). Biogas: Production, properties, applications, economic and challenges: A review. *Results in Chemistry, 7*(101549), 1–30. https://doi.org/10.1016/j.rechem.2024.101549

Janajreh, I., Alshehi, A., & Elagroudy, S. (2020). Anaerobic co-digestion of petroleum hydrocarbon waste and wastewater treatment sludge. *International Journal of Hydrogen Energy, 45*(20), 11538–11549. https://doi.org/10.1016/j.ijhydene.2018.05.100

Janesch, E., Pereira, J., Neubauer, P., & Junne, S. (2021). Phase separation in anaerobic digestion: A potential for easier process combination? *Frontiers in Chemical Engineering, 3*, 1–16. https://doi.org/10.3389/fceng.2021.711971

Jasim, N. (2020). The design for wastewater treatment plant (WWTP) with GPS X modelling. *Cogent Engineering, 7*(1). https://doi.org/10.1080/23311916.2020.1723782

Jeganathan, J., Nakhla, G., & Bassi, A. (2006). Long-term performance of high-rate anaerobic reactors for the treatment of oily wastewater. *Environmental Science & Technology, 40*(20), 6466–6472. https://doi.org/10.1021/es061071m

Jeihanipour, A., Aslanzadeh, S., Rajendran, K., Balasubramanian, G., & Taherzadeh, M. J. (2013). High-rate biogas production from waste textiles using a two-stage process. *Renewable Energy, 52*, 128–135. https://doi.org/10.1016/j.renene.2012.10.042

Jendželovský, N., & Uhlířová, L. (2021). Dynamic analysis of biogas tank. *7th World Multidisciplinary Earth Sciences Symposium (WMESS 2021), 906*(012049), 1–6. https://doi.org/10.1088/1755-1315/906/1/012049

Jiang, C. (2011). In R. Lindegren (Ed.), *A general investigation of Shanghai sewerage treatment system)* [Master Thesis Programme of Applied Environmental Science, School of Business and Engineering, Halmstad University]. https://www.diva-portal.org/smash/get/diva2:427129/FULLTEXT01.pdf

Jinadasa, K. B. S. N. (2003). In W. J. Ng & M. A. Aziz (Eds.), *Acidogenic biotreatment of wastewater containing 2-nitroaniline and copper* [Thesis for the Degree of Master of Engineering, Department of Civil Engineering, National University of Singapore]. https://core.ac.uk/download/pdf/48625486.pdf

Jones, J. G., Simon, B. M., & Gardener, S. (1982). Factors affecting methanogenesis and associated anaerobic processes in the sediments of a stratified Eutrophic Lake. *Journal of General Microbiolog, 128*(1), 1–11. https://doi.org/10.1099/00221287-128-1-1

Joshi, D. L. (1998). High rate anaerobic treatment of industrial wastewater in tropics. *Thammasat International Journal of Science and Technology, 3*(1), 1–7. https://www.thaiscience.info/journals/Article/TSTJ/10480586.pdf

Kabeyi, M. J. B., & Olanrewaju, O. A. (2022a). Biogas production and applications in the sustainable energy transition. *Journal of Energy, 2022*(750221), 1–43. https://doi.org/10.1155/2022/8750221

Kabeyi, M. J. B., & Olanrewaju, O. A. (2022b). Smart grid technologies and application in the sustainable energy transition: A review. *International Journal of Sustainable Energy, 42*(1), 685–758. https://doi.org/10.1080/14786451.2023.2222298

Kanat, G., & Yeter, D. (2020). Optimization and reduction of industrial load in a wastewater basin. *Desalination and Water Treatment, 205*, 316–327. https://doi.org/10.5004/dwt.2020.26420

Kannan, A., & Singaram, J. (2012). Anaerobic sequencing batch reactors and its influencing factors: an overview. *Journal of Environmental Science & Engineering, 54*(2), 317–322. https://www.researchgate.net/publication/261767299

Kapdi, S. S., Vijay, V. K., Rajesh, S. K., & Prasad, R. (2005). Biogas scrubbing, compression and storage: Perspective and prospectus in Indian context. *Renewable Energy, 30*(8), 1195–1202. https://doi.org/10.1016/j.renene.2004.09.012

Kataki, S., Hazarika, S., & Baruah, D. C. (2018). Recycling of bioenergy by-products as crop nutrient: Application in different phases for improvement of soil and crop. *Environmental Progress & Sustainable Energy, 38*(4), 1–11. https://doi.org/10.1002/ep.13099

Kavan, K. V, Mahendiran, R., Subramanian, P., & Karthikeyan, S. (2024). Upflow anaerobic sludge blanket (UASB) reactors for bio-methane production from limed tannery fleshings: Lab and pilot scale reactors. *Sustainable Chemistry One World*, *2*(100006). https://doi.org/10.1016/j.scowo.2024.100006

Kazimierowicz, J., & Dębowski, M. (2022). Aerobic granular sludge as a substrate in anaerobic digestion—Current status and perspectives. *Sustainability, 14*(10904), 1–24. https://doi.org/10.3390/su141710904

Kechrist, O. (2022). Economic analysis of biogas production via biogas digester made from composite material. *ChemEngineering, 6*(67), 1–12. https://doi.org/10.3390/chemengineering6050067

KENPRO. (2023). *Major donor agencies funding biogas projects in Africa.* KENPRO; KENPRO ICT Team. https://www.kenpro.org/major-donor-agencies-funding-biogas-projects-in-africa/

Kesharwani, N., & Bajpai, S. (2024). Anaerobic digestion of wastewater and resource recovery. In M. M. Ghangrekar, S. Yadav, & R. N. Yadava (Eds.), *Biological and hybrid wastewater treatment technology* (pp. 257–266). Springer, Cham. https://doi.org/10.1007/978-3-031-63046-0_10

Khalid, M. (2024). Smart grids and renewable energy systems: Perspectives and grid integration challenges. *Energy Strategy Reviews, 51*, 1–26. https://doi.org/10.1016/j.esr.2024.101299

Khalil, E. F., Whitmore, T. N., Gamal-El-Din, H., El-Bassel, A., & Lloyd, D. (1988). The effects of detergents on anaerobic digestion. *Applied Microbiology and Biotechnology, 29*, 517–522. https://doi.org/10.1007/BF00269079

Khan, I., Gupta, D. D., & Gupta, A. (2024). Sewage treatment by Kolkata's natural wetland system. *Nature Environment and Pollution Technology*, *23*(3), 1801–1816. https://doi.org/10.46488/NEPT.2024.v23i03.053

Khandaker, N., Islam, T., & Rabbi, F. (2015). Biogas generation potential from textile wastewater: Aggregation analysis of hundred textile facilities in Greater Dhaka Bangladesh. In *International conference on sustainable and renewable energy development and design (SREDD2017), 3–5 April 2017* (pp. 20–22). https://www.ijsr.net/conf/SREDD2017/1071705.pdf

Khandaker, N. R., Rahman, M. M., & Diba, D. S. (2021). Acetogenic pretreatment as an energy efficient method for treatment of textile processing wastewater. In A. Sikora (Ed.), *Anaerobic digestion in built environments* (pp. 1–12). InTech Open. https://doi.org/10.5772/intechopen.99107

Kharpude, S. N., Sharma, D., & Samar, K. (2017). A computer-based expert system to design Deenbandhu Biogas Plant. *Energy and Environment*, 177–190. https://doi.org/10.1007/978-981-10-5798-4_18

Khotimah, C. K., Handayani, D. S., Sudarno, S., Hadiwidodo, M., & Wardhana, W. I. (2021). The efficiency of biological oxygen demand removal in domestic wastewater treatment using multi soil layering. *International Conference on Research Collaboration of Environmental Science, 802*, 1–6. https://doi.org/10.1088/1755-1315/802/1/012006

Kilingo, F. M., Bernard, Z., -Bin, C. H., & Mkangombe, K. F. (2021). The analysis of wastewater treatment system efficiencies in Kenya: A review. *International Journal of Scientific and Research Publications, 11*(5), 204–215. https://doi.org/10.29322/IJSRP.11.05.2021.p11322

Kim, B. J., Alleman, J. E., Gee, C. S., & Bandy, J. T. (1991). *Use of programmable logic controllers to automate control and monitoring of U.S. Army wastewater treatment systems*. US Army Corps of Engineers-Construction Engineering Research Laboratory [USACERL Technical Report N-91/27]. https://apps.dtic.mil/sti/tr/pdf/ADA241372.pdf

Kim, K. (2008). *The characteristic of the sequencing batch reactor (SBR), anaerobic sequencing batch reactor (ASBR) and sequencing batch biofilm reactor (SBBR)* (pp. 1–10). https://www.engineering.iastate.edu/~tge/ce421-521/koobum.pdf

Kim, M., Chul, P., Kim, W., & Cui, F. (2022). Application of data smoothing and principal component analysis to develop a parameter ranking system for the anaerobic digestion process. *Chemosphere, 299*(134444). https://doi.org/10.1016/j.chemosphere.2022.134444

Kong, Z., Li, L., Xue, Y., Yang, M., & Li, Y.-Y. (2019). Challenges and prospects for the anaerobic treatment of chemical-industrial organic wastewater: A review. *Journal of Cleaner Production, 231*, 913–927. https://doi.org/10.1016/j.jclepro.2019.05.233

Kothari, R., Azam, R., Singh, H. M., Kumar, P., Kumar, V., Singh, R. P., & Tyagi, V. V. (2024). Nutrients sequestration from slaughterhouse wastewater with kinetic model studies using *C. vulgaris* for lipid production and reduction in freshwater footprint: A synergistic approach. *Waste and Biomass Valorization, 15*, 1807–1818. https://doi.org/10.1007/s12649-023-02226-0

Kovács, E. (2016). Biomethane injection into the national grid-Lille, France. In L. Lavrič, L. Nordin, W. Rogalski, B. Barbarisi, & N. Merkel (Eds.), *Success stories: Anaerobic digestion of biodegradeable municipal solid waste in European Cities* (pp. 1–8). European Biogas Association (EBA). https://www.europeanbiogas.eu/wp-content/uploads/2016/03

Kozak, T., & Majchrzycka, A. (2009). Application of biogas for combined heat and power production in the rural region. In *Workshop: "Solar Systems" by Faculty of Mathematics & Natural Science—[FMNS 2009]* (pp. 184–189). https://inis.iaea.org/collection/NCLCollectionStore/_Public/47/013/47013704.pdf

Krajnakova, E., Svazas, M., & Navickas, V. (2019). Biomass blockchain as a factor of energetical sustainability development. *Entrepreneurship and Sustainability Issues, 6*(3), 1456–1467. https://doi.org/10.9770/jesi.2019.6.3(28)

Kumar, R., Thakur, R., & Chatterji, S. (2014). PLC based automation of biogas plant. *International Journal of Advanced Research (IJAR), 2*(9), 314–318. https://www.journalijar.com/uploads/952_IJAR-4015.pdf

Kumari, A., Maurya, N. S., Kumar, A., Yadav, R. K., & Kumar, A. (2023). Options for the disposal and reuse of wastewater sludge, associated benefit, and environmental risk. In B. K. Taşeli (Ed.), *Sewage management* (pp. 1–15). InTech Open. https://doi.org/10.5772/intechopen.109410

Laaksonen, P., Sinkkonen, A., Zaitsev, G., Mäkinen, E., Grönroos, T., & Romantschuk, M. (2017). Treatment of municipal wastewater in full-scale on-site sand filter reduces BOD efficiently but does not reach requirements for nitrogen and phosphorus removal. *Environmental Science and Pollution Research, 24*, 11446–11458. https://doi.org/10.1007/s11356-017-8779-x

Lam, K. L., Zlatanović, L., & Van der Hoek, J. P. (2020). Life cycle assessment of nutrient recycling from wastewater: A critical review. *Water Research, 173*(115519), 1–16. https://doi.org/10.1016/j.watres.2020.115519

Lamb, J. J., Bernard, O., Sarker, S., Lien, K. M., & Hjelme, D. R. (2019). Perspectives of optical colourimetric sensors for anaerobic digestion. *Renewable and Sustainable Energy Reviews, 111*, 87–96. https://doi.org/10.1016/j.rser.2019.04.083

Lansing, S., Maile-Moskowitz, A., & Eaton, A. (2017). Waste treatment and energy production from small-scale wastewater digesters. *Bioresource Technology, 245*(Part A), 801–809. https://doi.org/10.1016/j.biortech.2017.08.215

Lettinga, G., Rebac, S., Parshina, S., Nozhevnikova, A., van Lier, J. B., & Stams, A. J. M. (1999). High-rate anaerobic treatment of wastewater at low temperatures. *Applied and Environmental Microbiology, 65*(4), 1696–1702.

Levavasseur, F., Martin, L., Boros, L., Cadiou, J., Carozzi, M., Martin, P., & Houot, S. (2023). Land cover changes with the development of anaerobic digestion for biogas production in France. *GCB-Bioenergy, 15*(5), 630–641. https://doi.org/10.1111/gcbb.13042

Levey, M. (2016). Methods to control odors from anaerobic digester plants and prevent nuisance to nearby communities. *Guest Column*, X–XX. https://www.wateronline.com/doc

Li, B. (2011). *Optimization of acidogenic anaerobic wastewater treatment with the potential for water reclamation* [Thesis Project 2008CT177B]. https://ctiwr.media.uconn.edu/

Li, P., Wang, J., Peng, H., Li, Q., Wang, M., Yan, W., Boboua, S. Y. B., Li, W., Sun, Y., Zheng, G., & Zhang, H. (2022). The effect of heat pre-treatment on the anaerobic digestion of high-solid pig manure under high organic loading level. *Frontiers in Bioengineering and Biotechnology, 10*(972361). https://doi.org/10.3389/fbioe.2022.972361

Li, K., Xu, P., Chen, X., Li, P., & Pu, Y. (2022). A study of a composite biofilm reactor for the treatment of mariculture wastewater: Performance and microbial communities. *Sustainability, 14*(5743). https://doi.org/10.3390/su14105743

Li, Y., Kontos, G. A., Cabrera, D. V., Avila, N. M., Parkinson, T. W., Viswanathan, M. B., Singh, V., Altpeter, F., Labatut, R. A., & Guest, J. S. (2023). Design of a high-rate wastewater treatment process for energy and water recovery at biorefineries. *ACS Sustainable Chemistry & Engineering, 11*(9), 3861–3872. https://doi.org/10.1021/acssuschemeng.2c07139

Li, Y., Qi, C., Zhang, Y., Li, Y., Wang, Y., Li, G., & Luo, W. (2021). Anaerobic digestion of agricultural wastes from liquid to solid state: Performance and environ-economic comparison. *Bioresource Technology, 332*(125080), 1–10. https://doi.org/10.1016/j.biortech.2021.125080

Lindmark, J., Thorin, E., Fdhila, R. B., & Dahlquist, E. (2014). Effects of mixing on the result of anaerobic digestion: Review. *Renewable and Sustainable Energy Reviews, 40*, 1030–1047. https://doi.org/10.1016/j.rser.2014.07.182

Lingaraju, A. K., Niranjanamurthy, M., Bose, P., Acharya, B., Gerogiannis, V. C., Kanavos, A., & Manika, S. (2023). IoT-based waste segregation with location tracking and air quality monitoring for smart cities. *Smart Cities, 6*, 1507–1522. https://doi.org/10.3390/smartcities6030071

Lippman, L., Perez, L., & Lopez, S. R. (2023). Fifty-year review of effluent discharge trends to the Southern California Bight from large municipal wastewater treatment facilities from 1971 to 2020. *Water, 15*(3876). https://doi.org/10.3390/w15223876

Lisiak-Zielińska, M., Jałoszyńska, S., Borowiak, K., Budka, A., & Dach, J. (2023). Perception of biogas plants: A public awareness and preference—A case study for the agricultural landscape. *Renewable Energy, 217*(119212), 1–10. https://doi.org/10.1016/j.renene.2023.119212

Litvinov, V., Daumova, G., Shaikhov, M., & Sergeyeva, N. (2022). Analysis of the composition of municipal wastewater sludge from small settlements in East Kazakhstan. *Journal of Ecological Engineering (JEE), 23*(7), 105–112. https://doi.org/10.12911/22998993/149896

Liu, J., Liu, T., Chen, S., Yu, H., Zhang, Y., & Quan, X. (2020). Enhancing anaerobic digestion in anaerobic integrated floating fixed-film activated sludge (An-IFFAS) system using novel electron mediator suspended biofilm carriers. *Water Research, 175*(115697). https://doi.org/10.1016/j.watres.2020.115697

Liu, X., Chen, Q., & Zhu, L. (2016). Improving biodegradation potential of domestic wastewater by manipulating the size distribution of organic matter. *Journal of Environmental Sciences, 47*, 174–182. https://doi.org/10.1016/j.jes.2016.02.004

Liu, Y., Zhang, Y., Quan, X., Li, Y., Zhao, Z., Meng, X., & Chen, S. (2012). Optimization of anaerobic acidogenesis by adding FeO powder to enhance anaerobic wastewater treatment. *Chemical Engineering Journal, 192*, 179–185. https://doi.org/10.1016/j.cej.2012.03.044

Loboichenko, V., Iranzo, A., Casado-Manzano, M., Navas, S. J., Pino, F. J., & Rosa, F. (2024). Study of the use of biogas as an energy vector for microgrids. *Renewable and Sustainable Energy Reviews, 200*(114574), 1–15. https://doi.org/10.1016/j.rser.2024.114574

Logan, M., Safi, M., Lens, P., & Visvanathan, C. (2019). Investigating the performance of internet of things based anaerobic digestion of food waste. *Process Safety and Environmental Protection, 127*, 277–287. https://doi.org/10.1016/j.psep.2019.05.025

Lohani, S. P., & Havukainen, J. (2018). Anaerobic digestion: Factors affecting anaerobic digestion process. In S. Varjani, E. Gnansounou, B. Gurunathan, D. Pant, & Z. Zakaria (Eds.), *Waste bioremediation* (pp. 343–359). Springer, Singapore. https://doi.org/10.1007/978-981-10-7413-4_18

Lohani, S. P., Chhetri, A., & Khanal, S. N. (2015). A simple anaerobic system for onsite treatment of domestic wastewater. *African Journal of Environmental Science and Technology, 9*(4), 292–300. https://doi.org/10.5897/AJEST2014.1848

Lopez-Arenas, T., Pulis, A., Mulas, M., & Baratti, R. (2004). A software sensor for a wastewater treatment plant. *IFAC Proceedings Volumes, 37*(1), 251–256. https://doi.org/10.1016/S1474-6670(17)38740-2

López-Estrada, F. R., Gonzáles, J. H. C., Ponsart, J. C., Theilliol, D., & Astorga-Zaragoza, C. M. (2015). Robust sensor fault detection and isolation of an anerarobic bioreactor modeled as a descriptor-LPV system. *IFAC-Papers OnLine, 48*(26), 176–181. https://doi.org/10.1016/j.ifacol.2015.11.133

Lu, J., Li, D., Chen, L., Kitamura, Y., Jiang, W., & Li, B. (2013). Simultaneous pretreatment and acidogenesis of solid food wastes by a rotational drum fermentation system with methanogenic leachate recirculation and andesite porphyry addition. *Bioresource Technology, 138*, 101–108. https://doi.org/10.1016/j.biortech.2013.03.166

Lu, M., Zhou, C., Wang, C., Jackson, R. B., & Kempes, C. P. (2022). *Worldwide scaling of waste generation in urban systems* (pp. 1–26). https://doi.org/10.48550/arXiv.2208.07917

Luo, H., Sun, Y., Yin, Y., & Wang, Z.-W. (2022). Odor generation, identification, and control during anaerobic digestion of organic solid wastes. In *Advances in bioenergy* (Vol. 7, pp. 139–165). Elsevier. https://doi.org/10.1016/bs.aibe.2022.04.002

Luste, S., & Luostarinen, S. (2010). Anaerobic co-digestion of meat-processing by-products and sewage sludge—Effect of hygienization and organic loading rate. *Bioresource Technology, 101*(8), 2657–2664. https://doi.org/10.1016/j.biortech.2009.10.071

Lv, Y. (2023). Transitioning to sustainable energy: Opportunities, challenges, and the potential of blockchain technology. *Frontiers in Energy Research, 11*, 1–20. https://doi.org/10.3389/fenrg.2023.1258044

Mabonga, M. K., Maingi, S. M., & Kirubi, G. (2020). Wastewater quality parameters and methane production in Dandora Estate Sewerage Treatment Works, Nairobi City County, Kenya. *International Journal of Environmental Sciences, 3*(1), 1–20. https://www.iprjb.org/journals/index.php/IJES/article/view/1032

Macura, B., Johannesdottir, S. L., Piniewski, M., & Haddaway, N. R. (2019). Effectiveness of ecotechnologies for recovery of nitrogen and phosphorus from anaerobic digestate and effectiveness of the recovery products as fertilisers: A systematic review protocol. *Environmental Evidence, 8*(29), 1–9. https://doi.org/10.1186/s13750-019-0173-3

Madukasi, E. I., & Zhang, G. (2010). Microaerobic biodegradation of high organic load wastewater by phototrophic bacteria. *African Journal of Biotechnology, 9*(25), 3852–3860. http://www.academicjournals.org/AJB

Magdalena, J. A., Greses, S., & González-Fernández, C. (2019). Impact of organic loading rate in volatile fatty acids production and population dynamics using microalgae biomass as substrate. *Scientific Reports, 9*(18374), 1–11. https://doi.org/10.1038/s41598-019-54914-4

Mahendra, B. G., & Patil, S. (2013). Anaerobic digestion of domestic wastewater. *International Journal of Research in Engineering and Technology (IJRET)*, 283–287. https://ijret.org/volumes/2013v02/i13/IJRET20130213051.pdf

Mahmoud, A. S., Farag, R. S., & Elshfai, M. M. (2020). Reduction of organic matter from municipal wastewater at low cost using green synthesis nano iron extracted from black tea: Artificial intelligence with regression analysis. *Egyptian Journal of Petroleum, 29*(1), 9–20. https://doi.org/10.1016/j.ejpe.2019.09.001

Mainardis, M., Buttazzoni, M., & Goi, D. (2020). Up-flow Anaerobic Sludge Blanket (UASB) technology for energy recovery: A review on state-of-the-art and recent technological advances. *Bioengineering, 7*(43). https://doi.org/10.3390/bioengineering7020043

Makhtar, M. M. Z., Don, M. M., & Tajarudin, H. A. (2018). Microbial Fuel Cell (MFC) development from anaerobic digestion system. In N. Horan, A. Z. Yaser, & N. Wid (Eds.), *Anaerobic digestion processes-green energy and technology* (pp. 9–31). Springer, Singapore. https://doi.org/10.1007/978-981-10-8129-3_2

Makowska, M., & Spychała, M. (2014). Organic compounds fractionation for domestic wastewater treatment modeling. *Polish Journal of Environmental Studies, 23*(1), 131–137. https://www.pjoes.com/

Maleki-Ghelichi, E., & Sharifi, M. (2017). Prioritize and choose the best process of anaerobic digestion to produce energy from biomass using analytic hierarchy process (AHP). *Geology Ecology and Landscapes, 1*(1), 1–6. https://doi.org/10.1080/24749508.2017.1389451

Mandhadi, K., & Iyer, R. (2012). Remote control monitoring of bioreactors. In *2012 ASEE annual conference & exposition [AC 2012-4815]* (pp. 1–10). https://doi.org/10.18260/1-2--21872

Martínez, R., Vela, N., El Aatik, A., Murray, E., Roche, P., & Navarro, J. M. (2020). On the use of an IoT integrated system for water quality monitoring and management in wastewater treatment plants. *Water, 12*(1096), 1–21. https://doi.org/10.3390/w12041096

Martínez-Sibaja, A., Alvarado-Lassman, A., Astorga-Zaragoza, C. M., Adam-Medina, M., Posada-Gómez, R., & Rodríguez-Jarquin, J. P. (2011). Volumetric gas meter for laboratory-scale anaerobic bioreactors. *Measurement, 44*(10), 1801–1805. https://doi.org/10.1016/j.measurement.2011.08.018

Maruthi, Y. A., Das, N. L., Hossain, K., Sarma, K. S. S., Rawat, K. P., & Sabharwal, S. (2011). Disinfection and reduction of organic load of sewage water by electron beam radiation. *Applied Water Science, 1*, 49–56. https://doi.org/10.1007/s13201-011-0008-z

Matheson, R. (2016). In D. Burton (Ed.), *Reduction of the organic load of industrial wood preserving wastewater via chemical oxidation* [Degree of Master of Science, Dalhousie University]. http://hdl.handle.net/10222/72285

Mathur, R., Sharma, M. K., Loganathan, K., Abbas, M., Hussain, S., Kataria, G., Alqahtani, M. S., & Rao, K. S. (2024). Modeling of two-stage anaerobic onsite wastewater sanitation system to predict effluent soluble chemical oxygen demand through machine learning. *Scientific Reports, 14*(1835), 1–14. https://www.nature.com/articles/s41598-023-50805-x

Mazza, S. A., Abdurahman, H. N., Said, N., Azwina, Z., & Mohamed, B. S. (2014). Optimization of ultrasonicated membrane anaerobic system (UMAS) for sewage sludge treatment using response surface methodology (RSM). *International Journal of Research in Engineering and Technology (IJRET), 3*(2), 1–5. https://ijret.org/volumes/2014v03/i02/IJRET20140302001.pdf

Mazzanti, M., Modica, M., & Rampa, A. (2021). *The biogas dilemma: An analysis on the social approval of large new plants* (02/2021; SEEDS Working Paper). https://www.sustainability-seeds.org/papers/RePec/srt/wpaper/0221.pdf

McDonnell, D., Burke, M., Dowdall, J., Foster, P., & Mahon, K. (2018). In M. Cremin (Ed.), *Guidelines for anaerobic digestion in Ireland* (pp. 1–72). Cré-Composting & Anaerobic Digestion Association of Ireland. http://www.cre.ie/

Meidiana, C., Ramadhani, Z. G., & Dinanti, D. (2017). Measuring public acceptance value of rural biogas development through logistic regression and willingness to pay. In S. Sabah (Ed.), *Consumer behavior—Practice oriented perspectives* (pp. 67–88). InTech Open. https://doi.org/10.5772/intechopen.69191

Meidiana, C., Hidayah, A., & Wijayanti, W. P. (2019). Factors influencing the biogas acceptance in rural area. In *IOP conference series: Earth and environmental science [ICRESBE 2019]* (pp. 1–6). https://doi.org/10.1088/1755-1315/520/1/012006

Menacho, W. A., Mazid, A. M., & Das, N. (2022). Modelling and analysis for biogas production process simulation of food waste using Aspen Plus. *Fuel, 309*(122058). https://doi.org/10.1016/j.fuel.2021.122058

Mensah, K. A., & Forster, C. F. (2003). An examination of the effects of detergents on anaerobic digestion. *Bioresource Technology, 90*(2), 133–138. https://doi.org/10.1016/S0960-8524(03)00126-3

Menzel, T., Neubauer, P., & Junne, S. (2020). Role of microbial hydrolysis in anaerobic digestion. *Energies, 13*(5555). https://doi.org/10.3390/en13215555

Mertins, A., & Wawer, T. (2022). How to use biogas?: A systematic review of biogas utilization pathways and business models. *Bioresources and Bioprocessing, 9*(59), 1–18. https://doi.org/10.1186/s40643-022-00545-z

Meyer, T., & Edwards, E. A. (2014). Anaerobic digestion of pulp and paper mill wastewater and sludge. *Water Research, 65*, 321–349. https://doi.org/10.1016/j.watres.2014.07.022

Mikkonen, L., Rämö, J., Keiski, R. L., & Pongrácz, E. (2013). Heat recovery from wastewater: Assessing the potential in northern areas. In *Conference: Water research at the University of OuluAt: Oulu, Finland* (pp. 161–164). https://doi.org/10.13140/2.1.1728.8326

Miruka, A. C., Kariuki, D. K., Yusuf, A. O., & Onyatta, J. O. (2018). Efficiency of a wastewater treatment plant in the City of Nairobi. *International Journal of Sciences: Basic and Applied Research (IJSBAR), 38*(1), 152–162. https://www.gssrr.org/index.php/JournalOfBasicAndApplied/article/view/8751

Moedinger, F., Ast, F., Ragazzi, M., Foladori, P., Rada, E. C., & Binnig, R. (2012). Innovative biogas multi-stage biogas plant and novel analytical system. *Energy Procedia, 18*, 672–680. https://doi.org/10.1016/j.egypro.2012.05.082

Moeller, L., & Görsch, K. (2015). Foam formation in full-scale biogas plants processing biogenic waste. *Energy, Sustainability and Society, 5*(1). https://doi.org/10.1186/s13705-014-0031-7

Mohan, S. V. (2008). Fermentative hydrogen production with simultaneous wastewater treatment: Influence of pretreatment and system operating conditions. *Journal of Scientific & Industrial Research, 67*(11), 950–961. https://www.researchgate.net/publication/228680756

Morán, A., Alonso, M. E., Carrillo-Peña, D., & Gómez, X. (2023). Feasibility of integrating small-scale anaerobic digestion in urban areas: Analyzing energy efficiency. In A. Sayigh (Ed.), *Mediterranean architecture and the green-digital transition* (pp. 453–466). Springer, Cham. https://doi.org/10.1007/978-3-031-33148-0_35

Morero, B., Vicentin, R., & Campanella, E. A. (2017). Assessment of biogas production in Argentina from co-digestion of sludge and municipal solid waste. *Waste Management, 61*, 195–205. https://doi.org/10.1016/j.wasman.2016.11.033

Mou, A., Yu, N., Yang, X., & Liu, Y. (2024). Enhancing methane production and organic loading capacity from high solid-content wastewater in modified granular activated carbon (GAC)-amended up-flow anaerobic sludge blanket (UASB). *Science of The Total Environment, 906*(167609). https://doi.org/10.1016/j.scitotenv.2023.167609

Mouele, E. S. M., Tijani, J. O., Badmus, K. O., Pereao, O., Babajide, O., Zhang, C., Shao, T., Sosnin, E., Tarasenko, V., Fatoba, O. O., Laatikainen, K., & Petrik, L. F. (2021). Removal of pharmaceutical residues from water and wastewater using dielectric barrier discharge methods—A review. *International Journal of Environmental Research and Public Health, 18*(1683). https://doi.org/10.3390/ijerph18041683

Mould, K., Da Silva, F. R., Knott, S. F., & Regan, B. O. (2022). A comparative analysis of biogas and hydrogen, and the impact of the certificates and blockchain new paradigms. *International Journal of Hydrogen Energy, 47*(93), 1–16. https://doi.org/10.1016/j.ijhydene.2022.09.107

MPCA. (2023). *Regulatory requirements for anaerobic digesters*. Minnesota Pollution Control Agency (MPCA). https://www.pca.state.mn.us/sites/default/files/wq-f8-54.pdf

Mukawa, J., Pajak, T., Rzepecki, T., & Banas, M. (2022). Energy potential of biogas from sewage sludge after thermal hydrolysis and digestion. *Energies, 15*(5255), 1–15. https://doi.org/10.3390/en15145255

Mulat, D. G., Dibdiakova, J., & Horn, S. J. (2018). Microbial biogas production from hydrolysis lignin: Insight into lignin structural changes. *Biotechnology for Biofuels, 11*(61), 1–16. https://doi.org/10.1186/s13068-018-1054-7

Muñoz, R., Meier, L., Díaz, I., & Jeison, D. (2015). A review on the state-of-the-art of physical/chemical and biological technologies for biogas upgrading. *Reviews in Environmental Science and Bio/technology, 14*, 727–759. https://doi.org/10.1007/s11157-015-9379-1

Musolino, V. M. (2016). In A. Bertucco, E. Sforza, & E. Barbera (Eds.), *Anaerobic digestion for nutrient recovery in industrial microalgae cultivation: Experiments and process simulation* [Master's Degree Course Thesis in Chemical and Industrial Process Engineering, Department of Industrial Engineering, University of Padova]. https://core.ac.uk/reader/41988210

Mwaniki, J., Mwazighe, F. M., Gitura, D. W., Gakuo, S. M., & Kamau, G. N. (2021). Useful products from microalgae: Case study of Ruai Wastewater Treatment Plant in Nairobi. *Journal Kenya Chemical Society, 9*(1), 15–34. http://erepository.uonbi.ac.ke:8080/handle/11295/100678

Nagpal, H., Spriet, J., Murali, M. K., & McNabola, A. (2021). Heat recovery from wastewater—A review of available resource. *Water, 13*(1274). https://doi.org/10.3390/w13091274

Naguib, A. M., Abdel-Gawad, S. A., & Mahmoud, A. S. (2024). Reduction of organic contaminants from industrial effluent using the advanced oxidation process, chemical coagulation, and green nanotechnology. *Scientific Reports, 14*(15221), 1–13. https://doi.org/10.1038/s41598-024-65162-6

Nallamothu, R. B., Teferra, A., & Rao, B. V. A. (2013). Biogas purification, compression and bottling. *Global Journal of Engineering, Design & Technology, 2*(6), 34–38. https://www.build-a-biogas-plant.com/PDF/BIOGASCompression_gjedt_2013.PDF

Nan, L. B., Trably, E., Santa-Catalina, G., Bernet, N., Delgenès, J.-P., & Escudié, R. (2020). Biomethanation processes: New insights on the effect of a high H_2 partial pressure on microbial communities. *Biotechnology for Biofuels, 13*(141), 1–17.

Nandy, T., & Kaul, S. N. (2001). Anaerobic pre-treatment of herbal-based pharmaceutical wastewater using fixed-film reactor with recourse to energy recovery. *Water Research, 35*(2), 351–362. https://doi.org/10.1016/S0043-1354(00)00263-3

Narayanan, C. M., & Narayan, V. (2019). Biological wastewater treatment and bioreactor design: A review. *Sustainable Environment Research, 29*(33), 1–17. https://doi.org/10.1186/s42834-019-0036-1

Newsom, G., West, A., & Dunbar, J. (2022). In A. Edalati, K. Uy, J. Steinbuck, & D. Bohan (Eds.), *Optimizing food waste anaerobic digestion at water resource recovery facilities*. https://www.energy.ca.gov/sites/default/files/2022-10/CEC-500-2022-016.pdf

Ngema, L., Madondo, I. N., Mguni, L. L., Sathiyah, D., Tetteh, E. K., & Rathilal, S. (2023). Optimizing anaerobic digestion of a local South African municipal wastewater using Box-Behnken Design of Response Surface Methodology. In *39th Johannesburg international conference on "chemical, biological and environmental engineering" (JCBEE-23) Nov. 16–17, 2023* (pp. 45–49). https://doi.org/10.17758/IICBE5.C1123020

Nguyen, B., Bui, X.-T., Hien, V. T. D., Cao, T. N.-D., Dang, B.-T., Tra, V.-T., Tran, H. T., Luu, T. Le, & Ngo, H. H. (2020). Anaerobic membrane bioreactors for industrial wastewater treatment. In H. H. Ngo, W. Guo, H. Y. Ng, G. Mannina, & A. Pandey (Eds.), *Current developments in biotechnology and bioengineering: advanced membrane separation processes for sustainable water and wastewater management—Anaerobic membrane bioreactor processes and technologies* (pp. 167–196). Elsevier. https://doi.org/10.1016/B978-0-12-819852-0.00007-5

Nguyen, L. N., Kumar, J., Vu, M. T., Mohammed, J. A. H., Pathak, N., Commault, A. S., Sutherland, D., Zdarta, J., Tyagi, V. K., & Nghiem, L. D. (2021). Biomethane production from anaerobic co-digestion at wastewater treatment plants: A critical review on development and innovations in biogas upgrading techniques. *Science of The Total Environment, 765*(142753). https://doi.org/10.1016/j.scitotenv.2020.142753

Nijman, J. (2022). *Support schemes for biogas and biomethane in Flanders*. Biogas-E. https://www.biogas-e.be/englishsupportscheme

Nwazor, N. O., Ibrahim, N., & Seun, O. K. (2019). Development of a smart, automated waste management system. *European Journal of Engineering and Technology, 7*(2), 46–55. https://www.idpublications.org/

O'Donovan, P., Coburn, D., Jones, E., Hannon, L., Glavin, M., Mullins, D., & Clifford, E. (2015). A cloud-based distributed data collection system for decentralised wastewater treatment plants. *Procedia Engineering, 119*(1), 464–469. https://doi.org/10.1016/j.proeng.2015.08.865

O'Dwyer, E., Chen, K., Wang, H., Wang, A., Shah, N., & Guo, M. (2020). Optimisation of wastewater treatment strategies in eco-industrial parks: Technology, location and transport. *Chemical Engineering Journal, 381*(122643). https://doi.org/10.1016/j.cej.2019.122643

Obaideen, K., Abdelkareem, M. A., Wilberforce, T., Elsaid, K., Sayed, E. T., Maghrabie, H. M., & Olabi, A. G. (2022). Biogas role in achievement of the sustainable development goals: Evaluation, challenges, and guidelines. *Journal of the Taiwan Institute of Chemical Engineers, 131*(104207). https://doi.org/10.1016/j.jtice.2022.104207

Offie, I., Piadeh, F., Behzadian, K., Alani, A. M., Yaman, R., & Campus, L. (2022). Real-time monitoring of decentralised anaerobic digestion using artificial intelligence-based framework. In *Conference: Real-time operation of decentralised anaerobic digestion using artificial intelligence framework* (pp. 1–9). http://repository.uwl.ac.uk/id/eprint/9368/

Offie, I. C. (2023). In K. Behzadian, J. Rizzuto, & L. Campos (Eds.), *Development of artificial intelligence systems for anaerobic digestion operations* [Degree of Doctor of Philosophy Thesis, University of West London]. https://doi.org/10.36828/xvqy2136

Offie, I., Piadeh, F., Behzadian, K., Campos, L. C., & Yaman, R. (2023). Development of an artificial intelligence-based framework for biogas generation from a micro anaerobic digestion plant. *Waste Management, 158*, 66–75. https://doi.org/10.1016/j.wasman.2022.12.034

Olawade, D. B., Fapohunda, O., Wada, O. Z., Usman, S. O., Ige, A. O., Ajisafe, O., & Oladapo, B. I. (2024). Smart waste management: A paradigm shift enabled by artificial intelligence. *Waste Management Bulletin, 2*(2), 244–263. https://doi.org/10.1016/j.wmb.2024.05.001

Olayanju, O. K., Oladepo, K. T., Azeez, O., Adebusuyi, A. O., & Obisakin, O. (2024). Review of anaerobic baffled reactor for treatment of industrial wastewater in Nigeria. In B. K. Taşeli, E. Jacob-Lopes, L. Q. Zepka, & M. C. Deprá (Eds.), *Wastewater Treatment—Past and future perspectives* (pp. 1–15). InTech Open. https://doi.org/10.5772/intechopen.1004565

Oliveira, S. M. A. C., Parkinson, J. N., & von Sperling, M. (2006). Wastewater treatment in Brazil: Institutional framework, recent initiatives and actual plant performance. *International Journal of Technology Management & Sustainable Development, 5*(3), 241–256. https://doi.org/10.1386/ijtm.5.3.241/1

Onu, P., Mbohwa, C., & Pradhan, A. (2023). Artificial intelligence-based IoT-enabled biogas production. *International Conference on Control, Automation and Diagnosis (ICCAD 2023), Rome, Italy*. https://doi.org/10.1109/ICCAD57653.2023.10152349

Owuor, O. B. (2020). In R. K. Kimilu & A. A. Aganda (Eds.), *Use of biogas from brewery waste water: A case study of Kenya Breweries Limited* [Degree of Master of Science in Energy Management, Department of Mechanical Engineering, University of Nairobi]. https://erepository.uonbi.ac.ke/handle/11295/152940

Palatsi, J., Viñas, M., Guivernau, M., Fernandez, B., & Flotats, X. (2011). Anaerobic digestion of slaughterhouse waste: Main process limitations and microbial community interactions. *Bioresource Technology, 102*(3), 2219–2227. https://doi.org/10.1016/j.biortech.2010.09.121

Pan, S., Wen, C., Liu, Q., Chi, Y., Mi, H., Li, Z., Du, L., Huang, R., & Wei, Y. (2019). A novel hydraulic biogas digester controlling the scum formation in batch and semi-continuous tests using banana stems. *Bioresource Technology, 286*(121372), 1–9. https://doi.org/10.1016/j.biortech.2019.121372

Panchenko, V. A., Kovalev, A. A., Kovalev, D. A., & Litty, Y. V. (2023). Review of modern methods and technologies for using of solar energy in the operation of anaerobic digestion systems. *International Journal of Hydrogen Energy, 48*(53), 20264–20278. https://doi.org/10.1016/j.ijhydene.2023.02.109

Pang, N., Bergeron, A. D., Gu, X., Fu, X., Dong, T., Yao, Y., & Chen, S. (2020). Recycling of nutrients from dairy wastewater by extremophilic microalgae with high ammonia tolerance. *Environmental Science and Technology, 54*(23), 15366–15375. https://doi.org/10.1021/acs.est.0c02833

Pani, D. F., Dos Reis, J. A. T., Rodrigues, M. B., Mendonça, A. S. F., Marques, S. M., & Da Silva, F. G. B. (2024). Prospects for reducing the contribution of organic load in a water basin with significant urban occupation. *Engenharia Sanitária e Ambiental, 29*, 1–9. https://doi.org/10.1590/S1413-415220230079

Parawira, W. (2004). In A. K. Kivaisi (Ed.), *Anaerobic treatment of agricultural residues and wastewater—Application of high-rate reactors* [Media-Tryck, Lund University]. https://lup.lub.lu.se/search/ws/files/5606523/1472236.pdf

Parawira, W., Kudita, I., Nyandoroh, M. G., & Zvauya, R. (2005). A study of industrial anaerobic treatment of opaque beer brewery wastewater in a tropical climate using a full-scale UASB reactor seeded with activated sludge. *Process Biochemistry, 40*(2), 593–599. https://doi.org/10.1016/j.procbio.2004.01.036

Pastor, D. J. (2020). The effects of renewables portfolio standards on renewable energy generation. *Economics Bulletin, 40*(3), 2121–2133. https://ideas.repec.org/a/ebl/ecbull/eb-19-01081.html

Pennington, M. (2019). In D. L. Forman, A. Milbrandt, C. Carusiello, C. Fabiano, E. Goldstein, E. Meyer, & A. Schnitzer (Eds.), *Anaerobic digestion facilities processing food waste in the United States in 2015*. https://www.epa.gov/sites/default/files/2018-08/documents/ad_data_report_final_508_compliant_no_password.pdf

Pereda, R. O. C., Rivera-Muñoz, E. M., & Ruiz, G. H. (2010). Automatic volumetric gas flow meter for monitoring biogas production from laboratory-scale anaerobic digester. *Sensors and Actuators B Chemical, 147*(1), 10–14. https://doi.org/10.1016/j.snb.2010.03.053

Pereira, C. P., Goldenstein, J. P. N., & Bassin, J. P. (2021). Industrial wastewater contaminants and their hazardous impacts. In R. Selvasembian & P. Singh (Eds.), *Biosorption for wastewater contaminants*. https://doi.org/10.1002/9781119737629.ch1

Pereira, J., De Melo, M. M. R., Silva, C. M., Lemos, P. C., & Serafim, L. S. (2022). Impact of a pretreatment step on the acidogenic fermentation of spent coffee grounds. *Bioengineering, 9*(362). https://doi.org/10.3390/bioengineering9080362

Petre, E., Selisteanu, D., & Sendrescu, D. (2013). Adaptive and robust-adaptive control strategies for anaerobic wastewater treatment bioprocesses. *Chemical Engineering Journal, 217*(5), 363–378. https://doi.org/10.1016/j.cej.2012.11.129

Pfau, S. F., Hagens, J. E., & Hagens, J. E. (2017). Biogas between renewable energy and bio-economy policies—Opportunities and constraints resulting from a dual role. *Energy, Sustainability and Society, 7*(17), 1–15. https://doi.org/10.1186/s13705-017-0120-5

Picavet, M., & Alves, M. (2010). *A compact high-rate anaerobic reactor configuration for the treatment of effluents with high lipid content* (pp. 1–8). https://core.ac.uk/download/pdf/55627563.pdf

Pilarska, A. A., Kulupa, T., Kubiak, A., Wolna-Maruwka, A., Pilarski, K., & Niewiadomska, A. (2023). Anaerobic digestion of food waste—A short review. *Energies, 15*(5742), 1–23. https://doi.org/10.3390/en16155742

Power, A. J. (2023). *Green bitcoin mining: Harnessing methane from animal waste for a sustainable crypto future*. Compass Mining Inc. https://education.compassmining.io/education

Prabakar, D., Subha, S. K., Manimudi, V. T., Mathimani, T., Kumar, G., Rene, E. R., & Pugazhendhi, A. (2018). Pretreatment technologies for industrial effluents: Critical review on bioenergy production and environmental concerns. *Journal of Environmental Management, 218*, 165–180. https://doi.org/10.1016/j.jenvman.2018.03.136

Pradt, L. A., & Meidl, J. A. (1993). *Purification of non-biodegradable industrial wastewaters* (US3977966A). https://patents.google.com/patent/US3977966A/en

Pratap, V., Kumar, S., & Yadav, B. R. (2024). Optimization of biogas production from thermal-alkali pre-treated sludge using response surface methodology and random forest regressor model. *Journal of the Taiwan Institute of Chemical Engineers, 105571*. https://doi.org/10.1016/j.jtice.2024.105571

Qian, J., Xu, C., Song, H., Zhou, W., Toda, T., Li, H., Takayama, Y., Sekine, M., Koga, S., Li, J., & Liu, J. (2023). Enhancing algal growth and nutrient recovery from anaerobic digestion piggery effluent by an integrated pretreatment strategy of ammonia stripping and flocculation. *Frontiers in Bioengineering and Biotechnology, 11*(1219103), 1–8. https://doi.org/10.3389/fbioe.2023.1219103

Qiu, L., Wang, L., Yang, P., Jia, H., & Chen, L. (2005). *Study on expert system for the fault diagnosis of rural biogas project*. https://www.researchgate.net/publication/290164285

Qu, J., Sun, Y., Awasthi, M. K., Liu, Y., Xu, X., Meng, X., & Zhang, H. (2021). Effect of different aerobic hydrolysis time on the anaerobic digestion characteristics and energy consumption analysis. *Bioresource Technology, 320*(Part A). https://doi.org/10.1016/j.biortech.2020.124332

Qu, D. (2022). Application of artificial intelligence in waste classification management at University. In *Proceedings of the International Conference on Intelligent Vision and Computing (ICIVC 2021)* (pp. 330–343). https://doi.org/10.1007/978-3-030-97196-0_27

Raboni, M., & Urbini, G. (2014). Production and use of biogas in Europe: a survey of current status and perspectives. *Ambiente & Água - An Interdisciplinary Journal of Applied Science, 9*(2), 191–202. https://www.redalyc.org/pdf/928/92831117002.pdf

Rachbauer, L., Granda, C. B., Shrestha, S., Fuchs, W., Gabauer, W., Singer, S. W., Simmons, B. A., & Urgun-Demirtas, M. (2024). Energy and nutrient recovery from municipal and industrial waste and wastewater—A perspective. *Journal of Industrial Microbiology and Biotechnology*, 1–49. https://doi.org/10.1093/jimb/kuae040

Radcliffe, J. C., & Page, D. (2020). Water reuse and recycling in Australia—History, current situation and future perspectives. *Water Cycle, 1*, 19–40. https://doi.org/10.1016/j.watcyc.2020.05.005

Radetic, B. (2022). Municipal sludge thickening technologies. *Handbook of Water and Used Water Purification*, 1–25. https://doi.org/10.1007/978-3-319-66382-1_138-1

Rahimzadeh, A., Ranjbarrad, S., Ein-Mozaffari, F., & Lohi, A. (2024). Application of machine learning models in coaxial bioreactors: Classification and torque prediction. *ChemEngineering, 8*(42), 1–16. https://doi.org/10.3390/chemengineering8020042

Rahman, M., Yeoh, S. L., Tamiri, F. M., & Melvin, G. J. H. (2018). Anaerobic digestion of food waste. In N. Horan, A. Yaser, & N. Wid (Eds.), *Green energy and technology* (pp. 105–122). Springer Nature Singapore Pte Ltd. https://doi.org/10.1007/978-981-10-8129-3_7

Rajakumaran, G., Usharani, S., Vincent, C., & Sujatha, M. (2023). Smart waste management: Waste segregation using machine learning. *Recent Advances in Wireless Communications & Emerging Technologies (RAWCET 2022), 2471*(012030), 1–13. https://doi.org/10.1088/1742-6596/2471/1/012030

Raju, C. S., Løkke, M. M., Sutaryo, S., Ward, A. J., & Møller, H. B. (2012). NIR monitoring of ammonia in anaerobic digesters using a diffuse reflectance probe. *Sensors, 12*(2), 2340–2350. https://doi.org/10.3390/s120202340

Rana, R. S., Singh, P., Kandari, V., Dobhal, R., & Gupta, S. (2017). A review on characterization and bioremediation of pharmaceutical industries' wastewater: An Indian perspective. *Applied Water Science, 7*, 1–12. https://doi.org/10.1007/s13201-014-0225-3

Rasouli, M., Dini, M., & Ataeiyan, B. (2023). Anaerobic co-digestion of sewage sludge and Cladophora green algae: Investigation of synergistic effects and optimization of substrate composition ratio. *Environmental Engineering Research, 28*(5). https://doi.org/10.4491/eer.2021.516

Raudkivi, M., & Tenno, T. (2021). In M. Andreeva, D. Frank-Kamenetsky, L. Herman, A. Ilola, S. Rettig, K. Riiko, & K. Rosqvist (Eds.), *Palette of solutions for nutrient recycling in the Baltic Sea Region*. https://helcom.fi/wp-content/uploads/2021/07/

Ravendran, R. R., Abdulrazik, A., & Zailan, R. (2019). Aspen Plus simulation of optimal biogas production in anaerobic digestion process. In *1st ProSES symposium 2019 4 September 2019, Kuantan, Pahang, Malaysia* (Vol. 702, pp. 1–9). https://doi.org/10.1088/1757-899X/702/1/012001

Ravi, P. P., Lindner, J., Oechsner, H., & Lemmer, A. (2018). Effects of target pH-value on organic acids and methane production in two-stage anaerobic digestion of vegetable waste. *Bioresource Technology, 247*, 96–102. https://doi.org/10.1016/j.biortech.2017.09.068

Ray, N. H. S., Mohanty, M. K., & Mohanty, R. C. (2016). Biogas compression and storage system for cooking applications in rural households. *International Journal of Renewable Energy Research (IJRER), 6*(2), 593–598. https://www.ijrer.org/ijrer/index.php/ijrer/article/view/3449

Rea, J. (2014). In A. Slocum & A. Hosoi (Eds.), *Kinetic modeling and experimentation of anaerobic digestion* [Degree of Bachelor of Science Thesis in Mechanical Engineering, Department of Mechanical Engineering, Massachusetts Institute of Technology]. https://dspace.mit.edu/handle/1721.1/92070

Reardon, K. F. (2021). Practical monitoring technologies for cells and substrates in biomanufacturing. *Current Opinion in Biotechnology, 71*, 225–230. https://doi.org/10.1016/j.copbio.2021.08.006

Reymond, P., Abdel Wahaab, R., Moussa, M. S., & Lüthi, C. (2018). Scaling up small scale wastewater treatment systems in low- and middle-income countries: An analysis of challenges and ways forward through the case of Egypt. *Utilities Policy, 52*, 13–21. https://doi.org/10.1016/j.jup.2018.03.008

Rodero, M. R., Muñoz, R., González-Sánchez, A., Ruiz, H. A., & Quijano, G. (2024). Membrane materials for biogas purification and upgrading: Fundamentals, recent advances and challenges. *Journal of Environmental Chemical Engineering, 12*(5), 1–21. https://doi.org/10.1016/j.jece.2024.114106

Romero, M. P., Sánchez-Braza, A., Salvador-Ponce, J., & Sánchez-Labrador, N. (2017). An overview of feed-in tariffs, premiums and tenders to promote electricity from biogas in the EU-28. *Renewable and Sustainable Energy Reviews, 73*, 1366–1379. https://doi.org/10.1016/j.rser.2017.01.132

Romero-Güiza, M. S., Mata-Alvarez, J., Chimenos, J. M., & Astals, S. (2016). Nutrient recovery technologies for anaerobic digestion systems: An overview. *Revista ION, 29*(1), 7–26. https://doi.org/10.18273/revion.v29n1-2016001

Rosas-Mendoza, E. S., Méndez-Contreras, J. M., & Sibaja, A. M. (2018). Anaerobic digestion of citrus industry effluents using an anaerobic hybrid reactor. *Clean Technologies and Environmental Policy, 20*(3), 1–11. https://doi.org/10.1007/s10098-017-1483-1

Saddoud, A., & Sayadi, S. (2007). Application of acidogenic fixed-bed reactor prior to anaerobic membrane bioreactor for sustainable slaughterhouse wastewater treatment. *Journal of Hazardous Materials, 49*(3), 700–706. https://doi.org/10.1016/j.jhazmat.2007.04.031

Sadiku, M. N. O., Musa, S. M., & Musa, O. M. (2019). Big data in chemical engineering. *International Journal of Advances in Scientific Research and Engineering (IJASRE), 5*(6), 100–102. https://doi.org/10.31695/IJASRE.2019.33254

Saha, B., Kumar, S., Verma, D. K., Nag, A., Bhattacharya, P., Dutta, S. K., Kumar, V., Kumari, S., Anjum, M., Kumari, S., Saha, S., Sahay, S., Sathyanarayana, K., & Vats, A. K. (2024). Microbial biodegradation of the agricultural wastes for environmental sustainability. In S. P. Saha, D. Mazumdar, S. Roy, & P. Mathur (Eds.), *Agro-waste to microbe assisted value added product: Challenges and future prospects* (pp. 157–174). Springer, Cham. https://doi.org/10.1007/978-3-031-58025-3_7

Saha, B., Khwairakpam, M., & Kalamdhad, A. S. (2021). Thermal pre-treatment—A prerequisite for the reduction of hydrolysis stage during anaerobic digestion of *Ageratum conyzoides*. *Materials Science for Energy Technologies, 4*, 34–45. https://doi.org/10.1016/j.mset.2020.12.002

Sahm, H. (2005). Anaerobic wastewater treatment. In *Advances in biochemical engineering/biotechnology*. Springer, Berlin, Heidelberg. https://doi.org/10.1007/Bfb0000691

Saini, J. K., Lohchab, R. K., Nain, A., & Kumari, M. (2019). Designing of anaerobic hybrid reactor to treat household sewage for decentralised system. *Pollution Research, 38*(1), 204–210. https://www.researchgate.net/publication/331976765

Salem, R. M. M., Saraya, M. S., & Ali-Eldin, A. (2022). An industrial cloud-based IoT system for real-time monitoring and controlling of wastewater. *IEEE Access, XX*, 1–13. https://doi.org/10.1109/ACCESS.2022.3141977

Sam, A., Bi, X., & Farnsworth, D. (2017). How incentives affect the adoption of anaerobic digesters in the United States. *Sustainability, 9*(1221), 1–14. https://doi.org/10.3390/su9071221

Sanathkumar, G., Nagesh, K. J., Hadimani, G., Laxman, Charanraj, B. R., & Honnavalli, P. B. (2021). Smart waste segregation. In *2021 IEEE 9th region 10 humanitarian technology conference (R10-HTC), Bangalore, India* (pp. 1–6). https://doi.org/10.1109/R10-HTC53172.2021.9641526

Sánchez-Contreras, I., Morales-Arrieta, S., Okoye, P. U., Guillén-Garcés, R. A., Sebastian, P. J., & Arias, D. M. (2021). Recycling industrial wastewater for improved carbohydrate-rich biomass production in a semi-continuous photobioreactor: Effect of hydraulic retention time. *Journal of Environmental Management, 284*(112065). https://doi.org/10.1016/j.jenvman.2021.112065

Sanders, W. T. M. (2001). In G. Lettinga & G. Zeeman (Eds.), *Anaerobic hydrolysis during digestion of complex substrates* [Doctor of Philosophy, Wageningen University]. https://edepot.wur.nl/198997

Santos, R. A., Teles, E. O., & Freires, F. G. M. (2024). Applying the circular economy framework to blockchain agricultural production. *Sustainability, 26*(8004), 1–17. https://doi.org/10.3390/su16188004

Sarathkumar, D., Srinivasan, M., Stonier, A. A., Kumar, S., & Vanaja, D. S. (2021). A review on renewable energy based self-healing approaches for smart grid. In *International conference on advancements in electrical, electronics, communication, computing and automation (ICAECA), 08–09 October 2021*. https://doi.org/10.1109/ICAECA52838.2021.9675495

Sarkar, S., Ghosh, P., Mukherjee, K., Sil, A. K., & Saha, T. (2009). Sewage treatment in a single pond system at East Kolkata Wetland, India. *Water Science & Technology, 60*(9), 2309–2317. https://doi.org/10.2166/wst.2009.673

Sarto, S., Hildayati, R., & Syaichurrozi, I. (2019). Effect of chemical pretreatment using sulfuric acid on biogas production from water hyacinth and kinetics. *Renewable Energy, 132*, 335–350. https://doi.org/10.1016/j.renene.2018.07.121

Schmidt, J. E., & Ahring, B. K. (1997). Treatment of wastewater from a multi-product food processing company, in upflow anaerobic sludge blanket (UASB) reactors: The effect of seasonal variation. In D. L. Wise (Ed.), *Global Environmental Biotechnology* (pp. 245–252). Springer, Dordrecht. https://doi.org/10.1007/978-94-017-1711-3_21

Sebola, M., Muzenda, E., & Tesfagiorgis, H. (2015). Effect of particle size on anaerobic digestion of different feedstocks. *South African Journal of Chemical Engineering, 20*, 11–26. https://api.semanticscholar.org/CorpusID:100559732

Selişteanu, D., Popescu, I. M., Petre, E., Roman, M., Şendrescu, D., & Popa, B. (2018). Distributed control systems for a wastewater treatment plant: Architectures and advanced control solutions. In T. Yonar (Ed.), *Wastewater and water quality* (pp. 153–181). InTech Open. https://doi.org/10.5772/intechopen.74827

Sforza, E., Barbera, E., Girotto, F., Cossu, R., & Bertucco, A. (2017). Anaerobic digestion of lipid-extracted microalgae: Enhancing nutrient recovery towards a closed loop recycling. *Biochemical Engineering Journal, 121*, 139–146. https://doi.org/10.1016/j.bej.2017.02.004

Sforza, E., Kumkum, P., Barbera, E., & Kumar, S. (2020). Bioremediation of industrial effluents: How a biochar pretreatment may increase the microalgal growth in tannery wastewater. *Journal of Water Process Engineering, 37*(101431), 1–12. https://doi.org/10.1016/j.jwpe.2020.101431

Shah, F. A., Mahmood, Q., Shah, M. M., Pervez, A., & Asad, S. A. (2014). Microbial ecology of anaerobic digesters: The key players of anaerobiosis. *Scientific World Journal, 183752*. https://doi.org/10.1155/2014/183752

Shahidul, M. I., Malcolm, M. L., Eugene, J. J., & Mamunur, R. (2018). Optimization of factors affecting biogas production from POME. *Science International, 30*(6), 851–859. https://www.sci-int.com/pdf/636788212165565332

Shen, Y., Linville, J. L., Urgun-Demirtas, M., Mintz, M. M., & Snyder, S. W. (2015). An overview of biogas production and utilization at full-scale wastewater treatment plants (WWTPs) in the United States: Challenges and opportunities towards energy-neutral WWTPs. *Renewable and Sustainable Energy Reviews, 50*, 346–362. https://doi.org/10.1016/j.rser.2015.04.129

Sher, F., Smječanin, N., Hrnjić, H., Karadža, A., Omanović, R., Šehović, E., & Sulejmanović, J. (2024). Emerging technologies for biogas production: A critical review on recent progress, challenges and future perspectives. *Process Safety and Environmental Protection, 188*, 834–859. https://doi.org/10.1016/j.psep.2024.05.138

Siddique, M., Sarkinbaka, Z. M., Abdul, A. Z., Asif, M., & Elboughdiri, N. (2023). Municipal solid waste to energy strategies in Pakistan and its air pollution impacts on the environment, landfill leachates: A review. *FMDB Transactions on Sustainable Energy Sequence, 1*(1), 38–48. https://www.fmdbpub.com/user/journals/article_details/FTSES/58

Silva, J., Gonçalves, J. C., Rocha, C., Vilaça, J., & Madeira, L. M. (2024). Biomethane production from biogas obtained in wastewater treatment plants: Process optimization and economic analysis. *Renewable Energy, 220*(119469), 1–14. https://doi.org/10.1016/j.renene.2023.119469

Simwambi, A., Yamba, F., Hibler, S., & Mulenga, K. (2020). Renewable energy potential of sewage in Zambia. *Open Journal of Applied Sciences, 10*(6), 328–350. https://doi.org/10.4236/ojapps.2020.106025

Sinduja, B., & Kumar, T. (2024). An intelligent app-based system for waste segregation and collection. *Procedia Computer Science, 235*, 2843–2856. https://doi.org/10.1016/j.procs.2024.04.269

Singh, A., & Kumar, V. (2021). Recent developments in monitoring technology for anaerobic digesters: A focus on bio-electrochemical systems. *Bioresource Technology, 329*(124937), 1–11. https://doi.org/10.1016/j.biortech.2021.124937

Smetana, G., & Grosser, A. (2024). The application of an Upflow Anaerobic Sludge Blanket reactor in the treatment of brewery and dairy wastewater: A critical review. *Energies, 17*(1504), 1–30. https://doi.org/10.3390/en17061504

Soehartanto, T., Pratama, I. P. E. W., Tanaya, J. F., & Sumirman, E. (2023). Digital monitoring system for anaerobic bioreactor conditions based on information on biogas volume, biogas pressure and methane gas content. *International Conference on Biomass and Bioenergy (2022), 1187*(012025), 1–8. https://doi.org/10.1088/1755-1315/1187/1/012025

Song, X., Liu, Q., Yu, Y., & Lv, X. (2016). *Using PLC programmable logic controller to realize the automation controller of anaerobic fermentation of biogas engineering*. https://www.researchgate.net/publication/311928016

Sowndharya, V., Savitha, P., & Selvaraj, H. J. R. (2019). Smart waste segregation and monitoring system using IoT. *International Research Journal of Multidisciplinary Technovation, 1*(2), 1–10. https://doi.org/10.34256/irjmt1921

Srivichai, P., & Chavalparit, O. (2020). Optimization of biogas production from co-digestion of waste activated sludge and modified tapioca starch sludge. *International Journal of GEOMATE, 18*(67), 148–155. https://doi.org/10.21660/2020.67.9356

Stampfli, J. J. (2020). In I. W. Hunter, N. C. Hogan, & M. Yang (Eds.), *Design of a two-stage multi-state experimental biogas reactor* [Degree of Bachelor of Science in Mechanical Engineering, Department of Mechanical Engineering, Massachusetts Institute of Technology (MIT)]. https://dspace.mit.edu/bitstream/handle/1721.1/127871/1196830605-MIT.pdf?sequence=1&isAllowed=y

Steenkamp, J. (2023). In A. Callaghan (Ed.), *Balfour wastewater treatment plant | resource consent application and assessment of environmental effects*. REP Balfour WWTP Resource Consent Application. https://www.es.govt.nz/repository/libraries/id:26gi9ayo517q9stt81sd

Stewart, W. C. (2014). *Three stage, multiple phase anaerobic digestion system and method* (US8765449B2). https://patents.google.com/patent/US8765449B2/en

Steyer, J.-P., Buffière, P., Rolland, D., & Moletta, R. (1999). Advanced control of anaerobic digestion processes through disturbances monitoring. *Water Research, 33*(9), 2059–2068. https://doi.org/10.1016/S0043-1354(98)00430-8

Steyer, J.-P., Genovesi, A., & Harmand, J. (2001). Advanced monitoring and control of anaerobic wastewater treatment plants: Fault detection and isolation. *Water Science & Technology, 43*(7), 183–190. https://doi.org/10.2166/wst.2001.0420

Suhan, B. K., Shuchi, S. B., Abir, R. A. R., & Hasan, R. (2019). Innovative multi-stage anaerobic modelling of a 10 MW biogas plant using Aspen HYSYS. In *International conference on petroleum engineering (ICPE 2019), Bangladesh University of Engineering & Technology (BUET), Dhaka-1000, Bangladesh* (pp. 1–17)

Sundh, I., Carlsson, H., Nordberg, Å., Hansson, M., & Mathisen, B. (2003). Effects of glucose overloading on microbial community structure and biogas production in a laboratory-scale anaerobic digester. *Bioresource Technology, 89*(3), 237–243. https://doi.org/10.1016/S0960-8524(03)00075-0

Sutthiprapa, S., Towprayoon, S., Chiemchaisri, C., Chaiprasert, P., & Wangyao, K. (2024). Optimizing an anaerobic hybrid reactor series for effective high-strength fresh leachate treatment and biogas generation. *Sustainability, 16*(3076). https://doi.org/10.3390/su16073076

Taherdoost, H. (2024). Blockchain integration and its impact on renewable energy. *Computers, 13*(107), 1–19. https://doi.org/10.3390/computers13040107

Taron, A., Majumder, A., Bodach, S., & Agbefu, D. (2023). Public-private partnerships for the circular bio-economy in the Global South: lessons learned. In *Resource Recovery and Reuse Series 22* (pp. 1–56). International Water Management Institute (IWMI). https://doi.org/10.22004/ag.econ.339137

Tassew, F. A., Bergland, W. H., Dinamarca, C., Kommedal, R., & Bakke, R. (2019). Granular sludge bed processes in anaerobic digestion of particle-rich substrates. *Energies, 12*(2940), 1–20. https://doi.org/10.3390/en12152940

Teixeira, P. F. S., Moura, L. F., & Lima, S. W. S. (2017). Development of a low-cost data acquisition system for biodigester. *Journal of Sustainable Bioenergy Systems, 7*(3), 117–137. https://doi.org/10.4236/jsbs.2017.73009

Thakor, D., & Patel, B. (2019). P2S_DLB: Pluggable to scheduler dynamic load balancing algorithm for distributed computing environment: Proceedings of ICETEAS 2018. In V. Rathore, M. Worring, D. Mishra, A. Joshi, & S. Maheshwari (Eds.), *Emerging trends in expert applications and security. Advances in intelligent systems and computing* (pp. 347–355). Springer, Singapore. https://doi.org/10.1007/978-981-13-2285-3_41

Thakur, S. V, & Barjibhe, R. B. (2019). Effect of canola oil to reduce the scum formations. *International Journal of Engineering and Advanced Technology (IJEAT), 9*(1), 7136–7139. https://doi.org/10.35940/ijeat.A1813.109119

Thakur, S., Barjibhe, R., Nitnaware, P. T., Jawanjal, P. B., & Suryawanshi, D. (2019). Additives for scum reduction in biogas plant: A simplified testing methedology. *International Journal of Innovative Technology and Exploring Engineering (IJITEE), 8*(11), 1798–1801. https://doi.org/10.35940/ijitee.K1775.0981119

Tharmarajah, N., Shahbaz, K., & Baroutian, S. (2024). Textile waste pretreatment for anaerobic digestion: A review and technology feasibility study. *Journal of Chemical Technology and Biotechnology, 99*(11), 2213–2227. https://doi.org/10.1002/jctb.7728

Thirugnanasambandham, K., Sivakumar, V., & Maran, J. P. (2014). Modeling and optimization of biogas production from rice mill effluent using up flow anaerobic sludge blanket reactor. *Journal of Renewable Sustainable Energy, 6*(2). https://doi.org/10.1063/1.4873400

Thomaz, I. P. L., Mahler, C. F., & Calôba, L. P. (2023). Artificial Intelligence (AI) applied to waste management: A contingency measure to fill out the lack of information resulting from restrictions on field sampling. *Waste Management Bulletin, 1*(3), 11–17. https://doi.org/10.1016/j.wmb.2023.06.002

Thrän, D., Deprie, K., Dotzauer, M., Kornatz, P., Nelles, M., Radtke, K. S., & Schindler, H. (2023). The potential contribution of biogas to the security of gas supply in Germany. *Energy, Sustainability and Society, 13*(12), 1–15. https://doi.org/10.1186/s13705-023-00389-1

Tian, X., Song, Y., Shen, Z., Zhou, Y., Wang, K., Jin, X., Han, Z., & Liu, T. (2020). A comprehensive review on toxic petrochemical wastewater pretreatment and advanced treatment. *Journal of Cleaner Production, 245*(118692), 1–68. https://doi.org/10.1016/j.jclepro.2019.118692

Tolman, A., Rubin, A., Akbari, A., Ferguson, A., McCarthy, A., Kukreja, A., Duplechien, A., Shore, B., Gerber, B., Sheff, B., Lang, B., Parmentier, B., Pleima, B., KillKelley, B., Nudelbacher, B., Kopman, C., Jones, D., Darr, D., Wells, D., … Oosthuizen, W. (2024). *Carbon accounting methodology for biogas*. TPR Enterprises, LLC, and Developed by EcoEngineers for the American Biogas Council. https://americanbiogascouncil.org/

Tomczak, W., Gryta, M., Daniluk, M., & Żak, S. (2024). Biogas upgrading using a single-membrane system: A review. *Membranes, 14*(80), 1–21. https://doi.org/10.3390/membranes14040080

Tompkins, D. (2021). *The impacts on human health and environment arising from the spreading of sewage sludge to land (CR/2016/23): Sewage sludge processing systems in Scotland* (pp. 1–61). James Hutton Institute, Project Number: 594. https://www.gov.scot/binaries/content/documents/govscot/publications

Topić, D., Barukčić, M., & Mezei, C. (2020). Optimization model for biogas power plant feedstock mixture considering feedstock and transportation costs using a Differential Evolution Algorithm. *Energies, 13*(1610), 1–22. https://doi.org/10.3390/en13071610

Trilita, M. N., Hendrasarie, N., & Wahjudijanto, I. (2016). Removal of organic load in communal wastewater by using the six stage Anaerobic Baffle Reactor (ABR). In *The 3rd Bali international seminar on science & technology (BISSTECH 2015), 58*(01023), 1–4. https://doi.org/10.1051/matecconf/20165801023

Tsapekos, P., Kougias, P. G., & Angelidaki, I. (2018). Mechanical pretreatment for increased biogas production from lignocellulosic biomass; predicting the methane yield from structural plant components. *Waste Management, 78*, 903–910. https://doi.org/10.1016/j.wasman.2018.07.017

Tsegaye, D., & Leta, S. (2022). Optimization of operating parameters for biogas production using two-phase bench-scale anaerobic digestion of slaughterhouse wastewater: Focus on methanogenic step. *Bioresources and Bioprocessing, 9*(125), 1–16. https://doi.org/10.1186/s40643-022-00611-6

Tsiakiri, E. P., Mpougali, A., Lemonidis, I., Tzenos, C. A., Kalamaras, S. D., Kotsopoulos, T. A., & Samaras, P. (2021). Estimation of energy recovery potential from primary residues of four municipal wastewater treatment plants. *Sustainability, 13*(7198). https://doi.org/10.3390/su13137198

Udaeta, M. E. M., Medeiros, G. A. S., Da Silva, V. O., & Galvão, L. C. R. (2019). Basic and procedural requirements for energy potential from biogas of sewage treatment plants. *Journal of Environmental Management, 15*(236), 380–387. https://doi.org/10.1016/j.jenvman.2018.12.110

Udomsom, S., Budwong, A., Wongsa, C., Sangngam, P., Baipaywad, P., Manaspon, C., Auephanwiriyakul, S., Theera-Umpon, N., & Paengnakorn, P. (2022). Automatic programmable bioreactor with pH monitoring system for tissue engineering application. *Bioengineering, 9*(187), 1–12. https://doi.org/10.3390/bioengineering9050187

UN. (2020). In Danfoss (Ed.), *Energy efficiency and environmental protection in wastewater treatment at the Marselisborg Wastewater Treatment Plant*. https://www.un.org/sites/un2.un.org/files/2020/09/

USDA. (2017). *Conservation practice standards for anaerobic digester*. United States Department of Agriculture (USDA). https://www.nrcs.usda.gov/sites/default/files/2022-08

USEPA. (2006). *Biosolids technology fact sheet: Multi-stage anaerobic digestion*. https://www.epa.gov/sites/default/files/2018-11/documents/multistage-anaerobic-digestion-factsheet.pdf

Usman, M. N., Suleiman, M. A., & Binni, M. I. (2021). Anaerobic digestion of agricultural wastes: A potential remedy for energy shortfalls in Nigeria. *Journal of Waste Management and Disposal, 4*(1), 1–13. https://article.scholarena.com/Anaerobic-Digestion-of-Agricultural.pdf

Uwamungu, J. Y., Kumar, P., Alkhayyat, A., Younas, T., Capangpangan, R. Y., Alguno, A. C., & Ofori, I. (2022). Future of water/wastewater treatment and management by Industry 4.0 integrated nanocomposite manufacturing. *Journal of Nanomaterials*. https://doi.org/10.1155/2022/5316228

van der Berg, D. J., Teke, G. M., Görgens, J. F., & van Rensburg, E. (2024). Predicting commercial-scale anaerobic digestion using biomethane potential. *Renewable Energy, 235*(121304), 1–9. https://doi.org/10.1016/j.renene.2024.121304

van Lier, J. B., Rebac, S., & Lettinga, G. (1997). High-rate anaerobic wastewater treatment under psychrophilic and thermophilic conditions. *Journal Water Science and Technology, 10*(10), 199–206. https://doi.org/10.1016/S0273-1223(97)00202-3

Vanrolleghem, P. A. (1999). Sensors for anaerobic digestion: An overview. *CiteseerX*, 1–7. https://citeseerx.ist.psu.edu/document?repid

Varjani, S., Joshi, R., Srivastava, V. K., Ngo, H. H., & Guo, W. (2020). Treatment of wastewater from petroleum industry: Current practices and perspectives. *Environmental Science and Pollution Research, 27*, 27172–27180. https://doi.org/10.1007/s11356-019-04725-x

Vasco-Correa, J., Manandhar, A., & Shah, A. (2018). *Economic implications of anaerobic digestion for bioenergy production and waste management*. Agriculture and Natural Resources FABE-661.1; College of Food, Agricultural and Environnmental Sciences, The Ohio State University. https://ohioline.osu.edu/factsheet/fabe-6611

Vaz, A., & Mattos, A. P. (2020). Energy and carbon credits generation from the production of biogas from the ethanol stillage of corn and sugar cane [ENC-2020–9999], November 16–20. In *18th Brazilian congress of thermal sciences and engineering* (pp. 1–11). https://doi.org/10.26678/ABCM.ENCIT2020.CIT20-0129

Verbeeck, K., Buelens, L. C., Galvita, V. V., Marin, G. B., Van Geem, K. M., & Rabaey, K. (2018). Upgrading the value of anaerobic digestion via chemical production from grid injected biomethane. *Energy & Environmental Science, 11*(7), 1788–1802. https://doi.org/10.1039/C8EE01059E

Vergara-Araya, M., Hilgenfeldt, V., Steinmetz, H., & Wiese, J. (2022). Combining shift to biogas production in a large WWTP in China with optimisation of nitrogen removal. *Energies, 15*(2710), 1–13. https://doi.org/10.3390/en15082710

Vigneswaran, S., Sundaravadivel, M., & Chaudhary, D. S. (2004). Sequencing batch reactors: Principles, design/operation and case studies. In *Encyclopedia of life support systems (EOLSS)* (pp. 1–21). Water and Wastewater Treatment Technologies. https://www.eolss.net/sample-chapters/c07/E6-144-11.pdf

Visali, K., Chitra, M., & Pappa, N. (2017). Automation and design of real-time controllers for a laboratory scale bioreactor. In *International conference on innovations in control, communication and information systems [ICICCI 2017], 12–13 August 2017 Greater Noida, India* (pp. 1–6). https://doi.org/10.1109/ICICCIS.2017.8660815

Viswanath, P., Devi, S. S., & Nand, K. (1992). Anaerobic digestion of fruit and vegetable processing wastes for biogas production. *Bioresource Technology, 40*(1), 43–48. https://doi.org/10.1016/0960-8524(92)90117-G

Vítězová, M., Kohoutová, A., Vítěz, T., Hanišáková, N., & Kushkevych, I. (2020). Methanogenic microorganisms in industrial wastewater anaerobic treatment. *Processes, 8*(1546), 1–27. https://doi.org/10.3390/pr8121546

Vrbová, V., & Ciahotný, K. (2017). Upgrading biogas to biomethane using membrane separation. *Energy & Fuels, 31*(9), 9393–9401. https://doi.org/10.1021/acs.energyfuels.7b00120

Vysokomornaya, O. V., Kurilenko, E. Y., & Shcherbinina, A. A. (2015). Major contaminants in industrial and domestic wastewater. *MATEC Web of Conferences, 23*(01041), 1–3. https://doi.org/10.1051/matecconf/20152301041

Wang, B., Wang, Z., Chen, T., & Zhao, X. (2020). Development of novel bioreactor control systems based on smart sensors and actuators. *Frontiers in Bioengineering and Biotechnology, 8*(7), 1–15. https://doi.org/10.3389/fbioe.2020.00007

Wang, Y., Zang, B., Gong, X., Liu, Y., & Li, G. (2017). Effects of pH buffering agents on the anaerobic hydrolysis acidification stage of kitchen waste. *Waste Management, 68*, 603–609. https://doi.org/10.1016/j.wasman.2017.06.041

Wanke, C. P., & Gallagher, R. (2024). Orange county sanitation district wastewater refunding revenue obligations. In S. Faessel, C. Marick, & J. Ahn (Eds.), *Preliminary official statement* (p. 266). https://arizent.brightspotcdn.com/85/07

Wei, X., Zhang, S., Han, Y., & Wolfe, F. A. (2019). Treatment of petrochemical wastewater and produced water from oil and gas. *Water Environment Research, 91*(10), 1025–1033. https://doi.org/10.1002/wer.1172

Westerholm, M., Castillo, M. P., Andersson, A. C., Nilsen, P. J., & Schnürer, A. (2019). Effects of thermal hydrolytic pre-treatment on biogas process efficiency and microbial community structure in industrial- and laboratory-scale digesters. *Waste Management, 95*, 150–160. https://doi.org/10.1016/j.wasman.2019.06.004

Wid, N., & Horan, N. J. (2016). Anaerobic digestion of wastewater screenings for resource recovery and waste reduction. *International Conference on Chemical and Bioprocess Engineering 9–12 December 2015, Kota Kinabalu, Malaysia, 36*(012017), 1–7. https://doi.org/10.1088/1755-1315/36/1/012017

Wid, N., & Horan, N. (2018). Anaerobic digestion of screenings for biogas recovery. In N. Horan, A. Yaser, & N. Wid (Eds.), *Anaerobic digestion processes* (pp. 85–103). Springer, Singapore. https://doi.org/10.1007/978-981-10-8129-3_6

Wolfson, S. J., Porter, A. W., Campbell, J. K., & Young, L. Y. (2018). Naproxen is transformed via acetogenesis and syntrophic acetate oxidation by a methanogenic wastewater consortium. *Microbial Ecology, 76*, 362–371. https://doi.org/10.1007/s00248-017-1136-2

WSDR. (2018). *Tax incentives for converting biogas into marketable coproducts and manufacturing wood biomass fuel*. Washington State Department of Revenue. https://dor.wa.gov/

Yadav, N. K., & Singh, A. K. (2018). Factors affecting biogas production during anaerobic digestion—A review. *International Journal of Creative Research Thoughts (IJCRT), 6*(2), 1276–1282. https://ijcrt.org/papers/IJCRT1812589.pdf

Yamaji, D. M., Amâncio-Vieira, S. F., Fidelis, R., & Contani, E. A. D. R. (2024). Decision-making in biogas production projects: Paradigms and prospection. *Journal of the Air & Waste Management Association, 74*(6), 416–438. https://doi.org/10.1080/10962247.2024.2338747

Yan, P., Gai, M., Wang, Y., & Gao, X. (2021). Review of soft sensors in anaerobic digestion process. *Processes, 9*(1434), 1–21. https://doi.org/10.3390/pr9081434

Yang, C. (2020). In Y. Dong (Ed.), *Petrochemical wastewater and its treatment* [Thesis-Centria University of Applied Sciences Environmental Chemistry and Technology]. https://www.theseus.fi/bitstream/handle/10024/344161/Chengyi_Yang.pdf?sequence=2

Yang, G., Zhang, P., Zhang, G., Wang, Y., & Yang, A. (2015). Degradation properties of protein and carbohydrate during sludge anaerobic digestion. *Bioresource Technology, 192*, 126–130. https://doi.org/10.1016/j.biortech.2015.05.076

Yang, L., Ge, X., Wan, C., Yu, F., & Li, Y. (2014). Progress and perspectives in converting biogas to transportation fuels. *Renewable and Sustainable Energy Reviews, 40*, 1133–1152. https://doi.org/10.1016/j.rser.2014.08.008

Yangın-Gömeç, Ç. (2006). Behavior of the anaerobic CSTR in the presence of scum during primary sludge digestion and the role of pH. *Journal of Environmental Science and Health Part A Toxic/hazardous Substances & Environmental Engineering, 41*(6), 1117–1127. https://doi.org/10.1080/10934520600620519

Yin, Y., Chen, S., Li, X., Jiang, B., Zhao, J. R., & Nong, G. (2021). Comparative analysis of different CHP systems using biogas for the cassava starch plants. *Energy, 232*(121028). https://doi.org/10.1016/j.energy.2021.121028

Yoshida, K., Kametani, K., & Shimizu, N. (2020). Adaptive identification of anaerobic digestion process for biogas production management systems. *Bioprocess and Biosystems Engineering, 43*(1), 1–10. https://doi.org/10.1007/s00449-019-02203-9

Yu, H.-Q., Zheng, X.-J., Hu, Z.-H., & Gu, G.-W. (2003). High-rate anaerobic hydrolysis and acidogenesis of sewage sludge in a modified upflow reactor. *Water Science & Technology, 48*(4), 69–75. https://doi.org/10.2166/wst.2003.0224

Yuan, T., Zhang, Z., Lei, Z., Shimizu, K., & Lee, D.-J. (2022). A review on biogas upgrading in anaerobic digestion systems treating organic solids and wastewaters via biogas recirculation. *Bioresource Technology, 344*(Part B). https://doi.org/10.1016/j.biortech.2021.126412

Zacharof, M.-P., Mandale, S. J., Oatley-Radcliffe, D., & Lovitt, R. W. (2019). Nutrient recovery and fractionation of anaerobic digester effluents employing pilot scale membrane technology. *Journal of Water Process Engineering, 31*(100846), 1–10. https://doi.org/10.1016/j.jwpe.2019.100846

Zahmatkesh, S., Gholian-Jouybari, F., Klemeš, J. J., Bokhari, A., & Hajiaghaei-Keshteli, M. (2023). Sustainable and optimized values for municipal wastewater: The removal of biological oxygen demand and chemical oxygen demand by various levels of geranular activated carbon- and genetic algorithm-based simulation. *Journal of Cleaner Production, 417*(137932). https://doi.org/10.1016/j.jclepro.2023.137932

Zaiat, M., Rodrigues, J. A. D., Ratusznei, S. M., De Camargo, E. F. M., & Borzani, W. (2001). Anaerobic sequencing batch reactors for wastewater treatment: A developing technology. *Applied Microbiology and Biotechnology, 55*(1), 29–35. https://doi.org/10.1007/s002530000475

Zarina, R., & Mezule, L. (2024). Enzymatic hydrolysis of waste streams originating from wastewater treatment plants. *Biotechnology for Biofuels and Bioproducts, 17*(104). https://doi.org/10.1186/s13068-024-02553-x

Zeeman, G., & Lettinga, G. (1999). The role of anaerobic digestion of domestic sewage in closing the water and nutrient cycle at community level. *Water Science and Technology, 39*(5), 187–194. https://doi.org/10.1016/S0273-1223(99)00101-8

Zeppilli, M., Cristiani, L., Ferretti, J., Majone, M., & Villano, M. (2022). Acetogenic inoculum selection for acetate production from waste biomasses via thermal shock treatment. *Chemical Engineering Transactions (CEt), 92*, 13–18. https://doi.org/10.3303/CET2292003

Zhang, L., Loh, K.-C., Lim, J. W., & Zhang, J. (2018). Bioinformatics analysis of metagenomics data of biogas-producing microbial communities in anaerobic digesters: A review. *Renewable and Sustainable Energy Reviews, 100*(C), 110–126. https://doi.org/10.1016/j.rser.2018.10.021

Zhang, J., Lv, C., Liu, J., Liu, J., Yu, D., Wang, Y., Chen, M., & Wei, Y. (2016). Optimization and microbial community analysis of anaerobic co-digestion of food waste and sewage sludge based on microwave pretreatment. *Bioresource Technology, 200*, 253–261. https://doi.org/10.1016/j.biortech.2015.10.037

Zhou, H., Xing, D., Ma, J., Su, Y., & Zhang, Y. (2022). Electrifying anaerobic granular sludge for enhanced waste anaerobic digestion and biogas production. *Separation and Purification Technology, 295*(121300), 1–10. https://doi.org/10.1016/j.seppur.2022.121300

Zwain, H. M., Barghash, H., Vakili, M., Majdi, H. S., & Dahlan, I. (2022). Modeling and optimization of process parametric interaction during high-rate anaerobic digestion of recycled paper mill wastewater using the response surface methodology. *Water Reuse, 12*(1), 78–91. https://doi.org/10.2166/wrd.2022.088

Chapter 6
Nutrient Recovery Technologies: Advanced Strategies for Nitrogen Removal and Recovery

L. R. Pooja, Mamta Bisht, Renu Singh, Rajdeep Das, M. Anjaly, and Manish Kumar

Abstract The growing population and increased dependence on industries have increased the production of wastewater which contains useful nutrients like nitrogen (N) and phosphorous (P), posing sever environmental threats. This chapter focuses on the advanced technologies for nitrogen removal and recovery in wastewater while considering their economic viability and environmental impact. Traditional methods, which are widely used, are scrutinized for their potential environmental setbacks like the emission of nitrous oxide (N_2O), a potent greenhouse gas. However, cutting-edge methods such as partial nitritation-anaerobic ammonium oxidation are emphasized because of their beneficial economics and ability, under proper management, to lessen environmental issues. Further, in this chapter opportunities for nitrogen recovery from wastewater, which contains high concentrations of nitrogen, which is suitable for processes like ammonia stripping are also, explored which results in less dependence on fertilizer industries and encourages a circular economy. This chapter also emphasizes the importance of the need for improvement in both biological and physico-chemical processes in the removal of nitrogen. It also explains emerging

L. R. Pooja (✉) · R. Singh · R. Das · M. Anjaly
Division of Environment Science, ICAR-Indian Agricultural Research Institute, New Delhi, India
e-mail: poojalr2913@gmail.com; pooja_12043@iari.res.in

R. Singh
e-mail: renusingh_env@iari.res.in

R. Das
e-mail: rajdeepdas611@gmail.com

M. Anjaly
e-mail: anjalykolaroad@gmail.com

M. Bisht
ICAR–Indian Institute of Water Management (IIWM), Bhubaneswar, India
e-mail: mambisht10@gmail.com

M. Kumar
School of Engineering and Sciences, Tecnológico de Monterrey, Monterrey, Nuevo Leon, Mexico
e-mail: manish@tec.mx

S. Dhiman and G. Mukherjee (eds.), *Resource Resurgence*, Springer Water,
https://doi.org/10.1007/978-3-032-02750-4_6

technologies in the recovery of nitrogen from wastewater such as electrochemical systems, denitrifying anaerobic methane oxidation (DAMO), and direct ammonia oxidation. Recovery of valuable byproducts other than nitrogen such as struvite reinforces the circular economy. The chapter presents the current status and promising nitrogen recovery methods through a study of recent technical breakthroughs and a detailed analysis of economic and environmental implications.

Keywords Circular economy · Wastewater · Denitrifying anaerobic methane oxidation · Membrane filtration · Ion exchange · Sustainable management

6.1 Introduction—Overview of Nitrogen Removal and Recovery

Nitrogen is a very essential nutrient for all life forms and it plays a crucial role in agricultural productivity (Mohamed et al., 2022). However, the excessive amount of nitrogen in wastewater poses significant environmental challenges which include water pollution, eutrophication, and greenhouse gas emissions such as nitrous oxide (N_2O) (Ahmed et al., 2022). This problem mostly arrives in densely populated areas and extensive industrial areas because of the discharge of high levels of nitrogen frequently in natural water bodies. Nitrogen primarily occurs in two forms in wastewater: Ammonia and organic nitrogen compounds. These forms of nitrogen must be converted into more stable forms or removed to prevent harmful effects on aquatic ecosystems. The nitrogen cycle in wastewater treatment comprises several biological and chemical processes that transform nitrogen compounds into less reactive or recoverable forms. Traditionally, nitrification–denitrification has been the backbone of nitrogen removal, converting ammonia into nitrate (NO_3^-) and then into nitrogen gas (N_2), which is harmless when released into the atmosphere (Rajta et al., 2020). However, these conventional biological processes come with many drawbacks. They are not only energy-intensive but also, they may result in the emission of N_2O which is a potent greenhouse gas that contributes to global warming. Moreover, they do not focus on recovering nitrogen, a valuable resource that can be reused, especially in agriculture.

The conventional methods of nitrogen removal, including nitrification–denitrification, are effective but suffer from major drawbacks in terms of energy-intensive utility consumption, complex operation, and adverse environmental impact. Major limitations include: (1) Nitrification requires significant aeration, which consumes large amounts of energy. (2) Denitrification may also produce nitrous oxide, a long-lived gas with a greenhouse effect that is 300 times greater than carbon dioxide (CO_2). (3) These processes are primarily driven by the removal of nitrogen. Losses of potentially recoverable nitrogen will be lost because it can be reused in agriculture to limit the circular economy approach.

To address these challenges, advanced nitrogen removal and recovery methods such as Partial Nitritation-Anammox (PN-A) which bypasses the traditional nitrification–denitrification process and converts ammonia into nitrogen gas directly, Denitrifying Anaerobic Methane Oxidation (DAMO) approach that can both do nitrogen removal and mitigate methane emissions, Electrochemical Systems, Ammonia Stripping, Struvite Precipitation method have been developed. These technologies aim not only to remove nitrogen efficiently but also to recover it as a valuable resource (Chen et. al., 2021a). Nitrogen recovery from wastewater not only helps to fight environmental pollution but is also beneficial in terms of economic outcomes because it provides a sustainable source of nitrogenous fertilizers. With the increasing global demand for nitrogen-based fertilizers and the environmental costs of their synthetic production, recovering nitrogen from wastewater can reduce the dependency on chemical fertilizers, lower greenhouse gas emissions, and promote a more circular and sustainable economy (Rahimi et al., 2020).

6.2 Importance in Wastewater Treatment and Environmental Sustainability

The rise in food production led to the high demand for nitrogen and phosphorus fertilizers. The Haber–Bosch process was used for the production of nitrogen-based fertilizers by the fixation of N_2 into NH_3. In wastewater treatment nitrogen is mainly lost as N_2 gas to the atmosphere. Many technologies are used for nitrogen recovery from wastewater treatment like vacuum membrane filtration, struvite precipitation, the treatment of digester reject water by air stripping, hydrophobic membrane filtration, and treatment of air from thermal sludge drying. Separate urine collection was found to be higher nitrogen recovery (60%), but that needs a completely new infrastructure for wastewater collection and treatment (Van der Hoek et al., 2018). It is important to investigate more sustainable nitrogen cycling routes which comprise direct recovery and reuse of nitrogen from wastewater. As yet only small interventions have been made in the field of full-scale nitrogen recovery and reuse from wastewater. To make wastewater treatment more sustainable, there are several programs beyond nitrogen recovery and reuse, and most of them concentrate on recovering resources. One adequate solution from conventional wastewater treatment to sustainable wastewater treatment plants by the reuse of resources from wastewater. To increase sustainability, radical adjustments in wastewater treatment and the use of different methods of nitrogen recovery are needed. This needs a separate collection of urine and its treatment.

An ion exchange/adsorption-based process that gives concentrated streams of reactive nitrogen is used for nitrogen recovery. Nitrogen as NH_3 (g) or $(NH_4)_2SO_4$ is recovered efficiently and effectively by biochemical systems. The recovery of ammonia through air stripping from anaerobic digestate has been observed to recover 70–92% of nitrogen. 99–100% recovery with no secondary contamination in the permeate was yielded from membrane separation (Zielińska & Bułkowska, 2025). Most of the wastewater treatment plants use energy-consuming nitrogen removal methods which don't have additional benefits. As an alternative, nitrogen recovery aims to both wastewater treatment and collection of concentrated ammonia products which generate a circular economy solution. By using a small amount of energy, gas-permeable membrane technology can recover ammonia efficiently and supply a small income with the sale of the product (Beckinghausen et al., 2020). Nitrogen recovery from wastewater, which manages nutrient entry into the waterbodies, and the recovered nitrogen can be used for transforming into fertilizers.

6.3 Scope and Objectives

This chapter provides a comprehensive discussion on the nitrogen removal and recovery technologies of wastewater treatment, judging them from their technical feasibility in their role, environmental impacts created, and economic implications. First, there is an emphasis on conventional nitrogen removal techniques, including shortfalls together with their environmental consequences, particularly in terms of greenhouse gases like nitrous oxide (N_2O). The advanced technologies include Partial nitritation-anammox, electrochemical systems, DAMO, and direct ammonia oxidation that come with improved efficiency and reduced environmental impacts. Another aspect is the recovering of nitrogen from wastewater streams: a high-nitrogen potential source such as ammonium stripping for reducing its reliance on synthetic fertilizers. More importantly, it gives a comparative evaluation of various processes of nitrogen removal and recovery based on their economic and ecological merits, aside from the advantages of the latest technical developments. The chapter thus concludes by summarizing the current challenges alongside the future research agenda and addressing the need to assimilate both biological and physico-chemical steps to further improve the handling of nitrogen from wastewater effluents.

6.4 Nitrogen in Wastewater: Sources and Challenges

The major challenge with environmental nitrogen emanates from agricultural runoff, industrial effluents, domestic sewage, and urban stormwater, which originate largely from wastewater. Agriculture leads to approximately 60–80% of nitrogen pollution in freshwater ecosystems in almost every corner of the globe because of intensive fertilizer application, thereby causing nitrate (NO_3^-) to leak into water sources. Industrial sources, such as food, chemical, and pharmaceutical productions, contribute roughly 15–20% to nitrogen in wastewater systems, thus making treatment a hassle. Source of domestic water in urban areas accounts for around 10–15% due to organic waste and detergents used at home, whereas the 20% contribution during heavy rainfall is accounted for by stormwater runoff carrying fertilizers and waste directly into water systems (Huang et al., 2023). Nitrate in wastewater poses some challenges. High nitrates contribute to the eutrophication of lakes and rivers, resulting in harmful algal blooms and hypoxic conditions that kill aquatic life. In addition, the chemical stability of nitrogen compounds ammonia (NH_4^+) and nitrate renders them very hard to remove using conventional methods. Partial treatment of the compounds is likely to result in the emission of nitrous oxide into the water; being a potent greenhouse gas causes degradation of air quality and climate change (Garcia et al., 2024).

Moreover, changes in concentration due to seasonal agricultural activities, changes in industrial output, and population growth further complicate the unpredictability of the composition of wastewater and nitrogen. These pose challenges for responsive and adaptive treatment processes that will not only remove nitrogen effectively but not introduce harm to the environment. Therefore, dealing with nitrogen in wastewater offers yet another exciting challenge toward enhancing water quality protection, public health, and ecosystem resilience.

6.4.1 Characteristics of Nitrogen in Wastewater

There are four major forms of nitrogen in wastewater, including organic nitrogen, ammonia (NH_4^+), nitrite (NO_2^-), and nitrate (NO_3^-). Their relative proportions depend upon the type of wastewater, its origin, and treatment stages, and hence the requirements for treatment and environmental impacts.

Organic Nitrogen

Organic nitrogen is that which exists in the form of proteins, amino acids, and other organic compounds. It primarily occurs in municipal wastewater through human excreta and food wastes, as well as in industrial wastewater from processing food, pharmaceutical industry, etc. Organic nitrogen converts into ammonia through biological processes, which makes its treatment cumbersome as ammonia requires additional treatment for better management.

Ammonia (NH_4^+)

Ammonia is the most abundant nitrogen species in wastewater, particularly from domestic sewage, runoff from livestock, and certain industrial effluents. Ammonia nitrogen typically comprises the largest fraction, accounting for 73.5–77.2% of total Kjeldahl nitrogen in raw and settled wastewater (Młyńska & Chmielowski, 2024). It is highly soluble in water and is toxic to aquatic organisms if left untreated. It is also a precursor to nitrous oxide, which is a potent greenhouse gas formed in the treatment process. Due to its stability, it is difficult to remove and generally undergoes biological nitrification and subsequent denitrification during wastewater treatment processes.

Nitrite (NO_2^-)

Nitrite is generally an intermediate product of nitrogen produced during the biological oxidation of ammonia to nitrate in a process called nitrification. While normally present at much lower concentrations, nitrite is ecologically important because it may be harmful to aquatic animals, even at low concentrations. In wastewater treatment, the buildup of nitrite commonly indicates incomplete nitrification and should be carefully watched for potential environmental discharge.

Nitrate (NO_3^-)

The most common and oxidized form of nitrogen present in wastewater is nitrate, especially where nitrification has occurred. The main nitrogen species in agricultural runoff and some industrial effluents is nitrate, which is largely of agricultural origin. High nitrate levels in discharged effluent lead to eutrophication, which causes oxygen depletion in the receiving water bodies and the production of harmful algal blooms.

6.4.2 Environmental and Economic Impacts of Nitrogen Pollution

Nitrogen pollution poses significant environmental, economic, and health challenges globally. It contributes to air and water pollution, climate change, and stratospheric ozone depletion. Agriculture is a major source of nitrogen pollution, and improving nitrogen management could potentially benefit farmers, the fertilizer industry, and the environment. Nitrogen pollution has secondary implications for increased emission of greenhouse gases, including nitrous oxide, with a global warming potential that is nearly 300 times that of carbon dioxide (Viancelli & Michelon, 2024). Wastewater treatment process effluents from dealing with high loads of nitrogen release N_2O to the atmosphere and contribute to accelerating climate change. Besides, ammonia volatilization and nitrate leaching from nitrogen-rich wastewater are harmful activities that cause air and water pollution. This leads to health issues, including respiratory problem and methemoglobinemia, popularly known as "blue baby syndrome." However, the environmental benefits of improved nitrogen management significantly outweigh the economic impacts on farmers and the fertilizer industry. Reducing nitrogen pollution requires a multifaceted approach, including optimizing fertilizer use and reducing emissions from transport and agriculture. Economically, nitrogen pollution imposes substantial costs. Water treatment facilities must invest in advanced technologies to remove nitrogen from wastewater, driving up operational expenses. In the U.S. alone, the costs of mitigating nitrogen pollution in water bodies are estimated to be in the billions of dollars annually. Additionally, nutrient pollution damages fisheries and tourism industries, as degraded water quality reduces fish populations and deters recreational activities.

6.4.3 Regulatory Frameworks and Standards for Nitrogen Discharge

Regulatory frameworks for nitrogen discharge vary across jurisdictions, with the EU Nitrates Directive serving as a key policy in Europe. Member states implement Action Programs, which may include nitrogen application standards, closed periods for manure application, and guidelines for fertilizer use. In the United States, ballast water regulations have moved towards concentration-based numeric discharge limits, with variations between federal and state standards (Wang et al., 2021). Denmark has implemented a strict regulatory regime, achieving significant reductions in nitrogen leaching, but faces challenges in addressing site-specific ecological demands (Nygaard et al., 2022). The regulation of nitrogen pollution in Europe has co-evolved with political integration, recognizing the transboundary nature of environmental effects. Scientific initiatives have played a role in informing strategies to mitigate pollution levels, though the interaction between institutional settings and scientific efforts remains complex (Yang et al., 2022).

6.5 Traditional Nitrogen Removal Technologies

Traditional nitrogen removal technologies from wastewater encompass both biological and physico-chemical processes. Biological methods include conventional nitrification–denitrification as well as novel approaches like simultaneous nitrification–denitrification, ANAMMOX, CANON, SHARON, and OLAND (Rodríguez-de la Garza et al., 2023). These processes leverage microorganisms to convert nitrogen compounds into harmless forms. Physico-chemical technologies comprise air and steam stripping, vacuum distillation, struvite precipitation, ion exchange, breakpoint chlorination, adsorption, and chemical reduction. Advanced oxidation processes, including ozonation, peroxone oxidation, and catalytic wet air oxidation, are emerging as promising alternatives. The choice of treatment method depends on factors such as temperature, dissolved oxygen, nitrate concentration, salinity, and pH. Effective nitrogen removal is crucial for preventing eutrophication, protecting human health, and meeting increasingly stringent environmental regulations.

6.5.1 Biological Nitrification–Denitrification Processes

Biological nitrification–denitrification is the fundamental two-step process applied for removing nitrogen from wastewater. In this process, specific microbial pathways are used to oxidize ammonia (NH_4^+) to nitrogen gas (N_2). A separate population of the microbial populations catalyzes nitrification in a two-step process at variable aerobic and anoxic conditions.

Nitrification involves the oxidation of ammonia by sequential steps involving a series of autotrophic bacteria under aerobic conditions followed by oxidation. The first step Eq. (6.1) is catalyzed by species of *Nitrosomonas* that catalyze ammonia oxidation into nitrite.

$$NH_4^+ + 1.5O_2 \rightarrow NO_2^- + 2H^+ + H_2O \quad (6.1)$$

Nitrobacter species further oxidize nitrite to nitrate (NO_3^-) in an oxygen-dependent reaction Eq. (6.2):

$$NO_2^- + 0.5O_2 \rightarrow NO_3^- \quad (6.2)$$

This oxidation reaction is very sensitive to parameters like temperature, pH, and concentrations of dissolved oxygen concentration, all being regulated carefully to maximize microbial activity.

Denitrification occurs in anoxic zones where heterotrophic bacteria, including *Pseudomonas* and Paracoccus, reduce nitrate to nitrogen gas. This is carried through a series of intermediate compounds Eq. (6.3), among which are nitric oxide (NO) and nitrous oxide (N_2O):

$$NO_3^- \rightarrow NO_2^- \rightarrow NO \rightarrow N_2O \rightarrow N_2 \quad (6.3)$$

An external source of carbon must be provided as an electron donor for denitrification to complete the reduction pathway efficiently. The process is also sensitive to the levels of oxygen: even trace amounts of oxygen can inhibit the reduction pathway, thereby leading to incomplete denitrification and potential nitrous oxide emission, an important greenhouse gas. Biological nitrification–denitrification is a process of effective removal of nitrogen, although it needs control over aerobic and anoxic stages, along with mitigating greenhouse gas emissions for sustainable water resource management.

6.5.2 Chemical Precipitation and Ammonia Stripping

Chemical precipitation and ammonia stripping are two widely used physicochemical methods for nitrogen removal in wastewater treatment, each targeting different nitrogen forms and applicable under specific wastewater conditions. Chemical precipitation is commonly applied to remove ammonium and other dissolved nitrogen forms by transforming them into insoluble compounds. A prevalent approach involves dosing wastewater with magnesium and phosphate to precipitate ammonium as magnesium ammonium phosphate (MAP), or struvite. This reaction typically occurs under alkaline conditions and is represented by Eq. (6.4):

$$NH_4^+ + Mg_2^+ + PO_4^{3-} \rightarrow MgNH_4PO_4 \cdot 6H_2O(\text{struvite}) \quad (6.4)$$

The struvite formed is insoluble and can be easily removed through filtration or sedimentation. This process not only reduces nitrogen concentrations but also recovers valuable nutrients, as struvite is often used as a slow-release fertilizer. However, chemical precipitation requires careful pH control, typically around 8.5 to 9.5, to optimize struvite formation and reduce operational costs.

Ammonia stripping is an effective technique for removing free ammonia from wastewater, especially in streams with high ammonia concentrations. This process involves increasing the pH to around 11, converting ammonium ions to free ammonia gas. Air or steam is then bubbled through the wastewater in a stripping tower, transferring the ammonia gas from the liquid phase to the gas phase. The reactions Eq. (6.5) involved are:

$$NH_4^+ + OH^- \rightarrow NH_3 + H_2O \quad (6.5)$$

The gaseous ammonia is subsequently captured in an acid solution, often as ammonium sulfate or ammonium nitrate, enabling its reuse as a fertilizer component. Ammonia stripping is advantageous for high-strength ammonia wastewaters but requires significant energy input to maintain elevated pH levels and can incur high costs associated with air or steam stripping systems.

6.5.3 Limitations and Environmental Concerns

Traditional nitrogen removal technologies in wastewater treatment face limitations and environmental concerns. Biological processes, while prevalent, raise concerns due to potential N_2O emissions (Karmann et al., 2023). Biological nitrogen removal in wastewater treatment primarily relies on nitrification and denitrification processes. However, these conventional methods face limitations such as low carbon availability, nitrite and nitrous oxide accumulation, and ammonium production. Recent research has focused on novel approaches to overcome these challenges, including anaerobic ammonium oxidation (anammox), complete ammonia oxidation (comammox), and heterotrophic nitrification-aerobic denitrification (HN-AD) (Arumugham et al., 2024). HN-AD, in particular, offers advantages by achieving one-stage nitrogen removal under aerobic conditions. Other innovative processes like SHARON and simultaneous nitrification–denitrification (SND) have also been explored. Despite these advancements, further research is needed to address the impact of operating parameters, minimize nitrous oxide emissions, and apply next-generation sequencing to study microbial community diversity. Additionally, the application of biodegradable polymer substrates and heterotrophic-autotrophic denitrification processes shows promise in mitigating some limitations. In chemical precipitation and ammonia stripping, struvite precipitation can efficiently remove phosphorus and nitrogen from anaerobic digestate, with optimal performance at pH 9.5 and a $Mg:PO_4$ ratio of at least 1:1. However, the process is sensitive to pH and $Mg:PO_4$ ratio variations. Ammonia stripping, while effective, has limited evidence base and requires further research (Lorick et al., 2020). Integrating these technologies can improve efficiency and economic benefits, but challenges remain, such as optimizing struvite collection and addressing non-preferential adsorption in combined systems. Environmental concerns include the use of chemicals like $Ca(OH)_2$ for pH adjustment and the potential for residual ammonia in treated wastewater. Despite these limitations, these technologies offer promising solutions for nutrient recovery from various wastewater sources. Ongoing research aims to optimize these processes and explore new technologies to address the limitations of traditional methods.

6.6 Advanced Nitrogen Removal Technologies

Advanced nitrogen removal technologies for wastewater treatment encompass biological, physical, and chemical methods. Biological processes include simultaneous nitrification–denitrification (SND), anammox, and emerging technologies like SHARON, CANON, and OLAND. These processes can achieve high total nitrogen removal efficiencies, often exceeding 90% (Anjali & Sabumon, 2024). Physicochemical methods such as air stripping, ion exchange, and advanced oxidation processes are also effective for nitrogen removal. Recent developments in bioreactor designs, including suspended growth and moving biofilm systems, have improved treatment efficiency and resilience to varying conditions (Mahto & Das, 2022). Operational parameters, process control strategies, and economic considerations are crucial for optimizing nitrogen removal processes. Ongoing research focuses on enhancing microbial interactions, developing cost-effective solutions, and exploring novel technologies like microwave pretreatment and high-frequency electric fields to improve ammonia removal.

6.6.1 Partial Nitritation-Anammox

Partial Nitritation-Anammox is an innovative nitrogen removal technology for wastewater treatment, offering cost and energy efficiency advantages over conventional methods. PN/A is a two-step process for nitrogen removal from wastewater with low COD/N ratios. In the first step (Fig. 6.1), partial nitritation oxidizes ammonium to nitrite, typically in a granular sequencing batch reactor (SBR). The second step involves anaerobic ammonium oxidation to convert the remaining ammonium and nitrite to nitrogen gas. This process offers advantages over conventional nitrification–denitrification, including reduced oxygen requirements and sludge production. PN/A performance is influenced by operational conditions such as temperature, pH, dissolved oxygen, and retention times. Microbial communities in PN reactors are dominated by ammonium-oxidizing bacteria like *Nitrosomonas*, while anammox reactors contain Planctomycetes such as *Brocadia* (Ribeiro et al., 2022). PN/A has been successfully applied to treat livestock manure digester liquor and municipal

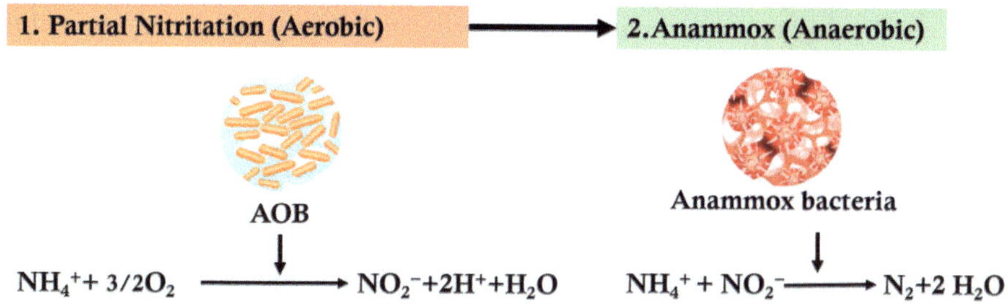

Fig. 6.1 Schematic of the partial nitritation-anammox process for nitrogen removal

wastewater, demonstrating its potential for various high-strength nitrogen wastewaters. Recent studies have demonstrated successful pilot-scale implementations of PN-A systems for municipal wastewater treatment. Kowalski et al. (2019) reported effective nitrogen removal using a two-stage MBBR/IFAS configuration, while Hoekstra et al. (2019) achieved stable PN-A performance in both summer and winter temperatures. These studies highlight the potential for mainstream PN-A application in municipal wastewater treatment plants. However, the impact of organics on PN-A systems is complex and requires careful consideration. Many studies suggest that optimum influent organics can improve nitrogen removal and operational stability. Despite challenges, PN-A technology shows promise for transforming wastewater treatment plants into energy-self-sufficient facilities by allowing more organic matter to be converted to biogas.

6.6.2 Membrane Filtration Technologies

Membrane filtration technologies play a crucial role in nitrogen removal and recovery from wastewater. These technologies include membrane bioreactors (MBRs), which combine activated sludge processes with membrane filtration (Table 6.1). Bio-membrane based integrated systems (BMISs) have emerged as promising solutions, incorporating osmotic membrane bioreactors, bioelectrochemical systems, and membrane photobioreactors. Gas permeable membranes (GPMs) and their combinations with other technologies have shown particular promise in terms of energy efficiency and recovery rates. Membrane separation processes can achieve ammonia nitrogen removal rates exceeding 99% under optimal conditions (Jin et al., 2024). The focus of research has shifted from nutrient concentration to producing marketable products using improved membrane materials and designs. These technologies offer significant potential for addressing nitrogen pollution and meeting regulatory discharge limits while simultaneously recovering valuable resources from wastewater streams. The field is evolving towards developing hybrid systems and exploring new processes like microbial fuel cells and anaerobic osmotic membrane bioreactors for improved efficiency and cost-effectiveness. Future research should focus on aligning recovery products with fertilizer industry needs and improving communication between researchers and end-users.

6.6.3 Electrochemical Processes

Nitrogen removal from wastewater is crucial to prevent environmental pollution, with various technologies being explored to improve efficiency and reduce energy consumption. Bioelectrochemical systems (BESs) have emerged as promising alternatives, combining electrodes and microorganisms to remove or recover nitrogen through oxidation, reduction, and membrane transport processes (Wan et al., 2023). Physico-chemical methods, including air stripping, ion exchange, and advanced oxidation processes, offer additional options for nitrogen removal. For mine and quarry wastewaters in cold environments, a combination of geochemical sorption and electrochemical techniques shows potential for treating large volumes with low nitrogen concentrations (Ryskie et al., 2021). Conventional biological processes like nitrification, denitrification, and anammox, along with novel bioelectrochemical methods such as microbial fuel cells and microbial electrolysis cells, are being studied for their effectiveness in nitrogen removal and recovery. These diverse approaches aim to address the challenges of nitrogen contamination while exploring opportunities for resource recovery.

6.6.4 Denitrifying Anaerobic Methane Oxidation (DAMO)

DAMO is an emerging technology for nitrogen removal from wastewater, utilizing methane as an electron donor for denitrification. This process can be coupled with anammox in membrane biofilm reactors, achieving high nitrogen removal rates without nitrite accumulation. Model-based studies have shown that optimizing the NO_2^-/NH_4^+ ratio and methane supply can lead to over 99% total nitrogen removal in such systems (Chen et al., 2016). DAMO-centered technologies offer potential for sustainable operation of wastewater treatment plants (WWTPs), addressing energy concerns by utilizing methane produced on-site from anaerobic digesters. The microbial community in these systems typically consists of DAMO bacteria, DAMO archaea, and anammox bacteria, working synergistically to remove nitrogen. Further research is needed to fully understand the mechanisms and optimize the process for practical applications in WWTPs.

Table 6.1 Overview of membrane filtration technologies for nitrogen recovery in wastewater treatment

Membrane technology	Membrane type	Characteristics	Advantages	Limitations	Applications in nitrogen recovery	References
Microfiltration (MF)	Polymer (e.g., PVDF, PES), Ceramic	Pore size: 0.1–10 μm; operates at low pressure; removes suspended solids and microorganisms	Low energy requirement; effective pre-treatment step	Does not remove dissolved nitrogen species	Pre-treatment to remove large particulates, enabling further processes	Castro & Abejón, 2024
Ultrafiltration (UF)	Polymer (e.g., PVDF, PES), Ceramic	Pore size: 0.01–0.1 μm; moderate pressure; retains bacteria and macromolecules	Removes pathogens; moderate fouling potential	Limited removal of small dissolved nitrogen compounds	Supports nitrification and denitrification by separating larger organics	Castro & Abejón, 2024
Nanofiltration (NF)	Thin-film composite	Pore size: 0.001–0.01 μm; moderate-to-high pressure; retains divalent ions, organic molecules	Removes organic nitrogen; partially removes ammonia	Higher energy costs; moderate fouling	Concentration of ammonium ions and organic nitrogen removal	Castro & Abejón, 2024
Reverse osmosis (RO)	Polyamide, Thin-film composite	Pore size: < 0.001 μm; high-pressure process; removes most dissolved ions and small molecules	High rejection rate of ammonia and nitrate	High energy requirement; significant fouling potential	Concentrates nitrogen compounds, effective in ammonia and nitrate removal	Kim et al., 2021

(continued)

Table 6.1 (continued)

Membrane technology	Membrane type	Characteristics	Advantages	Limitations	Applications in nitrogen recovery	References
Membrane bioreactor (MBR)	Hollow fiber, Flat sheet	Integrates biological treatment with membrane filtration; removes suspended solids and dissolved organics	Combines biological and filtration processes; compact design	Expensive setup; membrane fouling	Facilitates nitrification and denitrification in wastewater with high nitrogen	Cairone et al., 2024
Forward osmosis (FO)	Cellulose triacetate, Thin-film	Osmotic-driven; low energy; low fouling tendency	Energy-efficient; minimal chemical usage	Limited ammonia removal; requires draw solution regeneration	Recovery of ammonia and nitrate under low energy requirements	Liu et al., 2023
Electrodialysis (ED)	Ion-exchange membranes	Ion-selective transport driven by electric field; efficient in separating and concentrating ions	Selective ion recovery; low chemical requirement	Limited to ionized nitrogen forms; high operational complexity	Recovers ammonium and nitrate; suitable for nitrogen concentration in reuse applications	Belhamidi et al., 2022
Membrane distillation (MD)	Hydrophobic membranes (e.g., PTFE)	Utilizes thermal gradient; removes volatile nitrogen compounds like ammonia	High ammonia removal efficiency; minimal fouling	Energy-intensive; affected by temperature and salinity	Effective for ammonia recovery, especially in high-temperature streams	Carmona & Abejón, 2023

(continued)

Table 6.1 (continued)

Membrane technology	Membrane type	Characteristics	Advantages	Limitations	Applications in nitrogen recovery	References
Anion exchange membrane (AEM)	Ion-selective polymer	Selectively transports anions like nitrate under an applied electric field	High selectivity for nitrate recovery	Requires integration with other nitrogen recovery methods	Suitable for nitrate recovery in high-salinity wastewater	Jiang et al., 2021
Cation exchange membrane (CEM)	Ion-selective polymer	Selectively transports cations like ammonium in an applied electric field	Selective recovery of ammonium	Requires energy for electrical field application	Ammonium recovery; is often used in combination with ED	Jiang et al., 2021

6.6.5 Direct Ammonia Oxidation

Recent advancements in nitrogen removal technologies have focused on autotrophic processes that can efficiently treat high-strength ammonia wastewater without additional carbon sources. These include SHARON, ANAMMOX, CANON, OLAND, and other similar systems (Paul & Banerjee, 2022). While these biological methods show promise, their full-scale implementation remains limited due to various challenges. Physicochemical processes like air stripping, ion exchange, and advanced oxidation are also being explored for nitrogen removal. A novel approach utilizing microbial electrolysis cells (MECs) has emerged, where electroactive biofilms oxidize ammonium using polarized electrodes as electron acceptors, offering the potential for low C/N environments (Fan et al., 2025). Despite these innovations, each technology faces unique obstacles, and further research is needed to optimize its performance and expand its practical applications in wastewater treatment.

6.7 Nitrogen Recovery Technologies

Nitrogen is an essential and abundant nutrient in the atmosphere which exists mainly as nitrogen gas (N_2). N_2 form is largely unusable by most organisms except for certain nitrogen-fixing bacteria where N_2 is transformed into reactive forms such as nitrate (NO_3^-), nitrite (NO_2^-), ammonium (NH_4^+), and ammonia (NH_3). However, the Haber–Bosch process where atmospheric N_2 is converted into ammonia. This process is highly energy-intensive and contributes substantially to global energy consumption and greenhouse gas emissions. Therefore, a wide variety of N_2 recovery techniques are available such as filtration, stripping, adsorption, membrane processes, advanced, hybrid techniques etc. (Beckinghausen et al., 2020). Table 6.2 presents nitrogen recovery efficiency with different nitrogen recovery processes.

6.7.1 Ion Exchange and Adsorption Techniques

Ion exchange and adsorption techniques are promising methods for nitrogen recovery from wastewater. These processes can be implemented using fixed-bed column systems or dispersed throughout wastewater. Various adsorbents, including clinoptilolite, biochar, and synthetic resins, have been studied for ammonium recovery from urine, with synthetic resins showing high regeneration efficiencies and adsorption capacities (Tarpeh et al., 2017). Ion exchange technology demonstrates high nitrogen removal-recovery efficiency and competitive sustainability scores compared to conventional nitrification–denitrification and Anammox processes. Zeolites, a type of natural mineral, have been extensively researched for ammonium ion exchange in wastewater treatment. Factors influencing ammonium adsorption during the loading

Table 6.2 Application of different nitrogen recovery techniques

Technique	Source	Form	pH	NH_4^+-N (mg/L)	TN (mg/L)	N Recovery (%)	References
SLS/SE/RO	Industrial	Biogas plant digestate	7.8	3500	9910	100	Stopp et al., 2017
SMDC	Synthetic	Digestate	7.8	2400	87.9	100	Serra Toro, 2020
Struvite precipitation w/ LSSA	Agricultural	Pig slurry digestate	9.5	1727	91.95	100	Sena & Hicks, 2018
Ammonia stripping/biogas scrubbing	Agricultural	Dairy manure digestate	7.5	362	88.9	98–99	Adghim, 2023
BES	Synthetic	Digestate	–	504	83	100	Qin et al., 2017
Gas-permeable membrane	Agricultural	Swine effluent digestate	8.6	1465–2097	–	96–98	Vanotti et al., 2017
Thermal stripping/acid absorption	Agricultural	Cattle manure digestate	8.58	1100	96.5–99	100	Yang et al., 2017
Vacuum membrane distillation	Domestic	Anaerobic digestate supernatant	7.7	70	85	100	Karanasiou et al., 2023
Vacuum thermal stripping/acid absorption	Domestic	High ammonia food waste digestate	–	6400	54.45	100	Ukwuani and Tao, 2016
Liquid–liquid membrane contactors	Domestic	Zeolite regeneration solution	12.2	2400	85–98	95–98	Sancho et al., 2017
Stripping/acid absorption	Domestic urine	Undiluted urine	9.6	–	4500	76	Tian et al., 2019
Bio-electrodialysis	Domestic wastewater	Primary clarifier effluent	7.8	70	–	80	Serra Toro, 2020

(continued)

Table 6.2 (continued)

Technique	Source	Form	pH	NH_4^+-N (mg/L)	TN (mg/L)	N Recovery (%)	References
EDI Electrodeionization	Synthetic wastewater	Synthetic domestic	–	28,700	–	100	Zheng and Zhang 2023
MEC aeration	Domestic wastewater	Landfill leachate	–	4540	–	53	Zou et al., 2017
MEC-FO	Synthetic wastewater	Synthetic side stream concentrate	–	1000		100	Zou et al., 2017
MPC-IE	Domestic wastewater	Raw sewage after solid separation		27.4		37.5	Gong et al., 2017
Struvite recycling	Agricultural wastewater	Swine wastewater	7.7	378	–	91	Huang et al., 2016
Vacuum membrane distillation	Industrial wastewater	Carbonated downstream effluent from IX regeneration process	6.1	167	–	100	Yang et al., 2017
Vacuum thermal stripping/acid absorption	Domestic wastewater	Retentate of landfill leachate	10	167	–	100	Yang et al., 2017
Stripping/acid absorption	Domestic urine	Undiluted urine	9.6	–	4500	28–46	Swami et al., 2024

phase are well-understood, and both chemical and biological regeneration methods have been developed to optimize the process. Despite their potential, full-scale applications of ion exchange for nitrogen recovery in wastewater treatment plants remain limited, indicating room for further development and optimization.

6.7.2 *Ammonia Stripping*

Ammonia stripping is a chemical technique that extracts ammonium (NH_4^+) from wastewater by transforming it into ammonia gas (NH_3). This procedure comprises four primary stages i.e., conversion of ammonium ion (NH_4^+) to ammonia gas (NH_3), diffusion of NH_3 to the air–water contact, emission of NH_3 into the atmosphere at the interface, and diffusion of ammonia (NH_3) into the atmosphere above the sea surface. However, the efficiency of this process is influenced by factors such as pH, temperature, and the mass transfer area. Generally, elevated pH levels and temperatures facilitate the conversion of NH_4^+ to NH_3, hence optimizing the stripping process (Limoli et al., 2016). Ammonia gas is typically absorbed using phosphoric acid or sulfuric acid to avert its discharge into the atmosphere, where it may exacerbate greenhouse gas emissions. The absorption leads to the production of ammonium sulfate, which serves as a fertilizer with 40–60% ammonium sulfate concentration and negligible organic impurities (Shen et al., 2017). Ammonia stripping functions in continuous or batch modes, allowing for adaptability to various sizes and configurations. Nonetheless, it is restricted to the elimination of NH_4^+ and does not affect the concentrations of phosphorus or chemical oxygen demand (COD) in wastewater, both of which are essential for water quality (Cao et al., 2019).

6.7.3 *Struvite Precipitation*

Struvite precipitation is an efficient chemical method for the recovery of nitrogen and phosphorus. Struvite, or magnesium ammonium phosphate ($MgNH_4PO_4{\cdot}6H_2O$), is a white crystalline solid that functions as a beneficial slow-release fertilizer (Chen et al., 2021a, 2021b). The formation of $MgNH_4PO_4{\cdot}6H_2O$ requires magnesium (Mg^{2+}), phosphate (PO_4^{3-}), and ammonium ions (NH_4^+) to increase the solubility product of struvite and lead to make crystal formation (Cao et al., 2019). However, the stoichiometric requirements of struvite precipitation pose challenges when applying this technology to high-total ammoniacal nitrogen sources like pig slurry. TAN levels are typically high in such waste streams, but they lack sufficient magnesium and phosphate, necessitating their external addition to achieve effective struvite recovery (Guan et al., 2023). The addition of commercial-grade magnesium and phosphate sources, including $MgCl_2$, Mg $(OH)_2$, and MgO for magnesium, or H_3PO_4, Na_2HPO_4, and K_2HPO_4 for phosphate, has been commonly applied. However, these reagents add significant cost to the struvite precipitation process, reducing its overall economic viability (Astals et al., 2021). To address these cost constraints, researchers have been investigating alternative, low-cost sources of magnesium that can support struvite formation economically. Additionally, naturally occurring minerals such as magnesite, wood ash, and low-grade magnesium oxide (LG-MgO), which results from the calcination of magnesite, offer potential as alternative magnesium supplies for struvite precipitation (Chen et al., 2021a, 2021b).

6.7.4 Adsorption Using Zeolites

Zeolite is a naturally porous material that can swap cations very well and is very good at choosing ammonium ions (Fig. 6.2). Because of this, it is perfect for recovering nitrogen through adsorption. NH_4^+ is adducted onto the zeolite surface in a process that can be undone. The material then works as a physical capture agent. After being filled with nitrogen, the zeolite can be restored by treating it with a solution of NaCl or NaOH. This creates a concentrated ammonium or ammonia solution that can have amounts between 2 and 6 g/L. The zeolite that is filled with ammonium can also be used directly to improve the soil, working as a slow-release nitrogen fertilizer. This use avoids the need for removal and lowers the overall cost of treatment, making it an environmentally friendly way for farmers to recycle nitrogen (Godifredo et al., 2023).

6.7.5 Bio-Electrochemical Systems (BES)

BES provides a novel method for nitrogen recovery from wastewater which emphasizes the extraction of ammonia and ammonium from waste water. This technique is based on the biological oxidation of organic matter in the anodic compartment of the wastewater system (Chen et al., 2021a, 2021b). Bio-electrochemical systems offer a promising approach to nitrogen recovery, leveraging electrochemically active microorganisms that facilitate energy transformations between chemical and electrical forms. BES uses specialized microbial communities at either the anode or cathode of a cell, where they drive redox reactions essential for nitrogen removal and recovery. Key types of BES used for nitrogen recovery include microbial fuel cells (MFCs), microbial electrolysis cells (MECs), and microbial desalination cells (MDCs) (Beckinghausen et al., 2020). This oxidation process increases energy recovery and promotes the passage of ammonia/ammonium ions from the wastewater to the cathode compartment. Utilizing a cation exchange membrane, ammonia/ammonium ions migrate to the cathode, where they are subjected to a high pH environment. The primary impetus for the migration of ammonia/ammonium over the ion exchange membrane is the electric field generated inside the bioelectrochemical system. This configuration facilitates the effective extraction of ammonia and ammonium from wastewater (Karmann et al., 2023). BES can also facilitate ammonium

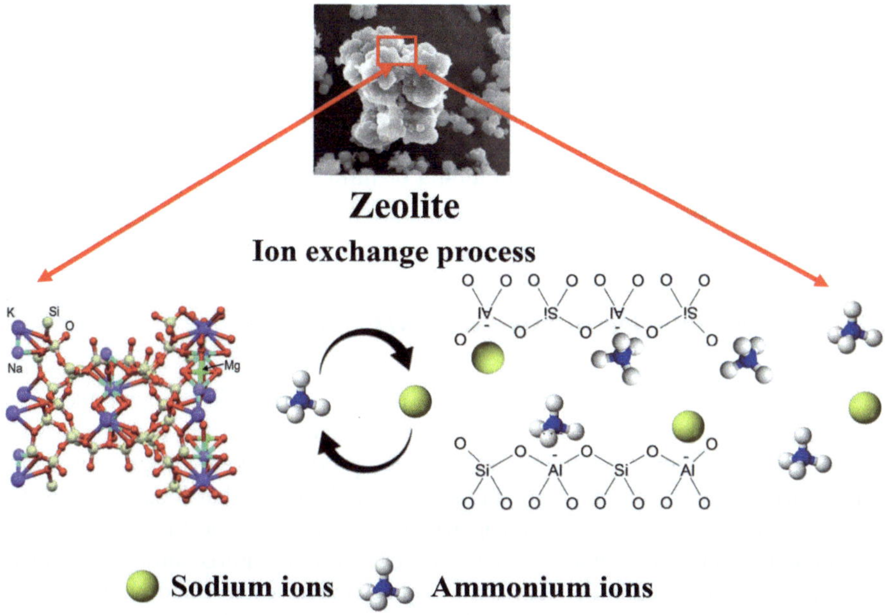

Fig. 6.2 Ion exchange of ammonium ions (NH_4^+) on zeolite, where NH_4^+ ions replace existing cations in the zeolite's lattice, enabling effective nitrogen removal from wastewater

oxidation in the cathodic compartment. Nonetheless, despite their potential applicability, BES technologies have several operational problems. The electrolytes in these systems frequently comprise different cations and anions in addition to hydrogen (H^+) and hydroxide (OH^-) ions (Rahimi et al., 2020). The translocation of these ions across the ion exchange membranes influences the pH equilibrium between the anode and cathode compartments, in addition to the membrane potential. The transfer of ammonium ions (NH_4^+) facilitates nitrogen recovery; nevertheless, pH gradients and the concentration gradients of other ions may cause considerable energy losses across the cation exchange membranes. This inefficiency constrains the overall efficacy of bioelectrochemical reactors (Astals et al., 2021).

6.7.6 Microbial Fuel Cells (MFCs)

Microbial fuel cells are systems that generate electrical energy from a thermodynamically favorable reaction occurring at the anode. In an MFC, microorganisms at the anode oxidize organic matter in wastewater, releasing electrons that travel through an external circuit to the cathode. Oxygen is often used as the electron acceptor at the cathode, completing the circuit and generating power. The main advantage of MFCs is that they recover energy directly as electricity, making them efficient for simultaneous wastewater treatment and power generation. In the context of nitrogen recovery, MFCs can facilitate nitrification and denitrification processes, where ammonia is first oxidized to nitrate (nitrification) and then reduced to nitrogen gas (denitrification), effectively removing nitrogen from the system (Boas et al., 2022). However, while MFCs are effective in nitrogen removal, their application for nitrogen recovery (in the form of ammonia or nitrate) remains limited due to the focus on energy recovery.

6.7.7 *Microbial Electrolysis Cells (MECs)*

Microbial electrolysis cells operate under a thermodynamically unfavorable reaction, requiring an external power source to drive reactions at the cathode. MECs are specifically designed to produce value-added products like hydrogen gas rather than electrical power, making them suitable for applications that focus on resource recovery (Cheng et al., 2022). For nitrogen recovery, MECs can promote the recovery of ammonia by applying a small voltage that drives the reduction of nitrate or nitrite to ammonium ions at the cathode. This ammonium can then be harvested as a nutrient or fertilizer. MECs are also advantageous for environments with high organic loading, as they support effective wastewater treatment while recovering nitrogen compounds as usable products (Zheng & Zhang, 2023).

6.7.8 *Microbial Desalination Cells (MDCs)*

Microbial desalination cells are a hybrid BES technology combining MFC functionality with desalination capability. MDCs utilize microbial oxidation of organic matter to generate electrons at the anode, similar to MFCs, but also feature a middle chamber that allows ion exchange, facilitating salt removal from water. This technology is effective in saline or brackish water treatment and is often used for water purification and desalination rather than direct nitrogen recovery. In nitrogen recovery applications, MDCs have the potential for simultaneous nutrient and salt removal, particularly in saline wastewater. This can be beneficial for agricultural use where both desalted water and recovered nitrogen nutrients are required (Chen et al., 2021a, 2021b).

6.7.9 *Microalgae and Cyanobacteria*

Microalgae and cyanobacteria are naturally present in freshwater and marine ecosystems. In wastewater systems, microalgae and cyanobacteria generate oxygen via photosynthesis, which is then employed by bacteria to decompose organic waste into simpler inorganic nutrients. These bacteria contribute to the elimination of surplus nutrients during tertiary treatment before wastewater release (Delgadillo-Mirquez et al., 2016). Nitrogen and phosphorus are vital for the growth of microalgae and cyanobacteria, rendering wastewater an optimal nutrition supply for these microorganisms. They assimilate nitrogen (in forms like nitrate and ammonia) and transform it into biomass, rather than releasing it into the environment as nitrogen gas (N_2) via dissimilatory nitrate reduction. Species such as *Chlorella* sp. have been extensively researched for nitrogen removal, sometimes in combination with other microbes or bacteria to enhance effectiveness. Certain research has investigated the nitrogen recovery potential of specific microalgal species. *Dunaliella tertiolecta, Neochloris oleobundans,* and *Chlorella vulgaris* exhibit nitrogen absorption rates of 155, 150, and 103 mg L^{-1} d^{-1}, respectively (Rahimi et al., 2020). These findings indicate that microalgae and cyanobacteria can function as efficient mechanisms for nutrient recovery in sustainable wastewater treatment.

6.7.10 Electrodialysis (ED)

ED is an ion-exchange membrane (IEM) technique that use electric fields to segregate ions, including cation and anion exchange membranes divided by compartments for dilute and concentrated solutions. The ions are transported in opposing directions, often dictated by their charges, across the ion exchange membranes (IEMs). In wastewater applications, electrodialysis is particularly advantageous for ammonia recovery. This procedure entails the removal and absorption of concentrated ammonia ions (Al-Amshawee et al., 2020). Xu et al. (2022) demonstrated an ammonia transfer efficiency of up to 96% by the integration of electrodialysis with stripping and absorption methods. Nonetheless, it was observed that the presence of other ions, such as sodium (Na^+) and potassium (K^+), might impede the effectiveness of ammonia recovery. During electrolysis, hydroxide ions produced at the cathode can inherently elevate the pH, possibly diminishing the necessity for additional chemicals. Nonetheless, there are several practical obstacles associated with electrodialysis, including membrane fouling. This fouling frequently transpires when salts precipitate on the cathode, especially during extended electrochemical deposition operations. Under situ pH management is advised to enhance ammonia recovery; however, appropriate configurations for this method remain under development (Chen et al., 2021a, 2021b).

6.7.11 Electrodeionization (EDI)

It is an advanced method that combines electrodialysis (ED) and ion exchange (IEX) processes, enhancing the efficiency of deionization by integrating both concentration polarization and chemical regeneration within EDI stacks (Chen et al., 2021a, 2021b). EDI has structural parallels to ED, however integrates ion exchange resins in the dilute compartment. These resin beds, which provide high ionic conductivity, mitigate concentration polarization by enabling continuous ion transfer. In the application of an electric field, ions in the dilute compartment are extracted by a process including adsorption onto the resin and translocation across ion-exchange membranes. A notable characteristic of EDI is the in-situ regeneration of the resin beds resulting from water splitting on the membrane and resin surfaces. This reaction produces protons and hydroxyl ions, maintaining the activity of the resin beds and facilitating continuous ion removal. When ions in the resin beds attain saturation, they provide effective transport pathways, reducing ion electro-migration resistance and facilitating optimum separation in low-conductivity solutions (Pan et al., 2018).

6.7.12 Continuous Electrodeionization (CEDI)

Continuous Electrodeionization (CEDI) is acknowledged as a novel form of conventional Electrodeionization (EDI), providing an efficient method for ion collection and discharge. In contrast to traditional EDI, which functions in batch mode by circulating diluted and concentrated solutions until the separation process concludes, CEDI employs a single-pass flow system. CEDI is ideally suited for applications that need the treatment of low-conductivity water (about 40 μS/cm), ion separation, continuous desalination, and the generation of high-purity water (Jiang et al., 2018). CEDI has demonstrated considerable adaptability, showing efficacious in the remediation of heavy metal ions (e.g., copper, nickel, and chromium), water softening, and the production of ultrapure water. The performance of this method is affected by two parameters i.e., flow rate and applied current. Nonetheless, despite its efficacy in several domains, CEDI has been little investigated for ammonia recovery (Chen et al., 2021a, 2021b).

6.7.13 Capacitive Deionization (CDI)

Capacitive Deionization is a novel ion separation process that utilizes electrodes that undergo cyclic charging and discharging to extract ionic species from aqueous solutions. This approach utilizes porous carbon electrodes that adsorb ions into the electrical double layer on the electrode surface during charging, and then release these ions back into the solution during discharge, so successfully renewing the system (Dykstra et al., 2016). In CDI technology, a crucial element is the uninterrupted circulation of aqueous solutions between the electrodes, facilitating the efficient adsorption of ions throughout the charging cycle. The efficacy of CDI is significantly affected by the specific surface area, pore architecture, and pore size distribution of the carbon electrodes since these attributes directly influence ion electro-sorption capacity (Chen et al., 2021a, 2021b). This approach has several uses, including water desalination, hard water softening, metal ion removal, and the treatment of various wastewater types (Fan et al., 2025). CDI technology has demonstrated potential in ammonia recovery, however it remains an emerging subject. Improving the efficiency of CDI for ammonia elimination and recovery can be gained by modified CDI (MCDI) configurations. Therefore, MCDI technology may provide an alternative to conventional methods for ammonia recovery from wastewater (Rihami et al., 2020).

6.7.14 Integrated System or Hybrid Method

Traditional techniques for the removal of ammonia or N2 from wastewater, such as stripping and membrane stripping, have concentrated on achieving regulatory compliance for ammonia–nitrogen discharge by concentrating ammonia and soaking it in acidic solutions to produce ammonium salts (Limoli et al., 2016). Hybrid methods that include electrochemical and stripping procedures for ammonia recovery have gained prominence to mitigate these constraints. Electrochemical techniques utilize an electrical driving force to facilitate the movement of ammonium ions for elimination, presenting the possibility for concurrent removal and recovery within a unified system. Advancements in hybrid ammonia recovery systems concentrate on enhancing operating efficiency. In previous study the nitrogen recovery rate of 94.3% by the integration of a microbial fuel cell, microbial electrolysis cell, and a stripping column for nitrogen extraction from pig slurry, achieving a maximum nitrogen flow of 25.5 g-N/d/m^2 (Chen et al., 2021a, 2021b). Likewise, the integration of membrane stripping technology with electrochemical systems enhances nitrogen recovery Kuntke et al. (2018) devised a system that integrates membrane electrodes with microbial electrolysis, attaining ammonia flow rates of 36.2 g-NH_3-N/m^2/d while markedly decreasing energy consumption to 1.61 kWh/NH_3-N, demonstrating a 70–90% enhancement in efficiency compared to conventional approaches. Subsequent developments encompassed a hydrogen recycling electrochemical system (HRES) for total ammonia nitrogen (TAN) recovery, with TAN removal efficiencies of up to 83% with energy reductions of 23.4 kJ/g-N.

6.8 Environmental and Economic Benefit of Ammonia Recovery

Nitrogen-containing wastewater derived from human and animal waste as well as industrial discharges, has raised significant environmental concerns, particularly in urban regions. Ammonia nitrogen emissions in wastewater are increasingly problematic, contributing to water pollution and associated health hazards (Palakkal et al., 2020). To address these trends, decentralized nitrogen recovery from urban wastewater sources has emerged as a sustainable solution, aiming to recycle nitrogen at a local level to minimize urban nitrogen discharges and environmental pollution.

Nitrogen recovery from wastewater helps limit eutrophication, which can otherwise lead to species loss and habitat degradation. Studies like those by Mishra et al. (2022) illustrate the relationship between nitrogen removal from wastewater and reduced eutrophication risks; with scenarios highlighting that, a total nitrogen effluent of 10 mg/L can be achieved using advanced biological nitrogen removal methods. By integrating nitrogen recovery with wastewater treatment, facilities can enhance water quality while reducing nutrient runoff, positively impacting rivers, lakes, and coastal ecosystems. Efficient nitrogen recovery methods, such as the Anammox process and struvite precipitation, offer a more environmentally friendly alternative to traditional ammonia management practices, which are energy-intensive and emit substantial greenhouse gases (GHGs) (Zhang et al., 2020a).

The environmental impact of nitrogen recovery is a major focus of LCA studies, which assess everything from resource extraction to disposal impacts of products and processes. For example, Kavvada et al. (2017) used spatial modelling and life cycle assessments to evaluate the environmental impact of ion-exchange technologies for nitrogen recovery in source-separated urine in San Francisco. Findings showed that fertilizer co-products from the recovery process contributed to energy savings and a reduction in greenhouse gases, adding to its appeal as a sustainable urban nitrogen management strategy (Chen et al., 2021b). Environmental benefits can thus be maximized by optimizing nitrogen recovery technologies to operate within a minimal environmental footprint. Economic Feasibility of Ammonia Stripping and Air Absorption Ammonia recovery technologies utilizing air stripping coupled with acid absorption are highly adaptable to various wastewater sources. They capture ammonia in volatile form to produce ammonium salts, which can be sold as fertilizers or industrial additives. The process is economically efficient, with potential profits from products like ammonium sulfate and ammonium citrate (Zhang et al., 2020a, 2020b).

6.9 Emerging and Future Technologies

With new technologies that develop nitrogen recovery, efficient wastewater management is innovated to be sustainable. Electrochemical systems include electrochemical oxidation, electrocoagulation, and ammonia stripping, allowing for the direct removal of nitrate with minimal chemical inputs and the possibility of conversion of nitrogen compounds into reusable products such as ammonium salts. This fact endows a new level of environmental benefits by reducing greenhouse gas emissions including N_2O. Other processes are also under development in the recovery of nitrogen, complete ammonia oxidation (comammox), which is another method that simplifies nitrification, by direct conversion of ammonia to nitrate in one step, increasing efficiency, while DAMO (denitrifying anaerobic methane oxidation) uses methane as the electron donor to remove nitrate—an energy-efficient low-carbon option. Hybrid approaches are particularly promising: coupling electrochemical and biological systems achieves greater efficiency. Examples include MBRs together with electrochemical cells, increasing the removal efficiencies of nitrate and reducing fouling problems and integrated anaerobic–aerobic systems that will maximize the biofilm attributes of different mixed populations of microorganisms for treating high nitrogen-containing wastewaters. It is proved by the practical application that these technologies are feasible because partial nitritation-anammox is used today throughout Europe and Asia for concentrated wastewater treatment with savings of the additional production of bioreactive sludge and energy. Recovery of ammonia from nitrogen-also utilized in the development of other industries that manage landfill leachate-can be regarded as a viable economic opportunity for capturing that substance as reused. This relates to broader changes in wastewater management toward greater environmental sustainability and circularity.

6.10 Challenges and Future Directions

Technical and operational challenges put severe restrictions on the full-scale implementation of nitrogen recovery technologies in wastewater management. One of the important issues is that advanced treatments, such as electrochemical and partial nitritation-anammox, require a high energy expenditure, which in turn influences their economic feasibility, especially at a larger scale of application. Another issue with variability in the composition of wastewater, with fluctuations that call for adaptation measures to obtain a maintained level of consistency regarding nitrogen removal efficiency. In addition, membrane fouling, a problem that often occurs in hybrid systems, and the need for exact pH control in ammonia stripping processes make it even more operative and demanding usually requiring more resources for maintenance and infrastructure. This is also an environmental issue that must be addressed through the appropriate management of secondary pollution, for example, disposal of residual by-products such as sludge or concentrated brine. New studies are nowadays focused on improvement and adaptability, promising lines of development. For instance, the optimization of microbial processes—both comammox and anaerobic methane oxidation—could treat greater wastewater profile diversity. AI-driven monitoring and control technologies promise great scope for real-time optimization, which would increase stability and minimize intervention by humans. Future innovations in biodegradable polymers for membranes and microbial community engineering will help to overcome persistent operational problems and further improve the overall sustainability of efforts in nitrogen recovery. Further work through targeted research and cross-disciplinary innovation in new directions will be important to move nitrogen-recovery technologies toward sustainable large-scale solutions for wastewater treatment.

6.11 Conclusions

Advanced nutrient recovery technologies for nitrogen removal and recovery thus represent a new significant shift towards the sustainable management of wastewater, and the traditional methods for treating nitrogen may be effective, but they often lack efficiency and give rise to a series of problems in the environment such as nitrous oxide emission and high sludge production. Further increasing efficiency with minimal environmental impact, the partial nitritation-anammox, electrochemical systems, and hybrid methods efficiently use energy for nitrogen removal while minimizing greenhouse gas emissions. The biological innovation coming from comammox and anaerobic methane oxidation to an integrated approach using both biological and physicochemical methods helped handle variables involved in wastewater conditions optimally recover nitrogen. Areas for continued research and development are necessary to continue forward in addressing remaining operational challenges and to further the broad usage of these sustainable technologies, primarily concerning AI-driven monitoring systems and next-generation materials. Improved strategies exist that can boost nitrogen management, promote better water quality, and support circular resource use during wastewater treatment.

Acknowledgements The authors sincerely acknowledge the institutional support provided by the ICAR–Indian Agricultural Research Institute, New Delhi—110012, India, which was instrumental in the successful preparation of this chapter.

References

Adghim, M. (2023). *Post-hydrolysis ammonia stripping as a new approach to enhance methane potential of high nitrogen feedstock* [Doctoral dissertation, Université d'Ottawa/University of Ottawa].

Ahmed, S. F., Kumar, P. S., Kabir, M., Zuhara, F. T., Mehjabin, A., Tasannum, N., & Mofijur, M. (2022). Threats, challenges and sustainable conservation strategies for freshwater biodiversity. *Environmental Research, 214*, 113808.

Al-Amshawee, S., Yunus, M. Y. B. M., Azoddein, A. A. M., Hassell, D. G., Dakhil, I. H., & Hasan, H. A. (2020). Electrodialysis desalination for water and wastewater: A review. *Chemical Engineering Journal, 380*, 122231.

Anjali, G., & Sabumon, P. C. (2024). Revolutionizing Wastewater Treatment: Cutting-Edge Technologies for Biological Nutrient Removal. In *Biological and Hybrid Wastewater Treatment Technology: Recent Developments in India* (pp. 125–158). Cham: Springer Nature Switzerland.

Astals, S., Martínez-Martorell, M., Huete-Hernández, S., Aguilar-Pozo, V. B., Dosta, J., & Chimenos, J. M. (2021). Nitrogen recovery from pig slurry by struvite precipitation using a low-cost magnesium oxide. *Science of the Total Environment, 768*, 144284.

Beckinghausen, A., Odlare, M., Thorin, E., & Schwede, S. (2020). From removal to recovery: An evaluation of nitrogen recovery techniques from wastewater. *Applied Energy, 263*, 114616.

Belhamidi, S., El-Ghzizel, S., Taky, M., & Elmidaoui, A. (2022). Nitrate removal of groundwater by reverse osmosis, nanofiltration and electrodialysis: Performances and cost comparison. *Desalination and Water Treatment, 262*, 338–346.

Boas, J. V., Oliveira, V. B., Simões, M., & Pinto, A. M. (2022). Review on microbial fuel cells applications, developments and costs. *Journal of Environmental Management, 307*, 114525.

Cairone, S., Mineo, A., Pollice, A., Belgiorno, V., Mannina, G., & Naddeo, V. (2024, June). Innovative Membrane Bioreactors for Advanced and Sustainable Wastewater Treatment. In *IWA Regional Membrane Technology Conference* (pp. 120–126). Cham: Springer Nature Switzerland.

Cao, L., Wang, J., Xiang, S., Huang, Z., Ruan, R., & Liu, Y. (2019). Nutrient removal from digested swine wastewater by combining ammonia stripping with struvite precipitation. *Environmental Science and Pollution Research, 26*, 6725–6734. https://doi.org/10.1007/s11356-019-04153-x

Carmona, B., & Abejón, R. (2023). Innovative membrane technologies for the treatment of wastewater polluted with heavy metals: Perspective of the potential of electrodialysis, membrane distillation, and forward osmosis from a bibliometric analysis. *Membranes, 13*(4), 385.

Castro, K., & Abejón, R. (2024). Removal of heavy metals from wastewaters and other aqueous streams by pressure-driven membrane technologies: An outlook on reverse osmosis, nanofiltration, ultrafiltration and microfiltration potential from a bibliometric analysis. *Membranes, 14*(8), 180.

Chen, J., Wei, X. D., Liu, Y. S., Ying, G. G., Liu, S. S., He, L. Y., Su, H. C., Hu, L. X., Chen, F. R., Yang, Y. Q. (2016). Removal of antibiotics and antibiotic resistance genes from domestic sewage by constructed wetlands: Optimization of wetland substrates and hydraulic loading. *Science of the Total Environment, 565*, 240–248.

Chen, T. L., Chen, L. H., Lin, Y. J., Yu, C. P., Ma, H. W., & Chiang, P. C. (2021a). Advanced ammonia nitrogen removal and recovery technology using electrokinetic and stripping process towards a sustainable nitrogen cycle: A review. *Journal of Cleaner Production, 309*, 127369.

Chen, T. L., Chen, L. H., Lin, Y., Yu, C. P., Ma, H. W., & Chiang, P. C. (2021b). Advanced ammonia nitrogen removal and recovery technology using electrokinetic and stripping process towards a sustainable nitrogen cycle: A review. *Journal of Cleaner Production, 309*, 127369. https://doi.org/10.1016/j.jclepro.2021.127369

Cheng, D., Ngo, H. H., Guo, W., Chang, S. W., Nguyen, D. D., Zhang, S., Deng, S., An, D., & Hoang, N. B. (2022). Impact factors and novel strategies for improving biohydrogen production in microbial electrolysis cells. *Bioresource Technology, 346*, 126588.

Delgadillo-Mirquez, L., Lopes, F., Taidi, B., & Pareau, D. (2016). Nitrogen and phosphate removal from wastewater with a mixed microalgae and bacteria culture. *Biotechnology Reports, 11*, 18–26. https://doi.org/10.1016/j.btre.2016.04.003

Dykstra, J. E., Zhao, R., Biesheuvel, P. M., & van der Wal, A. (2016). Resistance identification and rational process design in capacitive deionization. *Water Research, 88*, 358–370.

Fan, S., Song, Y., Zheng, D., Peng, X., Li, S., Gao, P., & Li, D. (2025). Advancing autotrophic nitrogen removal in low C/N ratio wastewater: Innovative application of supercapacitor to enhance microbial electrolysis cells. *Journal of Environmental Sciences, 152*, 87–98.

Garcia, C., Pérez-Sánchez, M., Ramos, H. M., López-Jiménez, P. A., & Sánchez-Romero, F. J. (2024). New strategies in wastewater treatment plants to evaluate the achievement of sustainable development goals: A case study. *Urban Water Journal*, 1–12.

Godifredo, J., Ferrer, J., Seco, A., & Barat, R. (2023). Zeolites for nitrogen recovery from the anaerobic membrane bioreactor permeate: Zeolite characterization. *Water, 15*, 1007. https://doi.org/10.3390/w15061007

Gong, H., Wang, Z., Zhang, X., Jin, Z., Wang, C., Zhang, L., et al. (2017). Organics and nitrogen recovery from sewage via membrane-based pre-concentration combined with ion exchange process. *Chemical Engineering Journal, 311*, 13–19.

Guan, Q., Li, Y., Zhong, Y., Liu, W., Zhang, J., Yu, X., & Zeng, G. (2023). A review of struvite crystallization for nutrient source recovery from wastewater. *Journal of Environmental Management, 344*, 118383.

Hoekstra, M., Geilvoet, S. P., Hendrickx, T. L., van Erp Taalman Kip, C. S., Kleerebezem, R., & van Loosdrecht, M. C. (2019). Towards mainstream anammox: Lessons learned from pilot-scale research at WWTP Dokhaven. *Environmental Technology, 40*(13), 1721–1733

Huang, H., Zhang, P., Zhang, Z., Liu, J., Xiao, J., & Gao, F. (2016). Simultaneous removal of ammonia nitrogen and recovery of phosphate from swine wastewater by struvite electrochemical precipitation and recycling technology. *Journal of Cleaner Production, 127*, 302–310.

Huang, Y., Zhao, Y., Tang, C., Yadav, A. K., Abbassi, R., Kang, P., Cai, Y., Liu, A., Yang, A., Li, M. (2023). A glance of coupled water and wastewater treatment systems based on microbial fuel cells. *Science of the Total Environment, 892*, 164599.

Jiang, B., Zhang, X., Zhao, X., & Li, F. (2018). Removal of high-level boron in aqueous solutions using continuous electrodeionization (CEDI). *Separation Science and Technology, 192*, 297–301.

Jiang, S., Sun, H., Wang, H., Ladewig, B. P., & Yao, Z. (2021). A comprehensive review on the synthesis and applications of ion exchange membranes. *Chemosphere, 282*, 130817.

Jin, X. G., Tang, X., Ren, T. X., Wang, J., Zheng, P., Ma, X. H., & Xu, Z. L. (2024). Ionic liquids tailored ultra-permeable antifouling nanofiltration membranes for water purification. *Journal of Membrane Science, 696*, 122536.

Karanasiou, A., Angistali, K., Plakas, K. V., Kostoglou, M., & Karabelas, A. J. (2023). Ammonia recovery from anaerobic-fermentation liquid digestate with vacuum membrane distillation. *Separation and Purification Technology, 314*, 123602.

Karmann, C., Magrova, A., Jenicek, P., Bartacek, J., & Kouba, V. (2023). Advances in nitrogen removal and recovery technologies from reject water: Economic and environmental perspectives. *Bioresource Technology, 391*, 129888. https://doi.org/10.1016/j.biortech.2023.129888

Kavvada, O., Tarpeh, W. A., Horvath, A., & Nelson, K. L. (2017). Life-cycle cost and environmental assessment of decentralized nitrogen recovery using ion exchange from source-separated urine through spatial modeling. *Environmental Science & Technology, 51*, 12061–12071.

Kim, J., Lim, J., Park, K., & Hong, S. (2021). Fundamentals and application of reverse osmosis membrane processes. In *Osmosis Engineering* (pp. 17–52). Elsevier.

Kowalski, M. S., Devlin, T. R., di Biase, A., & Oleszkiewicz, J. A. (2019). Effective nitrogen removal in a two-stage partial nitritation-anammox reactor treating municipal wastewater–piloting PN-MBBR/AMX-IFAS configuration. *Bioresource Technology, 289*, 121742.

Kuntke, P., Sleutels, T., Rodriguez Arredondo, M., Georg, S., Barbosa, S. G., Ter Heijne, A., Hamelers, H. V. M., & Buisman, C. J. N. (2018). (Bio)electrochemical ammonia recovery: Progress and perspectives. *Applied Microbiology and Biotechnology, 102*, 3865–3878.

Limoli, A., Langone, M., & Andreottola, G. (2016). Ammonia removal from raw manure digestate by means of a turbulent mixing stripping process. *Journal of Environmental Management, 176*, 1–10. https://doi.org/10.1016/j.jenvman.2016.03.007

Liu, H., Wang, C., Sohn, W., Wang, Q., Shon, H. K., & Sun, P. (2023). Source-separated urine treatment based on forward osmosis technology: Performance, applications and future prospects. *Desalination, 565*, 116872.

Lorick, D., Macura, B., Ahlström, M., Grimvall, A., & Harder, R. (2020). Effectiveness of struvite precipitation and ammonia stripping for recovery of phosphorus and nitrogen from anaerobic digestate: A systematic review. *Environmental Evidence, 9*, 1–20.

Mahto, K. U., & Das, S. (2022). Bacterial biofilm and extracellular polymeric substances in the moving bed biofilm reactor for wastewater treatment: A review. *Bioresource Technology, 345*, 126476.

Mishra, S., Singh, V., Cheng, L., Hussain, A., & Ormeci, B. (2022). Nitrogen removal from wastewater: A comprehensive review of biological nitrogen removal processes, critical operation parameters and bioreactor design. *Journal of Environmental Chemical Engineering, 10*(3), 107387.

Młyńska, A., & Chmielowski, K. (2024). The changes of nitrogen content during sewage treatment: A study of a two-stages wastewater treatment plant. *Journal of Water and Land Development*, 106–114.

Mohamed, D., Fergany, M., Elhabbasha, E. F., & El-temsah, M. (2022). Productivity improvement of canola genotypes under salinity stress conditions by integration between mineral and nano-scale forms of nitrogen fertilizer. *Arab Universities Journal of Agricultural Sciences, 30*(2), 229–244.

Nygaard, K., Schaper, S., Jacobsen, B. H., & Hansen, B. (2022). Sustainable value propositions of a new technology for targeted nitrogen regulation. *Journal of Cleaner Production, 337*, 130496.

Palakkal, V. M., Valentino, L., Lei, Q., Kole, S., Lin, Y. J., & Arges, C. G. (2020). Advancing electrodeionization with conductive ionomer binders that immobilize ion-exchange resin particles into porous wafer substrates. *npj Clean Water, 3*.

Pan, S.-Y., Snyder, S. W., Lin, Y. J., & Chiang, P.-C. (2018). Electrokinetic desalination of brackish water and associated challenges in the water and energy nexus. *Environmental Science: Water Research & Technology, 4*, 613–638.

Paul, D., & Banerjee, A. (2022). Technologies for biological and bioelectrochemical removal of inorganic nitrogen from wastewater: A review. *Nitrogen, 3*(2), 298–313.

Qin, M., Liu, Y., Luo, S., Qiao, R., & He, Z. (2017). Integrated experimental and modeling evaluation of energy consumption for ammonia recovery in bioelectrochemical systems. *Chemical Engineering Journal, 327*, 924–931.

Rahimi, S., Modin, O., & Mijakovic, I. (2020). Technologies for biological removal and recovery of nitrogen from wastewater. *Biotechnology Advances, 43*, 107570.

Rajta, A., Bhatia, R., Setia, H., & Pathania, P. (2020). Role of heterotrophic aerobic denitrifying bacteria in nitrate removal from wastewater. *Journal of Applied Microbiology, 128*(5), 1261–1278.

Ribeiro, H., Wijaya, I. M. W., Soares-Santos, V., Soedjono, E. S., Slamet, A., Teixeira, C., & Bordalo, A. A. (2022). Microbial community composition, dynamics, and biogeochemistry during the start-up of a partial nitritation-anammox pathway in an upflow reactor. *Sustainable Environment Research, 32*(1), 18.

Rodríguez-de la Garza, J. A., Pérez-Rodríguez, P., González-Méndez, L. M., & Martínez-Amador, S. Y. (2023). Recent advances in biological wastewater treatment. *Current Status of Fresh Water Microbiology*, 205–240.

Ryskie, S., Neculita, C. M., Rosa, E., Coudert, L., & Couture, P. (2021). Active treatment of contaminants of emerging concern in cold mine water using advanced oxidation and membrane-related processes: A review. *Minerals, 11*(3), 259.

Sancho, I., Licon, E., Valderrama, C., de Arespacochaga, N., López-Palau, S., & Cortina, J. L. (2017). Recovery of ammonia from domestic wastewater effluents as liquid fertilizers by integration of natural zeolites and hollow fiber membrane contactors. *Science of the Total Environment, 584*, 244–251.

Sena, M., & Hicks, A. (2018). Life cycle assessment review of struvite precipitation in wastewater treatment. *Resources, Conservation and Recycling, 139*, 194–204.

Serra Toro, A. (2020). *Nitrogen recovery from urban wastewater by means of membranes technology*.

Shen, Y., Tan, M. T. T., Chong, C., Xiao, W., & Wang, C. H. (2017). An environmentally friendly animal waste disposal process with ammonia recovery and energy production: Experimental study and economic analysis. *Waste Management, 68*, 636–645. https://doi.org/10.1016/j.wasman.2017.07.027

Stopp, P., Weichgrebe, D., & Rosenwinkel, K. H. (2017). *Innovative digestate evaporation for nutrient recovery and post-processing*. Resources Recovery, Beijing, China.

Swami, S., Vyas, A. D., Malik, B., & Shrivastava, R. (2024). Agrochemical nutrients reclamation from human urine and wastewater via struvite ($MgNH_4PO_4{\cdot}6H_2O$) crystallization: A comprehensive review on recent developments. *Proceedings of the Indian National Science Academy, 90*(3), 638–663.

Tarpeh, W. A., Udert, K. M., & Nelson, K. L. (2017). Comparing ion exchange adsorbents for nitrogen recovery from source-separated urine. *Environmental Science & Technology, 51*(4), 2373–2381.

Tian, X., Gao, Z., Feng, H., Zhang, Z., Li, J., & Wang, A. (2019). Efficient nutrient recovery/removal from real source-separated urine by coupling vacuum thermal stripping with activated sludge processes. *Journal of Cleaner Production, 220*, 965–973.

Ukwuani, A. T., & Tao, W. (2016). Developing a vacuum thermal stripping–acid absorption process for ammonia recovery from anaerobic digester effluent. *Water Research, 106*, 108–115.

Van der Hoek, J. P., Duijff, R., & Reinstra, O. (2018). Nitrogen recovery from wastewater: Possibilities, competition with other resources, and adaptation pathways. *Sustainability, 10*(12), 4605.

Vanotti, M. B., Dube, P. J., Szogi, A. A., & García-González, M. C. (2017). Recovery of ammonia and phosphate minerals from swine wastewater using gas-permeable membranes. *Water Research, 112*, 137–146.

Viancelli, A., & Michelon, W. (2024). Climate change and nitrogen dynamics: Challenges and strategies for a sustainable future. *Nitrogen, 5*(3), 688–701.

Wan, Y., Li, R., Wang, X., & Liao, C. (2023). Recovery of reactive nitrogen from wastewater using bioelectrochemical systems. *Separation and Purification Technology, 327*, 125002.

Wang, Z. (2021). *Management of vessel-borne biological invasions: policy, economic, and technology assessment*. University of Delaware.

Xu, L., Liu, S., Zhao, S., Li, K., Cao, A., & Wang, J. (2022). A novel electrocoagulation-membrane stripping hybrid system for simultaneous ammonia recovery and contaminant removal. *Separation and Purification Technology, 296*, 121377.

Yang, A. L., Raghuram, N., Adhya, T. K., Porter, S. D., Panda, A. N., Kaushik, H., Jayaweera, A., Nissanka, S. P., Anik, A. R., Shifa, S., Sharna, S. C., Joshi, R., Watto, M. A., Pokharel, A., Shazly, A., Hassan, R., Bansal, S., Kanter, D., Das, S., Jeffery, R. (2022). Policies to combat nitrogen pollution in South Asia: gaps and opportunities. Environmental Research Letters, 17(2), 025007.

Yang, X., Pang, H., Zhang, J., Liubinas, A., & Duke, M. (2017). Sustainable wastewater deammonification by vacuum membrane distillation without pH adjustment: Role of water chemistry. *Chemical Engineering Journal, 328*, 884–893.

Zhang, C., Ma, J., & Waite, T. D. (2020a). The impact of absorbents on ammonia recovery in a capacitive membrane stripping system. *Chemical Engineering Journal, 382*, 122851.

Zhang, H., Wang, L., Van Herle, J., Maréchal, F., & Desideri, U. (2020b). Techno-economic comparison of green ammonia production processes. *Applied Energy, 259*, 114135.

Zheng, M., & Zhang, Y. (2023). Nutrient recovery from wastewater using advanced technologies: An overview and future prospects. *Science of the Total Environment, 770*, 145414.

Zielińska, M., & Bułkowska, K. (2025). Use of membrane techniques for removal and recovery of nutrients from liquid fraction of anaerobic digestate. *Membranes, 15*(2), 45.

Zou, S., Qin, M., Moreau, Y., & He, Z. (2017). Nutrient-energy-water recovery from synthetic sidestream centrate using a microbial electrolysis cell-forward osmosis hybrid system. *Journal of Cleaner Production, 154*, 16–25.

Chapter 7
Microalgae as a Biofactory: Integrating Wastewater Treatment with Renewable Energy Production

Charu Sharma, Swati, and Papiya Mukherjee

Abstract Microalgae are drawing significant attention for their role as renewable energy sources and their wide-ranging utility in both industry and agriculture. These single-celled autotrophs help produce oxygen while manufacturing essential biomolecules. Microalgae can survive in multiple environments and effectively remove nutrients from wastewater, positioning them as essential tools in wastewater treatment and circular economy systems. This chapter examines new harvesting techniques and methods to recover important organic compounds from wastewater, thereby emphasizing the combination of microalgae-based treatment systems with bioenergy production. Economically valuable products emerge from wastewater when microbiological processes convert it into sustainable solutions for water scarcity and pollution. This chapter illustrates the innovations in bioprocessing and wastewater treatment by highlighting the significant contributions of microalgae to environmental protection and renewable energy production. Microalgae play an essential role in the bio-circular economy where they support resource recovery and global sustainability initiatives through the ability to connect wastewater management and bioenergy production.

Keywords Microalgae · Wastewater treatment · Biomass · Bioenergy · Nanoparticles

7.1 Introduction

Scientists have been studying microalgae for over a hundred years, and even today modern biotechnology serves as a source of renewable energy as well as a supplier of industrial and agricultural products. Microalgae are single-celled autotrophic organisms that produce proteins, lipids, carbohydrates, colors along with other valuable molecules. These species are essential for oxygen production and require sunlight as

C. Sharma · Swati · P. Mukherjee (✉)
Department of Botany, Panjab University, Chandigarh, India
e-mail: pmukherjee@pu.ac.in

S. Dhiman and G. Mukherjee (eds.), *Resource Resurgence*, Springer Water,
https://doi.org/10.1007/978-3-032-02750-4_7

their only energy source by producing half of the oxygen we breathe. While thriving in many aquatic habitats, these photosynthetic bacteria have shown potential for environmental sustainability and generation of renewable energy. Efficiency in high nutrient removal has also been shown by algae growing in freshwater and many kinds of wastewater, including municipal wastewater, anaerobic digestate effluent, and cattle wastewater (Xu et al., 2015; Luo et al., 2019). The wastewater management concept is a comprehensive overview of the sustainable reuse of various categories of wastewater and the effective remediation of wastewater (Meena et al., 2019). Research aimed at reducing water scarcity mostly focuses on this as it is the pillar of the circular economy and helps to address water shortage (Hossain et al., 2020). Though in small amounts, the treated water has several commercially important organic chemicals that can be used in agriculture and horticulture (Rajasulochana & Preethy, 2016). The recovery of these elements from wastewater by employing microbiological techniques is now drawing attention. Physical, chemical, biological, and magnetic techniques have been applied to treat wastewater produced from several operations. Employing microorganisms in wastewater treatment and resource recovery facilitates the management of waste streams and closure of material cycles. Nonetheless, the technology and methods applied have to be chosen carefully by considering technological feasibility, simplicity, cost-effectiveness, cultural relevance, and political aspects. The production of economically useful products is facilitated by meticulous planning and research that are tailored to the availability of organic compounds, nutrient requirements of various microbial species, origin and elements of the effluent (Morais et al., 2021). The type of wastewater has enabled the development of different methods that combine wastewater treatment with the manufacturing of industrially relevant products. Nevertheless, the methods of wastewater management using microorganisms as well as the effective synthesis of biohydrogen, bioalcohol, biomethane, and bioelectricity are under progress (Arutselvan et al., 2022). The recent trends in the evolution of bioprocesses and wastewater treatment for the recuperation of value-added products have been critically analyzed. The focus of this chapter is to examine the potential of harvesting techniques to establish a self-sustaining "factory" by integrating microalgae-based wastewater remediation along with bioenergy production. This emphasizes the potential of microalgae as a major element to shape the future of sustainable energy and wastewater management by analysing its technological developments and obstacles. The acceleration of sustainable global development could be facilitated by advanced bio-circular economy processes. These processes rely on a deeper understanding and improvement of bioprocesses used for wastewater treatment, along with the large-scale synthesis of bioenergy.

7.2 Strategies for Harvesting Microalgae for Wastewater Treatment

7.2.1 Physical Methods

Mechanical or physical techniques are used in several ways for microalgae extraction from wastewater to separate microalgal biomass from the liquid media. In wastewater treatment, these techniques are vital since they allow microalgae to be recovered following absorption of nutrients and pollutants. The recovery percentage of microalgae in different species by filtration technique is summarized in Table 7.1.

7.2.1.1 Filtration

Filtration is a fundamental physical process used to separate solids from liquids, therefore allowing only liquids to pass. Microalgae are collected using several filtration techniques, including vacuum filtration, pressure filtration, microfiltration, and ultrafiltration. During the filtration process, several elements are crucial in controlling the microalgae harvesting. Membrane fouling is a major issue in the framework of expensive harvesting methods and low fluxes throughout the procedure. Usually, a recommended option to reduce fouling is surface-coating materials like hydrophilic polyvinyl alcohol (PVA) polymer. A nylon 6,6 nanofiber membrane clearly showed the improved filtration efficacy for *Chlorella vulgaris* due to its increased surface pore density, larger pore size, and resistance to fouling (Bilad et al., 2018). Particularly with larger patterns, Zhao et al. (2020) observed that membranes with higher polyethylene glycol and polysulfone content displayed different pattern forms, hence lowering fouling and increasing membrane flux. The most recent study integrated aeration with ultra-low-pressure membrane filtration to harvest *C. vulgaris* while noting significant changes in broth permeability during the process (Osman et al., 2021). The highest permeance was observed in patterned polysulfone membranes fabricated with 28% polyethylene glycol in pure water and microalgal suspensions (Zhao et al., 2021). Chen et al. (2021) designed superabsorbent polymer spheres to filter microalgal cultures. These beads were more easily separated and were reused during microalgal concentration with higher swelling ratios in microalgal suspensions. Therefore, multiple permeable membranes have been developed and used to filter microalgal debris in membrane filtration systems efficiently. Rapid decrease in permeance was shown by a nylon 6,6 nanofiber membrane with a 0.12 μm pore size, surface porosity of 25.82%, and water permeance (1018 L/m^{2}h bar). The permeance dropped gradually from 1018–528 L/m^{2}h bar and then further to 300 L/m^{2}h by filtration within 15 min (Bilad et al., 2018). In the harvesting of *C. vulgaris*, Ultra-low-pressure membrane filtering has been identified with low energy usage (4.4×10^{-1} kWh/m^1), and membranes having pore sizes 40–100 kDa were proved effective

Table 7.1 Recovery percentage of microalgae in different species by filtration technique

Microalgae	Filtration	Recovery (%)	Reference
Chlorella sp.	Crossflow membrane filtration-hydrophilic polyvinyl alcohol polymer	100%	Hwang et al., 2013
	Ultrafiltration, Fouling-resistant PVDF membranes	94%, 100%	Hwang et al., 2015
	Transparent exopolymeric particles-polycarbonate filter	97%	Discart et al., 2013
	Crossflow membrane filtration, UH050 membrane-hydrophilic polyethersulfone	100%	Elcik and Cakmakci, 2017
Chlorella vulgaris	Crossflow filtration, ultra-low-pressure filtration system	76%	Osman et al., 2021
	Coagulation-polyaluminum chloride and polytetrafluoroethylene membrane	31% lipid, 28% protein, 8% carbohydrate	Lee et al., 2012
	Nylon 6,6 nanofiber membrane, polyvinylidene fluoride phase-inverted membrane	Enhanced its competitiveness	Bilad et al., 2018
	Nylon 6,6 nanofiber membrane	–	Mat Nawi et al., 2020
	Combining the submerged membrane bioreactor microfiltration with centrifugation	–	Bilad et al. 2012
Chlorella pyrenoidosa	Diatomite dynamic membrane	–	Zhang et al. 2012
Nannochloropsis gaditana	Dynamic filtration, polyethersulfone membrane	–	Nurra et al. 2014
Dictyosphaerium sp.	Combination of patterned membrane filtration and flocculation at standardized chitosan dosage, crossflow filtration, polyethylene glycol	Highest stable membrane permeance	Zhao et al., 2021
	Combining the submerged membrane bioreactor microfiltration with centrifugation	–	Bilad et al. 2012
Phaeodactylum tricornutum	Dynamic filtration, polyethersulfone membrane	–	Nurra et al. 2014
Aurantiochytrium sp.	Dynamic filtration module, an FMX B-class	100%	Kim et al., 2015

for long-term use. Furthermore, compared to microfiltration membranes, ultrafiltration membranes show better permeability and fouling resistance (Osman et al., 2021).

7.2.1.2 Flotation

Recently, flotation is considered to be a major separation method used to remove microorganisms from suspension. The type of flotation method described in Table 7.2 determines the recovery percentage by different algae species. During the flotation process, air or gas changes into bubbles inside a solid or liquid solution. These bubbles let solid particles connect to the gaseous molecules and gather at the top surface. The size of the bubbles determines forms of the flotation process, ozonation dispersed flotation, electrolytic flotation, dissolved air flotation, and dispersed air flotation. Using chitosan and saponin in the previously described technique has improved *C. vulgaris* and *Scenedesmus obliquus* yield (Kurniawati et al., 2014). Shen et al., (2018) showed that the surfactant Cetyltrimethylammonium bromide (CTAB) offered better harvesting efficiency for *C. vulgaris* relative to tea saponin. While tea saponin changed the surface of microalgae from hydrophilic to hydrophobic, the neutralizing potential of the algae by CTAB resulted in developing the affinity between *C. vulgaris* and bubbles. Al-Humairi et al., (2023) claim that harvesting efficiency of *C. vulgaris* was enhanced by raising the concentration of cationic trimethylammonium bromide at the initial level in the continuous foam flotation procedure. Leite et al. (2020) found pH modification during dissolved air flotation (DAF) as an effective biomass collecting technique for *C. sorokyana from wasterwater*. The marine alga, *Nannochloropsis sp.,* was harvested efficiently from DAF method by using a tannin-based biopolymer flocculant (AFlok-BP1) (Fuad et al., 2021). This procedure generates surface-layered polymeric microspheres in *C. vulgaris* for achieving at a maximum harvesting efficacy of 98.43%. In *C. vulgaris,* combining traces of aluminium sulfate with sodium alginate microspheres showed a novel approach to achieve a substantially higher harvesting efficiency than both air flotation and conventional buoy-bead flotation methods (Zhang et al., 2022). Thermal pre-flocculation denatures lipids, carbohydrates, and proteins on microalgal cell surfaces, reducing electrostatic repulsion between cells and air bubbles, thereby enhancing harvesting efficiency. A novel additive-free harvesting method was developed for *Ochromonas danica,* leveraging its hydrophobic surface and larger size for efficient lipid production from waste. Optimized flotation with waste cooking oil and air flotation achieved over 90% recovery in a single stage, with > 98% recovery at higher cell concentrations or using multi-stage processing (Hosseini et al., 2016). In *Nannochloropsis sp.*, a bio-flotation technique using oil emulsion and moringa protein extract was developed to reduce chemical residues, achieving a harvesting efficiency of 86.5% (Kandasamy and Shaleh 2018).

Table 7.2 Microalgal biomass recovery percentages in different species by flotation technique

Microalgae	Flotation	Recovery (%)	References
Chlorella vulgaris	Dispersed air flotation	93	Kurniawati et al., 2014
	Dissolved air flotation	90	Koley et al., 2017
	Surfactant, hexadecyltrimethyl-ammonium bromide, and tea saponin	89.23	Shen et al., 2018
	Buoy-bead flotation, surface-layered polymeric microspheres	98.43	Xu et al., 2018
	Thermal flotation	91.96	Zou et al., 2019
	N, N'-bis (ethyl dimethyl)-1,4-butane diammonium dibromide	99.2	Huang et al., 2019
	Buoy-bead flotation	89.9	Wen et al., 2019
	Colloidal gas aphrons technology, surfactants—cationic hexadecyl trimethyl ammonium bromide, anionic sodium dodecylbenzene sulfonate, sodium dodecyl sulfate, and combinations of these surfactants	95	Pal et al., 2021
Chromochloris zofingiensis	Buoy-bead flotation, sodium alginate microspheres	93.78	Zhang et al., 2022
	Continuous foam flotation, cationic trimethyl-ammonium bromide	96	Al-Humairi et al., 2023
	Dissolved air flotation, dissolved organic matter, increasing Al^{3+} concentration	95.2	Zhang et al., 2012
Chlorella sp.	Al^{3+} and cetyltrimethylammonium bromide	98.73	Xia et al. 2017
Chlorella sorokiniana	Dissolved air flotation, pH modulation	96.5–97.9	Souza Leite et al., 2020
Scenedesmus obliquus	Dispersed air flotation	93	Kurniawati et al., 2014
	Dissolved air flotation	90	Koley et al., 2017
	Thermal flotation	91.96	Zou et al., 2019
Ochromonas danica	Oil and air flotation	89	Hosseini et al., 2016
Dunaliella salina	Flocculation/flotation	80	Besson et al. 2019
Arthrospira platensis	Surfactants—cationic hexadecyltrimethylammonium bromide, anionic sodium dodecylbenzene sulfonate, sodium dodecyl sulfate, and combinations of these surfactants	95	Pal et al., 2021
Nannochloropsis sp.	Flotation, *Moringa* protein–oil emulsion	86	Kandasamy and Shaleh, 2018

(continued)

Table 7.2 (continued)

Microalgae	Flotation	Recovery (%)	References
	Dissolved air flotation, tannin-based biopolymer flocculant, AFlok-BP1	–	Fuad et al., 2021
Nannochloropsis oculata	Surfactants—cationic hexadecyl trimethyl ammonium bromide, anionic sodium dodecylbenzene sulfonate, sodium dodecyl sulfate, and combinations of these surfactants	95	Pal et al., 2021

7.2.2 Chemical Method

As a result of chemical and bio-flocculants, the technique of chemical flocculation enables microscale algal biomass to be harvested at a different degree of efficiency. Different bio-flocculants result in varying success rates, but overall, this remains to be a productive method for microalgae extraction. Gerchman et al. (2017) tested the sedimentation of *C. vulgaris* cultures with the use of polydiallyldimethylammonium chloride, superfloc and chitosan. Among the three, polydiallyldimethylammonium chloride performed the best at 90% flocculation with 5 mg/L over 60 min at pH 10. This suggests that certain flocculence can be utilized to increase the chance of recovering microalgal biomass. Liu et al., (2020a, 2020b) confirmed this with their microwave-assisted flocculation approach that there is a decrease in the necessary flocculant dosage in cultures of *C. vulgaris.* Flocculation using ferric sulfate ($Fe_2(SO_4)_3$) and sulfuric acid (H_2SO_4) for harvesting *Chlorella sp.* KR-1 is done in a more economical and industrial friendly way according to Kim et al. (2017). This method solve the problems associated with contamination and coagulant reuse, as well as in the aiding ferric ion recovery. Additionally, the acid-treated biomass allows direct lipid extraction by achieving a 32% yield and 90% FAME conversion efficiency, making it highly suitable for microalgal biorefinery applications.

Koley et al. (2017) investigated many harvesting methods in *Scenedesmus obliquus*, reporting flocculation efficiency of 80.2% with $FeCl_3$, 95% with alum, and 91% with chitosan. Likewise, $FeCl_3 \cdot 6H_2O$ (0.50 mM) proved to be quite an efficient flocculant for *Dunaliella salina* (Khadim et al., 2018). Using $CaCl_2$ (Caetano et al., 2020), the biomass of *Arthrospira maximum* was effectively obtained at concentrations between 0.2 and 2.0 g/L. An inventive helical tube method was designed, Lin et al. (2021) to increase microalgal biomass recovery in *Chlorella.* A 90% recovery rate was obtained using an algal solution of 0.12 g/L, flocculating for 18 s, and then sedimenting for 15 min. Recent research on *C. vulgaris* showed that flocculants based on sulfates and chloride have efficiency ranging from 93.5–98.8%. Morphological observations of *Phaeodactylum tricornutum, Porphyridium purpureum*, and *C. vulgaris* have highlighted the significance of alkaline pH in the flocculation process

(Ma et al., 2022). Furthermore, adding polyacrylamide retained industrially important chemical components and enhanced harvesting efficiency. These results underline the requirement of optimizing flocculant type, concentration, and environmental factors, including pH, for effective microalgae harvesting with possible applications in biofuel generation and wastewater treatment.

7.2.3 Biological Method

Among the several flocculation techniques used to gather microorganisms, bioflocculation is especially interesting since it reduces the detrimental effects of chemical leftovers. Given this, a cationic polyelectrolyte, chitosan, is rather used to extract several algae species due to its biodegradable and non-toxic characteristics. Still, the high expense of chitosan has driven research on other bio-flocculants. The study explores bioflocculation as an efficient microalga harvesting method, enhancing flocculant effectiveness while reducing costs. Chitosan at a concentration of 10 mg/L was successful in achieving 89% efficiency, while walnut protein extract (WPE) alone reached 40%. However, co-flocculation with WPE improved chitosan's efficiency to 98% at a reduced concentration (6 mg/L), with optimal performance at neutral pH. This is a reasonably affordable and sustainable method for large-scale microalgal collection (Xu et al., 2021), according to structural investigation revealing a web-like interaction. The flocculation efficiency of cationic nanocrystalline cellulose was examined in *C. sorokiniana* under harvesting conditions. Microalgal cultures were observed to flocculate at concentrations both above and below their isoelectric point (Lopez-Exposito et al., 2019). By use of polyacrylamide and chitosan, Wang et al., (2022a, 2022b) enhanced the *C. vulgaris* harvesting. Without changing the chemical composition, chitosan (10 mg/L) showed greater flocculation efficiency than polyacrylamide (25 mg/L). Furthermore, included to improve the collection of microalgae and reduce biological contamination, a cationic biopolymer, α-poly-L-lysine was added, which possess inherent antibacterial qualities. The most successful way of producing pellets, according to Toscano et al. (2018), is by co-cultivating *Aspergillus flavus* and *C. sorokiniana* in nutrient supplemented media. Using *A. oryzae* pellets at 30 °C, 130 rpm, and a 1:1 fungal to algal ratio, Chu et al. (2021) likewise attained a maximum harvesting efficiency of 99.23% in *C. vulgaris*. The authors also proposed that the formation of metabolites in the medium could affect the properties of bioflocculation. Chen et al. (2018) assessed different coagulants for recovering microalgal biomass from anaerobic wastewater using a fungal spore and pellet-assisted approach using *Penicillium sp.* Cassini et al. (2017) reported a similar effectiveness of 99% for *Chlorella sp.* At pH 7–10, cationic starch, modified tannin, and aluminium sulfate attained > 90% recovery whereas cationic starch performed better over a larger pH range. *Hibiscus esculentus* and *Moringa oleifera* gum were less effective as natural coagulants. The efficiency of organic flocculants was also investigated. Tannin showed flocculation efficiency on Niemi and Gentili (2021) comparable to that of tested microalgal species. Yang et al. (2021) evaluated natural flocculants, *Moringa oleifera*, cationic starch, chitosan, Tanfloc for harvesting *C. vulgaris* and *Scenedesmus obliquus*. In comparison, Tanfloc exhibited the highest harvesting efficacy at 30 mg/L for C. vulgaris and 20 mg/L for S. obliquus.

7.2.4 Magnetic Method

Atomic or molecular aggregates with a range of 1–100 nm in size, nanoparticles (NPs) show different physicochemical characteristics than bulk materials (Saxena & Harish 2018). Among them, magnetic nanoparticles (MNPs) have attracted interest for collecting microalgae because of their minimal contamination risk, great efficiency, fast, automated, and scalable processing capacity. Their small scale, special surface qualities, and somewhat low manufacturing cost make them valuable. MNPs follow microalgal cells, allowing for quick, effective, and large-scale separation by use of an external magnetic field, hence facilitating easy recovery (Seo et al., 2017). Current research is focused on functionalizing them with tannins (Wang et al., 2022a, 2022b), ammonium quaternaries (Qin et al., 2021), and other hydrophobic chemicals (Kumar et al. 2022) thereby improving their durability and harvesting efficiency. Reusability of nanoparticles, convenience of biomass removal, simplicity of surface re-functionalization, and so many benefits make this method potentially more energy-saving than traditional centrifugation and filtration (Patel et al., 2022). Still, the demand for affordable mass manufacturing solutions and recovery of such micro and nanoparticles persists since most of the current techniques depend on solvents and bases that set microalgae apart from metallic components. Though methods for microalgae harvesting have advanced significantly, commercial uses for filtration, magnetic, and microfluidic approaches remain limited. Harvesting microalgae is the best by using traditional coagulation/flocculation methods. This approach has created guidelines which along with life cycle assessment studies, uses metallic ions or chemical compounds. Analysing the many coatings for magnetic nanoparticles to improve the efficiency and speed of microalgae harvesting, has lately attracted attention.

7.2.4.1 Nanoparticles and Polyethylenimine

Recent work shows how well polyethylenimine (PEI)-coated magnetic nanoparticles collect microorganisms. Emphasizing the great magnetic and amine-functionalized capabilities of these nanocomposites, Gerulová et al. (2018) showed that Fe_3O_4 nanoparticles functionalized with PEI achieved an astounding 97% harvesting efficiency for *C. ellipsoidea* at a low dosage of 20 mg/L. Likewise, a study of magnetic nanoparticles in algal biotechnology by Li et al. (2024) underlined their efficiency in quick biomass separation, therefore underlining the need of PEI-coated versions for optimal microalgae recovery. In harvesting *C. pyrenoidosa*, Liu et al. (2019)

further compared several functionalized magnetic nanoparticles finding that PEI-coated Fe_3O_4 outperformed both uncoated Fe_3O_4 and acrylamide-functionalized forms. Although these experiments validate the efficiency of PEI-functionalized nanoparticles, differences in microalgal species and nanoparticle modifications imply that optimal techniques should be customized to particular algal strains and intended uses.

7.2.4.2 Nanoparticles and Polyamidoamine

Polyamidoamine (PAMAM)-coated nanoparticles have lately shown great success in improving microalgae harvesting efficiency. Strong electrostatic interactions between PAMAM and microalgal cells are highlighted by Liu et al., (2020a, 2020b) reporting on amino-rich PAMAM-coated Fe_3O_4 nanoparticles obtaining a 95% harvesting efficiency for *Chlorella sp.* Similar surface-functionalized iron oxide nanoparticles (SF-IONPs) for magnetophoretic separation of *Chlorella sp.*, Zhang et al., (2023) confirmed their robust adhesion to microalgal cells and effective biomass recovery. While both studies confirm the potential of functionalized nanoparticles in microalgae harvesting, Liu et al., (2020a, 2020b) especially emphasized the advantages of PAMAM coatings, whereas Zhang et al., (2023) investigate surface-functionalized iron oxide nanoparticles without stating the polymer utilized.

7.2.4.3 Nanoparticles-Polyaluminium Chloride and Polyacrylamide

The application of polyacrylamide (PAM) and polyaluminum chloride (PAC) in nanoparticle-based materials has attracted interest for enhancing their features in several domains. Adding PAC to tin oxide (SnO_2) nanoparticle precursor solutions improves the dispersion of the nanoparticles more precisely and lowers their colloidal size in perovskite solar systems. Better electrical conductivity in SnO_2 films results from this, which could improve photovoltaic performance (Zhang et al., 2025). In a similar line, scientists have created gold nanoparticles and polymerized them under sunlight to synthesis polyacrylamide-based nanocomposite hydrogels. This method removes the need for extra initiators or crosslinking chemicals, therefore producing hydrogels with better mechanical strength and possible uses in biomedical sectors (Anguilar et al., 2025). These results underline the flexibility of PAC and PAM are in improving nanoparticle-based materials, hence advancing many different technical domains.

7.2.4.4 Nanoparticles and Chitosan

Natural biopolymer chitosan has shown great success in the harvesting of microalgae. Islam et al., (2023) found that chitosan extracted by demineralization, deproteination, and deacetylation had 97% solubility and 58% degree of deacetylation, considerable for flocculation. Its potential for biomass recovery reached 100% harvesting efficiency was at a flocculant dosage of 20 mg/L with a mixing rate of 120 rpm. In a similar line, Wang et al., (2023) investigated chitosan-functionalized magnetic nanoparticles and found that their adsorption qualities greatly enhanced microalgae separation efficiency. While both studies validate chitosan's importance in biomass harvesting, Islam et al., (2023) concentrated on its direct use as a flocculant, whereas Wang et al., (2023) investigated its function in nanoparticle functionalizing for improved separation.

7.2.4.5 Nanoparticles and Poly (Diallyldimethylammonium Chloride)

By improving the flocculation and sedimentation of microalgal cells, PDDA-coated magnetic nanoparticles helped Chan et al., (2023) to lower energy use in biomass collecting. Encapsulated with the cationic polymer PDDA, magnetic graphene oxide (GO) nanocomposites was known for their efficacy in capturing *Chlorella sp.* Their stable graphene base and strong positive charge enable GO-Fe_3O_4/PDDA nanocomposites to work effectively throughout a wide pH range (4–12), hence compatible for industrial usage. With a raw/reactivated mixed method, these nanomaterials also demonstrated great recycling stability and maintained high harvesting efficiency across five cycles (Liu et al., 2016).

7.2.4.6 Nanoparticles and Fumed Silica Particles

Flame-derived, silica-coated magnetic particles provide a cost-effective alternative to precipitated magnetite for improving the magnetic separation of microalgae. Fe_3O_4 was used as a carrier when Zhang et al., (2023) produced magnetic mesoporous silica nanoparticles and assessed their efficacy in gathering *Chlorella sp.* Good dispersion and steady harvesting efficiency over a broad pH range (4–12) were shown by silica-coated Fe_3O_4 microspheres, according to the study. Eliminating organic elements from the surface of the material enhanced its porosity structure, therefore improving harvesting efficiency by 17.8% (from 65.9% to 83.7%). Further a maximum harvesting efficiency of 84.2% was achieved by the molar ratio of the active agent cetyltrimethylammonium bromide (CTAB) to the stabilizer polyvinylpyrrolidone (PVP) when set at 1:0.092 (pH 4) with an algal concentration of 0.5 g/L. While the literature highlights the effectiveness of fumed silica nanoparticles in microalgae harvesting, this study focused on the stability of magnetic mesoporous silica, emphasizing improvements in porous structure.

7.2.4.7 Nanoparticles and Plant Polyphenol

Using magnetic particles coated in plant polyphenolic (PP), *C. vulgaris* was magnetically collected by Zhao et al., (2015). The Fe_3O_4-QPP (quaternized polyphenol) composite directly caught 91% of the cells from the culture medium at a concentration of 5.0 g/L, therefore producing a high harvesting efficiency. In an alkaline environment Fe_3O_4-QPP maintained an efficient harvesting percentage. Using ultrasound to effectively separate magnetic particles from Fe_3O_4-QPP, cell aggregates produced the recovery of 96% of the magnetite. With a harvesting efficiency of 76%, the recycled Fe_3O_4-QPP composite kept a good degree of performance despite the tenth reuse cycle. Polyphenols from *Sargassum wightii* have facilitated the biosynthesis of zinc oxide nanoparticles with a high zeta potential of − 49.39 mV, indicating notable stability for biomedical applications (Sundaresan & Kasi, 2024). These studies underscore the potential of plant polyphenols in the eco-friendly synthesis of nanoparticles, offering sustainable alternatives for environmental and biomedical applications.

7.3 Utilization of Microalgae Biomass for Resource Recovery

7.3.1 Biofuels

Offering a renewable substitute for fossil fuels, biofuels come from organic sources like plants, agricultural waste, and algae. They are a better energy source since these fuels are created by natural processes including photosynthesis, fermentation, and fat extraction. Microalgae stand out among other biofuel sources for their fast development, great production, and abundance of energy-dense molecules including carbs and lipids. Biogas, biodiesel, and bioethanol can all be made from microalgae. While carbohydrate-rich species are fermented to generate bioethanol, therefore strengthening its promise as a sustainable fuel source, lipid-rich strains are broken down into biodiesel by transesterification.

7.3.1.1 Biohydrogen

High energy yielding biohydrogen, 141.65 MJ/kg, is a sustainable and clean substitute for fossil fuels. Using diverse metabolic routes, several bacteria including algae, cyanobacteria, dark fermentative bacteria, and photosynthetic bacteria each generate it physiologically. As they can fix CO_2, acquire vital nutrients, and use sunlight via photosynthesis, microalgae show great promise for producing biohydrogen. Khetkorn et al. (2017) have looked closely at the processes in which microalgae generate biohydrogen. Two primary enzymes, hydrogenases (H_2ase) and nitrogenases (N_2ase), play key roles in biohydrogen production across microbial systems. Starting from an algal culture exposed to light following a dark and anaerobic acclimation phase, the biophotolysis process in microalgae directly links hydrogen generation. As oxygen generated during photosynthesis can inactivate the binding sites of the [FeFe]-hydrogenase enzyme, it works only in anaerobic conditions in wild-type algae. This calls for methods to keep cell viability while reducing photosynthesis. Using stored biological components under anaerobic circumstances generates hydrogen (H_2). Nutrient depletion (e.g., sulfur, nitrogen, phosphorous, and magnesium), acetate control, and coculturing with different bacterial strains have been investigated to meet these obstacles (Wirth et al., 2018). Some bacteria can oxidize oxygen, hence creating anaerobic conditions fit for H_2 generation in algal systems. Moreover, using bacterial strains with low hydrogen absorption capacity will help to improve H_2 yield even more.

7.3.1.2 Bioalcohol

Although first-generation feedstocks such as food crops have long been the source of bioalcohol, this raises questions regarding world food security. As such, a viable solution for sustainable bioalcohol generation is alternative and reasonably priced feedstocks like microalgae grown in wastewater. When fermenting, microalgae have better photosynthetic efficiency and favourable chemical composition than traditional crops. In turn, wastewater can supply vital nutrients for the growth of algae, therefore lowering environmental impact and production costs. For instance, Rizza et al. (2019) produced biomass with up to 60% carbohydrate content by dry weight by growing cyanobacteria able of absorbing ambient nitrogen using a semi-closed loop system. After biomass pretreatment with diluted sulfuric acid, fermentation with yeast produced 50 g/l of ethanol. Onay (2019) evaluated *Hindakia tetrachotoma* ME03 grown in varying wastewater concentrations as a bioethanol feedstock in a flat-photobioreactor. Enzymatic hydrolysis yielded the highest saccharification efficiency (92.3%), and fermentation with *Saccharomyces cerevisiae* achieved a maximum bioethanol yield of 11.2 g L^{-1} in 50% wastewater. Addressing contamination through filtration could enhance its feasibility for bioethanol production. With a yield of 1.31 g/l, Chavan and Mutnuri (2021) likewise treated household wastewater using microalgal consortia, removing lipids from the biomass before fermenting the leftover biomass to generate ethanol.

7.3.1.3 Biomethane

Combining wastewater treatment with renewable energy generation is shown by using wastewater-grown algal biomass for biogas generation, especially methane. Utilizing CO_2 and nutrients from wastewater, photosynthetic algae may generate important molecules including lipids, proteins, and carbohydrates. Through anaerobic digestion, this combined approach not only reduces wastewater treatment costs rather recovers nutrients and increases biogas generation. Driven by separate microbial groups, methane fermentation- the method used to produce biogas, occurs in four stages, hydrolysis, acidogenesis, acetogenesis, and methanation. Methane yield is affected by elements like biomass composition, pH, temperature, and retention duration. Important is the carbon to nitrogen (C/N) ratio since imbalances can either rapidly deplete nitrogen too quickly or lead to ammonia accumulation, both of which reduce methane generation. Nutrient levels change the composition of algal biomass; restricting nitrogen and phosphorus improves carbohydrate and lipid content, hence improving methane production (Perazzoli et al., 2016). Studies reveal that with their higher lipid content, nitrogen-starved algae create more methane than species rich in carbohydrates. By breaking down complicated organic components, pretreatment techniques like microwave, alkaline, and heat treatments allow easy bacterial break down and increase methane generation. By balancing methane-producing bacteria, co-digesting algal biomass with waste products such as sewage sludge or lipids increases methane yield even more.

7.3.1.4 Bioelectricity

Microorganisms breaking down biomass waste or wastewater generates a renewable energy source called bioelectricity. Microbial fuel cells (MFCs), which handle organic matter employing microbial metabolism to transform chemical energy into electricity, offer one way for producing bioelectricity from wastewater, is depicted in Fig. 7.1. MFCs can process several wastewater types without pretreatment, generate little silt, and require less energy (He et al., 2017). At a 50:50 DSW-SWW ratio, Nayak and Ghosh (2018) built an H-type MFC for distillery spent wash (DSW) and sewage wastewater (SWW), with 97% suspended particles removal and a maximum power density of 836.81 mW/m^2. Other investigations have looked at MFCs for various wastewater sources. Using a two-chamber MFC, Firdous et al., (2018) treated vegetable oil industry effluent generating 5839 mV and 90% COD removal at 35 °C. Developing a trickling MFC for brewery wastewater, Gao et al., (2020) removed 60% of COD with a hydraulic retention time of 1.6 h and produced an average power density of 0.27 W/m^2. Apart from bioelectricity, MFCs have been assessed for biomass generation and CO_2 sequestration employing household wastewater (Ratna et al., 2021). Developed employing photosynthetic organisms, including *Synechococcus sp.*, these MFCs with *Chlorococcum sp.* MFC technology may produce bioelectricity concurrently and lower BOD and COD in wastewater. Still, it is essential to improve electrode performance and control the physio-chemical environment for bacterial growth for broad uses.

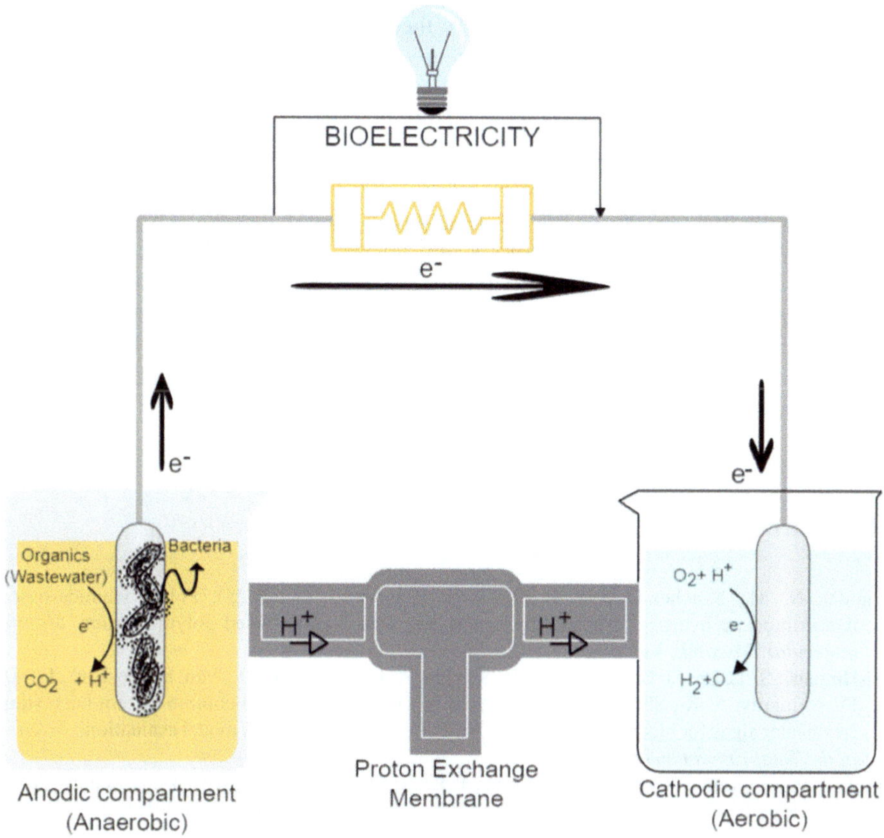

Fig. 7.1 The mechanism of generating electricity from wastewater

7.4 Conclusion and Future Prospects

The commercialization of microalgae and their products is an important process that needs evaluation of appropriate harvesting methods. Most studies show that *Chlorella sp.* have many industrial uses, and are among the well-investigated microalgae. There are several factors, including, microalgal species, chemical residues from post-harvesting, water body characteristics, pH, and medium composition, that influence the efficacy of microalgae biomass harvesting. Flocculation is a common method utilizing either natural or chemical flocculants in the growth of several types of microalgae. Bioflocculants are specially preferred and are sometimes known as natural organic flocculants, because of their low cost and lower contamination levels in the produced biomass. Combining two or more harvesting methods such as coagulation with flocculation or filtration with centrifugation has been shown to increase the effectiveness of harvesting in some cases. Employing microbial processes, wastewater can be transformed into economically valuable products such as biohydrogen, bioalcohol, biomethane, and bioelectricity, thereby offering sustainable solutions for pollution and water shortage. Despite recent advancements, many challenges remain yet to be solved, like the high energy requirements and the need for affordable harvesting techniques that could guarantee both purity and a high biomass output. The objective of future studies should be the improvement of these processes by genetically modifying the algal growth rates, and designing of economical, large-scale systems that can be seamlessly integrated into existing wastewater treatment systems. As microalgal biotechnology and harvesting methods keep developing, the notion of wastewater treatment plants as bioenergy factories gets more realistic. Adopting and investing in this technology can help us create a future whereby wastewater is turned from a liability to a profitable resource, thereby progressing us toward a sustainable, low-carbon world that would eventually protect ecosystems and empower communities.

References

Aguilar, N. M., Sanchez-Gaytan, B. L., & Soriano-Moro, G. (2025). Polyacrylamide-based nanocomposite hydrogel prepared by synergistic solar light-induced polymerization. *Macromolecular Research*, 1–7.

Al-Humairi, S. T., Lee, J. G., Harvey, A. P., Salman, A. D., Juzsakova, T., Van, B., Le, P. C., La, D. D., Mungray, A. K., Show, P. L., & Nguyen, D. D. (2023). A foam column system harvesting freshwater algae for biodiesel production: An experiment and process model evaluations. *Science of the Total Environment*, *862*, 160702.

Arutselvan, C., Kumar Seenivasan, H., Oscar, F. L., Ramya, G., Chi, N. T. L., Pugazhendhi, A., & Thajuddin, N. (2022). Review on wastewater treatment by microalgae in different cultivation systems and its importance in biodiesel production. *Fuel*, *324*, 124623

Bilad, M. R., Azizo, A. S., Wirzal, M. D. H., Jia, L. J., Putra, Z. A., Nordin, N. A. H. M., Mavukkandy, M. O., Jasni, M. J. F., & Yusoff, A. R. M. (2018). Tackling membrane fouling in microalgae filtration using nylon 6, 6 nanofiber membrane. *Journal of Environmental Management*, *223*, 23–28.

Bilad, M. R., Vandamme, D., Foubert, I., Muylaert, K., & Vankelecom, I. F. (2012). Harvesting microalgal biomass using submerged microfiltration membranes. *Bioresource Technology*, *111*, 343–352.

Caetano, N. S., Martins, A. A., Gorgich, M., Gutiérrez, D. M., Ribeiro, T. J., & Mata, T. M. (2020). Flocculation of Arthrospira maxima for improved harvesting. *Energy Reports*, *6*, 423–428.

Cassini, S. T., Francisco, S. A., Antunes, P. W. P., Oss, R. N., & Keller, R. (2017). Harvesting microalgal biomass grown in anaerobic sewage treatment effluent by the coagulation-flocculation method: Effect of pH. *Brazilian Archives of Biology and Technology*, *60*(00), e17160174.

Chan, K. S., Leung, S. K., Wong, S. S. Y., Chan, S. S., Suen, D. W. S., Tsang, C. W., & Chan, C. Y. (2023). Development of an energy-efficient rapid microalgal cell-harvesting method using synthesized magnetic nanocomposites. *Water*, *15*(3), 545.

Chavan, R., & Mutnuri, S. (2021). Domestic wastewater treatment by constructed wetland and microalgal treatment system for the production of value-added products. *Environmental Technology*, *42*(21), 3304–3317.

Chen, J., Leng, L., Ye, C., Lu, Q., Addy, M., Wang, J., Liu, J., Chen, P., Ruan, R., & Zhou, W. (2018). A comparative study between fungal pellet-and spore-assisted microalgae harvesting methods for algae bioflocculation. *Bioresource technology*, *259*, 181–190.

Chen, W., Wang, T., Dou, Z., & Xie, X. (2021). Microalgae harvesting by self-driven 3D microfiltration with rationally designed porous superabsorbent polymer (PSAP) beads. *Environmental Science & Technology, 55*(22), 15446–15455.

Chu, R., Li, S., Yin, Z., Hu, D., Zhang, L., Xiang, M., & Zhu, L. (2021). A fungal immobilization technique for efficient harvesting of oleaginous microalgae: Key parameter optimization, mechanism exploration and spent medium recycling. *Science of the Total Environment, 790*, 148174.

de Souza Leite, L., Dos Santos, P. R., & Daniel, L. A. (2020). Microalgae harvesting from wastewater by pH modulation and flotation: Assessing and optimizing operational parameters. *Journal of Environmental Management, 254*, 109825.

Discart, V., Bilad, M. R., Vandamme, D., Foubert, I., Muylaert, K., & Vankelecom, I. F. J. (2013). Role of transparent exopolymeric particles in membrane fouling: Chlorella vulgaris broth filtration. *Bioresource Technology, 129*, 18–25.

Elcik, H., & Cakmakci, M. (2017). Harvesting microalgal biomass using crossflow membrane filtration: critical flux, filtration performance, and fouling characterization. *Environmental Technology, 38(12)*, 1585–1596.

Firdous, S., Jin, W., Shahid, N., Bhatti, Z. A., Iqbal, A., Abbasi, U., Mahmood, Q., & Ali, A. (2018). The performance of microbial fuel cells treating vegetable oil industrial wastewater. *Environmental technology & innovation, 10*, 143–151

Fuad, N., Omar, R., Kamarudin, S., Harun, R., Idris, A., & Wakg, W. A. (2021). Harvesting marine microalgae Nannochloropsis sp. using dissolved air flotation (DAF) technique. *Sains Malaysiana, 50*(1), 73–83.

Gao, N., Fan, Y., Long, F., Qiu, Y., Geier, W., & Liu, H. (2020). Novel trickling microbial fuel cells for electricity generation from wastewater. *Chemosphere, 248*, 126058.

Gerchman, Y., Vasker, B., Tavasi, M., Mishael, Y., Kinel-Tahan, Y., & Yehoshua, Y. (2017). Effective harvesting of microalgae: Comparison of different polymeric flocculants. *Bioresource Technology, 228*, 141–146.

Gerulová, K., Bartošová, A., Blinová, L., Bártová, K., Dománková, M., Garaiová, Z., & Palcut, M. (2018). Magnetic Fe3O4-polyethyleneimine nanocomposites for efficient harvesting of Chlorella zofingiensis, Chlorella vulgaris, Chlorella sorokiniana, Chlorella ellipsoidea and Botryococcus braunii. *Algal Research, 33*, 165–172.

He, L., Du, P., Chen, Y., Lu, H., Cheng, X., Chang, B., & Wang, Z. (2017). Advances in microbial fuel cells for wastewater treatment. *Renewable and Sustainable Energy Reviews, 71*, 388–403.

Hossain, N., Bhuiyan, M. A., Pramanik, B. K., Nizamuddin, S., & Griffin, G. (2020). Waste materials for wastewater treatment and waste adsorbents for biofuel and cement supplement applications: A critical review. *Journal of Cleaner Production, 255*, 120261.

Hosseini, M., Starvaggi, H. A., & Ju, L. K. (2016). Additive-free harvesting of oleaginous phagotrophic microalga by oil and air flotation. *Bioprocess and Biosystems Engineering, 39*, 1181–1190.

Huang, Z., Cheng, C., Liu, Z., Luo, W., Zhong, H., He, G., & Fu, W. (2019). Gemini surfactant: A novel flotation collector for harvesting of microalgae by froth flotation. *Bioresource Technology, 275*, 421–424.

Hwang, T., Park, S. J., Oh, Y. K., Rashid, N., & Han, J. I. (2013). Harvesting of Chlorella sp. KR-1 using a cross-flow membrane filtration system equipped with an anti-fouling membrane. *Bioresource Technology, 139*, 379–382.

Hwang, T., Kotte, M. R., Han, J. I., Oh, Y. K., & Diallo, M. S. (2015). Microalgae recovery by ultrafiltration using novel fouling-resistant PVDF membranes with in situ PEGylated polyethyleneimine particles. *Water Research, 73*, 181–192.

Islam, M. B., Khalekuzzaman, M., Kabir, S. B., & Hossain, M. R. (2023, April). Characterization of chitosan extracted from shrimp shell waste and its utilization as a flocculant for harvesting of microalgae. In *AIP Conference Proceedings* (Vol. 2713, No. 1). AIP Publishing.

Kandasamy, G., & Shaleh, S. R. M. (2018). Flotation removal of the microalga Nannochloropsis sp. using Moringa protein–oil emulsion: A novel green approach. *Bioresource Technology, 247*, 327–331.

Khadim, S. R., Singh, P., Singh, A. K., Tiwari, A., Mohanta, A., & Asthana, R. K. (2018). Mass cultivation of Dunaliella salina in a flat plate photobioreactor and its effective harvesting. *Bioresource Technology, 270*, 20–29.

Khetkorn, W., Rastogi, R. P., Incharoensakdi, A., Lindblad, P., Madamwar, D., Pandey, A., & Larroche, C. (2017). Microalgal hydrogen production–a review. *Bioresource Technology, 243*, 1194–1206.

Kim, D. Y., Lee, K., Lee, J., Lee, Y. H., Han, J. I., Park, J. Y., & Oh, Y. K. (2017). Acidified-flocculation process for harvesting of microalgae: Coagulant reutilization and metal-free-microalgae recovery. *Bioresource Technology, 239*, 190–196.

Kim, K., Shin, H., Moon, M., Ryu, B. G., Han, J. I., Yang, J. W., & Chang, Y. K. (2015). Evaluation of various harvesting methods for high-density microalgae, Aurantiochytrium sp. KRS101. *Bioresource Technology, 198*, 828–835.

Koley, S., Prasad, S., Bagchi, S. K., & Mallick, N. (2017). Development of a harvesting technique for large-scale microalgal harvesting for biodiesel production. *RSC Advances, 7*(12), 7227–7237.

Kurniawati, H. A., Ismadji, S., & Liu, J. C. (2014). Microalgae harvesting by flotation using natural saponin and chitosan. *Bioresource Technology, 166*, 429–434.

Lee, D. J., Liao, G. Y., Chang, Y. R., & Chang, J. S. (2012). Coagulation-membrane filtration of Chlorella vulgaris. *Bioresource Technology, 108*, 184–189.

Li, L., Xue, S., Zhang, Y., Gao, Y., Yang, J., Zhang, X., & Zhang, W. (2024). A chemical-free magnetophoretic approach for recovering magnetic particles in microalgae removal through magnetic separation. *Journal of Cleaner Production, 467*, 143025.

Lin, Z., Li, C., Liu, J., Yang, Z., & Zhang, H. (2021). An effective process of harvesting Chlorella sp. biomass for bioresource by rapid flocculation in a helical tube. *Desalination and Water Treatment, 221*, 440–445.

Liu, P. R., Zhang, H. L., Wang, T., Yang, W. L., Hong, Y., & Hou, Y. L. (2016). Functional graphene-based magnetic nanocomposites as magnetic flocculant for efficient harvesting of oleaginous microalgae. *Algal Research, 19*, 86–95.

Liu, P., Wang, T., Yang, Z., Hong, Y., Xie, X., & Hou, Y. (2020). Effects of Fe3O4 nanoparticle fabrication and surface modification on Chlorella sp. harvesting efficiency. *Science of the Total Environment, 704*, 135286.

Liu, W., Cui, Y., Cheng, P., Huo, S., Ma, X., Chen, Q., Cobb, K., Chen, P., Ma, J., Gao, X., & Ruan, R. (2020). Microwave assisted flocculation for harvesting of Chlorella vulgaris. *Bioresource Technology, 314*, 123770.

Liu, Y., Jin, W., Zhou, X., Han, S. F., Tu, R., Feng, X., Jensen, P. D., & Wang, Q. (2019). Efficient harvesting of Chlorella pyrenoidosa and Scenedesmus obliquus cultivated in urban sewage by magnetic flocculation using nano-Fe3O4 coated with polyethyleneimine. *Bioresource Technology, 290*, 121771.

Lopez-Exposito, P., Campano, C., van de Ven, T. G., Negro, C., & Blanco, A. (2019). Microalgae harvesting with the novel flocculant hairy cationic nanocrystalline cellulose. *Colloids and Surfaces b: Biointerfaces, 178*, 329–336.

Luo, L., Zhang, Y., Xu, G., & Li, G. (2019). Microalgae-based nutrient recovery from anaerobic digestate: A review. *Bioresource Technology, 290*, 121790.

Ma, W., Feng, C., Guan, F., Ma, D., & Cai, J. (2022). Effective Chlorella vulgaris biomass harvesting through sulfate and chloride flocculants. *Journal of Marine Science and Engineering, 11*(1), 47.

Mat Nawi, N. I., Abd Halim, N. S., Lee, L. C., Wirzal, M. D. H., Bilad, M. R., Nordin, N. A. H., & Putra, Z. A. (2020). Improved nylon 6, 6 nanofiber membrane in a tilted panel filtration system for fouling control in microalgae harvesting. *Polymers, 12(2)*, 252.

Meena, R. A. A., Kannah, R. Y., Sindhu, J., Ragavi, J., Kumar, G., Gunasekaran, M., & Banu, J. R. (2019). Trends and resource recovery in biological wastewater treatment system. *Bioresource Technology Reports, 7*, 100235.

Morais, E. G., Cristofoli, N. L., Maia, I. B., Magina, T., Cerqueira, P. R., Teixeira, M. R., Varela, J., Barreira, L., & Gouveia, L. (2021). Microalgal systems for wastewater treatment: Technological trends and challenges towards waste recovery. *Energies, 14*(23), 8112

Nayak, J. K., & Ghosh, U. K. (2018). An innovative mixotrophic approach of distillery spent wash with sewage wastewater for biodegradation and bioelectricity generation using microbial fuel cell. *Journal of Water Process Engineering, 23*, 306–313.

Niemi, C., & Gentili, F. G. (2021). The use of natural organic flocculants for harvesting microalgae grown in municipal wastewater at different culture densities. *Physiologia Plantarum, 173*(2), 536–542.

Nurra, C., Clavero, E., Salvadó, J., & Torras, C. (2014). Vibrating membrane filtration as improved technology for microalgae dewatering. *Bioresource Technology, 157*, 247–253.

Onay, M. (2019). Bioethanol production via different saccharification strategies from H. tetrachotoma ME03 grown at various concentrations of municipal wastewater in a flat-photobioreactor. *Fuel, 239*, 1315–1323.

Osman, W. N. A. W., Nawi, N. I. M., Samsuri, S., Bilad, M. R., Khan, A. L., Hunaepi, H., Jaafar, J., & Lam, M. K. (2021). Ultra low-pressure filtration system for energy efficient microalgae filtration. *Heliyon, 7*(6).

Pal, P., Corpuz, A. G., Hasan, S. W., Sillanpää, M., & Banat, F. (2021). Microalgae harvesting using colloidal gas aphrons generated from single and mixed surfactants. *Chemosphere, 273*, 128568.

Patel, A. K., Kumar, P., Chen, C. W., Tambat, V. S., Nguyen, T. B., Hou, C. Y., Chang, J. S., Dong, C. D., & Singhania, R. R. (2022). Nano magnetite assisted flocculation for efficient harvesting of lutein and lipid producing microalgae biomass. *Bioresource technology, 363*, 128009.

Perazzoli, S., Bruchez, B. M., Michelon, W., Steinmetz, R. L., Mezzari, M. P., Nunes, E. O., & da Silva, M. L. (2016). Optimizing biomethane production from anaerobic degradation of Scenedesmus spp. biomass harvested from algae-based swine digestate treatment. *International Biodeterioration & Biodegradation, 109*, 23–28.

Qin, W., Zhang, S., Yang, S., Lin, F., Zhou, Q., Gong, H., Zhao, X., He, F., Zhou, Y., Feng, Y., Yu, G., Liu, Y., Li, Y., & Li, J. (2021). Highly dispersible magnetic nanoparticles modified by double quaternary ammonium surfactants for efficient harvesting of oleaginous microalgae. *Journal of Cleaner Production, 325*, 129297.

Rajasulochana, P., & Preethy, V. (2016). Comparison on efficiency of various techniques in treatment of waste and sewage water–A comprehensive review. *Resource-Efficient Technologies, 2*(4), 175–184.

Ratna, S., Rastogi, S., & Kumar, R. (2021). Current trends for distillery wastewater management and its emerging applications for sustainable environment. *Journal of Environmental Management, 290*, 112544.

Rizza, L. S., Coronel, C. D., Smachetti, M. E. S., Do Nascimento, M., & Curatti, L. (2019). A semi-closed loop microalgal biomass production-platform for ethanol from renewable sources of nitrogen and phosphorous. *Journal of Cleaner Production, 219*, 217–224.

Saxena, P., & Harish. (2018). Nanoecotoxicological reports of engineered metal oxide nanoparticles on algae. *Current Pollution Reports, 4*(2), 128–142.

Seo, J. Y., Kim, M. G., Lee, K., Lee, Y. C., Na, J. G., Jeon, S. G., Park, S. B., & Oh, Y. K. (2017). Multifunctional nanoparticle applications to microalgal biorefinery. *Nanotechnology for bioenergy and biofuel production*, 59–87.

Shen, Z., Li, Y., Wen, H., Ren, X., Liu, J., & Yang, L. (2018). Investigation on the role of surfactants in bubble-algae interaction in flotation harvesting of Chlorella vulgaris. *Scientific Reports, 8*(1), 3303.

Sundaresan, U., & Kasi, G. (2024). Synthesis of ZnO nanoparticles using Sargassum wightii ethanol extract and their antibacterial and anticancer applications. *Biomass Conversion and Biorefinery, 14*(20), 26173–26191.

Toscano, L., Ogden, K. L., Ogden, G., Cervantes, L., Steichen, S. A., Brown, C., Samaniego, B. Y., & Brown, J. K. (2018). Harvesting the microalga Chlorella sorokiniana by fungal-assisted pelletization. *Journal of Biobased Materials and Bioenergy, 12*(6), 493–505.

Wang, K., Zhang, F., Xu, K., Che, Y., Qi, M., & Song, C. (2023). Modified magnetic chitosan materials for heavy metal adsorption: A review. *RSC Advances, 13*(10), 6713–6736.

Wang, X., Liu, C., Qin, L., & Liang, W. (2022a). Self-assembly of Fe_3O_4 with natural tannin as composites for microalgal harvesting. *Fuel, 321*, 124038.

Wang, Y., Wang, J., Feng, C., Li, J., Wang, N., & Cai, J. (2022b). High-quality Chlorella vulgaris biomass harvesting through chitosan and polyacrylamid2e. *Environmental Science and Pollution Research, 29*(23), 34651–34658.

Wen, H., Zou, X., Xu, K., Shen, Z., Ren, X., & Li, Y. (2019). Buoy-bead flotation application for the harvesting of microalgae and mechanistic analysis of significant factors. *Bioprocess and Biosystems Engineering, 42(3),* 391–400.

Wirth, R., Lakatos, G., Böjti, T., Maróti, G., Bagi, Z., Rákhely, G., & Kovács, K. L. (2018). Anaerobic gaseous biofuel production using microalgal biomass–a review. *Anaerobe, 52*, 1–8.

Xia, L., Li, Y., Huang, R., & Song, S. (2017). Effective harvesting of microalgae by coagulation–flotation. *Royal Society Open Science, 4(11),* 170867.

Xu, K., Zou, X., Mouradov, A., Spangenberg, G., Chang, W., & Li, Y. (2021). Efficient bioflocculation of Chlorella vulgaris with a chitosan and walnut protein extract. *Biology, 10*(5), 352.

Xu, K., Zou, X., Wen, H., Xue, Y., Zhao, S., & Li, Y. (2018). Buoy-bead flotation harvesting of the microalgae Chlorella vulgaris using surface-layered polymeric microspheres: A novel approach. *Bioresource Technology, 267,* 341–346.

Xu, Y., Ibrahim, M. N. M., & Azam, M. M. (2015). Cultivation of microalgae in dairy farm wastewater for nutrient removal and biomass production. *Ecological Engineering, 84*, 472–478.

Yang, Z., Hou, J., & Miao, L. (2021). Harvesting freshwater microalgae with natural polymer flocculants. *Algal Research, 57*, 102358.

Zhang, H., Wen, H., Yin, H., Qin, W., Liu, X., Wang, Y., & Liu, Y. (2022). A novel approach for harvesting the microalgae Chlorella vulgaris with sodium alginate microspheres using buoy-bead flotation method. *Science of the Total Environment, 851*, 158418.

Zhang, X., Amendola, P., Hewson, J. C., Sommerfeld, M., & Hu, Q. (2012). Influence of growth phase on harvesting of Chlorella zofingiensis by dissolved air flotation. *Bioresource Technology, 116,* 477–484.

Zhang, X., Yao, D., Dong, H., Han, J., Zheng, W., Zhou, B., Tian, N., Peng, Y., Zheng, G., & Long, F. (2025). Regulating the crystallization of tin oxide nanoparticles by an environmentally friendly poly-aluminum chloride additive for a high-quality electron transport film in perovskite solar cells. *Journal of Materials Chemistry C, 13*(12), 6354–6363.

Zhang, Y., Liu, P., & Hong, Y. (2023). Fabrication of magnetic silica nanomaterials and their effects on algal harvesting. *Water, 15*(15), 2823.

Zhao, Y., Liang, W., Liu, L., Li, F., Fan, Q., & Sun, X. (2015). Harvesting Chlorella vulgaris by magnetic flocculation using Fe3O4 coating with polyaluminium chloride and polyacrylamide. *Bioresource Technology, 198*, 789–796.

Zhao, Z., Ilyas, A., Muylaert, K., & Vankelecom, I. F. (2020). Optimization of patterned polysulfone membranes for microalgae harvesting. *Bioresource Technology, 309*, 123367.

Zhao, Z., Muylaert, K., & Vankelecom, I. F. (2021). Combining patterned membrane filtration and flocculation for economical microalgae harvesting. *Water Research, 198*, 117181.

Zou, X., Xu, K., Wen, H., Xue, Y., Qu, Y., & Li, Y. (2019). Efficient microalgae harvesting using a thermal flotation method with response surface methodology. *Water Science and Technology, 80(3),* 426–436.

Chapter 8
Innovative Sensing and Monitoring Technologies: Real Time Monitoring for Efficient Resource Extraction and Sensor Technologies for Process Optimisation

Navneet Kaur, Devendra Sillu, Vagish Dwibedi, and Gurleen Kaur Sodhi

Abstract Innovative sensing and monitoring technologies are redefining how resources are managed and processes are optimized across industries. This chapter explores the critical role of real-time monitoring and advanced sensor systems in enhancing efficiency and sustainability in resource extraction, agriculture, environmental monitoring, and wastewater recycling. The chapter begins by contextualizing the importance of sensing technologies, emphasizing the value of real-time data acquisition for informed decision-making. It provides an overview of sensor types, including physical, chemical, and biological sensors, along with their working principles such as optical, electrochemical, and piezoelectric mechanisms. The integration of IoT-based monitoring platforms and data analytics tools is highlighted for enabling precise and actionable insights. Key applications of smart sensors in agriculture, such as soil health assessment, moisture tracking, and crop growth optimization, are discussed, demonstrating their potential in precision farming. Nano-sensors for environmental monitoring offer innovative solutions for detecting trace pollutants in air, soil, and water, contributing to improved environmental quality. The chapter also delves into the role of sensing technologies in wastewater recycling, including monitoring membrane filtration, sludge treatment, and ensuring the quality of treated water for reuse. Finally, emerging trends such as green sensors, IoT integration, and AI advancements are identified as pivotal to the future of sensing technologies. By

N. Kaur
Department of Biotechnology, Thapar Institute of Engineering and Technology, Patiala, Punjab, India

D. Sillu
Department of Environmental Science and Engineering, Guangdong Technion—Israel Institute of Technology, Shantou, People's Republic of China

V. Dwibedi · G. K. Sodhi (✉)
University Institute of Biotechnology, Chandigarh University, Mohali, Punjab, India
e-mail: gurleen.e17702@cumail.in

S. Dhiman and G. Mukherjee (eds.), *Resource Resurgence*, Springer Water,
https://doi.org/10.1007/978-3-032-02750-4_8

focusing on sustainable innovations and interdisciplinary approaches, this chapter underscores the transformative potential of advanced sensing systems in optimizing resource management processes for a sustainable future.

Keywords Real time monitoring · Precision agriculture · Green approach · Smart farming · Sustainability

8.1 Introduction

Over time, the extraction, management and optimisation of resources using sensing and monitoring technologies has seen immense advancements. To ensure sustainability and profitability of these processes it is eminent to introduce monitoring techniques in the industry. Be it mining, oil extraction, treatment of wastewater, the precise nature of real time sensing enables efficient process optimisation (Ally et al., 2023). Furthermore, integration of Artificial Intelligence (AI) and Internet of Things (IoT) further enhances the impact of sensing technologies. Unlike traditional methods, it has enabled reduction in resource wastage as well as degradation of the environment (Cairone et al., 2024). Some real-life applications such as detection of toxic gases in occupational work place by real-time gas monitoring can help prevent hazards (Cho et al., 2023). Moreover, various types of sensors are available which can be tailored for specific industry and application.

For instance, for detection of physical attributes such as vibrations, a piezoelectric sensor can be employed. Likewise optical sensors help in detection of infrastructure structural deformations. Other type of sensors, such as electrochemical sensors are widely utilised to detect contamination of environmental pollutants like heavy metals. Additionally, biological components like antibodies or enzymes could be detected using biosensors. Detection of other components like microbial contaminants and soil pathogens is also made possible by incorporation of such biosensors (Alygizakis et al., 2020; Huisman et al., 2024; Laghrib et al., 2020). The mentioned sensors work on a wide range of principles. For example, the piezoelectric sensors generate electrical charges, electrochemical sensors record electrical signal and optical sensor use absorption and reflection of light to identify changes. With time, the integration of IoT-based monitoring has enabled the simultaneous collection of data, which can be sent for real-time analysis to a cloud-based system. This not only enables real-time monitoring but also allows remote monitoring, optimization, and performance prediction in the industrial ecosystem. The IoT sensors have been crucial in predicting flow rate and efficiency of systems in chemical and wastewater treatment industries. On further integration of AI-driven analytics, predictive maintenance and anomaly detection can be achieved. Along with operational accuracy, it also enables sustainable regulatory compliance (Cairone et al., 2024; Laghrib et al., 2020).

Previous research has highlighted wide array of application based on the utilisation of sensors. Some aspects include, assessment of soil health, microbial activity, soil degradation, water requirements and plant health. Likewise, sensors related to wastewater recycling and detecting environmental pollutants allow timely intervention against early warnings of contamination. Such sensors can detect fouling, microbial activity in sludge and quality assurance of wastewater (Massoud et al., 2019; Mubarak et al., 2021). Over the years, development of sustainable biosensors and advance algorithms has garnered much attention. This chapter delves into exploration of nano sensors for tracing air, soil and water pollutants. The monitoring of water parameters and soil health using nano sensors is also discussed. In addition, the implementation of advance sensing in waste water treatment is also discussed. In conclusion, it would provide insights for researchers and stakeholders working in the area for future advancements.

8.2 Nano-sensors for Environmental Monitoring

The preservation of vital environmental resources and their sustainable utilization can be managed through systematic detection, monitoring and assessment of organic and inorganic pollutants released into the water, air, and soil (Chakraborty et al., 2021; Mahmoud et al., 2021). The advancement of nanotechnology, which has indispensable applications in almost all areas of our lives such as daily use gadgets, cosmetics, water disinfection, agriculture, food processing, industries, medical, defense and security (Chauhan et al., 2024; Dhiman et al., 2019, 2024; Dubey et al., 2022; Sillu et al., 2022; Wu et al., 2023a), can also assist in the direction of environmental detection. Significant research efforts have been directed towards leveraging the unique chemical and physical properties of nanostructures and nanomaterials for the development of nanosensors. With significant advantages, including enhanced interaction with analytes, superior sensitivity and selectivity, portability, rapid response, and facile operation, nanosensors trump over conventional sensors (Dimitrievska et al., 2023). Nanosensors from metal/metal-oxide, carbon and polymer based nanomaterials have been fabricated and utilized for the detection of various analytes such as pathogens, toxic gases, organic chemicals, heavy metals and pesticides from the environmental samples (Darwish et al., 2024; Dimitrievska et al., 2023). The amalgamation of nanotechnology and biotechnology has led to the development of even more selective and efficient nanobiosensors. The integration of nanosensors and nanobiosensors into environment management systems will facilitate the remote monitoring and identification of pollution issues, as well as ensure adherence to regulations and policies concerning its quality and conservation (Mahmoud et al., 2021). In this section, we have discussed various types of nanosensors based on the area of application in environmental monitoring.

8.2.1 Nano-sensors in Water Quality Monitoring

Mercury (Hg), lead (Pb), arsenic (As), chromium (Cr) and cadmium (Cd) are considered highly toxic heavy metal contaminants of aquatic systems, presenting human poisoning even at trace level. Therefore, sensitive and selective detection of metal ions in water through nanosensors has attracted considerable attention. The first carboxyl-rich carbon dots (CDs) based mercury sensor with a photoluminescence (PL) decay, for ultra-low level, rapid detection of Hg^{2+} is reported by Zdražil et al. (2023). The CDs sensor displayed record sensitivity, selectivity, and stability with high resistance to photobleaching, pH and ionic strength. Additionally, this sensor with economical precursors (0.14€ per reaction) and limit of detection (LOD) of 0.7 ppb (3.5 nM) for sensing Hg^{2+} in real water samples, showcased considerable commercial potential. Sahu et al. (2024) created a silver hydrosol exhibiting metal-enhanced fluorescence sensor, with quinone-capped silver nanoparticles (Q2HBAg) effectively removing methyl blue (MB) dye and detecting Pb^{2+} and Fe^{3+} ions. The unique fluorescence behavior of Q2HBAg, quenched by MB, enabled selective detection of Pb^{2+} with LOD of 5×10^{-8} M. Ballen et al. (2021) developed a graphene oxide/urease nanobiosensor for detecting Cd^{2+} in river water, exhibiting high sensitivity (0.0147 nm/ppt), a LOD of 18 ppt., and a reliable response. A study by Huang et al. (2024) presented an IoT-enabled and automated environmental monitoring platform that integrated a self-powered triboelectric nanosensor (TENS) with a robotic hand for real-time, on-site detection of heavy metal ions. The robotic hand, wirelessly controlled (by exo-hand) to avoid exposure to toxic chemicals, utilized copper oxide nanowires (CuO NWs)-based TENS for selective detection of Pb^{2+}, Cr^{6+}, and As^{3+} with LOD values of 5, 10, 10 nM, respectively.

Organic pollutants such as phenolic compounds, dyes, surfactants, and pesticides/herbicides in water are a cause of serious concern due to their toxicity, high bioaccumulation, and non-biodegradability (Dimitrievska et al., 2023). Dhiman et al. (2022) developed tyrosinase-gold nanoparticles (Ty-AuNPs) for the ultrasensitive detection of phenolic compounds, achieving an exceptionally low LOD of 0.01 ppb. Sadrolhosseini et al. (2023) developed a surface plasmon resonance (SPR) sensor featuring a nAu layer modified with NiCo-layered double hydroxide for detecting methylene blue and methylene orange. It achieved a LOD of 0.005 ppm and a response time of 268 s, indicating its efficiency and rapid performance. A rapid and sensitive colorimetric nanosensor developed from Sumac gum and copper nanoparticles (CuNPs@SG) enabled glyphosate quantification in aqueous models with a calibration range of 5–140 mg L^{-1}, and a LOD of 4.2 mg L^{-1} (Zeebaree et al., 2025).

Abrupt rise of unregulated emerging contaminants (ECs) such as PFAS, cyanotoxins, micro/nano-plastics, as well as various bacteria, viruses, and pathogens, has posed a significant challenge for maintaining drinking water quality (Ally et al., 2023; Wang et al., 2024) as these are often overlooked in existing monitoring systems. Therefore, development of nanosensors in trace detection of ECs in environmental matrices is imperative. Ma et al. (2025) demonstrated a colorimetric sensor using cyclodextrin-modified gold nanoparticles (CD@AuNPs) for visual perfluorooctanoic acid (PFOA) detection. The host–guest interaction between CD and PFOA induced AuNPs aggregation, causing a visible color change from red to blue-purple. The sensor displayed LOD of 170 nM, 156 nM, and 204 nM for α-CD@AuNPs, β-CD@AuNPs, and γ-CD@AuNPs, respectively, with high selectivity and recovery rates of 98% to 129% in spiked lake water samples. A sensing platform for microcystins was synthesized with imprinted nanohybrids (CDs-FITC-SiO_2@MIP) combining metformin as a template, carbon dots (CDs) and fluorescein isothiocyanate (FITC) for response and reference signals, respectively. This highly sensitive and selective ratiometric fluorescence nanosensor displayed a linear detection and LOD range of 0.5–500 μg L^{-1}; 0.013–0.022 μg L^{-1}, respectively (Li et al., 2023). A H_2O driven visible-light-sensitive microrobot, Au/TiO_2/Au/R-Fe_3O_4 (ATAR), was developed by Geng et al. (2024) for dynamic capture and in situ surface-enhanced Raman scattering (SERS) detection of nanoplastics. It demonstrated quantitative analysis of polystyrene (PS) and polymethyl methacrylate (PMMA) with LOD of 1.27 μg mL^{-1} and their accurate size differentiation using machine learning algorithms (MLAs).

Kumari et al. (2025b) presented a Förster resonance energy transfer (FRET)-based "turn-on" fluorescent nanosensor, where novel nanoconjugate of copper nanoclusters–encapsulated metal organic frameworks with MnO_2 nanospheres (CuNCs@ZIF-8@ MnO_2) was employed for rapid and sensitive detection of *Escherichia coli* in water. The presence of bacteria triggers a redox reaction, restoring fluorescence, allowing detection across a wide concentration range (5×10^1 to 5×10^5 CFU mL^{-1}) with a limit of 8 CFU mL^{-1} in 50 min and high recovery rates (94.3–106.5%) in real water samples. Chen et al. (2022) designed a chemiluminescence system using mercaptophenylboronic acid (MPBA)-functionalized CuSe nanoprobes (CuSeNPs@MPBA) for accurate and sensitive detection of *Staphylococcus aureus* and *Escherichia coli*, with LODs of 1.25 CFU mL^{-1} and 1.01 CFU mL^{-1}, respectively. Magnetic nanoparticles (MNPs) have been employed in nucleic acid detection and quantification, demonstrating effectiveness in detecting pathogens such as Hepatitis B virus and SARS-CoV-2, with a LOD of 50 IU mL^{-1} and 500 copies mL^{-1}, respectively (Fang et al., 2023).

8.2.2 Nano-sensors in Air Quality Monitoring

Gaseous pollutants, such as nitrogen dioxide (NO_2), nitric oxide (NO), ammonia (NH_3), sulphur dioxide (SO_2), carbon dioxide (CO_2), carbon monoxide (CO), hydrogen sulfide (H_2S), and ozone (O_3) are recognized as major environmental threats to public health and significant contributors to global climate change (Yi et al., 2015). Consequently, there is a significant drive to create gas nanosensors that can track such contaminants. Wu et al. (2023b) fabricated a flexible NO_2 electrochemical sensor with zinc trifluoromethanesulfonate ($Zn(OTf)_2$)/polyacrylamide

(PAM)-carbon hydrogel. It exhibited ultrahigh sensitivity (1.92%/ppb), ultralow LOD of 0.1 ppb, outstanding recovery, under different deformations in subzero temperatures and high humidity. A deep learning (DL) technology for early in situ detection of mixed chemical vapors using AI-enabled arrayed nanoelectrodes with multiplexed DNA-functionalized graphene was presented by Hwang et al. (2023). A single chip with seven graphene-based ssDNA-sensor arrays detected H_2S, NH_3, NO, NO_2, and their mixtures. Gas classification utilized the Boruta algorithm, achieving over 99% recognition in low humidity and over 98% in humid conditions. A one-dimensional convolutional neural network (1D CNN) built with PyTorch 1.10.0 and Python 3.7 provided near-perfect vapor composition discrimination across varying humidity levels. A study presented a room-temperature UV-light-assisted chemiresistive sensor for SO_2 detection, utilizing polyaniline (PANI) and silver nanoparticle-modified tin dioxide nanostructures (Ag/PANI/SnO_2) (Xu et al., 2021). The sensor facilitated by UV irradiation at 365 nm demonstrated a detection range of 20.1–50 ppm SO_2, with response and recovery times of 110 s and 100 s, respectively. A field-effect transistor (FET) sensor utilizing a Zn-doped In_2O_3 single nanowire was demonstrated for detecting gas concentrations of 1–5 ppm for CO, NO, and NO_2, with highest sensitivity to CO at ambient temperature (Dhall et al., 2021). A chemiresistive sensor based on $CuCo_2O_4$ nanomaterials was developed for O_3 detection, operating at 90 °C by Zhao et al. (2023). The sensor exhibited a response of 11 to 1 ppm O_3 at 0% relative humidity (RH), with the response increasing to 24 and 27 at 50% and 70% RH, respectively.

Volatile organic compounds (VOCs) are chemicals with high vapor pressure at room temperature and atmospheric pressure, making them complex and challenging to detect. VOCs such as alcohols, aldehydes, aromatic hydrocarbons, oxygenated organics, and halogenated hydrocarbons etc. may originate from both natural processes and anthropogenic activities. Their hazardous nature has made VOC sensors essential for environmental monitoring. Capman et al. (2022) aimed to enhance the selectivity of electronic noses by functionalizing graphene with various chemical receptors to develop a machine-learning (ML)-enabled gas sensor. The system employed graphene-based varactor arrays with 108 sensors, each functionalized with 36 chemical receptors, to achieve cross-selectivity which further enabled rapid detection of multiple VOCs, with data acquisition completed in seconds. Utilizing ML-based analysis (Bagging Random Forest), the sensor distinguished between octane, methyl ethyl ketone, toluene, hexanal, and ethanol with 98% accuracy. Wang et al. (2023) demonstrated that MXene- and Indium(iii)oxide-based nanocomposites ($In_2O_3/Ti_3C_2T_x$) showed superior formaldehyde (HCHO) sensing performance that increased from 66.06 to 106.8% as compared to pristine metal oxide. A single μLED-embedded photoactivated ultra-low-power (μLP) metal oxide gas sensor with time-variant illumination, integrated with DL analysis was developed by Cho et al. (2023). This sensor, consuming only 0.53-mW, achieved over 96.99% classification accuracy for NO_2, acetone, methanol, and ethanol, offering precise real-time gas mixture predictions for practical applications. Jeong et al. (2023) reported a CeO_2/Rh–SnO_2 bilayer chemiresistive sensor array, enhanced by pattern recognition, for distinguishing between aromatic and non-aromatic gases. The high sensing ability for volatile aromatic hydrocarbons (VAHs) was attributed to the CeO_2 overlayer, which facilitated the catalytic oxidation of reactive interfering gases into less reactive forms.

Airborne pathogens, including bacteria, fungi etc., emitted from ventilation, air conditioning, and humidifier systems, pose health risks, reduce productivity, and cause economic losses. Varvařovská et al. (2024) introduced a polyacrylonitrile nanofiber bioreceptor functionalized with antibodies targeting bacterial antigens for the selective capture of bacterial cells directly from the air. The antibody-functionalized nanofiber membranes demonstrated a statistically significant improvement in model organisms, *E. coli* (41%) and *S. aureus* (36%), compared to unmodified nanofibers. The area density of the functionalized membranes for bacterial detection was reduced to 2.5 g m^{-2}, thereby minimizing mechanical interception. A colorimetric sensor for detecting *Aspergillus niger* using AuNPs functionalized with a specific spore-binding peptide was developed by Lee et al. (2021). Peptide-functionalized AuNPs rapidly bound to *A. niger* spores over common airborne bacteria, causing a visible color change in the supernatant after spore sedimentation. This approach achieved high sensitivity, detecting $\sim$ 50 spores within 10 min, facilitated by a smartphone-based image analysis application.

8.2.3 Nano-sensors in Soil Quality Monitoring

Effective nutrient management is essential for maximizing crop yields, reducing environmental pollution, and preserving soil health. Nanosensors have been created to enable real-time monitoring of soil nutrients, including nitrogen (N), and phosphorus (P). Kundu et al. (2022) fabricated electrochemical nanosensor using nanocomposite of carbon nanotubes (CNT), hematite nanostructures along with immobilization of nitrate reductase (NiR) enzyme and bovine serum albumin (BSA) for NO_3^- detection in soil extracts. The BSA/NiR/CNT-α-Fe_2O_3/SPE nanosensor demonstrated excellent performance, with high sensitivity (63.87 μA/log (mg L^{-1}) cm^{-2}), 0.09 mg L^{-1} LOD, and excellent reproducibility (RSD = 0.43%), specificity, accuracy, and a 5-week shelf life. The detection of phosphate ions was monitored through the formation of silver phosphate (Ag_3PO_4) using a nozzle-jet printed silver/reduced graphene oxide (Ag/rGO) composite-based ISFET sensor, as reported in Bhat et al. (2019).

Soil pH and moisture are critical factors influencing nutrient availability, microbial activity, and plant growth (Buss et al., 2018; Robinson et al., 2016). A conductometric microsensor modified with polyaniline doped with dodecylbenzene sulfonic acid (PANI-DBSA) and an SU-8 (epoxy resin) nanocomposite film for soil pH measurement was developed by Patil et al. (2019). The nanocomposite, spin-coated on Au-IDE patterned on Si/SiO_2, demonstrated excellent pH responsiveness across standard buffer solutions (pH 2–10), red and bentonite soils (with varying moisture content), and calcium chloride ($CaCl_2$) solutions. Sensitivities of 41 and 52 μS/pH for red and bentonite soils, respectively, were achieved, with response and recovery times of 10 s and 30 s. A moisture sensor comprised of urea modified Cu-doped $LaFeO_3$

nanoparticles exhibited superior performance, with response and recovery times of 5.4 s and 3.8 s, respectively when operated across 11%–90% RH. The efficacy of the fabricated sensor to monitor real-time humidity observed via IoT on a laptop/smartphone ameliorate its potential to detect soil moisture for agriculture application (Kumari et al., 2025a).

Heavy metal contamination in soil poses a severe threat to human health, ecosystem stability, and food safety (Alloway, 2012). Nanosensors enabling early detection and remediation of contaminated sites have been encouraged for real-time monitoring of heavy metals in soil. QD-based sensors have been used for multiplexed detection of mercury (Hg), arsenic (As), and chromium (Cr) ions in soil, with LODs of 0.1, 0.5, and 1 nM, respectively (Parameswari et al., 2024). In a study by Das and Sharma (2019), a Bi_2O_3/Fe_2O_3-doped graphene oxide nanocomposite was synthesized for the square wave voltammetric detection of cadmium ions (Cd(II)). The nanocomposite exhibited high selectivity for Cd(II) amidst other metal ions and biomolecules, with recovery rates of 97–99% in Cd(II)-spiked soil samples. A miniaturized cable-type electrochemical sensor utilizing carbon nanotube-immobilized cellulose yarn detected Zn(II) in real-time, in the range of 0.1−500 ppm, with high sensitivity (< 1 ppm) and with deviations of < 6% compared to atomic absorption techniques (Mondal et al., 2019).

Soil contamination by organic pollutants, including petroleum hydrocarbon spills, explosives, plastics, pharmaceuticals, herbicides/pesticides, has become a significant environmental issue. Nurdin et al. (2025) introduced a novel TiO_2-ZnO/graphene (GTZ) hybrid nanocomposite electrode for ultra-sensitive electrochemical detection of fipronil. Cyclic voltammetry demonstrated a linear response to fipronil at concentrations as low as 0.0023 $\mu g\ L^{-1}$, with high reproducibility (RSD 1.77%). An electrochemical sensing platform was developed for quantifying the high-risk 2-nitroaniline (2-NA) using a bismuth molybdate (Bi_2MoO_6, BMO)-functionalized carbon nanofiber (f-CNF)-modified GC electrode with recoveries ranging from 97.8 to 99.4% in soil samples (Krishnapandi et al., 2021). CNT-based sensors have been used to detect organophosphate pesticides, such as parathion and malathion, with LODs of 0.1 and 0.5 nM, respectively (Parameswari et al., 2024). Komikawa et al. (2020) developed a fluorescent TNT chemosensor by conjugating TNT-binding peptides to fluorescent CdTe/CdS quantum dots (QDs) via thiol groups. The sensor exhibited a LOD of approximately 375 nM in fluorescence mode.

8.3 Innovative Applications in Wastewater Recycling

Fouling is one of the primary challenges hindering the consistent operation of the membrane process in wastewater treatments plants (WWTPs). Increment in fouling raises energy consumption, lowers permeability, and shortens the lifespan of membrane modules, which ultimately affects the overall performance and efficiency (Du et al., 2020; Iorhemen et al., 2016; Meng et al., 2017; Zuthi et al., 2012). Consequently, novel approaches for in situ fouling monitoring have been proposed, with some capable of collecting data in real time. Several studies have implemented AI and ML algorithms along with spectroscopic, tomographic techniques to observe membrane fouling by predicting fouling indicators (increased transmembrane pressure or decreased flux) in relation to input parameters and wastewater properties (Dalmau et al., 2015; Li et al., 2020; Zhong et al., 2022).

Ren et al. (2023) introduced an innovative approach utilizing a cascade neural network (CAS-NN) to characterize fouling in a real membrane bioreactor (MBR), which demonstrated high accuracy in predicting membrane permeability, integrity, and lifespan. Cámara et al. (2023) achieved accurate fouling predictions using filtration times and flux as input variables using long-short term memory (LSTM) networks, a category of recurrent neural network (RNN). A hybrid ANN-RSM (artificial neural networks-response surface methodology) model to simulate the forward osmosis process and predict membrane flux was developed by Jawad et al. (2021). The prediction performance, indicated by R^2 values, was 0.98036 and 0.9408 for the ANN and RSM models, respectively. In an insightful study, Kovacs et al. (2022) applied random forest (RF), ANN, and LSTM algorithms to predict transmembrane pressure using 80,000 data points from a four-year full-scale treatment plant operation. While all models performed well, RF achieved the best results based on root mean squared error and the coefficient of determination. Ba-Alawi and co-authors introduced a system combining explainable AI (XAI) with a multi-sensor fusion-based automated data reconciliation and imputation (MSF-ARI) method for monitoring membrane fouling in MBRs, with autonomous handling of sensor malfunctions. MSF-ARI achieved a 100% detection rate for fault groups, enabling early prevention of fouling caused by sludge cake formation while effectively balancing energy consumption and operational performance (Ba-Alawi et al., 2023).

Optical coherence tomography (OCT) is an in situ real-time monitoring technique that captures 2D cross-sectional images of fouling accumulation on membranes with real-time analysis of the thickness and morphology of the fouling layer (Cairone et al., 2024; Jeong, & Kim, 2023). Recently, Im et al. (2021) introduced a novel method for real-time fouling monitoring in forward osmosis (FO) by combining OCT with a convolutional neural network (CNN), an ANN specifically designed for processing grid-like data, including images. Wong et al. (2021) developed OCT application from 2 to 3D in membrane distillation (MD) process, enabling the quantification of fouling by distinguishing the fouling layer from the membrane. OCT has emerged as a non-invasive method for monitoring membrane fouling (Elcik et al., 2021; Fortunato et al., 2020; Huisman et al., 2024; Ranieri et al., 2024). Integrating OCT with AI models, such as CNNs, offers valuable insights to enhance advanced fouling control strategies. Quartz crystal microbalance with dissipation monitoring (QCM-D) allows in situ study of adsorptive fouling on model membranes. Rudolph-Schöpping et al. (2023) utilized QCM-D to analyse ultrafiltration fouling through flux measurements, frequency and dissipation changes, and surface characterization. Rudolph et al. (2021) utilized in situ QCM-D to identify fouling on ceramic tubular microfiltration membranes used for treating thermomechanical pulping process water. These studies suggested QCM-D, combined with other analytical methods, can offer valuable insights into membrane fouling in industrial wastewater applications.

Electrical impedance spectroscopy (EIS) enables real-time, in situ detection of electrical variations on membrane surfaces by measuring permeate flux and impedance, enabling fouling and wetting observation in the MD process. A study explored the use of EIS to monitor biofilm development on gold-coated membranes and verified that early detection of fouling is possible before flux reduction and electrical conductivity increase (Zhang et al., 2022). Early fouling detection enabled more effective mitigation strategies, such as applying intermittent cathodic potential (− 1.5 V) or cross-flow flushing. EIS sensitivity was consistently recovered across four mechanical fouling removal cycles, demonstrating that electrically conductive membranes (ECMs) serve dual roles in EIS-enabled biofouling detection and mitigation. Transmembrane electric potential (TMEP), indicative of the voltage drop across the membrane, reflects changes in membrane resistance associated with the extent of fouling. Xiang et al. (2021) proposed a mechanism for ion exchange membrane (IEM) fouling induced by anionic polyacrylamide (APAM) using TMEP and EIS.

Regular monitoring aims to stabilize the anaerobic digestion (AD) process and enhance biogas output. However, controlling such processes is challenging due to their extremely non-linear nature and unstable kinetic parameters (Cruz et al., 2021; Ghofrani-Isfahani et al., 2020). Selecting appropriate early-warning indicators is crucial for effective process monitoring and control. Various offline and online parameters, including pH, alkalinity, volatile fatty acids (VFAs) concentrations, biogas composition, methane yield, specific intermediate metabolites etc. have been proposed as stability indicators (Awhangbo et al., 2020a). Numerous studies have evaluated these parameters to develop efficient diagnostic methods for AD process issues. Near-infrared (NIR) spectroscopy, a common spectral analysis technique, requires developing a correlation model using chemometrics for sample composition prediction (Wu et al., 2019a). Several studies have utilized NIR to predict parameters such as volatile fatty acids (VFA), total and volatile solids, ammonium (NH_4^+), and ammonia (Awhangbo et al., 2020b; Finzi et al., 2015). In this context, Awhangbo et al. (2020c) demonstrated the effectiveness of the Sequential and Orthogonalized Partial Least Squares (SO-PLS) approach in estimating AD parameters using two NIR sources on digestate, highlighting a synergistic effect for optimal monitoring.

Electrochemical sensors measure signals from samples via electrodes, establishing a correlation to determine sample concentrations during monitoring (Wu et al., 2019a). These sensors have been extensively used in both offline and online monitoring of AD processes (Cruz et al., 2019; Laghrib et al., 2020). Rasheed et al. (2020) evaluated the real-time performance of a large-scale plant using an online biogas flowmeter (BF-3000) to measure gas pressure and flow rate, and a gas analyzer (GA 2000) to determine CH_4 and CO_2 levels in the biogas. This system enabled the authors to identify the optimal co-digestion ratio for each substrate. Recently, electrochemical biosensors have gained attention as a facile, online, cost-effective, and efficient method for VFA monitoring in AD (Do et al., 2020). Two prominent bioelectrochemical systems (BES) used in AD are microbial fuel cells (MFCs) and microbial electrolysis cells (MECs) (Wu et al., 2019b). Sun et al. (2019) demonstrated the viability of an MFC as a biosensor for selective acetate detection. The biosensor, comprising a conventional double-chamber design with an anion exchange membrane (AEM), exhibited high selectivity and was unaffected by temperature variations. A novel dual-chamber air–cathode MFC developed by Sun et al. (2021), displayed nonlinear response at higher VFA concentrations (2.8–112 mM) and linear at lower concentrations (< 14 mM) within 2–5 h. Its effectiveness was validated with AD reactor effluents, showing a strong correlation with gas chromatographic measurements. Zhao et al. (2018) proposed an MEC biosensor for online ammonia monitoring in AD and observed a linear relationship between ammonia levels and current–voltage, unaffected by pH or external voltage variations.

Soft sensors are models that estimate hard-to-quantify process variables using easily detectable data and can be integrated with measurement tools to provide real-time estimates of process variables as a cost-effective alternative (Kabugo et al., 2020). Some studies have applied soft sensors in AD, such as Wang et al. (2018), who used an Artificial Neural Network (ANN) based soft sensor to monitor real-time alkalinity in an anaerobic digester. Kazemi et al. (2020) compared several data-driven method-RF, ANN, extreme learning machine (ELM), support vector machine (SVM), and genetic programming (GP)—for monitoring VFA values in AD. Using synthetic data, they found that the GP soft sensor was the most accurate and easiest to implement. Similarly, Asadi et al. (2020) applied data-driven methods, including ANN and adaptive network-based fuzzy inference system (ANFIS), to predict biogas production rates at a Canadian wastewater plant affected by annual temperature fluctuations (40 to 30 °C). The ANFIS-FCMC model provided more precise estimates and could assist in enhancing biogas production. Likewise, IoT-based wireless monitoring for wastewater treatment processes shows potential as Xiu and Dong (2019) proposed its use in a sewage treatment plant to monitor and control output in real-time. However, IoT relies on complementary technologies, such as electrochemical or soft sensors, to detect environmental changes (Gopikumar et al., 2021).

In response to increasing water scarcity, treated wastewater reuse is considered a reliable alternative for sustainable water management and agricultural development (Marteleira et al., 2014; Massoud et al., 2019). Significant progress has been made in producing safe effluents through the removal of nutrients, metals, and minimizing chemical oxygen demand (COD) (Fatta-Kassinos et al., 2011; Kümmerer et al., 2018; Piña et al., 2020; Rizzo et al., 2013). Research on novel monitoring methods and advanced analytical tools aims to mitigate the adverse effects of chemical and biological factors such as emerging contaminants (ECs) and antibiotic resistance genes (ARGs) (Corno et al., 2019; Turner et al., 2019). An agarose hydrogel diffusion-based passive sampler coupled with liquid chromatography tandem MS/MS (LC-QQQ), and liquid chromatography high-resolution mass spectrometry (LC-HRMS) was deployed at the outlet of an MBR-WWTP in Cyprus. This enabled detection of ecotoxicologically ECs (antibiotics-azithromycin, clarithromycin, erythromycin, ofloxacin) and ARGs (except blaCTX-M-32) indicating potential chemical and biological hazards in wastewater reuse (Alygizakis et al., 2020). A high-throughput effect-directed analysis (HT-EDA) platform, combining downscaled luminescent Ames and reporter gene assays with high-resolution fraction collection and UPLC-QTOF MS, was developed for WWTP effluent analysis

(Zwart et al., 2020). It identified mutagens and endocrine disruptors, including 1,2,3-benzotriazole and androstenedione, highlighting its potential for routine water quality monitoring and detection of emerging contaminants.

Yang et al. (2022) proposed a dynamic principal component analysis with a nonlinear autoregressive model with exogenous inputs (PCA-NARX) model for predicting effluent quality and enabling real-time adjustments in WWTP operations. The dynamic model outperformed static ANN models in effluent quality prediction. El-Rawy et al. (2021) introduced Deep Cascade-Forward Backpropagation (DCB) and Deep Learning Time Series Forecasting (DLTSF) models to predict effluent water quality and assess the treatment performance of the El-Berka WWTP. Xie et al. (2022) combined an improved Feedforward Neural Network (IFFNN) with Genetic Algorithm (GA) to predict real-time effluent water quality in a Jiangsu Province WWTP. The IFFNN-GA model improved prediction performance by 52.3% for COD and 72.6% for total nitrogen (TN) compared to the traditional FFNN model. A novel ML-based framework, incorporating RF models, Deep Neural Networks (DNN), Variable Importance Measure (VIM), and Partial Dependence Plot (PDP) analysis, was developed for effluent quality control at Umeå WWTP in Sweden (Wang et al., 2021). The study focused on Total Suspended Solids (TSS) and Phosphate (PO_4) as effluent parameters, with VIM identifying key influencing variables and PDP clarifying their specific impacts on TSS and PO_4 levels, respectively. Mubarak et al. (2021) applied three data-driven models—Hammerstein-Weiner (HW), Support Vector Machine (SVM), and Feedforward Backpropagation Neural Network (FFBPNN)—along with traditional multi-linear regression, as well as two non-linear ensemble techniques to predict effluent COD at the new Nicosia WWTP. The ensemble techniques (HW-ensemble and FFBPNN-ensemble) improved model performance during the verification phase by 12% for HW-E and 19% for FFBPNN-E. Another study by Sharafati et al. (2020) utilized three ensemble machine learning models—AdaBoost Regression (ABR), Gradient Boost Regression (GBR), and Random Forest Regression (RFR)—to predict effluent quality parameters Total Dissolved Solids (TDS), BOD, and COD at an industrial wastewater treatment plant in Qom, Iran. Results showed that the ABR model excelled in predicting TDS (CC = 0.962, RMSE = 30.3 mg L^{-1}), while GBR provided better accuracy for BOD (CC = 0.9, RMSE = 4.6 mg L^{-1}) and COD (CC = 0.75, RMSE = 9.6 mg L^{-1}).

8.4 Future Directions in Sensing Technologies

The evolution of sensing technologies is poised to shape the future of resource management and process optimization significantly. Emerging trends such as green sensors, the integration of the Internet of Things (IoT), and advancements in artificial intelligence (AI) are transforming the landscape of real-time monitoring. Green sensors, designed with eco-friendly materials and sustainable manufacturing processes, are gaining traction as industries prioritize environmental stewardship. These innovations aim to reduce energy consumption, eliminate hazardous substances, and ensure long-term usability, aligning with global sustainability goals. IoT integration continues to revolutionize the field by enabling seamless communication between sensors and centralized monitoring systems. This connectivity allows for real-time data aggregation and analysis, empowering decision-makers with actionable insights. Advanced IoT networks also promise to improve sensor efficiency, reliability, and adaptability in dynamic operational environments. AI advancements are unlocking unprecedented possibilities in data interpretation and predictive analysis. Machine learning algorithms can process vast datasets generated by sensors, identifying patterns and anomalies that may be imperceptible to human observers. AI-driven models are also enhancing system automation, enabling proactive maintenance, fault prediction, and optimal resource utilization. Looking forward, research is likely to focus on the development of multifunctional and miniaturized sensors capable of addressing complex challenges across industries. Innovations in nanotechnology, material science, and bioengineering could pave the way for ultra-sensitive sensors tailored to specific applications. Furthermore, interdisciplinary collaborations will play a critical role in bridging technological gaps and fostering the widespread adoption of these advancements. Overall, the future of sensing technologies lies in creating smarter, more sustainable systems that balance efficiency, environmental impact, and cost-effectiveness, ultimately contributing to global resource optimization efforts.

References

Alloway, B. J. (2012). *Heavy metals in soils: Trace metals and metalloids in soils and their bioavailability* (Vol. 22). Springer Science & Business Media.

Ally, N., & Gumbi, B. (2023). A review on metal nanoparticles as nano-sensors for environmental detection of emerging contaminants. *Materials Today: Proceedings*, 1–5. https://doi.org/10.1016/j.matpr.2023.08.032

Alygizakis, N. A., Urík, J., Beretsou, V. G., Kampouris, I., Galani, A., Oswaldova, M., Berendonk, T., Oswald, P., Thomaidis, N. S., & Slobodnik, J. (2020). Evaluation of chemical and biological contaminants of emerging concern in treated wastewater intended for agricultural reuse. *Environment International, 138*, 1–10. https://doi.org/10.1016/j.envint.2020.105597

Asadi, M., Guo, H., & McPhedran, K. (2020). Biogas production estimation using data-driven approaches for cold region municipal wastewater anaerobic digestion. *Journal of Environmental Management, 253*, 1–10. https://doi.org/10.1016/j.jenvman.2019.109708

Awhangbo, L., Bendoula, R., Roger, J.-M., & Béline, F. (2020a). Detection of early imbalances in semi-continuous anaerobic co-digestion process based on instantaneous biogas production rate. *Water Research, 171*, 1–12. https://doi.org/10.1016/j.watres.2019.115444

Awhangbo, L., Bendoula, R., Roger, J.-M., & Béline, F. (2020b). Multi-block data analysis for online monitoring of anaerobic co-digestion process. *Chemometrics and Intelligent Laboratory Systems, 205*, 1–17. https://doi.org/10.1016/j.chemolab.2020.104120

Awhangbo, L., Bendoula, R., Roger, J.-M., & Béline, F. (2020c). Multi-block SO-PLS approach based on infrared spectroscopy for anaerobic digestion process monitoring. *Chemometrics and Intelligent Laboratory Systems, 196*, 1–21. https://doi.org/10.1016/j.chemolab.2019.103905

Ba-Alawi, A. H., Nam, K., Heo, S., Woo, T., Aamer, H., & Yoo, C. (2023). Explainable multisensor fusion-based automatic reconciliation and imputation of faulty and missing data in membrane bioreactor plants for fouling alleviation and energy saving. *Chemical Engineering Journal, 452*, 1–13. https://doi.org/10.1016/j.cej.2022.139220

Ballen, S. C., Ostrowski, G. M., Steffens, J., & Steffens, C. (2021). Graphene oxide/urease nanobiosensor applied for cadmium detection in river water. *IEEE Sensors Journal, 21*(8), 9626–9633. https://doi.org/10.1109/JSEN.2021.3056042

Bhat, K. S., Nakate, U. T., Yoo, J.-Y., Wang, Y., Mahmoudi, T., & Hahn, Y.-B. (2019). Nozzle-jet-printed silver/graphene composite-based field-effect transistor sensor for phosphate ion detection. *ACS Omega, 4*(5), 8373–8380. https://doi.org/10.1021/acsomega.9b00559

Buss, W., Shepherd, J. G., Heal, K. V., & Mašek, O. (2018). Spatial and temporal microscale pH change at the soil-biochar interface. *Geoderma, 331*, 50–52. https://doi.org/10.1016/j.geoderma.2018.06.016

Cairone, S., Hasan, S. W., Choo, K.-H., Li, C.-W., Zorpas, A. A., Ksibi, M., Zarra, T., Belgiorno, V., & Naddeo, V. (2024). Enhancing membrane fouling control in wastewater treatment processes through artificial intelligence modeling: Research progress and future perspectives. *Euro-Mediterranean Journal for Environmental Integration*, 1–19. https://doi.org/10.1007/s41207-024-00659-0

Cámara, J. M., Diez, V., & Ramos, C. (2023). Neural network modelling and prediction of an Anaerobic Filter Membrane Bioreactor. *Engineering Applications of Artificial Intelligence, 118*, 1–14. https://doi.org/10.1016/j.engappai.2022.105643

Capman, N. S., Zhen, X. V., Nelson, J. T., Chaganti, V. S. K., Finc, R. C., Lyden, M. J., Williams, T. L., Freking, M., Sherwood, G. J., & Bühlmann, P. (2022). Machine learning-based rapid detection of volatile organic compounds in a graphene electronic nose. *ACS Nano, 16*(11), 19567–19583. https://doi.org/10.1021/acsnano.2c10240

Chakraborty, U., Kaur, G., & Chaudhary, G. R. (2021). Development of environmental nanosensors for detection monitoring and assessment. *New frontiers of nanomaterials in environmental science*, 91–143. https://doi.org/10.1007/978-981-15-9239-3_5

Chauhan, A., Agnihotri, S., & Vasundhara, M. (2024). Enhanced solar light-driven photocatalysis of norfloxacin using Fe-doped TiO_2: RSM optimization, DFT simulations, and toxicity study. *Environmental Science and Pollution Research, 31*(35), 47991–48013. https://doi.org/10.1007/s11356-024-34080-5

Chen, X., Wang, X., Fang, Y., Zhang, L., Zhao, M., & Liu, Y. (2022). Long-lasting chemiluminescence-based POCT for portable and visual pathogenic detection and *in situ* inactivation. *Analytical Chemistry, 94*(23), 8382–8391. https://doi.org/10.1021/acs.analchem.2c00877

Cho, I., Lee, K., Sim, Y. C., Jeong, J.-S., Cho, M., Jung, H., Kang, M., Cho, Y.-H., Ha, S. C., & Yoon, K.-J. (2023). Deep-learning-based gas identification by time-variant illumination of a single micro-LED-embedded gas sensor. *Light: Science & Applications, 12*(1), 1–12.

Corno, G., Yang, Y., Eckert, E. M., Fontaneto, D., Fiorentino, A., Galafassi, S., Zhang, T., & Di Cesare, A. (2019). Effluents of wastewater treatment plants promote the rapid stabilization of the antibiotic resistome in receiving freshwater bodies. *Water Research, 158*, 72–81. https://doi.org/10.1016/j.watres.2019.04.031

Cruz, I. A., Andrade, L. R., Bharagava, R. N., Nadda, A. K., Bilal, M., Figueiredo, R. T., & Ferreira, L. F. (2021). An overview of process monitoring for anaerobic digestion. *Biosystems Engineering, 207*, 106–119. https://doi.org/10.1016/j.biosystemseng.2021.04.008

Cruz, I. A., de Melo, L., Leite, A. N., Sátiro, J. V. M., Andrade, L. R. S., Torres, N. H., Padilla, R. Y. C., Bharagava, R. N., Tavares, R. F., & Ferreira, L. F. R. (2019). A new approach using an open-source low cost system for monitoring and controlling biogas production from dairy wastewater. *Journal of Cleaner Production, 241*, 1–9. https://doi.org/10.1016/j.jclepro.2019.118284

Dalmau, M., Atanasova, N., Gabarrón, S., Rodriguez-Roda, I., & Comas, J. (2015). Comparison of a deterministic and a data driven model to describe MBR fouling. *Chemical Engineering Journal, 260*, 300–308. https://doi.org/10.1016/j.cej.2014.09.003

Darwish, M. A., Abd-Elaziem, W., Elsheikh, A., & Zayed, A. A. (2024). Advancements in nanomaterials for nanosensors: A comprehensive review. *Nanoscale Advances*, 1–32. https://doi.org/10.1039/D4NA00214H

Das, T. R., & Sharma, P. K. (2019). Sensitive and selective electrochemical detection of Cd^{2+} by using bimetal oxide decorated graphene oxide (Bi_2O_3/Fe_2O_3@ GO) electrode. *Microchemical Journal, 147*, 1203–1214. https://doi.org/10.1016/j.microc.2019.04.001

Dhall, S., Mehta, B., Tyagi, A., & Sood, K. (2021). A review on environmental gas sensors: Materials and technologies. *Sensors International, 2*, 1–10. https://doi.org/10.1016/j.sintl.2021.100116

Dhiman, N. K., Agnihotri, S., & Shukla, R. (2019). Silver-based polymeric nanocomposites as antimicrobial coatings for biomedical applications. *Nanotechnology in Modern Animal Biotechnology: Recent Trends and Future Perspectives*, 115–171. https://doi.org/10.1007/978-981-13-6004-6_4

Dhiman, J., Vaid, K., Johns, T., Maurya, R., Arora, M., Negi, A., Gupta, R., Misra, M., Kim, K.-H., & Kumar, V. (2022). Tyrosinase-functionalized gold nanoparticle-tailored ultrasensitive nanosensing probe for hazardous and nutritional phenolic compounds. *Sensors and Actuators B: Chemical, 371*, 1–15. https://doi.org/10.1016/j.snb.2022.132434

Dhiman, N. K., Agnihotri, S., & Reddy, M. S. (2024). High permeation, biodegradable hydrogel based flow-through reactor for water disinfection with controlled silver release. *Chemical Engineering Journal, 489*, 1–14. https://doi.org/10.1016/j.cej.2024.151335

Dimitrievska, I., Paunovic, P., & Grozdanov, A. (2023). Recent advancements in nano sensors for air and water pollution control. *Materials Science and Engineering, 7*, 113–128. https://doi.org/10.15406/mseij.2023.07.00214

Do, M. H., Ngo, H. H., Guo, W., Chang, S. W., Nguyen, D. D., Liu, Y., Varjani, S., & Kumar, M. (2020). Microbial fuel cell-based biosensor for online monitoring wastewater quality: A critical review. *Science of the Total Environment, 712*, 1–50. https://doi.org/10.1016/j.scitotenv.2019.135612

Du, X., Shi, Y., Jegatheesan, V., & Haq, I. U. (2020). A review on the mechanism, impacts and control methods of membrane fouling in MBR system. *Membranes, 10*(2), 1–33. https://doi.org/10.3390/membranes10020024

Dubey, S. K., Dey, A., Singhvi, G., Pandey, M. M., Singh, V., & Kesharwani, P. (2022). Emerging trends of nanotechnology in advanced cosmetics. *Colloids and Surfaces B: Biointerfaces, 214*, 1–30. https://doi.org/10.1016/j.colsurfb.2022.112440

Elcik, H., Fortunato, L., Vrouwenvelder, J. S., & Ghaffour, N. (2021). Real-time membrane fouling analysis for the assessment of reclamation potential of textile wastewater processed by membrane distillation. *Journal of Water Process Engineering, 43*, 1–10. https://doi.org/10.1016/j.jwpe.2021.102296

El-Rawy, M., Abd-Ellah, M. K., Fathi, H., & Ahmed, A. K. A. (2021). Forecasting effluent and performance of wastewater treatment plant using different machine learning techniques. *Journal of Water Process Engineering, 44*, 1–23. https://doi.org/10.1016/j.jwpe.2021.102380

Fang, Y., Wang, Y., Zhu, L., Liu, H., Su, X., Liu, Y., Chen, Z., Chen, H., & He, N. (2023). A novel cartridge for nucleic acid extraction, amplification and detection of infectious disease pathogens with the help of magnetic nanoparticles. *Chinese Chemical Letters, 34*(8), 1–12. https://doi.org/10.1016/j.cclet.2022.108092

Fatta-Kassinos, D., Kalavrouziotis, I. K., Koukoulakis, P. H., & Vasquez, M. (2011). The risks associated with wastewater reuse and xenobiotics in the agroecological environment. *Science of the Total Environment, 409*(19), 3555–3563. https://doi.org/10.1016/j.scitotenv.2010.03.036

Finzi, A., Oberti, R., Negri, A., Perazzolo, F., Cocolo, G., Tambone, F., Cabassi, G., & Provolo, G. (2015). Effects of measurement technique and sample preparation on NIR spectroscopy analysis of livestock slurry and digestates. *Biosystems Engineering, 134*, 42–54. https://doi.org/10.1016/j.biosystemseng.2015.03.015

Fortunato, L., Ranieri, L., Naddeo, V., & Leiknes, T. (2020). Fouling control in a gravity-driven membrane (GDM) bioreactor treating primary wastewater by using relaxation and/or air scouring. *Journal of Membrane Science, 610*, 1–11. https://doi.org/10.1016/j.memsci.2020.118261

Geng, Z., Deng, T., Gu, B., Qian, X., Li, R., Duan, L., Li, J., Han, W., Qu, L., & Wei, K. (2024). Visible-light-sensitive microrobots using H_2O as fuel for highly efficient capture and precise detection of nanoplastics. *Journal of Hazardous Materials, 479*, 1–13. https://doi.org/10.1016/j.jhazmat.2024.135731

Ghofrani-Isfahani, P., Valverde-Pérez, B., Alvarado-Morales, M., Shahrokhi, M., Vossoughi, M., & Angelidaki, I. (2020). Supervisory control of an anaerobic digester subject to drastic substrate changes. *Chemical Engineering Journal, 391*, 1–41. https://doi.org/10.1016/j.cej.2019.123502

Gopikumar, S., Banu, J. R., Robinson, Y. H., Shanmuganathan, V., Kadry, S., & Rho, S. (2021). Novel framework of GIS based automated monitoring process on environmental biodegradability and risk analysis using Internet of Things. *Environmental Research, 194*, 1–13. https://doi.org/10.1016/j.envres.2020.110621

Huang, Y. T., Khan, A., Ganguly, A., Kaswan, K., Suresh, S., Cheng, Y. Y., Lee, K. M., Yu, J. H., & Lin, Z. H. (2024). Real-time wireless detection of heavy metal ions using a self-powered triboelectric nanosensor integrated with an autonomous thermoelectric generator-powered robotic system. *Advanced Science*, 1–12. https://doi.org/10.1002/advs.202410424

Huisman, K. T., Blankert, B., Horn, H., Wagner, M., Vrouwenvelder, J. S., Bucs, S., & Fortunato, L. (2024). Noninvasive monitoring of fouling in membrane processes by optical coherence tomography: A review. *Journal of Membrane Science, 692*, 1–12. https://doi.org/10.1016/j.memsci.2023.122291

Hwang, Y. J., Yu, H., Lee, G., Shackery, I., Seong, J., Jung, Y., Sung, S.-H., Choi, J., & Jun, S. C. (2023). Multiplexed DNA-functionalized graphene sensor with artificial intelligence-based discrimination performance for analyzing chemical vapor compositions. *Microsystems & Nanoengineering, 9*(1), 1–13. https://doi.org/10.1038/s41378-023-00499-y

Im, S. J., Viet, N. D., & Jang, A. (2021). Real-time monitoring of forward osmosis membrane fouling in wastewater reuse process performed with a deep learning model. *Chemosphere, 275*, 1–11. https://doi.org/10.1016/j.chemosphere.2021.130047

Iorhemen, O. T., Hamza, R. A., & Tay, J. H. (2016). Membrane bioreactor (MBR) technology for wastewater treatment and reclamation: Membrane fouling. *Membranes, 6*(2), 1–29. https://doi.org/10.3390/membranes6020033

Jawad, J., Hawari, A. H., & Zaidi, S. J. (2021). Modeling and sensitivity analysis of the forward osmosis process to predict membrane flux using a novel combination of neural network and response surface methodology techniques. *Membranes, 11*(1), 1–19. https://doi.org/10.3390/membranes11010070

Jeong, S., & Kim, H.-W. (2023). In situ real-time monitoring technologies for fouling detection in membrane processes *Current Developments in Biotechnology and Bioengineering* (pp. 43–64). Elsevier.

Jeong, S.-Y., Moon, Y. K., Wang, J., & Lee, J.-H. (2023). Exclusive detection of volatile aromatic hydrocarbons using bilayer oxide chemiresistors with catalytic overlayers. *Nature Communications, 14*(1), 1–13. https://doi.org/10.1038/s41467-023-35916-3

Kabugo, J. C., Jämsä-Jounela, S.-L., Schiemann, R., & Binder, C. (2020). Industry 4.0 based process data analytics platform: A waste-to-energy plant case study. *International Journal of Electrical Power & Energy Systems, 115*, 1–18. https://doi.org/10.1016/j.ijepes.2019.105508

Kazemi, P., Steyer, J.-P., Bengoa, C., Font, J., & Giralt, J. (2020). Robust data-driven soft sensors for online monitoring of volatile fatty acids in anaerobic digestion processes. *Processes, 8*(1), 1–17. https://doi.org/10.3390/pr8010067

Komikawa, T., Tanaka, M., Tamang, A., Evans, S. D., Critchley, K., & Okochi, M. (2020). Peptide-functionalized quantum dots for rapid label-free sensing of 2, 4, 6-trinitrotoluene. *Bioconjugate Chemistry, 31*(5), 1400–1407. https://doi.org/10.1021/acs.bioconjchem.0c00117

Kovacs, D. J., Li, Z., Baetz, B. W., Hong, Y., Donnaz, S., Zhao, X., Zhou, P., Ding, H., & Dong, Q. (2022). Membrane fouling prediction and uncertainty analysis using machine learning: A wastewater treatment plant case study. *Journal of Membrane Science, 660*, 1–12. https://doi.org/10.1016/j.memsci.2022.120817

Krishnapandi, A., Muthukutty, B., Chen, S.-M., Arul, K. T., Shiuan, H. J., & Selvaganapathy, M. (2021). Bismuth molybdate incorporated functionalized carbon nanofiber as an electrocatalytic tool for the pinpoint detection of organic pollutant in life samples. *Ecotoxicology and Environmental Safety, 209*, 1–12. https://doi.org/10.1016/j.ecoenv.2020.111828

Kumari, S., Nehra, M., Jain, S., Sheokand, A., Dilbaghi, N., Chaudhary, G. R., Kim, K.-H., & Kumar, S. (2025b). Luminescent Cu nanoclusters–encapsulated ZIF-8 as on–off–on fluorescent probe for efficient and selective quantification of *E. coli*. *Microchimica Acta, 192*(2), 1–17. https://doi.org/10.1007/s00604-024-06905-0

Kumari, M., Dhariwal, N., Yadav, P., Kumar, V., & Thakur, O. (2025a). Urea modified Cu-doped $LaFeO_3$ nano-particles for humidity sensing with contactless moisture detection for medical and agricultural application. *Sensors and Actuators B: Chemical, 423*, 1–13. https://doi.org/10.1016/j.snb.2024.136749

Kümmerer, K., Dionysiou, D. D., Olsson, O., & Fatta-Kassinos, D. (2018). A path to clean water. *Science, 361*(6399), 222–224. https://doi.org/10.1126/science.aau2405

Kundu, M., Krishnan, P., Chobhe, K. A., Manjaiah, K., Pant, R., & Chawla, G. (2022). Fabrication of electrochemical nanosensor for detection of nitrate content in soil extract. *Journal of Soil Science and Plant Nutrition, 22*(3), 2777–2792.

Laghrib, F., Bakasse, M., Lahrich, S., & El Mhammedi, M. (2020). Electrochemical sensors for improved detection of paraquat in food samples: A review. *Materials Science and Engineering: C, 107*, 1–9. https://doi.org/10.1016/j.msec.2019.110349

Lee, J. I., Jang, S. C., Chung, J., Choi, W.-K., Hong, C., Ahn, G. R., Kim, S. H., Lee, B. Y., & Chung, W.-J. (2021). Colorimetric allergenic fungal spore detection using peptide-modified gold nanoparticles. *Sensors and Actuators B: Chemical, 327*, 1–9. https://doi.org/10.1016/j.snb.2020.128894

Li, P., Fu, H., Bai, Z., Feng, X., Qi, J., Song, X., Hu, X., & Chen, L. (2023). A dummy molecularly imprinted ratiometric fluorescence nanosensor for the sensitive detection of guanidyl-microcystins in environmental water. *The Analyst, 148*(3), 573–582. https://doi.org/10.1039/D2AN01928K

Li, W., Li, C., & Wang, T. (2020). Application of machine learning algorithms in MBR simulation under big data platform. *Water Practice & Technology, 15*(4), 1238–1247. https://doi.org/10.2166/wpt.2020.095

Ma, J., Liu, C., Li, J., An, Z., Zhang, B., Hong, W., Ye, C., Li, M., & Guo, L.-H. (2025). Colorimetric visualization detection of perfluorooctanoic acid based on host–guest interactions with cyclodextrin-modified gold nanoparticles. *Environmental Science: Nano*, 1–12. https://doi.org/10.1039/D4EN01096E

Mahmoud, A. E. D., & Fawzy, M. (2021). Nanosensors and nanobiosensors for monitoring the environmental pollutants. *Waste Recycling Technologies for Nanomaterials Manufacturing*, 229–246. https://doi.org/10.1007/978-3-030-68031-2_9

Marteleira, R., Pinto, G., & Niza, S. (2014). Regional water flows—Assessing opportunities for sustainable management. *Resources, Conservation and Recycling, 82*, 63–74. https://doi.org/10.1016/j.resconrec.2013.10.016

Massoud, M. A., Terkawi, M., & Nakkash, R. (2019). Water reuse as an incentive to promote sustainable agriculture in Lebanon: Stakeholders' perspectives. *Integrated Environmental Assessment and Management, 15*(3), 412–421. https://doi.org/10.1002/ieam.4131

Meng, F., Zhang, S., Oh, Y., Zhou, Z., Shin, H.-S., & Chae, S.-R. (2017). Fouling in membrane bioreactors: An updated review. *Water Research, 114*, 151–180. https://doi.org/10.1016/j.watres.2017.02.006

Mondal, S., & Subramaniam, C. (2019). Point-of-care, cable-type electrochemical Zn^{2+} sensor with ultrahigh sensitivity and wide detection range for soil and sweat analysis. *ACS Sustainable Chemistry & Engineering, 7*(17), 14569–14579. https://doi.org/10.1021/acssuschemeng.9b02173

Mubarak, A., Esmaili, P., Ameen, Z., Abdulkadir, R., Gaya, M., Ozsoz, M., Saini, G., & Abba, S. (2021). Metro-environmental data approach for the prediction of chemical oxygen demand in new Nicosia wastewater treatment plant. *Desalination and Water Treatment, 221*, 31–40. https://doi.org/10.5004/dwt.2021.27049

Nurdin, M., Maulidiyah, M., & Nohong, N. (2025). Advanced TiO2-ZnO/graphene hybrid nanocomposite for ultra-sensitive electrochemical detection of fipronil pesticide. *Environmental Nanotechnology, Monitoring & Management, 23*, 101031.

Parameswari, P., Belagalla, N., Singh, B. V., Abhishek, G., Rajesh, G., Katiyar, D., Hazarika, B., & Paul, S. (2024). Nanotechnology-based sensors for real-time monitoring and assessment of soil health and quality: A review. *Asian Journal of Soil Science and Plant Nutrition, 10*(2), 157–173. https://doi.org/10.9734/ajsspn/2024/v10i2272

Patil, S., Ghadi, H., Ramgir, N., Adhikari, A., & Rao, V. R. (2019). Monitoring soil pH variation using Polyaniline/SU-8 composite film based conductometric microsensor. *Sensors and Actuators B: Chemical, 286*, 583–590. https://doi.org/10.1016/j.snb.2019.02.016

Piña, B., Bayona, J. M., Christou, A., Fatta-Kassinos, D., Guillon, E., Lambropoulou, D., Michael, C., Polesel, F., & Sayen, S. (2020). On the contribution of reclaimed wastewater irrigation to the potential exposure of humans to antibiotics, antibiotic resistant bacteria and antibiotic resistance genes—NEREUS COST Action ES1403 position paper. *Journal of Environmental Chemical Engineering, 8*(1), 1–11. https://doi.org/10.1016/j.jece.2018.01.011

Ranieri, L., Esposito, R., Nunes, S. P., Vrouwenvelder, J. S., & Fortunato, L. (2024). Biofilm rigidity, mechanics and composition in seawater desalination pretreatment employing ultrafiltration and microfiltration membranes. *Water Research, 253*, 1–21. https://doi.org/10.1016/j.watres.2024.121282

Rasheed, R., Tahir, F., Yasar, A., Ahmad, S. R., Tabinda, A. B., Sharif, F., & Su, Y. (2020). An industrial scale testing and analysis of waste-to-energy production from various substrates by employing a modern anaerobic digestion plant. *Biomass and Bioenergy, 138*, 1–10. https://doi.org/10.1016/j.biombioe.2020.105571

Ren, K., Jiao, Z., Wu, X., & Han, H. (2023). Multivariable identification of membrane fouling based on compacted cascade neural network. *Chinese Journal of Chemical Engineering, 53*, 37–45. https://doi.org/10.1016/j.cjche.2022.01.028

Rizzo, L., Manaia, C., Merlin, C., Schwartz, T., Dagot, C., Ploy, M., Michael, I., & Fatta-Kassinos, D. (2013). Urban wastewater treatment plants as hotspots for antibiotic resistant bacteria and genes spread into the environment: A review. *Science of the Total Environment, 447*, 345–360. https://doi.org/10.1016/j.scitotenv.2013.01.032

Robinson, D. A., Jones, S. B., Lebron, I., Reinsch, S., Domínguez, M. T., Smith, A. R., Jones, D. L., Marshall, M. R., & Emmett, B. A. (2016). Experimental evidence for drought induced alternative stable states of soil moisture. *Scientific Reports, 6*(1), 1–6. https://doi.org/10.1038/srep20018

Rudolph, G., Hermansson, A., Jönsson, A.-S., & Lipnizki, F. (2021). In situ real-time investigations on adsorptive membrane fouling by thermomechanical pulping process water with quartz crystal microbalance with dissipation monitoring (QCM-D). *Separation and Purification Technology, 254*, 1–9. https://doi.org/10.1016/j.seppur.2020.117578

Rudolph-Schöpping, G., Schagerlöf, H., Jönsson, A.-S., & Lipnizki, F. (2023). Comparison of membrane fouling during ultrafiltration with adsorption studied by quartz crystal microbalance with dissipation monitoring (QCM-D). *Journal of Membrane Science, 672*, 1–8. https://doi.org/10.1016/j.memsci.2022.121313

Sadrolhosseini, A. R., Ghasemi, E., Pirkarimi, A., Hamidi, S. M., & Ghahrizjani, R. T. (2023). Highly sensitive surface plasmon resonance sensor for detection of Methylene Blue and Methylene Orange dyes using NiCo-Layered Double Hydroxide. *Optics Communications, 529*, 1–7. https://doi.org/10.1016/j.optcom.2022.129057

Sahu, M., Ganguly, M., & Sharma, P. (2024). Highly fluorescent quinone-capped silver hydrosol for environmental remediation and sensing applications. *Spectrochimica Acta Part a: Molecular and Biomolecular Spectroscopy, 311*, 1–10. https://doi.org/10.1016/j.saa.2024.123981

Sharafati, A., Asadollah, S. B. H. S., & Hosseinzadeh, M. (2020). The potential of new ensemble machine learning models for effluent quality parameters prediction and related uncertainty. *Process Safety and Environmental Protection, 140*, 68–78. https://doi.org/10.1016/j.psep.2020.04.045

Sillu, D., Agnihotri, S., & Reddy, M. S. (2022). Magnetic halloysite nanotube/α-amylase based nanobiocatalytic transformation of food processing waste into an active fermentation ingredient. *ACS Sustainable Chemistry & Engineering, 10*(31), 10169–10180. https://doi.org/10.1021/acssuschemeng.2c01547

Sun, H., Xu, M., Wu, S., Dong, R., Angelidaki, I., & Zhang, Y. (2021). Innovative air-cathode bioelectrochemical sensor for monitoring of total volatile fatty acids during anaerobic digestion. *Chemosphere, 273*, 1–9. https://doi.org/10.1016/j.chemosphere.2021.129660

Sun, H., Zhang, Y., Wu, S., Dong, R., & Angelidaki, I. (2019). Innovative operation of microbial fuel cell-based biosensor for selective monitoring of acetate during anaerobic digestion. *Science of the Total Environment, 655*, 1439–1447. https://doi.org/10.1016/j.scitotenv.2018.11.336

Turner, R. D., Warne, M. S. J., Dawes, L. A., Thompson, K., & Will, G. D. (2019). Greywater irrigation as a source of organic micro-pollutants to shallow groundwater and nearby surface water. *Science of the Total Environment, 669*, 570–578. https://doi.org/10.1016/j.scitotenv.2019.03.073

Varvařovská, L., Kudrna, P., Sopko, B., & Jarošíková, T. (2024). The development of a specific nanofiber bioreceptor for detection of *Escherichia coli* and *Staphylococcus aureus* from air. *Biosensors, 14*(5), 1–12. https://doi.org/10.3390/bios14050234

Wang, D., Thunéll, S., Lindberg, U., Jiang, L., Trygg, J., Tysklind, M., & Souihi, N. (2021). A machine learning framework to improve effluent quality control in wastewater treatment plants. *Science of the Total Environment, 784*, 1–11. https://doi.org/10.1016/j.scitotenv.2021.147138

Wang, J., Zhu, W.-J., Zhang, J.-Y., Qi, B., Wang, J.-P., Gao, G.-G., Fan, L.-L., & Liu, H. (2023). Ratiometric response to formaldehyde by 3D silver SERS substrate with polyoxometalate as internal label. *Sensors and Actuators B: Chemical, 381*, 1–13. https://doi.org/10.1016/j.snb.2023.133450

Wang, W., Srivastava, S., & Vikesland, P. J. (2024). Overcoming barriers and embracing advances: Nanosensor implementation for practical water contaminant surveillance. *One Earth, 7*(8), 1351–1361. https://doi.org/10.1016/j.oneear.2024.07.006

Wang, X., Bai, X., Li, Z., Zhou, X., Cheng, S., Sun, J., & Liu, T. (2018). Evaluation of artificial neural network models for online monitoring of alkalinity in anaerobic co-digestion system. *Biochemical Engineering Journal, 140*, 85–92. https://doi.org/10.1016/j.bej.2018.09.010

Wong, P. W., Guo, J., Khanzada, N. K., Yim, V. M. W., & Kyoungjin, A. (2021). In-situ 3D fouling visualization of membrane distillation treating industrial textile wastewater by optical coherence tomography imaging. *Water Research, 205*, 1–11. https://doi.org/10.1016/j.watres.2021.117668

Wu, D., Li, L., Zhao, X., Peng, Y., Yang, P., & Peng, X. (2019a). Anaerobic digestion: A review on process monitoring. *Renewable and Sustainable Energy Reviews, 103*, 1–12. https://doi.org/10.1016/j.rser.2018.12.039

Wu, S., Sikdar, P., & Bhat, G. S. (2023a). Recent progress in developing ballistic and anti-impact materials: Nanotechnology and main approaches. *Defence Technology, 21*, 33–61. https://doi.org/10.1016/j.dt.2022.06.007

Wu, Y., Kovalovszki, A., Pan, J., Lin, C., Liu, H., Duan, N., & Angelidaki, I. (2019b). Early warning indicators for mesophilic anaerobic digestion of corn stalk: A combined experimental and simulation approach. *Biotechnology for Biofuels, 12*, 1–15. https://doi.org/10.1186/s13068-019-1442-7

Wu, Z., Wang, H., Ding, Q., Tao, K., Shi, W., Liu, C., Chen, J., & Wu, J. (2023b). A self-powered, rechargeable, and wearable hydrogel patch for wireless gas detection with extraordinary performance. *Advanced Functional Materials, 33*(21), 1–23. https://doi.org/10.1002/adfm.202300046

Xiang, W., Han, M., Dong, T., Yao, J., & Han, L. (2021). Fouling dynamics of anion polyacrylamide on anion exchange membrane in electrodialysis. *Desalination, 507*, 1–12. https://doi.org/10.1016/j.desal.2021.115036

Xie, Y., Chen, Y., Lian, Q., Yin, H., Peng, J., Sheng, M., & Wang, Y. (2022). Enhancing real-time prediction of effluent water quality of wastewater treatment plant based on improved feedforward neural network coupled with optimization algorithm. *Water, 14*(7), 1–16. https://doi.org/10.3390/w14071053

Xiu, C., & Dong, L. (2019). *Design of sewage treatment monitoring system based on internet of things*. Paper presented at the 2019 Chinese Control And Decision Conference (CCDC).

Xu, H., Li, J., Li, P., Shi, J., Gao, X., & Luo, W. (2021). Highly efficient SO_2 sensing by light-assisted Ag/PANI/SnO_2 at room temperature and the sensing mechanism. *ACS Applied Materials & Interfaces, 13*(41), 49194–49205. https://doi.org/10.1021/acsami.1c14548

Yang, Y., Kim, K.-R., Kou, R., Li, Y., Fu, J., Zhao, L., & Liu, H. (2022). Prediction of effluent quality in a wastewater treatment plant by dynamic neural network modeling. *Process Safety and Environmental Protection, 158*, 515–524. https://doi.org/10.1016/j.psep.2021.12.034

Yi, W. Y., Lo, K. M., Mak, T., Leung, K. S., Leung, Y., & Meng, M. L. (2015). A survey of wireless sensor network based air pollution monitoring systems. *Sensors, 15*(12), 31392–31427. https://doi.org/10.3390/s151229859

Zdražil, L., Panáček, D., Šedajová, V., Baďura, Z., Langer, M., Medveď, M., Paloncýová, M., Scheibe, M., Kalytchuk, S., & Zoppellaro, G. (2023). Carbon dots enabling parts-per-billion sensitive and ultraselective photoluminescence lifetime-based sensing of inorganic mercury. *Advanced Optical Materials, 11*(21), 1–8. https://doi.org/10.1002/adom.202300750

Zeebaree, A. Y. S., Sharaf, S. Y., Sultan, S. J., Rashid, R. F., Haji, O. I., Barani, S. S. I., Rumih, Z. K., Othman, I. I., Mohammad, K. S., & Albastky, N. M. S. (2025). Sustainable strategy and rapid naked eye performance of novel Cu@ SG nanosensor for detection of glyphosate in aquatic models. *Desalination and Water Treatment, 321*, 1–15. https://doi.org/10.1016/j.dwt.2024.100903

Zhang, N., Lee, H.-J., Wu, Y., Ganzoury, M. A., & de Lannoy, C.-F. (2022). Integrating biofouling sensing with fouling mitigation in a two-electrode electrically conductive membrane filtration system. *Separation and Purification Technology, 288*, 1–9. https://doi.org/10.1016/j.seppur.2022.120679

Zhao, N., Li, X., Jin, X., Angelidaki, I., & Zhang, Y. (2018). Integrated electrochemical-biological process as an alternative mean for ammonia monitoring during anaerobic digestion of organic wastes. *Chemosphere, 195*, 735–741. https://doi.org/10.1016/j.chemosphere.2017.12.139

Zhao, Z., Deng, Z., Zhang, R., Klamchuen, A., He, Y., Horprathum, M., Chang, J., Mi, L., Li, M., & Wang, S. (2023). Sensitive and selective ozone sensor based on $CuCo_2O_4$ synthesized by a facile solution combustion method. *Sensors and Actuators B: Chemical, 375*, 1–13. https://doi.org/10.1016/j.snb.2022.132912

Zhong, H., Yuan, Y., Luo, L., Ye, J., Chen, M., & Zhong, C. (2022). Water quality prediction of MBR based on machine learning: A novel dataset contribution analysis method. *Journal of Water Process Engineering, 50*, 1–17. https://doi.org/10.1016/j.jwpe.2022.103296

Zuthi, M., Ngo, H., & Guo, W. (2012). Modelling bioprocesses and membrane fouling in membrane bioreactor (MBR): A review towards finding an integrated model framework. *Bioresource Technology, 122*, 119–129. https://doi.org/10.1016/j.biortech.2012.04.090

Zwart, N., Jonker, W., Ten Broek, R., de Boer, J., Somsen, G., Kool, J., Hamers, T., Houtman, C. J., & Lamoree, M. H. (2020). Identification of mutagenic and endocrine disrupting compounds in surface water and wastewater treatment plant effluents using high-resolution effect-directed analysis. *Water Research, 168*, 1–34. https://doi.org/10.1016/j.watres.2019.115204

Chapter 9
Wastewater Recycling by Smart Technologies: Case Studies Around the World

Ranjan Pandey, Diksha Rana, Rahul Gupta, Ashima Awasthi, and Luiz Fernando Romanholo Ferreira

Abstract The rise in the global population has led to a significant pressure on the natural resources, specifically on water. The continuously growing volume of wastewater poses a risk of depletion of portable water and also pollution of the existing water bodies, giving way to new diseases. Research to tackle such issues is also gaining momentum, and one such economically and environmentally viable solution is the Circular Economy. Recycling conserves natural resources, reduces pollution, creates jobs, saves energy, lowers emissions, decreases consumption of raw materials, and promotes sustainability. With the fourth industrial revolution, nations are adapting to smart water and data-driven recycling technologies such as Automation systems, Digital meters, sensors, Internet of Things (IoT)-based wastewater management systems, and many more. These techniques improve the quality and quantity of the wastewater and conserve the latent energy, nutrients, and water

R. Pandey (✉)
Department of Sustainable Sciences, Delhi Skill and Entrepreneurship University, Govt. of NCT of Delhi, New Delhi, India
e-mail: ranjan.pandey@dseu.ac.in

D. Rana
Department of Environmental Science, Amity University, Mohali, Punjab, India
e-mail: dikshaevs@gmail.com

R. Gupta
Department of Civil Engineering, University of Calgary, Calgary, AB, Canada
e-mail: rahulgupta.9096@gmail.com

A. Awasthi
Department of Environmental Science, Sardar Patel University, Mandi, Himachal Pradesh, India
e-mail: aaashima.15@gmail.com

L. F. R. Ferreira
Genomic Sciences and Biotechnology Lab, Catholic University of Brasília, Brasília, Brazil

Lab. de Ciências Genômicas e Biotecnologia, PPG em Ciências Genomicas e Biotecnologia—Universidade Católica de Brasília, Brasília, DF, Brazil

L. F. R. Ferreira
e-mail: luiz.romanholo@p.ucb.br

S. Dhiman and G. Mukherjee (eds.), *Resource Resurgence*, Springer Water,
https://doi.org/10.1007/978-3-032-02750-4_9

embedded within it. It encourages water reuse, thus reducing the pressure on natural resources. This chapter attempts to develop an overview of wastewater recycling and discusses exemplary innovative recycling technologies practiced worldwide. It can help professionals working in the environment and sustainability fields to improve policies and planning by serving as a reference.

Keywords Circular economy · Smart techniques · Wastewater management · Machine learning · Digitization · Artificial intelligence · Internet of Things (IoT)

9.1 Water: An Essential Resource

Water is a necessity for survival and development. On earth, nearly 3% of this essential resource is fresh. It is used for drinking, agriculture, and manufacturing to recreational purposes. The changing climatic pattern and rise in the global population have led to significant pressure on the aquifers. The replenishment rate of fresh water ecosystem is much slower than the consumption rate, thus leading to uneven distribution and intense competition among the various sectors. As per the SDG 2022 report, around 2 billion people in the world don't have access to clean portable water, and close to 4 billion people face extreme water shortages, sometimes during the year (United Nations, 2022). These numbers are expected to go up. The effects of water shortage are evident in part of the Middle East, Africa, Asia, and Latin America. Apart from this, each year 380 billion m^3 of wastewater is produced worldwide, and 80% of this wastewater is released back into the natural aquatic ecosystem (Pratap et al., 2023). It is mainly from domestic, agricultural, or industrial sources. Many times, the effluents contain hazardous, i.e., dyes, heavy metals, oil and grease, hydrocarbons, pharmaceutical drugs, Poly-Chlorinated Biphenyls (PCBs), phthalates, microplastics, xenobiotics, and non-hazardous compounds, such as ions and nutrients (Masood et al., 2021). Often, the presence of viruses, bacteria, and protozoa is also detected in the discharged waste. The contaminant-loaded liquid waste finds its way into the ecosystem and contaminates the soil and freshwater resources. The shrinking of aquatic assets and continuously growing volume of wastewater is posing a risk to environmental safety, human survival, and giving way to new diseases. Thus, the world needs to explore all the available options for limiting the over-exploitation of freshwater resources.

Wastewater treatment technologies have been established over time in an attempt to reduce pollution in the ecosystem (Pandey et al., 2024). In developed countries, the use and treatment of water are both at a higher percentage when compared with the developing and lower-income countries. A country's economy and percentage of wastewater treatment are concomitant. High-income countries treat approximately 70% of the produced wastewater, and middle-income nations treat 38–28% of the wastewater produced (Pratap et al., 2023). However, countries with lower economies are able to treat only 8% of the total wastewater. The global industrial wastewater treatment market is poised to increase to the US$16.5 billion by 2026 with an average growth rate of 4.5% CAGR (Parde & Behera, 2023). However, treating the wastewater only reduces environmental pollution, and the pressure on the natural resources remains usual.

Over the past few decades, Sustainable Development has been gaining momentum. It explains the use of resources in such a way that the needs of the present generation are contended without compromising the ability of future generations to satisfy their own needs. It encourages the concept of circular economy. A circular economy is an economically and environmentally viable model that requires fewer resources for manufacturing materials and products (Yang et al., 2023). It is based on a redesigning model which reuses "waste" as a resource in production. It entails a market that promotes recycling of the products, rather than discarding them, and discourages pressure on new resources for manufacturing. Thus, it minimizes waste, reduces carbon footprints, and promotes sustainable use of natural resources. Thus, gradually, there is a paradigm shift of policies pushing the world to adopt measures for a better future. Circular Economy has pushed the recycling of treated wastewater for multipurpose. Globally, millions of smallholder farmers use treated wastewater to support their agricultural yields and livelihoods. Of the total wastewater generated worldwide every year, approx. 6.6% is used for agricultural purposes, and more than 10% of the global population consumes wastewater irrigated agricultural products (Yahaya et al., 2023). Barring agriculture, treated wastewater carries applicability in thermal power plants as coolant, construction activities, firefighting, groundwater recharging, lands such as park and golf course irrigation, toilet flushes, and vehicle washing.

Conventional water recycling technologies primarily encompass membrane filtration, chemical treatments, ion exchange, adsorption, activated sludge and biofilm, and advanced oxidation processes such as Fenton reactions, photocatalytic oxidation, and electrochemical oxidation reactions (Cairone et al., 2024). Such technologies have high operational and maintenance costs, often leading to sludge swelling and the possibility of secondary pollution. Traditional wastewater treatment focuses only on meeting the environmental discharge standards by removing the contaminants, but in this process, the latent energy, nutrients, and water embedded within, gets neglected. These techniques are time-consuming and lack real-time feedback. Thus, they are not sustainable on a large scale. Traditional wastewater treatment methods also lack the capability to work efficiently in response to variations in wastewater characteristics. Lack of real-time monitoring and longer water sampling and analysis often leads to delays in detecting defects and taking corrective action. Therefore, efficient technologies are required, which can minimize waste production and maximize the reuse of the resource.

With fast-paced development and constantly evolving industrial technologies, a fourth industrial revolution has occurred, i.e., Industry 4.0 (Fig. 9.1). Industry 4.0 is a change towards an intelligent industry with the integration of new technologies, including artificial intelligence (AI), cloud computing, and the Internet of Things (IoT) into manufacturing facilities and throughout operations (Narayanan et al., 2024). The application of sensors, software, and robotics often leads to increased

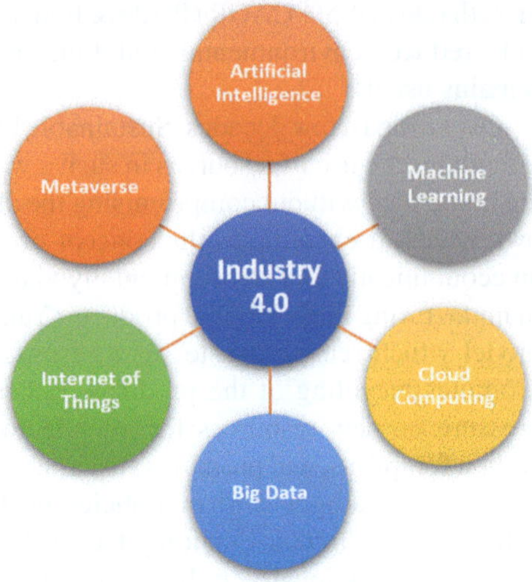

Fig. 9.1 Summary of Industry 4.0s

efficiency, automation, predictive maintenance, and self-optimization. This eventually helps to collect and analyse data for better decision-making. Industry 4.0 has the potential to achieve sustainable development goals by facilitating an efficient circular economy.

The established wastewater treatment and recycling technologies eliminate the compounds present in it and make the water reusable. It reduces the pressure on freshwater resources and decreases the volume of wastewater discharged into the environment. Digitalization plays an important role in supporting wastewater treatment systems to use information (big data) for real-time monitoring of the processes, collecting data on water quality and system functionality, thus, enhancing decision-making capability and operation efficiency (Nagpal et al., 2024). Incorporating cutting-edge technologies, including AI, machine learning, IoT, big data, and cloud computing in wastewater treatment systems can provide a definitive path for stringent wastewater treatment and disposal along with opportunities for resource recovery. Thus, the rise of smart water technology has the potential to transform traditional wastewater systems into instrumented, interconnected, and intelligent systems.

Lately, smart systems and data-driven innovations with information technology methods and internet communication techniques (ICT) are playing a critical role in managing water resources and changing the challenges of wastewater into opportunities (Alzahrani et al., 2023). Primarily smart recycling involves effective resource recovery, which limits the excessive waste of fresh water and monitors' water pollution. It is a stepping stone toward a brighter, greener, more sustainable future.

9.2 Smart Systems

Smart systems are a product of Industry 4.0, which consists of advanced technologies (software, hardware, analytics) that automate activities, use data to anticipate and make adaptive decisions to carry out actions, and make operations efficient and convenient (Khan & Razzak, 2021). Smart systems utilize advanced technologies including AI (artificial intelligence), IOT (internet of things), big data, advanced sensors, embedded software and robotics and data analytics to quickly track and record changes in soil, water or air (Rajak et al., 2023). This enables the smart system to detect malfunction or a leak and immediately fix the cause. Smart systems are used in industrial and civil industries and are successful in reducing costs, saving natural resources, and improving efficiency. One such industrial application where smart systems are being extensively used is wastewater management.

9.2.1 Wastewater Management and Resource Recovery

Smart water management system increases the efficiency of wastewater treatment processes by real-time monitoring, predictive maintenance, optimization of chemical dosage, and reduction in human error (Cairone et al., 2024). It helps in lowering the consumption of energy and heat and thus reduces carbon footprints. Further, smart water management systems are capable of measuring the performance parameters such as biological oxygen demand (BOD), chemical oxygen demand (COD), turbidity, hardness, and other physicochemical parameters. Many times, wastewater is comprised of resources like metals, chemicals, and nutrients. AI-based systems facilitate the treatment of such effluents along with the recovery of resources present in it (Zhang et al., 2023).

9.3 Data-Driven Approaches

Data-driven approach is the method of making decisions based on data analysis and interpretation. Data analytics analyze real-time data generated by connected devices (e.g., Sensors and digital meters). This helps businesses to track key performance indicators and improve processes (Matheny et al., 2023). Data-driven decisions make use of Machine Learning (ML) modeling. These models analyze the real-time data captured by sensors regarding weather conditions and soil conditions. In the agriculture sector, it aids in accurate timing for plantation, irrigation, and harvesting. Various natural conservation organizations also use big data for protection of endangered species (Worthington et al., 2020).

9.3.1 Wastewater Management and Resource Recovery

Sophisticated modeling of real-time data is a significant driver to increase efficiency and resource recovery in wastewater treatment operations (Khan & Javaid, 2021). The simulation models are based on the real-time monitoring of key parameters by sensors. Machine Learning (ML) creates a balance between the large volume of data collected by sensors and the complex statistical analysis (Durgun, 2024). The contemporary wastewater treatment systems are equipped with real-time sensing, analysis of data and online control of parameters, all supported by mathematical models (Alvi et al., 2023). Real-time data related to chemical concentrations, water quality indicators, and flow rates is fed into sophisticated models that analyze the data and predict the response of the treatment process to changing conditions (Durgun, 2024). This assists operators in making proactive and informed decisions regarding adjustments to treatment parameters. Simulation of different scenarios and prediction of potential issues through the models can significantly decrease downtime and repair costs.

A data-driven approach has been widely used to optimize resource recovery from wastewater. Machine learning algorithms are used to optimize the process for the proper disposal of biosolids in wastewater treatment plants and the creation of energy through biogas generation (Asadi & McPhedran, 2021). With the help of anaerobic digestion, wastewater plants convert biosolids to energy and other valuable products. Further, with the help of a data-driven solution, a controlled anaerobic co-digestion process produces fatty acids, which can be utilized to create sustainable chemicals for fertilizers.

9.4 Smart Technologies and Data-Driven Approaches: Together an Efficient System

Smart systems and Data-driven approaches together create a functional system that works impeccably. AI system mimics human abilities to perform complex tasks. Machine learning (ML) uses algorithms based on historical and real-time data to produce models to understand industrial operations (Wongburi & Park, 2021). In the context of the wastewater treatment system, historical data is related to the records preserved by the wastewater treatment plant over a period of time, including readings of flow rates, pH levels, temperature, wastewater conditions, loading rate, etc. On the other hand, real-time data related to wastewater quality is collected throughout ongoing operations and is tracked by sensors and digital meters. IoT devices transmit real-time data to a central monitoring system or a cloud-based computing system, where ML models analyse patterns and wastewater process behaviour and identify abnormalities throughout the everyday operation, by using real-time data collected from sensors (Essamlali et al., 2024). Further, ML models perform data cleaning, removing unreliable data to enhance output quality. Smart systems provide wastewater treatment workers with data related to increased pollution loading for immediate action. This helps workers quickly identify the contaminants and take appropriate actions to reduce pollution levels at the source (Alprol et al., 2024). With real-time insights regarding network conditions for the entire wastewater treatment plant, operators can forecast future outcomes and proactively plan and implement control strategies.

In the present chapter, many such technologies based on smart systems and data-driven approaches will be elucidated. Further, a few exemplary case studies related to each of these technologies, which are being practiced worldwide, will be discussed.

9.4.1 Digital Meters and Smart Sensor Networks

Smart meters and sensors are vital components of a smart system. These advanced technologies accurately measure real-time water quality parameters (Rodríguez-Alonso et al., 2024). Sensors and digital meters collect real-time data related to flow rate and water characteristics of influent and effluent and transmit them to cloud servers for operators' use (Fig. 9.2).

For instance, pH sensors monitor real-time acidity and alkalinity of the water. With control of pH, industries can reduce the levels of contaminants in the effluent for safe disposal into the environment (Ching et al., 2021). ORP sensors measure

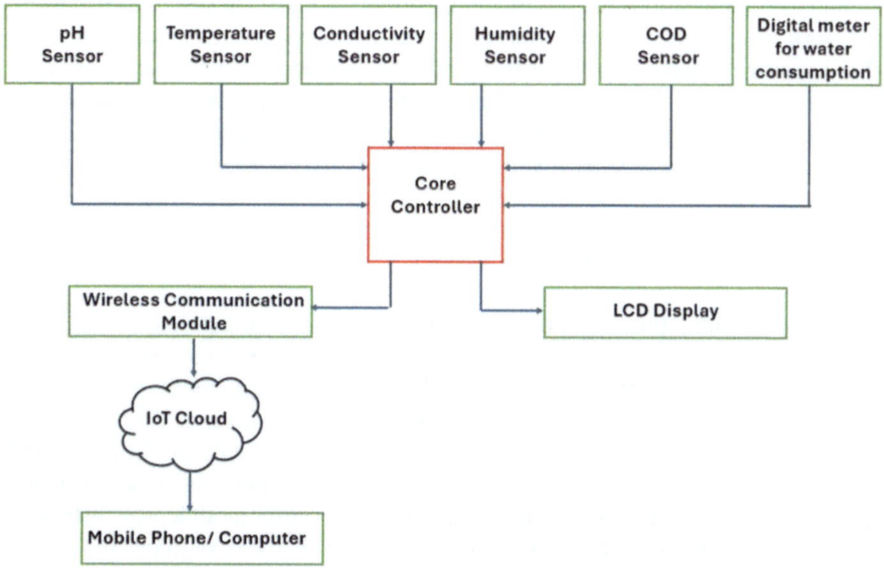

Fig. 9.2 Components of digital meters and smart sensors

the contamination in water. Turbidity sensors monitor the overall clarity of water by observing the amount of light scattered after hitting the surface of the water. Further, with the help of equipment asset management (EAM) and a computerized maintenance management system (CMMS), this real-time data, collected through sensors, can be used to monitor the performance of the machines (Martínez et al., 2020). In large factories, smart sensors can also control the temperature depending on the use of the operations and optimize the consumption of resources for improvement in the production (Martínez et al., 2020). It helps in tracking the life cycle of the water stream and prepares operators for scheduling preventive maintenance and calibration activities, along with scheduling meter readings and setting alarm conditions.

9.4.1.1 CASE STUDY 1: Malang, Indonesia

Earlier, a community wastewater treatment plant in Malang City, Indonesia, comprised an anaerobic (oxygen-free) tank to process organic material present in the wastewater, a phytoremediation pond that lowers the level of pollutants using plants like water hyacinth or vetiver, and finally, an aerobic filter to remove the solids in the wastewater. In the process of upgrading and enhancing the performance of the wastewater treatment plant, an online monitoring system was implemented to monitor four parameters (pH, turbidity, dissolved oxygen, and temperature). Sensors, electronic modules and an IoT platform were the key components of the online monitoring system (Soetedjo et al., 2022). The real-time data gathered by the sensors was transmitted to a cloud-based computing software with the help of a Global System for Mobile Communications (GSM) modem. Further, to make the system self-powered and work as a green energy monitoring system, a solar panel was installed at the top of the sensor module. Three sensor modules were placed at three locations. Location A and location B were located in the phytoremediation pond to monitor the influent and effluent parameters of the phytoremediation process. Location C was at the effluent of the wastewater treatment plant.

Configuration of Online Monitoring System

For each location, three sensor modules were placed. Each sensor module is comprised of sensors, a microcontroller with Wi-Fi, a GSM modem, and a solar panel and battery (Fig. 9.3). Each module had four sensors to measure parameters, i.e., pH, turbidity, dissolved oxygen and temperature (Soetedjo et al., 2022). Next, the microcontroller transmitted the data to the IoT cloud server with the help of a GSM modem. The IoT cloud server stored the historical data for future analysis and helped operators in the visualization of real-time data. Each sensor module was installed in a waterproof panel box. This arrangement was portable and easy to install.

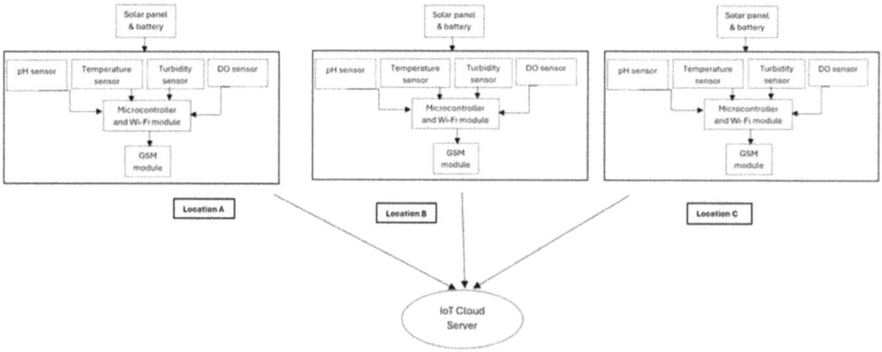

Fig. 9.3 Configuration of online monitoring system

Results

The online monitoring system efficiently monitored the real-time data of the parameters. The IoT cloud server displayed the real-time data collected through the sensors. The green colour indicated that values were within the quality standard, and the red colour indicated values going beyond the quality standard. The monitoring system achieved 81.4% reliability of the transfer of data with the period of transmission of data reduced to 1 min, which was sufficient for real-time monitoring at an interval of one day.

9.4.1.2 Case Study 2: Seosan City, Korea

The city of Seosan in Korea decided to install a smart water metering system for the local water supply system to measure drought (Kim & Kang, 2020). In response to regional and national drought reaction plans, smart meters were employed in the local water system with the aim to construct smart water management (SWM) systems. The focus of SWM was to reduce the leakages of water and improve the revenue water ratio with the help of remote metering, smart meters, and Information and Communication Technology (ICT). Previously, the meters installed in the local water supply systems were traditional. These meters required a person to check the data on meters manually and to physically remove from the flow line for calibration. Workers used to get in contact with process material while removing the meters. The newly installed smart meters provided hourly water usage data through a digital meter and wireless communication technology (ICT).

Results: As a result, the city witnessed a decrease in 190,000 m^3/year of water leakage along with a 20% improvement in the revenue-to-water ratio. It was projected that the new smart metering system would benefit USD 590,000 over the next 8 years.

9.4.2 Automation Systems

In wastewater treatment, automation systems may be referred to as advanced technology that can bring solutions that can be specifically applied to augment the accuracy, efficiency, and sustainability of the applied treatment (Gao & Liu, 2019). It comprises an integrated network of differing data platforms, actuators, controllers, and a network of sensors, which are used to regulate and monitor various parameters in real-time. Some of the important aspects, such as physical, chemical, biological, energy, nutrient, different operational, and water quality, are regularly tracked in order to optimize different treatment processes (Nivala & Kettunen, 2017). These systems are very much applicable at minimizing the effect on the environment and cost encoding for different operations by using a defined set of hardware and software that is able to monitor, control and optimize various operational parameters to ensure that wastewater is treated effectively (Wang et al., 2020).

Automation systems are comprised of different components. Sensors and instruments help calculate chemical and physical qualities like pH, turbidity, temperature, dissolved oxygen, strength of ammonia, pollution by heavy metals, and flow rate. When data is input into programmable logic controllers (PLCs), automation for exercising chemical dose, pumping, and aeration is accomplished (Khan & Razzak, 2021).

Chemical Dosing Automation assures correct delivery of disinfectants, pH regulators, and coagulants in response to pH and redox reactions to avoid overuse and maintain compliance. In parallel, aeration energy is established on levels of dissolved oxygen, further decreasing practical costs and not compromising the treatment for biological activity (Nassiri & Mohseni, 2021). Real-time data Analytics detects some patterns and possible failure probabilities. For instance, constant BOD and COD observations maintain efficient biochemical oxygen demands, and phosphate observations can prevent the abundant nutrients from provoking eutrophication (Huang & Cheng, 2020).

Energy management systems (EMS) control the air and pump energy utilization and check where costs can be cut. Several measures that quantify energy recovery capture biogas generation and utilization to ensure sustainability. The Human-Machine Interfaces (HMI)'s function is to depict system parameters in a visual form with the potential of remotely supervising several procedures for operators. These systems interconnect through communication networks that facilitate support, cooperation, and passing of information. Supervisory alarm and notification systems identify such conditions as highly increased sludge concentrations, failed equipment, or contamination events such as excessive heavy metals that warrant immediate corrective action (U.S. Environmental Protection Agency, 2021).

Another element of operational flexibility is improved by using key equipment monitoring for predictive maintenance and decreasing equipment idle time. You can also monitor and control this system through Mobile and Web Application Standards, making it easier to monitor from any place at any time (Khan & Mohammad, 2019).

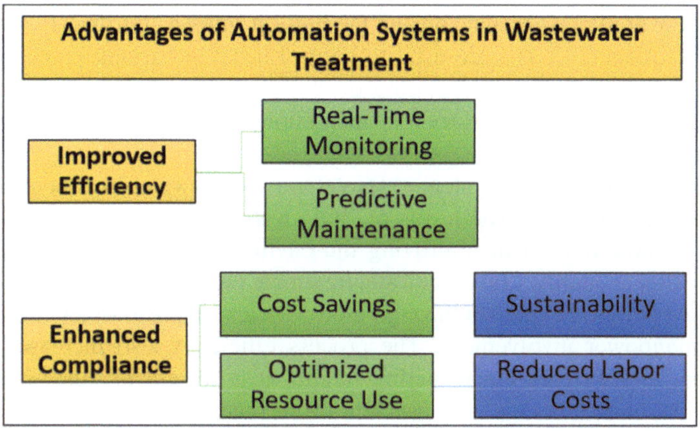

Fig. 9.4 Advantages of automation systems in wastewater treatment

Automation systems are now a standard in many new and existing wastewater treatment plants as they offer impressive solutions in terms of efficiency, reliability and especially on compliance (Fig. 9.4). Through real-time measurement and analysis, together with the use of control systems, these systems enable efficient treatment procedures, decreased expense, and improved effectiveness. Due to technology development, there remains a high chance of the continued growth of automation within the field of wastewater management.

9.4.2.1 CASE STUDY 1: Barcelona, Spain

In 2020, Barcelona's wastewater treatment facility was far from efficient, mainly because it relied on manual ways of analyzing and regulating the stages of treatment. It also made it hard for the facility to provide comparable treatment every time, independently of peak loads or changing conditions regarding the inflow. The system was inherently non-automated and thus yielded varying levels of BOD and nutrient removal and entailed a high probability of being labelled non-compliant with the stipulated regulatory benchmarks. However, these inefficiencies led to increased operations costs since the facility needed people to maintain the inconsistent treatment quality and manually regulate the chemical dosing quantity.

The facility implemented an automation system with automated flow monitoring, chemical dosing, and sludge management. The automated flow meters in the inflow line had a flow range from 0.1 to 10,000 L/min, up to an accuracy of ±0.5%. It ensured that inflow did not overwhelm the system to standardize the treatment results. The controlled dosing system guaranteed better usage of treatment chemicals, and it made effluent quality more stable, minimized chemical usage, and reduced the operational cost. The sludge removal automation enhanced the management of solid waste, reduced the handling of it by operators and technicians, and sustained the treatment's overall performance during the highest inflows. Further, Programmable Logic Controllers (PLCs) were used to control tasks in varying processes, as the control could be automated depending on the data from the sensors. The system incorporated a centralized interface for supervision and manipulation from a distance, which benefited the processes.

Results: The automation system at Barcelona's wastewater treatment facility led to notable improvements. The levels of BOD were reduced to 30%, and thus, the company became helpful in improving the environmental standard, which led to enhanced effluent quality. The wastewater treatment facility achieved a 20% improvement in the operation expenses through more efficient application of chemicals and minimal manpower involvement. The process efficiency also increased with the automation, as it offered equal quality treatment to all clients since it cannot be overtired like human beings, and was helpful in holding off large loads. Sustained real-time surveillance enabled more frequent predictive maintenance of the equipment and minimized cases of unscheduled breakdowns. Overall, the automation reduced the consumption of numerous chemicals used in the facility and enhanced sludge treatment methods.

9.4.2.2 CASE STUDY 2: Melbourne, Australia

In Melbourne wastewater treatment plants, difficulties in controlling high amounts of sludge were persistent and often produced high levels of inflow-overflow during its operation. The accumulation of sludge was causing further variability in the rate of treatment. Also, it required manual interference from operators to maintain it, raising the required manpower and operational expenses. Furthermore, poor sludge management handicapped the facility to provide the right environmental requirements, especially with the high loads. It sometimes threatened to violate compliance tests because of frequent spills and overflow.

To address these challenges, the facility implemented an automated sludge management system. It initiated continuous monitoring and controlling of the levels of sludge. The monitoring was made possible through ultrasonic sensors, which timely fed into the facility information on levels in the tanks in a timely manner and ensured that the facility maintained appropriate conditions of operations. Resistance Temperature Detector (RTD) sensors were introduced which help monitor sludge temperature as high temperatures lead to inaccuracy in the digestive sludge removal process. Since flow rates of sludge vary with time, automated pumping with Variable Frequency Drive (VFD) was applied to control pump speed to remove sludge effectively and, at the same time, reduce possible overflow. They also used historical data and machine learning technology for predicting sludge generation and allocating the number of removals to avoid high concentrations of sludge and to provide the necessary reaction to high loads.

Results: It increased processing capacity by at least 25%, decreasing cases of overflows and enhancing quality and efficiency in sludge treatment. Improvement in sludge shipment realized a 15% cost cut through technological advancements in energy use by the pumps and reduced human interference.

9.4.3 Supervisory Control and Data Acquisition (SCADA) Systems

SCADA is a sophisticated and incredibly adaptable technology used to automate, monitor, control, and regulate processes across industries, including wastewater and water treatment (González & Mora, 2021). In order to ensure real-time process management, optimize operations, and deliver practical knowledge, it integrates hardware (such as sensors and controllers) with software platforms (Stojanovic & Stojanovic, 2019). Fundamentally, SCADA is an industrial control system (ICS) that is based on three pillars, i.e., automation, centralized control, and continuous data collection. It collects data in real-time from distant sensors and machinery and transmits it to a central control platform. It allows operators to monitor important parameters like energy consumption, chemical dosage, equipment performance, and water quality (Zhou & Yang, 2020). By automating numerous operational choices and modifications, SCADA technology guarantees that processes remain dependable, efficient, and regulatory compliant. SCADA systems become quite essential in large, decentralized facilities with continuous monitoring and remote management (Stojanovic & Stojanovic, 2019).

9.4.3.1 CASE STUDY 1: Hamburg, Germany

Hygiene facilities of the Hamburg city's wastewater treatment plant were hampered by the lack of a monitoring system that helps detect an irregularity in treatment processes. A major consequence of this was that the operators could not dynamically monitor processes to settle any problems as they emerged. This led to low operational effectiveness and greater risks of failing to meet legal standards on environmental management. Thus, the facilities evolved as vulnerable to treatment fluctuations and feared it may lead to the breach of regulatory requirements and cause an impact on the environment.

The management team of the treatment plant deployed a SCADA system to enable real-time data processing and to have centralized control of the treatment processes. The system was interfaced with several sensors for data acquisition, i.e., electromagnetic flow sensors with a flow range of 0.5–5000 L/min and accuracy of $\pm 0.2\%$ and Piezoelectric pressure sensors with a range of 0–200 kPa and accuracy ± 1 kPa for measuring pressure levels across the system. Programmable Logic Controllers were connected with the SCADA system to control device outputs for response to sensor readings and for remote system control and supervision. Moreover, in order to enhance awareness, the existing SCADA software was incorporated with a user interface for real-time data, alarms, and alerts to prompt the occurrence of abnormality meant for the operators.

Results: Implementing the SCADA system led to substantial improvements in the facility's operations. With the help of the proposed facility response time was increased by 50% to treatment issues. Thus, it increased the availability and reduced the influence of any deviations from the treatment performance. The facility complied entirely with the requirements of the regulatory environmental discharge, indicating as a means of assisting the environment. The operational efficiency of the plant also improved. This was because remote control, real-time supervision, and monitoring led to more efficient utilization of existing resources and operational changes in the treatment processes. With the collected data, the facility could predict various problems that could lead to equipment malfunctions or disrupt the ordinary flow of operations. Therefore, various problems could be solved even before they caused such occurrences.

9.4.3.2 CASE STUDY 2: Phoenix, Arizona, USA

The operating cost of the wastewater treatment facilities in Phoenix, Arizona, USA, was increasing rapidly, which affected the organization greatly. The high operational cost was largely attributed to the absence of an integrated information management system. Operators lacked centralized monitoring system tools that would have enabled them to monitor the performance indicators and respond to treatment variation. Consequently, it led to increased labor requirements, constant exigent maintenance repairs, and shaky treatment quality, which culminated in the ineffectiveness of the applied wastewater management system. To overcome these challenges, the city implemented a SCADA system that centralized the data gathering and tracking at the different wastewater treatment plants/centers. In order to collect the data from different sensors embedded in the treatment process, some Remote Terminal Units (RTUs) were installed, which ensured that all the operational parameters were captured. Differential pressure flow sensors were also incorporated to give a flow measurement rate of 0.1–5000 L per minute with an accuracy of ±0.5%. Inline analyzers were incorporated to provide instantaneous reading of pH, turbidity, and ammonia levels, and in case of any variation from the normal set limit, an alarm was set to sound an alert to the operators. Further, a robust network infrastructure was created using wireless communications to enable live data streaming to the SCADA server for instant monitoring and controlling of treatment processes.

Results: The implemented changes led to an overall 25% operational cost reduction as more control was established, which reduced the frequency of maintenance and utilized resources most optimally. Integrated data management capabilities were around long-term planning, which allowed the city not to await the emergence of specific problems but to think about them and prevent their occurrence. The treatment efficiency of the Acrodex system (advanced wastewater treatment) increased as all the changes to treatment processes could be made centrally and more frequently. The success was measured by the overall reduction of emergency repairs due to more efficient monitoring and time response, leading to lower operational costs while providing a more efficient treatment system.

9.4.4 Geographic Information Systems (GIS) and Remote Monitoring

GIS stands for Geographic Information Systems; these are techniques that are applied for recording, storing, processing as well as depiction of geographic information. In this context, presenting data in the form of maps and other visuals, helps organizations to understand connections, correlations, and patterns. With the help of GIS, multiple data sources are employed in organizing resources and planning operations, and location-based problem-solving is efficient (Singh & Bhagat, 2022).

Remote monitoring can be defined as an approach to obtaining data from remote locations in real-time through the use of sensors, drones, and Internet of Things (IoT) devices. Thus, by transmitting this data to a common system, processes may be continuously supervised and controlled without physically overseeing them (Singh & Bhagat, 2022). As is described in this chapter, remote monitoring is more likely to ensure a quicker reaction to the issues and enhance the decision-making process due to providing real-time information (Schütze & Kuhl, 2021).

GIS, together with the remote monitoring system, is quite useful as it offers real-time tracking capability, enhanced visualization of collected information, and operational control from a distance. This has made them extremely beneficial in smart technology systems in various segments especially in waste and wastewater treatment processes (Duncan & Prichard, 2021).

9.4.4.1 CASE STUDY 1: Toronto, Canada

Toronto confronted important problems with its wastewater systems because the structures were old and were causing many combined sewer overflows (CSOs) during storms. The existing infrastructure could not accommodate the increased flow, leading to untreated sewage getting discharged into the rivers and other water bodies. Thus, it posed a considerable value threat to the environment as well as citizens' health. These overflows could not be managed with reliable monitoring and management tools, which delayed responses, worsened the problem, and raised operational expenses.

To tackle these issues, the city made several strategic enhancements. It established a geographic information system to organize and evaluate wastewater infrastructure. Integrating ArcGIS into detailed mapping and spatial analysis of the sewer system to improve the visual understanding of the infrastructural work. The management team digitized the records of the sensors, collected historical maintenance data along with rainfall data, and fed all the data into the GIS. Further, pressure sensors for pressures ranging from 0 to 100 kPa (accuracy of ± 0.5 kPa) and ultrasonic flow meters with an accuracy of $\pm 1\%$ were used for pressure and flow measurement of the sewer system in real-time. The use of mobile GIS applications on tablets were encouraged to aid in real-time data capturing during field inspection and thus increased data quality.

Results: The GIS-based monitoring system significantly improved the wastewater system. The change reduced both combined sewer overflow during storm events to a 40% overall discharge rate and the environmental effects. The increased resource deployment led to a reduction in emergency repairs by twenty-five percent, meaning that efficiency was improved. Since real-time and historical data were made available, planning and investment in the necessary infrastructure were made easier. The improvement facilitated the development of better relations between the city and the public and promoted new measures toward better wastewater management.

9.4.4.2 CASE STUDY 2: Singapore

Singapore has been experiencing high growth in the past few decades; this has brought about high concentrations of population and, therefore, higher volumes of wastewater than the initial sewage treatment plants were created to manage. Several of these sewage plants became outdated, hampering the efficiencies and increasing the facilities' costs. This situation led to questions as to whether the firm can continue to meet international and set high environmental compliance standards since the consequences of non-compliance have severe environmental and business impacts. Further, the live monitoring capacity was restricted, making it more challenging to counter treatment abnormality and impedes operational effectiveness. The situation, combined with the rising stigmatization of poor wastewater management and potential health impacts, encouraged the need for new approaches to improve the capacity and efficacy of Singapore's wastewater systems while protecting the public and the environment.

To overcome this difficult situation, the city implemented a remote monitoring system incorporating GIS which helped to process actual data from the treatment plants. Along with the GIS, multiple pH and dissolved oxygen sensors were installed. The collected real-time- data was transmitted to a monitoring system via wireless telemetry to allow immediate processing of the received information. QGIS was used for spatial analysis and visualization to detect wastewater trends and assess the efficiency of treatment.

Results: The implementation of the remote monitoring system resulted in noteworthy improvements. The change brought in a thirty percent enhancement of operational effectiveness resulting from enhanced central resource management and preventive action to treat pestering irregularities. Real-time monitoring and analysis provided practical benefits and saw compliance fines cut in half, showing better compliance. Improved availability of real-time data led to proper scheduling of infrastructure upgrades and strategies to pursue. Better management of the treatment processes was observed to have resulted in improvements in the environmental aspects and, hence, was helpful in enhancing sustainable management of urban water in Singapore.

9.4.5 *Machine Learning Algorithms*

One branch of artificial intelligence (AI) is known as machine learning, which deals with the development of statistical models and algorithms. It enables computer systems to perform certain activities without following a direct command by a person (Fig. 9.5). Instead, these systems acquire knowledge from the collected data and analyse trends that may exist within it. In layman's terms, machine learning is the ability of a computer to learn from previous experiences through various input processes and gradually improve its efficiency (Sen et al. 2022). Machine Learning has several advantages as discussed in Table 9.1.

9.4.5.1 CASE STUDY 1: New York City, USA

It was observed that due to the unanticipated breakdowns of equipment, New York City's wastewater treatment facilities were faced with numerous shortcomings. These failures led to operation interruptions and impeded the treatment process, followed by

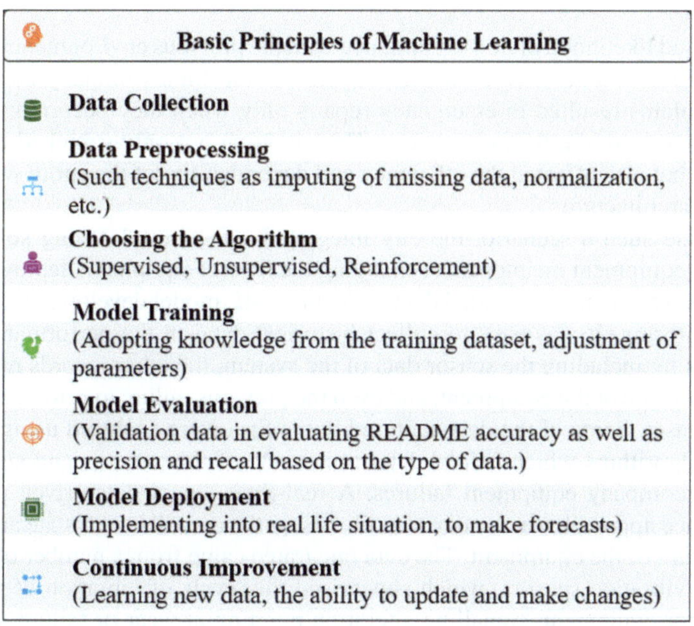

Fig. 9.5 Principles of machine learning

Table 9.1 Advantages of machine learning (ML) (Cheng & Zhang, 2020; Deng & Li, 2021; Lee & Kwon, 2023; Zhang & Wang, 2022)

Advantage	Description
Automation of tasks	ML reduces the need for manual interventions by automating processes like data entry, fraud detection, and customer support through chatbots
Improved decision making	ML analyses large datasets to provide insights, improving the accuracy and speed of decision-making
Real-time processing	ML systems process data and take actions in real-time, critical for applications like financial trading and predictive maintenance
Handling complex data	ML can process structured and unstructured data, including text, images, and videos, for improved analytics
Continuous improvement	ML models adapt over time, improving predictions and adapting to changing conditions with new data
Personalization	ML enables personalized recommendations by analysing user behaviour, such as Netflix and Amazon's recommendation engines
Scalability	ML solutions handle large datasets and scale seamlessly across industries with minimal adjustments
Predictive analytics	ML allows organizations to predict trends, risks, and failures, enhancing maintenance and inventory management
Cost efficiency	Automating workflows and processes reduces labour and operational costs over time
Enhanced accuracy	ML models often outperform traditional systems in areas like medical diagnostics, fraud detection, and image recognition

the increased likelihood of non-compliance with the previous environmental statutes. The proactive maintenance strategies adopted at the First Company ensured that many problems resulted in emergency repairs only when they occurred; however, equipment stayed idle for many hours. This situation not only stretched the cities' capacities but also affected the efficiency and dependability of the entire wastewater treatment architecture.

To tackle such a scenario, the city integrated a machine learning solution for predictive equipment maintenance, changing the flow of equipment deployment and usage. The key changes included data collection, ML model development, and real-time monitoring. It was easy to collect historical data on the performance of the equipment by including the sensor data of the system, the other records retained on the performance of the equipment, and even the previous failure logs for arraying the proper dataset. Some of the data captured in the firm were analyzed using machine learning algorithms which helped determine some of the patterns and signals that usually accompany equipment failures. A real-time sensor data-driven predictive maintenance approach was used to steadily assess the overall health state and predict future failure of the equipment. The data integrated came from a number of sources, including vibration sensors, which supervised the levels of vibration of particular machines or equipment, signalling alteration pointing at wear or failure, and pressure and temperature sensors, which delivered real-time productivity data that would enable the evaluation of equipment performance against set benchmarks. Applied decision tree and support vector machine algorithms for monitoring equipment failures led to a high accuracy of results. For the convenience of operators, a dashboard was designed to show these alerts and administrations' recommendations based on the predictions.

Results: Downtime Reduction was achieved by up to 30% in the unintended shutdown due to maintenance intervention. There was also a 20% reduction in maintenance costs since there were fewer instances of emergency repairs and improved planning of the maintenance activities. That way, the operation has been more streamlined and the legal guidelines have been followed—all in an effort that fuels New York City's environmentally-friendly vision towards wastewater treatment.

9.4.5.2 CASE STUDY 2: Copenhagen, Denmark

Copenhagen's wastewater treatment plants have reported having several issues in enhancing the treatment processes, especially during periods of varying inflow due to storms and seasons. The many adjustments to treatment processes made manually were usually done in response to system breakdowns hence there were many inefficiencies and variabilities in the treatment processes. Consequently, these manual interventions created challenges that led to low quality of effluents, which created challenges to the company in achieving environmental compliance and health standards. This was made worse because much of the data was not integrated in real-time, thereby limiting operators' ability to adapt to new conditions swiftly.

The city adopted a machine learning approach to optimize treatment processes by predicting the best operational parameters based on real-time data. Key changes included a Data Analytics Framework, which established good collections and analysis of data from the various treatment processes; Predictive Algorithms, which estimated the dosages of applied chemicals and respective changes in aeration according to measured influent quality with the aid of widely used active machine learning algorithms; and Feedback Loop, which implemented the use of an ongoing loop that allowed for regular enhancement of the prediction models as the actual treatment outcomes streamed in. Sensor Integration employs a field of sensors to gather information on influent quality and operational parameters. Important Parameters like biological oxygen demand, chemical oxygen demand, and total suspended solids were measured repeatedly. Attached real-time data on aeration rates, chemical dosing, and flow rates were considered. ML models used regression analysis and neural networks to enhance the presence of chemical dosing and aeration according to the input characteristics of the influent. A specifically developed set of interfaces should connect the ML models with the corresponding control systems to automate changes in the plant according to the predictions made by the models.

Results: The innovative use of a machine learning-based optimization system increased the effectiveness of treatment. Effluent quality improved by 25% under different influent conditions. Also, reacting quickly to control chemical dosing and aeration settings reduced operational costs by 15%, mainly from lesser use of chemicals and enhanced utilization of the available resources. This progressive approach summed up the company to the environmental standards and established Copenhagen as a model in proper sanitary sewer procedures.

9.4.6 Internet of Things (IOT)-Based Wastewater Management System

One process of controlling and managing wastewater through IoT technology is gathering information and transforming it into substance (Kumar & Singh, 2020). As for the period, the IoT system allows monitoring and evaluation of the process of wastewater treatment through connected devices or sensors (Zhou & Jiang, 2020). Because of this technology, management of wastewater facilities becomes easy because flow rates, chemical and equipment conditions, among other factors, are well established (Mavromatis & Koutitas, 2019). The attempt to combine the Internet of Things within the management of wastewater could lead to further development of environmental sustainability and improvements within the management and in the resource management by the operators (Zhou & Jiang, 2020).

Some of the fundamentals of IoT-Based Wastewater Treatment are:

Data collection: The system obtains data relating to factors such as water quality, temperature, and flow rates within facilities that handle wastewater by using these sensors (Kumar & Singh, 2020; Zhou & Jiang, 2020).

Connectivity: The collected data can then be transmitted to a centralised platform via the Internet for processing. Many communications solutions, such as satellite, cellular, Wi-Fi, etc., are available for this (Alavi & Mavrommatis, 2021; Moussa & Khedher, 2021).

Data Processing and Analysis: The centralised system analyzes the collected data to look for patterns, irregularities, or problems in the wastewater treatment process (Zhao & Zhang, 2021).

Visualization: Data received through the system can be presented on the graphical displays, and the operators can monitor the system in real-time (Moussa & Khedher, 2021; Zhou & Jiang, 2020).

Automation and Control: Based on the above data, it is possible to have automated control systems that make real adjustments to the treatment process, such as changing the degree of aeration or adjusting the chemical usage (Mavromatis & Koutitas, 2019).

Notifications and Alerts: To some extent, where there are anomalies or serious issues, it can alert operators so that the best action can be taken (Zhao & Zhang, 2021; Zhou & Jiang, 2020).

Advantages of IoT-Based Wastewater Management

Real-Time Monitoring

Ongoing monitoring of parameters in wastewater makes it easier to identify problems as they occur.

Improved Efficiency

Use of automation in the process also brings about a change in the consumption of resources like energy and chemicals hence improving the cost.

Data-Driven Decisions

Availability of actual time information can be used in decision making therefore increasing the operational efficiency.

Environmental Compliance

It helps in the evaluation of the corporation's compliance with the set legal complements into environmental matters.

Predictive Maintenance

By so doing, IoT systems can help to prevent failures from happening in equipment hence lowering on equipment downtimes and repair expenses.

Fig. 9.6 Advantages of IoT based wastewater management system

Ongoing Enhancement: The data set accumulates and is used to change the system's behaviour in a more efficient manner of handling wastewater at every time scale (Alavi & Mavrommatis, 2021; Kumar & Singh, 2020).

The application of IoT in wastewater treatment is changing old conceptions as more and more IoT-based solutions are implemented. These technologies help maximise operations, slash costs, and ensure compliance with environmental regulations through the use of real-time data acquisition, analysis, and control (Fig. 9.6). Therefore, this results in a more sustainable management system of water resources.

9.4.6.1 CASE STUDY 1: São Paulo, Brazil

Wastewater treatment plants in São Paulo experienced major issues mainly associated with the lack of automated monitoring and control systems. This shortcoming resulted in slow reaction times to treatment irregularities, affecting the efficiency of the general treatment procedures. Also, the lack of the ability to monitor the operational parameters allowed the poor management of resources, and, therefore, it became difficult for operators to intervene in a way that fixed any possible problems on their own. The conduct of the study essentially involved the use of manual methods of data collection, which worsened these difficulties and affected the ability to make timely interventions.

To improve monitoring and management, the smart, sustainable city set up an IoT-based system for the live monitoring of the area. Many IoT sensors were deployed across the treatment facilities to track flow rates, water quality, and practically all aspects of the equipment. This meant that the operators would always have real-time information concerning operations. The collected data were also moved to a central cloud platform where all the treatment processes could be analyzed using various visualization dashboards to enable a better understanding of the operation performance. An alerting mechanism was designed to warn operators of any peculiarities to support real-time monitoring of the system. This capability provided the required timeliness to intervene, enhancing response time to problems as and when they occurred. IoT sensor types included ultrasonic flow sensors with values of 0.1–10,000 L/min with an accuracy level of ±0.5% used for flow metering; and water quality sensors with different probes; and pH, turbidity, and dissolved oxygen probes were used to measure these parameters, to observe the quality of water. For data storage and data analysis purposes, a cloud-based solution was considered for accessibility and constant utilization. An application was created for operators to receive the alarm and monitor the operational functions of the plant from the comfort of their mobile devices making the function more flexible.

Results: This led to an improvement of the response time on treatment issues by 50% with the adoption of the IoT-based monitoring system. Further, real-time collection of operations data provided deeper visibility of the business to effect better decisions and optimize the use of resources. Such a management approach was not only used to fine-tune treatment processes but also adapted São Paulo's wastewater facilities for increased compliance with legislations and countering of environmental issues.

9.4.6.2 CASE STUDY 2: Tokyo, Japan

Tokyo faced some of its biggest problems in handling a vast network of wastewater, as the specificity of large urban environments influenced the latter. The current monitoring tools were incapable of giving an effective snapshot of the network's performance. This lack of data clarity resulted in problems in managing operational processes, and it took a long time to recognize and respond to various irregularities in treatment. The lack of monitoring capability of some elements of the wastewater network made it hard to intervene early, hence increasing operational risks and maintenance costs.

To boost wastewater management, Tokyo recently implemented IoT intelligent water management solution. Key changes included smart metering, data analytics integration, and automated controls. Citizens of the city employed IoT in their wastewater network where smart meters were placed to monitor flow rates and quality of water. This created more efficient data collection since it allowed the assessment to be more accurate and faster. The data from the smart meters was fed into an analytical platform where trends and even predictive analysis were performed to prevent problems from turning into bigger problems. To work in response to the data obtained from all the smart meters installed, automated control systems were created to manage different treatment processes for dynamic efficiency. Smart meters equipped with flow monitors were used to accurately measure the flow using Electromagnetic based on function meters with Measuring Range (0.5–5000 L/min) at ± 1 (%) accuracy. The operationalization of advanced sensors capable of performing multi-parameter measurements such as ammonia, nitrate, and turbidity comply with the on-site water quality directives. A combination of the sensor data (analytics platform) was analyzed to derive operational insights that led to quicker decision support. A robust communication network utilizing LoRaWAN technology was setup to have low-power long range data transportation of IoT devices across various nodes within the network.

Results

Monitoring and control improved overall network efficiency by 30%. Manual oversight was decreased, and operational costs were reduced by about 25% by optimising resource management.

9.4.7 AI Based Bioelectrochemical Systems (BES)

AI based Bio- electrochemical systems (BES) are a category of smart systems, which are being widely adopted (Fig. 9.7). These are based on the promising technology of recovery of resources such as hydrogen and methane and the efficient treatment of wastewater. Two primarily used bioelectrochemical systems, i.e., microbial fuel cells (MFC) and microbial electrolysis cells (MEC) (Hasany et al., 2016). Microbial fuel cell (MFC) catalyzes electrochemical reactions with the help of microorganisms whereas in MEC, the energy content from dissolved organic matter can be extracted as renewable energy in the form of hydrogen or methane by providing the cathode with anaerobic conditions (Kurniawan et al., 2022). MFC and MEC are also capable of recovering other valuable nutrients like phosphorus and nitrogen along with metals. To support this system, microbial electrochemical sensors are used in wastewater treatment plants, which enables the monitoring of biological oxygen demand of the plant and monitor the toxicity in the system. Further, these sensors can provide real-time data on biodegradable organics coming into and exiting the activated sludge process. Sensors can also be used in anaerobic digestors to provide data regarding the volatile fatty acids (VFA) content. This will help wastewater treatment plants decide whether to accept new organic waste in addition to sludge.

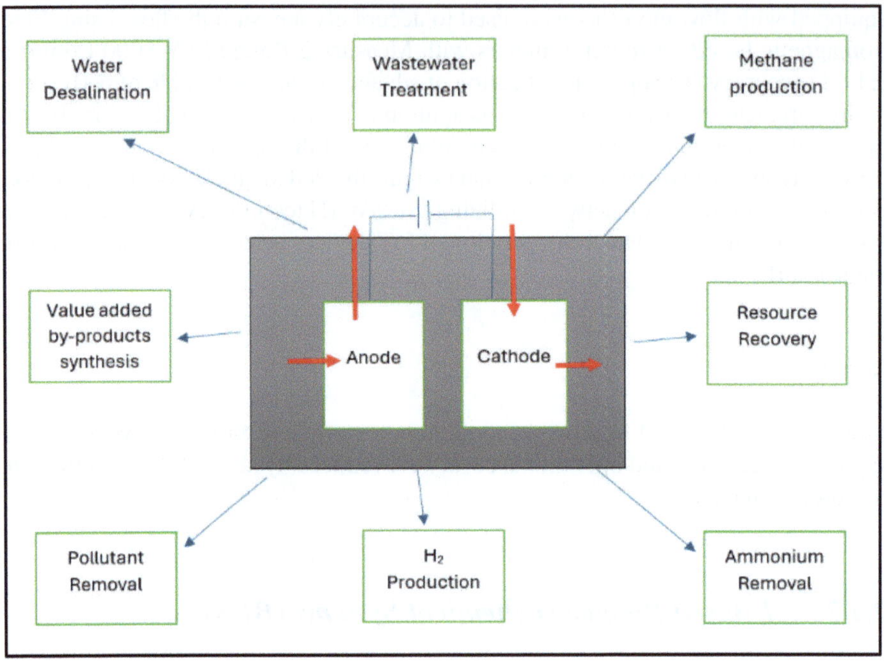

Fig. 9.7 Advantages of bioelectrochemical systems (BES) in wastewater treatment

9.4.7.1 Case Study: A Petroleum Refinery, China

A petroleum refinery's wastewater was used to study the impact of installing a combined microbial fuel cell-membrane bioreactor (MFC-MBR) system on wastewater treatment efficiency and potential for energy recovery. Further, the use of Microbial Fuel Cells (MFC) as sensors to enhance process control and energy recovery was also analyzed (Zhao et al., 2022).

Petroleum refinery wastewater (PRW) is comprised of a high pollutant load and complex composition of contaminants such as aromatic hydrocarbons, heavy metals, aliphatic hydrocarbons, phenol, sulfide, grease, and oil, along with extremely high chemical oxygen demand (COD).

Previously, the wastewater in the treatment plant was treated using the membrane bioreactor (MBR) technique. The MBR technique uses a membrane filtration process along with biodegradation by microorganisms. However, membrane fouling limits the widespread use of the MBR technique. Although many strategies have been used to decrease membrane fouling, including control of process parameters, pretreating substrates, modifying the material of membranes, and cleaning the module of the membrane, these strategies consumed high energy due to chemical and physical reagents. Further, the use of chemicals affected the life span of the membrane. Hence, an economical and efficient system was required to overcome this limitation.

The combination of MFC with MBR technique has been applied to treat Petroleum refinery wastewater (PRW). In MFC, electroactive bacteria are used to convert the chemical energy of organic contaminants into electrical energy. Organic matter is oxidized by bacteria in the anode chamber. This releases electrons and protons. Through an electric circuit, electrons travel toward the cathode and generate an electric current. Simultaneously, protons travel through a proton exchange chamber and combine with oxygen to form water.

A dual chamber system was installed for this study. The system is comprised of MFC and a hollow fiber MBR. An integrated system of MBR with MFC enhanced the treatment process as the microelectric field in the MFC positively impact on membrane fouling. Next, without any active monitoring of the wastewater, the performance of the MFC-MBR was affected, particularly the energy consumption. Hence, an MFC-based biosensor was used to quantify numerous analytes. The biosensors established a linear relationship between the substrate concentration and the voltage output. As COD is the critical parameter for MFC technology, biosensors were used to detect COD to manage the process between MFC and MBR in a two-stage combined treatment. The installed MFC-MBR system could handle a high hydraulic retention time (HRT) of 8–16 h, which is necessary for the degradation of the substrate along with the recovery of energy. Further, non-platinum air cathodes were used to increase scalability and reduce the cost of material. To analyze the system's performance under different conditions, varying loading rates (OLRs), 1.2 and 4.9 kg COD/m^3/day were applied.

Results: The removal of chemical oxygen demand (COD) showed a linear relationship with the output voltage of MFC (R2 = 0.9821). Hence, the MFC biosensor was used to control the combined system. The efficiency of removal of ammonium nitrogen (NH_4^+ -N), COD, and total nitrogen (TN) were 92.4%, 96.3%, and 86.6%, respectively. The transmembrane pressure was reduced by 50%, decreasing the membrane fouling in the treatment. The system achieved a maximum recovery of energy of 0.00258 kWh m^{-3}.

9.4.7.2 Case Study: Wastewater Treatment Plant, UK

For this case study, a microbial electrolysis cell (MEC) reactor with a capacity of 100 L was installed at Howdon wastewater treatment company, a commercial water utilities company. The wastewater treatment plant treated 250,000 m^3 (approximately) of domestic wastewater daily using activated sludge technology (Heidrich et al., 2014). The MEC reactor was operated for a 12-month period. The temperature range of raw domestic wastewater was from 1 to 22 °C.

The design of the MEC reactor consisted of a 120 L polypropylene tank. 6 separate MEC cells cassettes were placed in series within the wastewater tank. A Perspex lid was used to seal the top of the reactor to maintain anoxic conditions within the anode compartment. Each cathode compartment of every MEC cassette was connected with a 3 mm polyvinyl chloride tubing of 50 mm length to a 5 L gas bag for the collection of hydrogen gas. Each MEC cassette had 2 carbon anodes on each side of the cathode compartment, providing a today anode surface area of 16.4 m^2/m^3. All 6 cells were powered by DC power supplies, providing voltage to each cell.

With the help of a peristaltic pump, wastewater was pumped from the grit chamber to a holding tank (Heidrich et al., 2014). From the holding tank, wastewater was pumped into the MEC reactor with the help of a peristaltic pump. Further, Biosensors were used to detect the biological oxygen demand of the plant and monitor the toxicity in the system.

Results: The reactor produced an average of 0.6 L/day of hydrogen. The reactor could recover around 70% of the input electrical energy in the form of production of hydrogen gas. As the reactor was operated for 12 months, it was observed that the bacteria could work in different natural conditions. The reactor treated the wastewater within the range of 2 °C up to 21 °C. The MEC reactor could be easily retrofitted into the existing wastewater treatment plant. 2.3 KJ of energy was used to treat 1 g of COD using an MEC reactor. On the other hand, 2.5–7.5 kJ of energy is required to treat 1 g of COD using an activated sludge process. Self-sustaining MEC reactors were used to charge biosensors to monitor water quality.

9.5 Smart Technologies in the Nascent Stage of Development

Decades of research and development have established various smart systems and technologies in countries across the world today. These systems treat the wastewater sustainably and decrease the pressure on natural resources. However, several developments are still at the budding stage, to further improve the efficiency of the existing systems. This section outlines emergent technologies that are still in the early stages of development but show promise for future applications in wastewater treatment.

9.5.1 Smart Nanotechnology

One promising approach is the usage of nanotechnology, particularly nanomaterial-based treatment solutions in conjugation with artificial intelligence (AI). These AI driven nanostructured materials possess unique features, for instance high surface-to-volume ratios, exceptional reactivity and sensitivity, high adsorption capacity, and the ability to be easily functional (Rathod et al., 2024; Sharma et al., 2023). These features prove to be advantageous over the drawbacks of conventional treatment methods (Pandit et al., 2024). The application of smart nanotechnology holds significant potential for wastewater treatment and appears to have a bright future in this field (Rashid et al., 2025). However, achieving this potential requires genuine commitment and collaborative efforts from the scientific community and government organizations.

9.5.2 Hydroxyl Radical-Based Advanced Oxidation (HR-AOP)

Hydroxyl radical-based advanced oxidation processes have been explored for wastewater treatment across the past thirty years. The primary objective of HR-AOP is to eliminate resistant organic compounds, emerging contaminants, and specific inorganic pollutants. The effectiveness of the treatment largely depends on the types of AOPs used, the physical and chemical characteristics of the targeted pollutants, and the operating conditions. In recent times, sulfate radicals based AOP has emerged as an effective wastewater treatment method, as it can efficiently oxidize ammonia nitrogen in wastewater (He et al., 2024). Forthcoming studies should concentrate on investing the development of affordable AOPs.

9.5.3 AI Guided Chemical-Free Disinfection Technologies

In wastewater disinfection technologies, artificial intelligence could be used to minimise complexities as AI is applicable to predict the performance of water disinfection processes. Integration of AI in disinfection robots facilitates self-guided navigation, continuous environmental monitoring and flexibility in processes (Al Aani et al., 2019). These AI driven disinfection robots can adjust continuously to environmental changes, enhancing the disinfection efficiency and maintaining cleanliness standards (Khanna & Srivastava, 2022).

9.5.4 AI Driven Electrocoagulation

Electrocoagulation (EC) is a water treatment method that utilizes the micro-electrical currents to eliminate the water impurities. It works as an alternate to chemical coagulation (CC), which encompasses introduction of metallic coagulants into the water through the electro-chemical dissolution of electrodes (Das et al., 2022). However, this method can remove numerous pollutants, it may face practical challenges while focussing on novel recyclable and soluble biological contaminants. To address this problem, the process of EC can be embedded with diverse organic treatment technologies and AI techniques to augment contaminant removal proficiency (Mao et al., 2023; Shah et al., 2024). Research into enhancing effective parameters and electrode materials is advancing, to make electrocoagulation a more profitable alternative for widespread implementation.

9.5.5 *AI Driven Microbial System Modelling*

Traditional biological treatment methods face limitations in terms of adaptability and response time. Contemporary advancements in artificial intelligence and machine learning offer innovative possibilities for enhancing the productivity of biological treatment methods. These methods not only increase pollutant removal efficiency but also help in reducing functioning costs (Kamali et al., 2021). Furthermore, advanced machine learning techniques such as ensemble techniques and deep learning have yet to be comprehensively analysed (Cairone et al., 2024; Singh et al., 2023).

9.5.6 *Smart Adsorbent for Targeted Pollutant*

Several adsorbents such as agricultural residues, black carbon and synthetic polymers have been widely used for the removal of wastewater pollutants. Traditional adsorbents fail to achieve the necessary efficiency of pollutant removal. Therefore, stimulus-response adsorbents act as a novel variety of materials due to their adjustable and reversible physical and chemical characteristics (Yang et al., 2018). The development, synthesis, characterization, and performance analysis of these smart materials have unleashed new research prospects in wastewater treatment and environmental remediation (Sharma et al., 2024; Patel et al., 2025).

9.6 Conclusion

Water is the utmost significant resource on the planet Earth. Presently the world is facing key challenges associated with water quality, highlighting the urgent necessity to increase the wastewater treatment efforts to achieve Sustainable Development Goals. Effective management of water resources is highly important for achieving these goals as 11 out of the total 17 development goals are directly related to having access to clean and safe water. The rising demand for water resources at global level stimulated by demographic expansion, industrial growth and climate change, necessitates pioneering approaches in wastewater treatment and management.

The treatment of wastewater is significant for the protection of environment and public health, with emphasis on enhancing efficiency and sustainability in water treatment technologies. This chapter highlights how smart technologies in wastewater treatment and monitoring can revolutionize the way we deal with the wastewater problems. By integrating ML and AI with advanced systems, connected devices, data analytics, we can efficiently enhance the resource recovery, improve operational efficiency and reduce the effects of wastewater on environment and human health.

Through investigating numerous case studies, we have revealed the effectiveness and efficiency of different smart technologies, including automation systems, digital meters and IoT based wastewater management system. These smart technologies not only improve the wastewater management but also proposes real time insights that encourage the strategic decision-making. This chapter has stipulated a cross-section and investigation of widely used ML models along with smart technologies and AI techniques that have been utilised in water -based treatments. Shifting from conventional to smart technologies encourages water sustainability practices and highlights the substantial role of modernization in combating water scarcity and environmental pollution.

In the coming years it is evident that the wastewater treatment field is poised for further research and analysis, particularly technologies in their early stages of development. The integrated usage of ML, AI and IoT advances the development of smart water management technologies. The revolutionary potential of these technologies is not only limited to enhancing operational efficiency but also in advancing the broader efforts towards environmental protection, sustainability and smart water management systems.

9.7 Recommendations

While smart recycling technologies has shown substantial potential in advancing wastewater treatment processes, there are challenges and possible risks related to its application which need to be managed. To mitigate these challenges, there are numerous fields where advancements can be achieved.

1. Forthcoming research should focus on advancing the precision and robustness of smart technologies by integrating more comprehensive datasets addressing issues related to data availability and quality. Acquiring adequate data with high accuracy is challenging in water-treatment systems due to economic or technical restraints.
2. There is a need to address the research challenges related to data management, societal and legal considerations, consistency and accountability to further enhance the applicability of smart technologies in this field.
3. The use of AI and ML for real time monitoring is a growing development that could greatly enhance the flexibility and efficiency of wastewater treatment systems. Effective data from real wastewater treatment plants will enable the precise predictions of contaminant elimination.
4. In addition to predicting the efficiency of water treatment processes, smart technologies can be used to integrate all aspects of water management, including water transportation, water discharge, sludge treatment, environmental impact assessment, financial factors and policy formulations by incorporating data details such as geographical locations, sources and the framework.
5. Integration of energy storage devices and clean energy resources could play a critical role in abating challenges of energy consumption linked with smart water management systems.

6. Collaboration among stakeholders and knowledge sharing are vital for addressing various challenges related to technical expertise, energy usage, financial and compliance restrictions.
7. Harmonized efforts from various departments including government, industries and organizations are necessary for the successful implementation of smart wastewater treatment technologies. These synchronised efforts will help to minimise the environmental impacts of growing industrialisation and promote sustainable development.
8. Encourage international partnerships to share information, effective practices, and technologies related to wastewater treatment.
9. In the long term, effective local water management and preserving limited water resources are essential for developing sustainable wastewater treatment and reuse initiatives.
10. When selecting technology for wastewater treatment, it is important to consider that high-tech solutions are not always the most suitable. Ensuring that adequate financial and human resources are available to operate and maintain the facilities effectively is crucial.

References

Al Aani, S., Bonny, T., Hasan, S. W., & Hilal, N. (2019). Can machine language and artificial intelligence revolutionize process automation for water treatment and desalination? *Desalination, 458*, 84–96. https://doi.org/10.1016/j.desal.2019.02.005

Alavi, A. H., & Mavrommatis, K. (2021). IoT-based wastewater management systems: An overview. *Journal of Water Process Engineering, 43*, 102218. https://doi.org/10.1016/j.jwpe.2021.102218

Alprol, A. E., Mansour, A. T., Ibrahim, M. E. E.-D., & Ashour, M. (2024). Artificial intelligence technologies revolutionizing wastewater treatment: Current trends and future prospective. *Water, 16*(2), 314. https://doi.org/10.3390/w16020314

Alvi, M., Batstone, D., Mbamba, C. K., Keymer, P., French, T., Ward, A., Dwyer, J., & Cardell-Oliver, R. (2023). Deep learning in wastewater treatment: A critical review. *Water Research, 245*, 120518. https://doi.org/10.1016/j.watres.2023.120518

Alzahrani, A. I. A., Chauhdary, S. H., & Alshdadi, A. A. (2023). Internet of things (IoT)-based wastewater management in smart cities. *Electronics, 12*(12), 2590. https://doi.org/10.3390/electronics12122590

Asadi, M., & McPhedran, K. (2021). Biogas maximization using data-driven modelling with uncertainty analysis and genetic algorithm for municipal wastewater anaerobic digestion. *Journal of Environmental Management, 293*, 112875. https://doi.org/10.1016/j.jenvman.2021.112875

Cairone, S., Hasan, S. W., Choo, K.-H., Lekkas, D. F., Fortunato, L., Zorpas, A. A., Korshin, G., Zarra, T., Belgiorno, V., & Naddeo, V. (2024). Revolutionizing wastewater treatment toward circular economy and carbon neutrality goals: Pioneering sustainable and efficient solutions for automation and advanced process control with smart and cutting-edge technologies. *Journal of Water Process Engineering, 63*, 105486. https://doi.org/10.1016/j.jwpe.2024.105486

Cheng, X., & Zhang, Y. (2020). Applications of machine learning in data analytics: Challenges and opportunities. *Journal of Data Science, 18(3)*, 45–62.

Ching, P. M. L., So, R. H. Y., & Morck, T. (2021). Advances in soft sensors for wastewater treatment plants: A systematic review. *Journal of Water Process Engineering, 44*, 102367. https://doi.org/10.1016/j.jwpe.2021.102367

Das, P. P., Sharma, M., & Purkait, M. K. (2022). Recent progress on electrocoagulation process for wastewater treatment: A review. *Separation and Purification Technology, 292*, 121058. https://doi.org/10.1016/j.seppur.2022.121058

Deng, J., & Li, W. (2021). Adapting machine learning algorithms for continuous improvement in predictive modeling. *Artificial Intelligence Review, 34*(6), 1123–1140.

Duncan, H., & Prichard, L. (2021). Leveraging GIS and remote sensing for effective wastewater management in urban areas. *Sustainability, 13*(9), 5025. https://doi.org/10.3390/su13095025

Durgun, Y. (2024). Real-time water quality monitoring using AI-enabled sensors: Detection of contaminants and UV disinfection analysis in smart urban water systems. *Journal of King Saud University—Science, 36*(9), 103409. https://doi.org/10.1016/j.jksus.2024.103409

Essamlali, I., Nhaila, H., & El Khaili, M. (2024). Advances in machine learning and IoT for water quality monitoring: A comprehensive review. *Heliyon, 10*(6), e27920. https://doi.org/10.1016/j.heliyon.2024.e27920

Gao, J., & Liu, Y. (2019). Smart water management: A review of innovative technologies for wastewater treatment. *Water Research, 151*, 176–185. https://doi.org/10.1016/j.watres.2018.11.050

González, A., & Mora, R. (2021). Application of SCADA systems in water treatment plants. *Water Science and Technology, 83*(10), 2321–2330. https://doi.org/10.2166/wst.2021.063

Hasany, M., Mardanpour, M. M., & Yaghmaei, S. (2016). Biocatalysts in microbial electrolysis cells: A review. *International Journal of Hydrogen Energy, 41*(3), 1477–1493. https://doi.org/10.1016/j.ijhydene.2015.10.097

He, F., Chen, H., Li, J., Zhao, C., Zhang, J., & Wang, S. (2024). Photothermal-mediated advanced oxidation processes for wastewater purification. *Current Opinion in Chemical Engineering, 45*, 101039. https://doi.org/10.1016/j.coche.2024.101039

Heidrich, E. S., Edwards, S. R., Dolfing, J., Cotterill, S. E., & Curtis, T. P. (2014). Performance of a pilot scale microbial electrolysis cell fed on domestic wastewater at ambient temperatures for a 12-month period. *Bioresource Technology, 173*, 87–95. https://doi.org/10.1016/j.biortech.2014.09.083

Huang, Y., & Cheng, X. (2020). Data analytics in wastewater treatment: Applications and challenges. *Science of the Total Environment, 740*, 140072.

Sen, J., et al. (2022). *Introduction to machine learning: Concepts and applications*. Springer. https://doi.org/10.5772/intechopen.94615.

Kamali, M., Appels, L., Yu, X., Aminabhavi, T. M., & Dewil, R. (2021). Artificial intelligence as a sustainable tool in wastewater treatment using membrane bioreactors. *Chemical Engineering Journal, 417*, 128070. https://doi.org/10.1016/j.cej.2020.128070

Khan, A. A., & Razzak, S. A. (2021). The role of automation in improving the efficiency of wastewater treatment systems. *Journal of Water Process Engineering, 41*, 102014. https://doi.org/10.1016/j.jwpe.2021.102014

Khan, E. A., & Mohammad, F. (2019). Role of IoT in wastewater management. *Journal of Environmental Management, 245*, 85–92. https://doi.org/10.1016/j.jenvman.2019.05.012

Khan, I. H., & Javaid, M. (2021). Role of Internet of Things (IoT) in adoption of Industry 4.0. *Journal of Industrial Integration and Management, 7*(4), 2150006. https://doi.org/10.1142/s2424862221500068

Khanna, S., & Srivastava, S. (2022). The emergence of AI-based autonomous UV disinfection robots in pandemic response and hygiene maintenance. *International Journal of Applied Health Care Analytics, 7(11)*, 1–19.

Kim, H.-Y., & Kang, D.-W. (2020). South Korea's experience with smart infrastructure services: Smart water management. https://doi.org/10.18235/0002673

Kumar, P., & Singh, P. (2020). Application of Internet of Things (IoT) in wastewater management: A review. *Journal of Environmental Management, 271*, 110972. https://doi.org/10.1016/j.jenvman.2020.110972

Kurniawan, T. A., Othman, M. H. D., Liang, X., Ayub, M., Goh, H. H., Kusworo, T. D., Mohyuddin, A., & Chew, K. W. (2022). Microbial fuel cells (MFC): A potential game-changer in renewable energy development. *Sustainability, 14*(24), 16847. https://doi.org/10.3390/su142416847

Lee, H., & Kwon, J. (2023). Real-time processing with machine learning systems. *Journal of Advanced Computing, 31(4)*, 241–258.

Mao, Y., Zhao, Y., & Cotterill, S. (2023). Examining current and future applications of electrocoagulation in wastewater treatment. *Water, 15(8)*, 1455. https://doi.org/10.3390/w15081455

Martínez, R., Vela, N., el Aatik, A., Murray, E., Roche, P., & Navarro, J. M. (2020). On the use of an IoT integrated system for water quality monitoring and management in wastewater treatment plants. *Water, 12*(4), 1096. https://doi.org/10.3390/w12041096

Masood, F., Pandey, R., Singh, H. P., Gupta, A. S., Kaur, S., Batish, D. R., & Kohli, R. K. (2021). Cytotoxic and genotoxic assessment of agricultural soils from an industrial region. *Environmental Monitoring and Assessment, 193*, 1–11. https://doi.org/10.1007/s10661-021-09018-6

Matheny, M. E., Ohno-Machado, L., Davis, S. E., & Nemati, S. (2023). *Data-driven approaches to generating knowledge: Machine learning, artificial intelligence, and predictive modeling*, (pp. 217–255). Elsevier EBooks. https://doi.org/10.1016/b978-0-323-91200-6.00031-0.

Mavromatis, K., & Koutitas, C. (2019). Smart water management systems: Internet of Things in wastewater treatment. *Water, 11(10)*, 2036.

Moussa, A., & Khedher, L. (2021). Integrating IoT technologies in wastewater treatment: A review. *Sustainability, 13(4)*, 2030. https://doi.org/10.3390/su13042030

Nagpal, M., Siddique, M. A., Sharma, K., Sharma, N., & Mittal, A. (2024). Optimizing wastewater treatment through artificial intelligence: Recent advances and future prospects. *Water Science & Technology.* https://doi.org/10.2166/wst.2024.259

Narayanan, D., Bhat, M., Samuel Paul, N. R., Khatri, N., & Saroliya, A. (2024). Artificial intelligence-driven advances in wastewater treatment: Evaluating techniques for sustainability and efficacy in global facilities. *Desalination and Water Treatment, 320*, 100618. https://doi.org/10.1016/j.dwt.2024.100618

Nassiri, A., & Mohseni, M. (2021). Automation of wastewater treatment processes. *Water Research, 202*, 117446. https://doi.org/10.1016/j.watres.2021.117446

Nivala, J., & Kettunen, H. (2017). Automation in wastewater treatment plants: Impact on process control and environmental sustainability. *Environmental Science & Policy, 77*, 80–87. https://doi.org/10.1016/j.envsci.2017.08.013

Pandey, R., Krishnamurthy, B., Al Obaid, S., Ansari, M. J., Batish, D. R., & Singh, H. P. (2024). Glycolipid biosurfactant production and petroleum hydrocarbon degradation by a new strain of *Citricoccus zhacaiensis. Biocatalysis and Agricultural Biotechnology, 59*, 103248. https://doi.org/10.1016/j.bcab.2024.103248

Pandit, S., Yadav, N., Sharma, P., Prakash, A., & Kuila, A. (2024). Life cycle assessment and techno-economic analysis of nanotechnology-based wastewater treatment: Status, challenges and future prospectives. *Journal of the Taiwan Institute of Chemical Engineers*. https://doi.org/10.1016/j.jtice.2024.105567

Parde, D., & Behera, M. (2023). Challenges of wastewater and wastewater management. In *Sustainable industrial wastewater treatment and pollution control* (pp. 229–255). Springer Nature. https://doi.org/10.1007/978-981-99-3921-1_10

Patel, T. K. (2025). Biofiltration techniques for industrial effluent treatment. In *Biotechnology approaches to industrial and pharmaceutical wastewater treatment* (pp. 175–208). IGI Global Scientific Publishing. https://doi.org/10.4018/979-8-3693-8487-9.ch006

Pratap, B., Kumar, S., Nand, S., Azad, I., Bharagava, R. N., Ferreira, L. F. R., & Dutta, V. (2023). Wastewater generation and treatment by various eco-friendly technologies: Possible health hazards and further reuse for environmental safety. *Chemosphere, 313*, 137547. https://doi.org/10.1016/j.chemosphere.2022.137547

Rajak, P., Ganguly, A., Adhikary, S., & Bhattacharya, S. (2023). Internet of Things and smart sensors in agriculture: Scopes and challenges. *Journal of Agriculture and Food Research, 14*(14), 100776. https://doi.org/10.1016/j.jafr.2023.100776

Rashid, M., Arshad, M., Habib, M. S., & Haq, E. U. (2025). Conversion of smart nanomaterials to achieve sustainable goals. In *Functionalized cellulose materials: Sustainable manufacturing and applications* (pp. 229–251). Springer Nature. https://doi.org/10.1007/978-3-031-76953-5_10

Rathod, S., Preetam, S., Pandey, C., & Bera, S. P. (2024). Exploring synthesis and applications of green nanoparticles and the role of nanotechnology in wastewater treatment. *Biotechnology Reports, 8*, e00830. https://doi.org/10.1016/j.btre.2024.e00830

Rodríguez-Alonso, C., Pena-Regueiro, I., & García, Ó. (2024). Digital twin platform for water treatment plants using microservices architecture. *Sensors, 24*(5), 1568–1568. https://doi.org/10.3390/s24051568

Schütze, M., & Kuhl, T. (2021). Real-time monitoring and control of wastewater treatment plants: Case studies and challenges. *Water Science and Technology, 83*(4), 812–823. https://doi.org/10.2166/wst.2021.204

Shah, A. A., Walia, S., & Kazemian, H. (2024). Advancements in combined electrocoagulation processes for sustainable wastewater treatment: A comprehensive review of mechanisms, performance, and emerging applications. *Water Research, 121248*. https://doi.org/10.1016/j.watres.2024.121248

Sharma, A., Goel, H., Sharma, S., Rathore, H. S., Jamir, I., Kumar, A., et al. (2024). Cutting-edge technology for wastewater treatment using smart nanomaterials: Recent trends and futuristic advancements. *Environmental Science and Pollution Research, 1*–31.

Sharma, S., Dhingra, P., & Jain, S. (2023). Advancements & challenges of nanotechnology in wastewater treatment. *Materials Today: Proceedings, 80*, 18–23. https://doi.org/10.1016/j.matpr.2022.09.481

Singh, N. K., Yadav, M., Singh, V., Padhiyar, H., Kumar, V., Bhatia, S. K., & Show, P. L. (2023). Artificial intelligence and machine learning-based monitoring and design of biological wastewater treatment systems. *Bioresource Technology, 369*, 128486. https://doi.org/10.1016/j.biortech.2022.128486

Singh, R. K., & Bhagat, P. (2022). Application of GIS in wastewater management: A review. *Environmental Science and Pollution Research, 29*(30), 44805–44821. https://doi.org/10.1007/s11356-022-20635-y

Soetedjo, A., Hendriarianti, E., Wibowo, S. A., Novrian, J., Nugroho, A. B., Roby, M. F., Dewi, O. W., Apriliansyah, R. S., Mustofa, A., Sari, R. I., & Wijayanto, F. Y. (2022, June). Real-time implementation of wastewater monitoring system on the communal wastewater treatment plant using the IoT technology. *IOP Conference Series: Earth and Environmental Science, 1030*(1), 012006. https://doi.org/10.1088/1755-1315/1030/1/012006

Stojanovic, J., & Stojanovic, D. (2019). SCADA systems: An overview. *Journal of Automation and Control Engineering, 7*(1), 34–41. https://doi.org/10.18178/joace.7.1.34-41

U.S. Environmental Protection Agency (EPA). (2021). *Wastewater treatment: Automated systems.* Retrieved from https://www.epa.gov

United Nations. (2022). *The sustainable development goals report 2022.* https://unstats.un.org/sdgs/report/2022/The-Sustainable-Development-Goals-Report-2022.pdf

Wang, H., Wang, Y., & Zhang, Y. (2020). Advances in automation technologies for wastewater treatment: A review. *Journal of Environmental Management, 263*, 110368. https://doi.org/10.1016/j.jenvman.2020.110368

Wongburi, P., & Park, J. K. (2021). Big data analytics from a wastewater treatment plant. *Sustainability, 13*(22), 12383. https://doi.org/10.3390/su132212383

Worthington, T. A., Andradi-Brown, D. A., Bhargava, R., Buelow, C., Bunting, P., Duncan, C., Fatoyinbo, L., Friess, D. A., Goldberg, L., Hilarides, L., Lagomasino, D., Landis, E., Longley-Wood, K., Lovelock, C. E., Murray, N. J., Narayan, S., Rosenqvist, A., Sievers, M., Simard, M., & Thomas, N. (2020). Harnessing big data to support the conservation and rehabilitation of mangrove forests globally. *One Earth, 2*(5), 429–443. https://doi.org/10.1016/j.oneear.2020.04.018

Yahaya, S. M., Mahmud, A. A., & Abdu, N. (2023). The use of wastewater for irrigation: Pros and cons for human health in developing countries. *Total Environment Research Themes, 6*, 100044. https://doi.org/10.1016/j.toteres.2023.100044

Yang, M., Chen, L., Wang, J., Msigwa, G., Osman, A. I., Fawzy, S., Rooney, D. W., & Yap, P. S. (2023). Circular economy strategies for combating climate change and other environmental issues. *Environmental Chemistry Letters, 21*(1), 55–80. https://doi.org/10.1007/s10311-022-01419-8

Yang, Q., Wang, J., Chen, X., Yang, W., Pei, H., Hu, N., et al. (2018). The simultaneous detection and removal of organophosphorus pesticides by a novel Zr-MOF based smart adsorbent. *Journal of Materials Chemistry A, 6*(5), 2184–2192.

Zhang, S., Jin, Y., Chen, W., Wang, J., Wang, Y., & Ren, H. (2023). Artificial intelligence in wastewater treatment: A data-driven analysis of status and trends. *Chemosphere, 336*, 139163. https://doi.org/10.1016/j.chemosphere.2023.139163

Zhang, Y., & Wang, K. (2022). Machine learning in water quality prediction: A case study in wastewater treatment. *Journal of Environmental Management, 303*, 114146. https://doi.org/10.1016/j.jenvman.2021.114146

Zhao, C., & Zhang, X. (2021). Big data and machine learning in wastewater treatment: Current applications and future directions. *Environmental Technology & Innovation, 22*, 101384. https://doi.org/10.1016/j.eti.2021.101384

Zhao, S., Yun, H., Khan, A., Salama, E.-S., Redina, M. M., Liu, P., & Li, X. (2022). Two-stage microbial fuel cell (MFC) and membrane bioreactor (MBR) system for enhancing wastewater treatment and resource recovery based on MFC as a biosensor. *Environmental Research, 204*, 112089. https://doi.org/10.1016/j.envres.2021.112089

Zhou, H., & Yang, C. (2020). A review of SCADA systems and their applications. *Journal of Control Science and Engineering, 2020.* https://doi.org/10.1155/2020/1675180

Zhou, J., & Jiang, Z. (2020). Smart wastewater management system based on IoT technology. *Water, 12(5)*, 1286.

Chapter 10
Technology Integration and Synergy for Optimal Resource Recovery from Wastewater

Ashwani Kumar, Vikash Kumar, Jamilu Garba, and Sunny Dhiman

Abstract The chapter discusses how Artificial Intelligence functions with current wastewater treatment systems to generate sustainable resource extraction. Clean water along with essential nutrients and energy now functions as vital resources derived from wastewater since the worldwide need for sustainable water has increased. AI technology offers effective solutions for treatment processes through three aspects: real-time surveillance systems, predictive operational protocols and automated control mechanisms. AI projects implemented in wastewater facilities produce superior operational results by both reducing power needs and retrieving substantial resources. Machine learning algorithms and neural networks in AI enable the study of bio electrochemical system data patterns to generate better decision outcomes and minimized waste output. This chapter explores AI as an integration tool among municipal waste methods from microbial operations to chemical absorption systems to develop a self-reliant wastewater system. AI researchers study the entire wastewater management practice revolution through both practical assessments and technological discoveries that highlight economic and ecological improvements.

Keywords Artificial intelligence · Wastewater treatment · Bioelectrochemical systems · Resource recovery · Water management

A. Kumar (✉) · V. Kumar · S. Dhiman
Department of Biotechnology, University Institute of Biotechnology, Chandigarh University, Mohali 140413, Punjab, India
e-mail: ashwanibitian@gmail.com

J. Garba
Veterinary Medicine, Usmanu Danfodiyo University, Sokoto, Nigeria

S. Dhiman and G. Mukherjee (eds.), *Resource Resurgence*, Springer Water,
https://doi.org/10.1007/978-3-032-02750-4_10

10.1 Introduction to Resource Recovery from Wastewater

10.1.1 Importance of Wastewater as a Resource

Modern society faces two fundamental threats: climate change and environmental deterioration. As resources continue to diminish, it is essential to pursue sustainable development and work toward achieving climate neutrality. Water treatment facilities are essential for human health protection but generate increasing material disorder which causes environmental damage. Making resource recovery from wastewater into a priority over both destruction and removal proves effective for minimizing entropy production (Hao et al., 2022). Contaminant components in ponds, oceans, groundwater and canals are categorized as late-stage pollution. Human actions including industrial processes together with agricultural practices and improper waste disposal systems generate this type of contamination. Different types of pollutants existing chemically or physically or biologically in water environments create severe destructive outcomes. Waste water treatment plants (WWTPs) are now widely known to receive chemicals such as EDCs and pharmaceuticals and personal care products (PPCPs) and APIs or pharmaceutically active compounds (PACs) through their surface water systems (Kumar et al., 2023).

10.1.2 Global Challenges and Opportunities in Wastewater Management

The global transformation in wastewater management occurs from two main factors: environmental sustainability combined with decreasing resources which demands contemporary solution approaches. Population growth combined with industrial development patterns along with urbanization has overloaded existing infrastructure through an increase in wastewater amounts (Das et al., 2024). The previous methods of wastewater management through pollution control no longer serve today's expectations because modern solutions must be introduced. The present-day movement involves utilizing cutting-edge technology to recover vital resources from wastewater which creates a more sustainable cyclic process of wastewater management (Kumar & Choudhury, 2024). Worldwide wastewater operations face major difficulties in their current operations mechanism. The raw sewerage dumping at crucial points represents a primary environmental challenge. Global unregulated wastewater dumping to the environment reaches approximately 90% of all untreated wastewater discharges according to WWaP (Waste Water Pollutant) data. Developing countries face a serious urgent matter regarding wastewater management because they lack fundamental infrastructure together with technical skills as well as financial ability to implement proper wastewater treatment systems. The continuous pollution issue produces widespread water contamination that reduces quality and generates critical health hazards for wide areas (Akhtar et al., 2021). Modern wastewater treatment systems possess higher sophistication because they process three main pollutant groups: endocrine-disrupting chemicals (EDCs), pharmaceutical compounds, and microplastics. Traditional wastewater treatment methods face challenges when dealing with current emerging problems. Pollutants that stay in water bodies become major dangers for biological diversity while posing risks to human wellness (Kolawole & Iyiola, 2023). The use of substantial energy sources and resources by traditional wastewater facilities has earned them significant criticism. The present-day wastewater treatment systems face low sustainability because traditional treatment methods often result in loss of valuable by-products (Bdour et al., 2009). Traditional wastewater management systems face problems because rising energy expenses occur at a time when natural resource levels continue to decrease. Double hardships exist for under developing nations since they suffer from limited finances while trying to operate ancient infrastructure. The modern world currently experiences an evolving transition process because wastewater emerges as a valuable substance that provides multiple practical applications. Within wastewater systems exist important values that consist of energy-producing biogas together with clean water alongside nutrient resources containing phosphorus and nitrogen objects ready for industrial use (Shen et al., 2015). The establishment of resource recovery centres running under circular economy models within treatment plants yields combined environmental gains with economic advantages. Modern technological development enables this transformation through anaerobic digestion which serves as the main enabling factor. The process tests wastewater organic substances to convert them into biogas which operates as a sustainable green fuel (Kabeyi & Olanrewaju, 2022). The wastewater processing industry currently undergoes fundamental changes because of advancements in modern technology. Engineered microorganisms together with re-fined reactor designs resulted in efficient processes that converted waste into sustainable biogas for power generation (Speight, 2023). Modern wastewater treatment includes membrane applications that comprise MBRs (Membrane Bio Reactors) and FO (Forward Osmosis) and RO (Reverse Osmosis) technologies which efficiently recover quality water while eliminating contaminants (Abuhasel et al., 2021). Nanocomposites among other advanced materials have increased membrane strength and operation effectiveness to expand wastewater treatment limits. Electrochemical systems and photocatalysis as emerging technologies demonstrate high effectiveness in pharmaceutical pollutant removal and valuable metal recovery operations from industrial wastewater (Nabgan et al., 2022). Struvite precipitation technology enables wastewater treatment since it collects phosphorus and nitrogen for production into agricultural fertilizers that promote sustainable farming methods. The wastewater management sector of Earth uses Internet of Things devices and sensors together with artificial intelligence (AI) to bring about changes in the field (Rane et al., 2023). Smart systems technology enables real-time operational monitoring that simultaneously facilitates operational efficiency, resources recovery and expenses reduction. Combined systems which include these technologies improve the performance of resource recovery infrastructure as well as wastewater management waste reduction systems. Wastewater treatment acts as a platform to produce renewable energy through water-energy-nutrient nexus approaches that benefit sustainable agricultural practice (Mulier et al., 2022). The circular economy provides three designated systems through which organizations can achieve recycling and resource reuse leading to environmental benefit and reduced waste creation. Some nations have already proved the effectiveness of integrated wastewater management approaches. Wastewater treatment facilities in Amsterdam (Netherlands) follow circular economy methods by extracting biogas along with nutrients and clean water from their operations (van Leeuwen et al., 2018). Singapore received worldwide recognition for their NEWater initiative because they obtained ultra-clean water from treated wastewater through membrane filtration backed by ultraviolet disinfection processes (Sewerin et al., 2021). China manages industrial wastewater treatment within centralized facilities that both recovers chemical substances and produces clean water without producing environmental damage (Qu et al., 2022). The global expansion of such improvements requires strategic planning measures. Governments need to create new rules which offer motives for sustainable wastewater practices and resource recovery methods (Obaideen et al., 2022). Training programs together with technical workshops coupled with knowledge sharing initiatives build capability which develops specialized expertise primarily in regions that are developing (Cheng et al., 2023). Public support for resource recovery initiatives increases through community engagement activities which educate people about all the resource recovery advantages.

10.2 Overview of Technologies for Resource Recovery

10.2.1 Physical Methods: Sedimentation and Filtration

Physical wastewater treatment methods form the base of the entire wastewater water treatment system. They function as first-stage components which make wastewater ready for more advanced processes. Gravity controls sedimentation which operates as an easy yet functional waste treatment system. This process removes suspended solids from wastewater while heavy particles maintain separate from other water components by dropping to the bottom portion of the tank to create sludge. The sludge resulting from this process becomes eligible for advanced processing either to extract phosphorus nutrients or to prepare for direct soil condition utilization (Hušek et al., 2022). The low-cost sedimentation method helps eliminate disturbances that might cause blockages in subsequent operational stages. Modern high-rate settlers and lamella clarifiers enhance the process efficiency through improvements in settlement speed and take up less space. Sand filters have operated in the industry for many years until membrane filtration systems replaced them as an advancement in technology. High-purity water and most nutrients and organic matter can be retained when using microfiltration (MF), ultrafiltration (UF), and reverse osmosis (RO) membranes. The combination of biological treatment techniques with filtration in a Membrane bioreactor system creates a high-performance wastewater treatment system that performs water purification along with effective retention of desirable products (Ahmed et al., 2022). The core operational processes from physical methods make it possible for subsequent chemical and biological transformations to take place.

10.2.2 Chemical Methods: Precipitation and Chemical Recovery

The combination of biological treatment techniques with filtration in a Membrane bioreactor system creates a high-performance wastewater treatment system that performs water purification along with effective retention of desirable products. The core operational processes from physical methods make it possible for subsequent chemical and biological transformations to take place. Struvite works as a sustainable fertilizer solution that solves two environmental concerns: sustainable nutrient recovery and waterway eutrophication. Struvite treatment techniques have spread throughout municipal wastewater treatment facilities along with facilities serving industries (Sharifian & Wang, 2024). The advancement of electrochemical precipitation enhances both the recovery efficiency and diminishes chemical consumption rates. The combination of ion exchange and solvent extraction technologies serve as effective methods for fetching metals along with other essential compounds from sources. Through resin beads ion exchange achieves selective ion binding which effectively recovers copper metal as well as zinc metal and rare earth elements. Industrial applications benefit from solvent extraction which employs organic solvents to extract and focus valuable solutes through its separation method. Safety measures must be deployed to stop secondary waste production and maintain budget efficiency during the operation of these effective recovery processes.

10.2.3 Biological Methods: Anaerobic Digestion and Microbial Fuel Cells

The biological methods harness microorganism metabolic functions to recover energy and nutrients by using sustainable and eco-friendly process. Among established biological methods anaerobic digestion stands as one of the most proven techniques especially in the wastewater treatment facilities of municipalities. Organic matter breakdown through anaerobic microorganisms produces biogas by creating a renewable energy source whose main component is methane. Biogas from the process can undergo purification leading to two final uses for electricity generation as well as heating functions or alternative applications for vehicle fuel. The digestate by-product contains numerous nutrients including nitrogen and phosphorus which make it usable as biofertilizer. Thermophilic digestion and co-digestive operations in Anaerobic digestion systems boost both yield production and operational efficiency when applied to different organic waste streams (Paranjpe et al., 2023). The promising new technology known as microbial fuel cells (MFCs) provides a direct conversion mechanism which produces electricity from wastewater organic matter. Electroactive bacteria operating within MFCs transfer electrons to electrodes during organic matter decomposition thus producing an electric current (Yang et al., 2025). Due to their current developmental stage MFCs demonstrate great potential to serve as an innovative dual-function wastewater management solution which produces energy. Multiple academic investigations regarding electrode materials and reactor designs focus on scalability to establish these tools as significant promoters of sustainable resources.

10.2.4 Emerging Technologies: Nanotechnology and Advanced Oxidation Processes (AOPs)

The deployment of new technologies changes wastewater management into an accurate system for both efficient resource recovery and sustainable operations. Nanotechnology provides wastewater treatment facilities with fresh methods to achieve their goals. Nanomaterials successfully adsorb contaminants through their high ability to recover specific resources using carbon nanotubes graphene and metal–organic frameworks (MOFs). Nanomembranes provide both energy-efficient separation of tiny solutes and high separation performance. Engineered functionalized nanoparticles have the ability to detect particular compounds such as heavy metals or precious elements platinum and gold in industrial discharge materials. The high manufacturing expenses of nanomaterials together with their release-induced environmental hazards represent major barriers to sustainable environmental remediation operations. The reactive oxygen species used in AOPs generate hydroxyl radicals to decompose complex organic pollutants into less complex compounds. AOPs provide specific value to wastewater treatment when dealing with pharmaceuticals along with dyes and hard-to-break down substances (Estrada-Arriaga et al., 2016). AOPs function to enhance downstream recovery systems by removing obstacles from the water stream. Collaborative applications between AOPs and biological treatments result in full organic pollutant mineralization which improves both accessibility and effectiveness of nutrient and energy recovery systems.

10.3 Integration and Synergy Between Technologies

10.3.1 Combining Biological and Chemical Processes for Enhanced Recovery

Through their dual operation biological and chemical methods build a strong system for extracting energy alongside nutrients and useful substances from wastewater. The combined usage of biological and chemical processes leads to methods which enhance resource retrieval alongside the management of tough pollutants and the delivery of environmentally friendly answers.

10.3.1.1 Recovering Nutrients More Effectively

The main nutrients which agriculture needs can be found in wastewater supplies (Qadir et al., 2020). When nutrients exceed necessary levels, they can trigger eutrophic conditions that hurt the environment significantly. The combination of biological processes together with chemical ones helps achieve sustainable recovery operations and reduces environmental hazards. The organic matter breakdown process through microorganisms during Anaerobic Digestion (AD) generates renewable biogas as an energy product. After digestion occurs the leftover product contains high levels of ammonia together with phosphorus. The conversion of waste nutrients into struvite-crystalline fertilizer occurs through chemical processes known as struvite precipitation. The chemical compound struvite achieves high value in agricultural fertilizer production due to its controlled release capabilities which originate from its magnesium-ammonium-phosphate composition (Krishnamoorthy et al., 2021). The integrated process resolves two problems by generating energy and recovering valuable nutrients. Biological processes of nitrification-denitrification successfully transform nitrogen into environmentally harmless products. The combination of these methods with ammonia stripping produces usable nitrogen products including commercial fertilizer ammonium sulfate. Wastewater treatment plants become more economically sustainable due to the combined application of two operations.

10.3.1.2 Handling Tough Pollutants

Some pollutants consisting of industrial chemicals along with dyes and residual pharmaceuticals are unable to decompose by biological methods alone. State-of-the-art chemical approaches when combined with biological methods will produce higher success in treatments (Rashid et al., 2021). A Microbial Fuel Cell technology enables microbial breakdown of organic compounds which creates valuable energy through microbial processes. The application of advanced oxidation processes (AOPs) that employ reactive oxygen species (hydroxyl radicals) successfully converts complex compounds into less hazardous simpler structures for determined pollutants. The innovative pairing of techniques enables sustainable operation and high efficiency because it generates usable power while helping eliminate tough-to-remove pollutants (Rani et al., 2024).

10.3.1.3 Recovering Valuable Metals

End-of-pipe industrial wastewater consists of metals including copper as well as zinc and rare earth elements which serve essential yet non-renewable purposes. The implementation of biological with chemical approaches improves metal extraction efficiency. Heavy metals get attached to microorganisms as they perform biosorption operations through surface adsorption functions (Priya et al., 2022). After capture these metals undergo chemical extraction that leads to purification before industries use them in operations. The combined approach leads to reduced secondary waste output while helping the circular economy to restore scarce resource availability.

10.3.1.4 Energy and By-product Co-recovery

The integration of processes allows recovery systems to obtain both energy and useful by-products at the same time. The removal of carbon dioxide and hydrogen sulfide through chemical scrubber systems from biogas produces purified methane after anaerobic digestion. Biogas of high quality obtained from digester tanks can serve two purposes: powering cars and producing electricity. All by-product remnants undergo treatments which extract mineral nutrients before completion of this process. Chemical scrubbers implement purification of biogas by removing carbon dioxide and hydrogen sulfide from the process to produce clean methane (Kasulla et al., 2021). The high-quality biogas produced can function as vehicle fuel in addition to electricity generation operations. The rest of the digestate undergoes treatment to extract its phosphorus and nitrogen content thus eliminating waste in every step.

10.3.2 Role of Automation and AI in Optimizing Wastewater Treatment Systems

Wastewater treatment facilities need adaptable systems that operate efficiently because their systems grow increasingly complex. Automation and artificial intelligence functions as primary tools for handling complex systems to deliver operations which are smarter and both faster and sustainable as shown in Fig. 10.1.

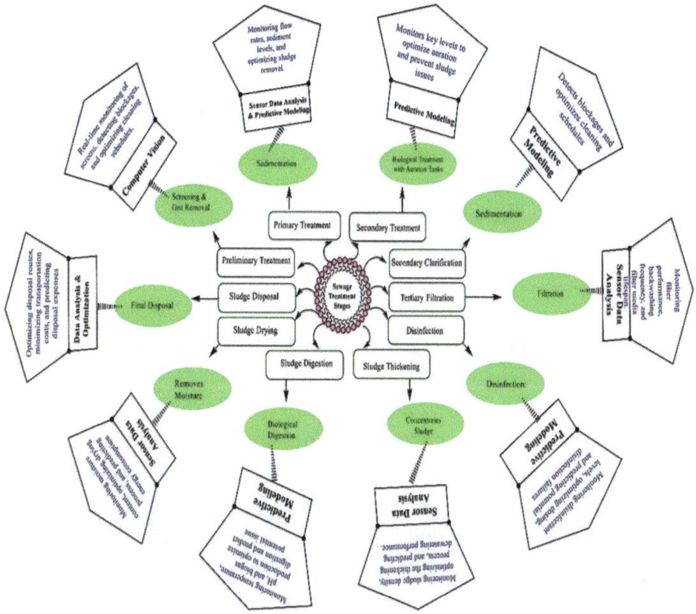

Fig. 10.1 AI based innovations in sewage water treatment

10.3.2.1 Real-Time Monitoring and Dynamic Adjustments

Wastewater facilities with modern technology employ smart sensors to track essential parameters consisting of pH, turbidity levels together with nutrient content and organic material measurement (Moretti et al., 2024). The data collected by sensors automatically activates AI systems to perform operations adjustments regularly. AI algorithms can examine nutrient levels which enables them to make immediate alterations to chemical dosing requirements. The system makes resource recovery more efficient by managing phosphorus and ammonia without chemical overuse so users save money while protecting the environment (Bohra et al., 2022). The biological treatment process needs aeration as its vital operational step but it requires significant energy input. Artificial intelligence systems precisely determine which amount of air is required for microbial activities and achieve efficient treatment with minimal energy usage.

10.3.2.2 Predictive Maintenance and Problem Prevention

Wastewater facilities use predictive technology provided by AI to monitor their systems constantly thus enabling them to detect upcoming problems before they happen (Mohanty et al., 2024). The system identifies the early warning indicators that acidification or overloading presents in anaerobic digester systems. Operating personnel can control stability and maximize biogas output through adjustments including feedstock modification or organic load reduction. Membrane fouling which affects filtration systems requires both extensive time and costly measures for remediation. AI technology detects membrane fouling patterns at an early stage thus allowing necessary preventive measures to be performed before operational interruptions occur (Abuwatfa et al., 2023).

10.3.2.3 Machine Learning for Continuous Improvement

Through its subset identity machine learning (ML) analyzes both previous and present data to boost wastewater treatment performance continuously. The use of ML determines the best selection of feedstocks that leads to higher methane production yields with reduced waste generation (Song et al., 2024). Using ML technology in wastewater infrastructure enables the optimal route selection for wastewater flow to maximize recovery efficiency together with cost reduction.

10.3.2.4 Simulating Scenarios with Digital Twins

A digital twin is a virtual model of a wastewater treatment facility that mirrors real-world operations. Digital twins allow operators to test changes and optimize processes without disrupting actual systems. These digital twins simulate the effects of different feedstocks on biogas production, helping operators choose the most efficient options (Kabeyi & Olanrewaju, 2022). Virtual models evaluate various chemical recovery strategies, reducing trial-and-error efforts and improving cost-effectiveness as depicted in Table 10.1.

10.4 Energy Recovery and Circular Economy

The concept of energy recovery from wastewater treatment plants (WWTPs) has emerged as a cornerstone of the circular economy, transforming what was once seen as waste into valuable resources. By leveraging advanced technologies for biogas production and utilization and integrating synergistic approaches, modern WWTPs are evolving into energy-positive facilities. These facilities utilize their energy needs and supply extra power to the electric grid alongside water and nutrient resource recovery technology for recycling purposes. This combined operational model simultaneously supports sustainability along with decreased environmental impact and it delivers better economic benefits.

Table 10.1 Integration of biogas and nutrient recovery with economic and environmental implications

Process	Technology used	Mechanism	Products	Environmental benefits	Economic benefits	Challenges
Biogas production	Anaerobic digestion (AD)	Microbial breakdown of organic matter in the absence of oxygen	Biogas (methane + CO_2); digestate	Reduces organic waste; mitigates GHG emissions	Renewable energy source; reduces energy costs	Requires optimized conditions (pH, temperature); odor management
Biogas purification	Chemical scrubbing; pressure swing adsorption	Removal of impurities like CO_2 and H_2S using chemical absorbents or pressure-based separation	Purified methane	High-purity fuel reduces reliance on fossil fuels	Enables vehicle fuel or electricity production	Cost of purification technologies; disposal of by-products
Digestate treatment	Nutrient recovery (e.g., struvite precipitation)	Extraction of phosphorus and nitrogen from digestate via chemical precipitation and ammonia stripping	Fertilizers (struvite, ammonium sulfate)	Prevents water eutrophication; supports circular nutrient economy	Value-added fertilizer products; meets agricultural demand	Investment in dual-processing infrastructure
Energy recovery enhancement	Combined heat and power (CHP); biogas storage	Conversion of purified biogas into electricity and heat or storage for peak energy demand	Electricity; thermal energy	Reduces dependency on grid power; minimizes waste	Generates secondary revenue streams	High upfront investment in CHP units
Co-digestion for optimization	Integration with food waste or industrial waste	Mixing multiple organic feedstocks to improve biogas yield and nutrient recovery	Increased biogas; enhanced nutrient value	Improves overall system efficiency	Higher biogas and nutrient recovery yield	Feedstock variability may affect consistency; requires careful monitoring

10.4.1 Technologies for Biogas Production and Utilization

Biogas production plays a fundamental role in wastewater energy recovery because of its dependence on anaerobic digestion (AD). The methodology depends on microorganisms to decompose organic materials under oxygen-free conditions which creates biogas together with digestate as the process's beneficial side product. The biogas produced from CH_4 and CO_2 mainly contains renewable energy for multiple applications.

1. **Anaerobic Digestion as a Foundation for Biogas Production:** Organic waste treatment adopts Anaerobic digestion as its primary standard-based method. Many operators use this technology to treat wastewater sludge as well as agricultural residues and food waste together with industrial effluents (Pascual et al., 2018). Trough co-digestion where several different organic materials are combined for digestion the biogas production increases significantly due to a balanced substrates mix. The combined operation between these processes results in optimized methane production while recovering nutrients which strengthens circular economy principle (Klemeš et al., 2019).
2. **Biogas Upgrading for Optimal Utilization:** The biogas produced as a raw substance needs to undergo processing and cleaning steps for increasing its methane content as well as its heating value. Pressure Swing Adsorption (PSA) technology uses compression along with adsorbents to upgrade biogas through selective CO_2, H_2S and water vapor adsorption at high pressure (Chen et al., 2021). These impurities leave the adsorbent after depressurization which allows for the adsorbent regeneration process. The Pressure Swing Adsorption method produces methane products which reach purities beyond 95% while recovering more than 90% of methane in the process. Biomethane emerges as a pure energy resource that can function in various applications through this process:
   (a) Electricity and Heat Generation: Biomethane is combusted in combined heat and power (CHP) units to produce electricity and thermal energy, supporting plant operations or external distribution.
   (b) Vehicle Fuel: Upgraded biomethane can substitute compressed natural gas (CNG) in transportation, offering a cleaner and sustainable alternative.
   (c) Pipeline Injection: Biomethane is compatible with natural gas infrastructure, allowing for direct injection into pipelines to meet regional energy demands.
3. **Digestate Valorization for Nutrient Recovery**: The digestate produced from AD is rich in nutrients like nitrogen, phosphorus, and potassium. This by-product is processed further into biofertilizers, reducing reliance on synthetic fertilizers and contributing to sustainable agriculture. By closing the nutrient loop, digestate utilization strengthens the economic and environmental value of wastewater treatment.

10.4.2 Synergies for Energy-Positive Treatment Plants

Huge amounts of electricity are needed to operate traditional wastewater treatment systems because aeration and sludge management require significant energy. WWTPs now generate more energy than they use through the combination of energy recovery elements with efficiency-boosting innovations. The facilities transform surplus energy into a total output greater than their energy requirements thus achieving goals of sustainable development.

1. **Maximizing Energy Production**
   (1) Co-Digestion for Enhanced Biogas Yield: Biogas production from wastewater sludge increases when combined with organic materials such as agricultural wastes or food residuals.
   (2) Innovative Digestion Techniques: Higher temperatures in operating digesters activate microorganisms which improves the speed of methane generation. The placement of Bitbial and methanogenic processes in different reactors produces optimal performance and boosts biogas production with simultaneous improvements in nutrient management (Sarker et al., 2019).

2. **Optimizing Energy Utilization**
   (1) Efficient Combined Heat and Power (CHP) Systems: The application of biogas in combined heat and power units enables generation of electricity together with thermal energy while retaining minimal energy waste (Dalpaz et al., 2020). WWTPs become self-operational through combined heat and power units that also generate financial benefit from excess energy production.
   (2) Integration with Renewable Energy Sources: Hybrid energy solutions are formed when energy recovery systems combine solar panels and wind turbines for additional carbon footprint reduction in plants (García et al., 2021).

3. **Minimizing Energy Consumption**
   (1) Smart Aeration Techniques: In biological treatment processes, aeration is a major energy consumer and is optimized through advanced monitoring systems and artificial intelligence (Gu et al., 2023). Real-time adjustment of aeration levels by AI algorithms maintains microbial health with a reduction in energy use.
   (2) Thermal Hydrolysis Pre-Treatment: Pre-treating sludge with thermal hydrolysis increases its biodegradability, allowing anaerobic microbes to extract more energy from the organic matter with less input energy (Serrano et al., 2016).
   (3) Circular Resource Flows: Treated wastewater is repurposed for irrigation, industrial cooling, or even potable use in water-scarce regions. This

practice conserves freshwater resources and reduces environmental strain (Ungureanu et al., 2020). Innovative techniques like struvite precipitation and ammonia stripping capture nutrients from wastewater, converting them into commercial-grade fertilizers for agricultural use. CO_2 produced during biogas upgrading can be captured and reused in industrial processes or as a raw material for synthetic fuels, aligning with global carbon neutrality goals (Kabeyi & Olanrewaju, 2022).

10.5 Advantages of Energy-Positive Treatment Plants

WWTPs that generate surplus energy reduce emissions by using renewable biogas as an alternative to fossil fuels. Both features decrease the flow of nutrients into water bodies which safeguards environmental ecosystems along with minimizing eutrophication risks. Operating energy generation at WWTPs leads to financial savings which boosts profitability in addition to reducing operational costs (Al-Dahidi et al., 2024). The business activities surrounding surplus energy marketing combined with water purification and biofertilizer supply both generate financial income which sustains plant operations. Water treatment systems maintain their ability to adjust their dimensions based on service characteristics and also handle various wastewater compositions including recovery requirements. These solutions demonstrate versatility which makes them applicable to different service settings across various locations and societal classes in Table 10.2.

Table 10.2 Future of energy-positive wastewater treatment

Category	Details	Benefits	Technologies
Circular resource cycles	Recover water, nutrients, and energy in a closed loop	• Reduces waste • Conserves resources	• Nanotech • AI-driven recovery • Smart sensors
Carbon neutral operations	Achieve carbon–neutral or positive energy efficiency and renewable	• Reduces emissions • Supports climate goals	• Digital twins • CO_2 utilization technologies
Global accessibility	Scalable solutions for urban and rural areas	• Improves water access • Supports development	• Modular systems • AI for optimization
Tech integration	Use AI, nanotech, and digital twins for better efficiency and scalability	• Drives innovation • Enhances decision-making	• Predictive analytics • Machine learning
Energy-positive WWTPs	Convert wastewater into energy, water, and nutrients, contributing surplus energy	• Supports circular economy • Increases viability	• Combined heat and power (CHP) systems produce electricity and heat from biogas • Advanced nutrient recovery

10.6 Case Studies and Real-World Applications

10.6.1 Successful Examples of Integrated Systems Globally

The Wastewater Treatment Plant located in Malmo Sweden demonstrates outstanding positive energy wastewater treatment methods (Brundin, 2018). The energy-positive wastewater treatment system implements three components which include anaerobic digestion together with co-digestion along with combined heat and power (CHP) systems for wastewater energy recovery. The plant receives wastewater from municipalities combined with food scraps and agricultural residues to maximize biogas production.

10.6.2 Energy-Positive Wastewater Treatment: The Malmo Wastewater Treatment Plant, Sweden

The Wastewater Treatment Plant located in Malmo Sweden demonstrates outstanding positive energy wastewater treatment methods. The energy-positive wastewater treatment system implements three components which include anaerobic digestion together with co-digestion along with combined heat and power (CHP) systems for wastewater energy recovery (Sarpong et al., 2020) The plant receives wastewater from municipalities combined with food scraps and agricultural residues to maximize biogas production. Anaerobic digestion breaks down organic matter in the absence of oxygen, producing biogas (mostly methane), which is then used to generate electricity and heat through CHP units. The co-digestion process, which includes food waste, maximizes methane production. The plant becomes energy self-sufficient by utilizing the biogas to power its operations and provide thermal energy, even generating surplus energy that is sold back to the grid (Augustyn et al., 2021). The Malmo plant is one of the first energy-positive wastewater treatment plants, contributing to the local community by supplying surplus electricity and reducing greenhouse gas emissions, making it a leading example of resource recovery and energy optimization.

10.6.3 Nutrient Recovery and Resource Recycling: The Blue Plains Advanced Wastewater Treatment Plant, Washington D.C., USA

The Blue Plains Advanced Wastewater Treatment Plant in Washington D.C. is a pioneering facility in nutrient recovery and resource recycling (Alley & Alley, 2022). The plant employs biological nutrient removal (BNR) in combination with chemical precipitation methods to remove nitrogen and phosphorus from wastewater, both of which are essential for plant growth but harmful when released in excess into water bodies. The struvite precipitation process denies phosphorus from biological treatment before creating struvite as an essential fertilizer. After ammonia recovery the plant converts it into ammonium sulfate which stands as a common agricultural fertilizer. The Blue Plains plant operates as a complete nutrient recycling facility because it organizes nitrogen and phosphorus resources which minimizes dependence on artificial fertilizers for agricultural needs (Grant & Flaten, 2019). The facility generates biosolids which undergo treatment before being applied to agricultural lands to aid reclamation projects and fertilizer agricultural purposes through a complete resource management system.

10.6.4 Water and Energy Recovery: The Pampa del Infierno Wastewater Treatment Plant, Argentina

The Pampa del Infierno Wastewater Treatment Plant serves as an example in Argentina of joining wastewater treatment with water and energy recovery practices (Sánchez & Martins, 2021). Through anaerobic digestion the facility processes wastewater sludge into biogas before converting the gas into electricity. Biogas production at the plant is accompanied by solar energy implementation that decreases dependency on outside energy suppliers. Minimizing energy usage goes beyond electricity production because the plant uses efficient pumping systems along with its energy recovery efforts to decrease its overall carbon emissions. Renewable power resources including solar power along with biogas have made the facility self-powered while enabling the excess electrical energy to supply the local power network (Obiora et al., 2024). The power-positive framework decreases operational costs while concurrently adding more electricity to the community power network thus making the facility a sustainability example for circula resource enhancement and power generation capabilities from wastewater treatment centers.

10.6.5 Global Synergy and Automation: The Xiangyin Wastewater Treatment Plant, China

The Xiangyin Wastewater Treatment Plant in China showcases the highest standard of international cooperation and automated water processing in modern wastewater facilities (Qu et al., 2022). The plant utilizes three pollution removal methods including biological treatment and membrane filtration and advanced oxidation processes (AOPs). The combination of biological wastewater treatment with membrane filtration through Membrane bioreactors (MBRs) produces water that reaches suitable standards for multiple reuse purposes. AOPs function as treatment methods to decompose difficult-to-remove pollutants specifically including pharmaceutical residues. Continuous real-time monitoring systems together with artificial intelligence (AI) allow the plant to enhance operational efficiencies. The plant implements AI systems for water quality parameter monitoring that trigger automated chemical distribution and aeration management and filter operations to decrease energetic requirements (Lowe et al., 2022). The processed water undergoes recycling for industrial and agricultural uses in a region with restricted access to freshwater leading to water preservation. Multiple process interactions at Xiangyin between chemicals, biology and filtration systems produce high quality effluents while maximizing resource recovery to establish new wastewater treatment performance standards.

10.7 Sustainability and Policy Frameworks

Current wastewater treatment methods express direct support for multiple United Nations Sustainable Development Goals through the implementation of advanced technologies (Table 10.3). Technological approaches in wastewater management achieve both the Sustainable Development Goals and handle crucial environmental concerns along with socioeconomic requirements. Three core UN Sustainable Development Goals match wastewater treatment technology implementations:

1. **SDG 6: Clean Water and Sanitation** The wastewater treatment efficiency can be improved through the use of modern water technologies including membrane bioreactors (MBRs) and biological nutrient removal (BNR) and advanced oxidation processes (AOPs) under the SDG 6 framework. Membrane bioreactors (MBRs) enable the removal of organic contaminants and pathogens and nutrients through their processes thus producing water at proper standards to support agricultural and industrial reuse and drinking purposes (Pervez et al., 2020). The

Table 10.3 Alignment of wastewater treatment technologies with UN SDGs

SDG	Technologies	Contribution
SDG 6	MBRs, BNR, AOPs	Improves water quality for reuse in agriculture, industry, and potable applications
SDG 7	AD, CHP	Converts wastewater into renewable energy, reducing grid reliance
SDG 9	AI, real-time monitoring	Optimizes treatment processes and infrastructure
SDG 12	Nutrient recovery, water recycling	Recovers nutrients and recycles water for industrial and agricultural uses
SDG 13	Biogas, solar/wind power	Reduces carbon footprint through renewable energy
SDG 11	Resource recovery technologies	Enhances sustainable urban infrastructure and water reuse

application of these technologies enables wastewater treatment to reach high standards for environmental release or reuse purposes thus actively generating more water and sanitation supplies.

2. **SDG 7: Affordable and Clean Energy** WWTPs now implement AD and CHP systems for sustainable energy generation from wastewater sources. The process of anaerobic digestion transforms wastewater organic matter to produce biogas that becomes a valuable source for electricity production combined with heat generation. The implementation of CHP systems enables plants to leverage biogas for power generation as well as operational plant supply and grid surplus power distribution. Through renewable energy systems WWTPs can generate excess electric power which makes them energy-positive systems able to supply electricity to local communities and Renewable power integration operates in line with SDG 7 (Rani et al., 2022).
3. **SDG 9: Industry, Innovation, and Infrastructure** Modern wastewater treatment facilities achieve better operational results by utilizing both artificial intelligence solutions real-time and data-based management procedures. Industrial waste management scenarios benefit from optimal resource use through automatic behaviour monitoring together with operational control adjustments and adjustment of chemical consumables as they happen (Cheah et al., 2022). AI alongside automation leads to infrastructure advancement in wastewater treatment facilities which creates secure adaptable plants with improved efficiency for growing water and wastewater needs.
4. **SDG 12: Responsible Consumption and Production** Wastewater treatment plants that integrate resource recovery technologies contribute to sustainable consumption and production patterns by reducing waste and promoting the reuse of materials (Diaz-Elsayed et al., 2019). Nutrients like nitrogen and phosphorus, which are essential for plant growth but harmful when released into water bodies, are recovered from wastewater through biological nutrient removal and chemical precipitation. These recovered nutrients can then be processed into valuable fertilizers, thus reducing the need for synthetic fertilizers. Furthermore, WWTPs that recycle water for industrial, agricultural, or even potable uses contribute to the responsible consumption of water resources.
5. **SDG 13: Climate Action** Many of the technologies used in energy-positive wastewater treatment plants are also aligned with climate action. Biogas production reduces methane emissions by capturing and using methane that would otherwise be released into the atmosphere, while the use of solar, wind, or hydroelectric power reduces the carbon footprint of wastewater treatment processes. Water treatment facilities gain reduced environmental impacts because their use of advanced technologies combined with energy-efficient equipment lowers their total energy footprint (Molinos-Senante & Sala-Garrido, 2018).
6. **SDG 11: Sustainable Cities and Communities** Global urbanization trends demand growing requirements for sustainable urban pipeline systems which manage water and wastewater distribution. Wastewater treatment plants gain resilience through the implementation of resource recovery technologies for their infrastructure development. These infrastructure facilities help cities become

more environmentally sustainable by lowering wastewater emissions together with maximizing water conservation benefits and creating renewable power generation (Obaideen et al., 2022). Treated water from these treatment plants becomes viable for industrial applications and agricultural use which extends the sustainability of both cities and communities.

10.7.1 Policy Recommendations for Adopting Integrated Approaches

A productive partnership between governments, industries and international institutions needs to develop policy platforms which promote integrated resource-based wastewater treatment methods to reach their maximum potential as shown in Table 10.4. The transition requires the following policies for implementation:

1. **Incentivize Research and Development (R&D):** The allocation of government funds needs to support research-based developments in wastewater treatment systems (Shi et al., 2021). Organization collaborations between public and private sectors enable better innovation characterizing the deployment of advanced technologies. Reserve funding from governments helps support development activities which create improved materials, processes and systems for wastewater resource recovery with cost reduction benefits and enhanced sustainability impact.

Table 10.4 Policy recommendations for wastewater treatment

Policy	Action plan	Purpose
Incentivize R&D	Fund R&D and support PPPs	Drive tech advancements and enhance resource recovery
Strengthen resource recovery	Set targets for nutrient, water, and energy recovery	Promote resource recovery and reduce environmental impact
Promote circular economy	Incentivize water/energy recovery and reuse	Promote sustainable resource recycling
Encourage PPPs	Engage the private sector in recovery projects	Speed up tech adoption and improve efficiency
Create education & training	Develop programs for operators and policymakers	Build expertise in handling complex treatment systems
Strengthen global cooperation	Foster global knowledge-sharing and collaboration	Ensure global access to wastewater treatment innovations
Implement data-driven systems	Support AI, big data, and machine learning use	Optimize treatment processes and improve efficiency
Conduct impact assessments	Require environmental and social impact evaluations	Identify risks and minimize negative effects

2. **Strengthen Regulations on Resource Recovery:** Policies aimed at fostering resource recovery technology adoption should establish quantitative goals for extracting water alongside nutrients and energy. Operation of WWTPs should bear legal requirements to extract specific nitrogen and phosphorus contents from wastewater to obtain marketable fertilizer products. The implementation of anaerobic digestion systems with CHP technologies must receive governmental support through measurement policies stipulating energy-positive facilities or offering financial incentives to convert surplus energy into grid-fed power.
3. **Promote Circular Economy Practices:** Water, energy and nutrient recycling activities should become the focus of policymakers for the wastewater sector through supportive measures that promote these circular economy initiatives (Guerra-Rodríguez et al., 2020). The implementation of tax breaks alongside grants and subsidies should be provided to municipalities and industries that build water and energy recovery systems. Public policies endorsing wastewater recycling for agricultural watering and industrial functions enable the creation of water cycle closures to minimize pressure on freshwater sources while increasing sustainability.
4. **Encourage Public–Private Partnerships (PPPs):** The implementation of sophisticated wastewater treatment systems needs joint efforts between public institutions and private enterprises because of their difficult deployment requirements and expensive character (Rodriguez et al., 2020). Public authorities need to develop systems that motivate private organizations to take part in wastewater project development alongside activities for resource recovery. Public–private partnerships advance municipality operations at reduced infrastructure expenses alongside faster new technology adoption.
5. **Create Education and Training Programs:** Advanced wastewater treatment systems need experts to perform design activities and handle operation and maintenance because their designs become progressively complex (Chen et al., 2020). National programs involving governmental support and industrial organizations must create training curricula to teach wastewater treatment personnel and policymakers about integrated resource recovery methods for proper implementation. The sector's future technological developments will need prepared personnel because of this initiative.
6. **Strengthen International Cooperation:** Our world faces a global problem with wastewater management which demands international cooperation for an effective solution. Governments need to work together with international organizations for exchanging knowledge along with technological expertise and best practice approaches. The implementation of cross-border projects aims to promote resource recovery technologies in regions without infrastructure so sustainable wastewater treatment systems become available worldwide.
7. **Implement Data-Driven Decision-Making:** Public authorities should establish policies to promote AI along with big data and machine learning for wastewater treatment facilities (Rane et al., 2023). Water treatment facilities use these computerized systems to improve their resilient resource recovery performance while decreasing operational costs while generating effluent of superior quality.

Speeding up the creation of baseball standards and evaluation methods for wastewater data alongside sharing protocols will make decisions better and water treatment more transparent.

8. **Implement Environmental and Social Impact Assessments:** Wastewater treatment projects requiring new technology should undergo complete environmental and social impact studies which evaluate the possible negative effects of system implementations. Such assessments detect possible threats that could affect local natural environments and communities along with economic systems thus enabling policymakers to develop mitigation plans that support societal advantages from technology deployment (Balogun et al., 2020).

10.8 Challenges and Future Directions

Wastewater treatment systems in the past concentrated mainly on wastewater remediation to maintain environmental requirements without prioritizing resource recovery from wastewater. Sustainable resource management needs have created rising pressure to shift wastewater treatment from traditional waste elimination practice into resource recovery practices. The complete realization of wastewater resource recovery technologies demands solution of various technological, economic and regulatory barriers. The fast development of wastewater resource recovery occurs through blockchain technology which operates with AI and IoT networks to generate efficient innovation processes.

10.8.1 Technological Barriers

Several technological barriers exist for implementing advanced wastewater treatments that enable resource recovery practice. The facilities at wastewater treatment plants require different processes that operate under distinctive conditions and requirements. The primary technological challenges included. Advanced resource recovery technologies such as membrane filtration, anaerobic digestion, and nutrient recovery systems need to work in synergy within a wastewater treatment plant. Each technology has different operating parameters (e.g., temperature, pressure, chemical dosing), and integrating them to work seamlessly can be difficult. For instance, combining energy recovery technologies (such as anaerobic digestion) with water reuse systems (such as reverse osmosis) demands a sophisticated system of controls, sensors, and real-time adjustments. While many advanced resource recovery technologies have shown promise in pilot studies or small-scale applications, scaling these technologies to large, municipal-level treatment plants remains a significant hurdle. Often, the cost of scaling is prohibitive, and the technologies do not always perform as effectively or efficiently in large-scale operations as they do in controlled environments. This scalability issue is a critical factor limiting widespread adoption. Many advanced treatment technologies, such as reverse osmosis or advanced oxidation processes (AOPs), are energy-intensive and require substantial operational costs. For resource recovery technologies to be economically viable, they must not only provide environmental benefits (e.g., reduced pollution) but also deliver a return on investment (ROI) (Velenturf & Jopson, 2019). This is especially true for energy recovery systems, where the cost of energy generation and storage must be carefully balanced against the recovery rate to make it financially sustainable. The implementation of multiple advanced systems (e.g., nutrient recovery, biogas production, water reclamation) requires effective monitoring and integration. The existing wastewater treatment facilities operate with outdated systems which cannot manage the necessary data volume needed to enhance operational optimization. Plant operators face implementation difficulties when trying to establish real-time data integration between all systems in existing plants because it requires thorough system upgrades.

10.8.2 Economic Barriers

Multiple complexities exist because of the substantial financial obstacles that prevent the adoption of sophisticated resource recovery methods. Starting advanced resource recovery systems with technologies including biogas generators and nutrient recovery systems as well as advanced filtration systems demands major initial financial investment (Diaz-Elsayed et al., 2019). Many treatment facilities in low- and middle-income countries face financially restricts to implement these solutions. The funds dedicated to infrastructure upgrades remain restricted because they aim to enhance basic treatment processes instead of implementing new resource recovery systems. A thorough return on investment (ROI) process associated with resource recovery technologies demands multiple years before achieving financial and environmental advantages. Energy recovery systems require multiple years to produce enough electricity that matches the original expenses spent during installation. Nutrient recovery systems together with water reuse operations require extended periods before establishing sustainability as a business model. Financial incentives together with obvious rapid return on investment enable wastewater treatment plants to adopt advanced systems but their absence may discourage implementation. Multiple regions lack appropriate financial programs that would drive the adoption of resource recovery systems. The financial framework which should support innovations such as biogas recovery and wastewater reuse (through resources as subsidies and tax incentives or resource recovery credits) remains underdeveloped (Rodriguez et al., 2020). The integrated value of reclaimed resources especially nutrients and water remains obscure in pricing systems thus creating obstacles for plants to establish recovery technology investments. Operation of sophisticated wastewater treatment systems increases overall business expenditures because specialized skills are needed for maintenance work. Membrane filtration systems alongside other resource recovery technologies have upkeep needs that demand periodic component maintenance or replacement leading to expenses. It is necessary for plants to manage their operational expenses without sacrificing the quality of effluent treatment while maintaining sustainability.

10.8.3 Regulatory Barriers

Historical wastewater treatment standards have centred their efforts on creating safe effluents for the environment as well as safeguarding public health. Existing regulations struggle to motivate or encourage the use of modern technologies because resource recovery has recently gained greater prominence. Current wastewater treatment rules have not been updated to consider innovations in resource recycling because they were designed before modern technologies emerged. Wastewater reuse regulations focus on effluent standards instead of evaluating the worth of extracted resources including phosphorus, nitrogen and water. Confusion together with delayed implementation of new systems is a result of unclear regulations. The circular economy depends on wastewater reuse but numerous countries maintain strict rules about using treated wastewater for tasks aside from drinking water including irrigation and industrial applications (Mannina et al., 2022). The regulatory restrictions which stem from public health and safety worries work as barriers that prevent municipal authorities from putting their water recycling systems into complete operation. There are no standardized water quality requirements for different applications within several existing wastewater reuse regulations. Multiple issues emerge in the standardization of efficiency evaluation methods for resource recovery technologies. Requirements for standards in performance create delays in regulatory approval for technology implementation on a broad scale. The lack of performance standards alienates businesses from introducing breakthrough innovations because they struggle to win customer trust. The regions that show inconsistent behavior in environmental regulation enforcement discourage investors from investing in advanced wastewater treatment systems. Without sufficient penalties or incentives for reaching high resource recovery municipalities show lower motivation for employing advanced technologies (Kehrein et al., 2020).

10.9 Future Trends: AI, IoT, and Blockchain in Wastewater Recovery

Various modern technological advancements of AI alongside IoT and blockchain technology enable the solution of various issues connected to wastewater handling and resource retrieval. The latest technological solutions provide new operational methods which lower expenses while enhancing wastewater system transparency as well as operational efficiency.

1. **AI for Process Optimization and Predictive Maintenance:** AI systems evaluate substantial wastewater information through their algorithms while automatically changing operational settings in real-time (Mohanty et al., 2024). High energy efficiency combined with superior nutrient removal occurs due to optimized control of chemical treatments together with air distribution systems and filter operations. The system decreases energy usage while enhancing the efficiency of resource recovery and it minimizes operational expenses. AI assesses equipment failure possibilities to deliver strategically planned equipment maintenance which prevents equipment breakdowns. By using sensors alongside pattern recognition AI discovers operational patterns human operators normally overlook and thus strengthens operational reliability across the plant.
2. **IoT for Real-Time Monitoring and Integration:** Off-site Hardware Monitoring (Off-HM) devices allow real-time tracking of wastewater treatment plant indicators including pH measurements together with chemical levels alongside flow measurements and energy monitoring through sensors and smart meters. The data provides instant opportunities to adjust operations for peak resource recovery system efficiency. The implementation of IoT-enabled devices allows different resource recovery systems (including nutrient recovery and biogas production and water reuse) to interoperate through shared operational platforms for operators (Rane et al., 2023). All plant systems function together harmoniously because of their integrated operational framework.
3. **Blockchain for Transparency and Resource Trading:** Through blockchain technology the wastewater treatment process can achieve secure transparent transaction recording starting from energy recovery and ending with water reuse operations (Hakak et al., 2020). Blockchain establishes transparent methods to develop user trust between regulation bodies and consumers which results in enhanced standards and organizational oversight responsibility. The implementation of blockchain technology allows the development of markets where municipalities together with private entities can trade recovered resources such as water energy and nutrients. Wastewater treatment plants can earn revenue by returning their surplus biogas to the power grid and swapping nutrient substances with farming businesses. The transactions on Blockchain preserve their integrity to establish a more efficient circular economy system.

10.10 Conclusion and Recommendations

Modern technological innovations in wastewater treatment hold immense promise—not only for purifying water, but for transforming waste into wealth by recovering energy and nutrients. By integrating physical, chemical, and biological processes such as anaerobic digestion, advanced oxidation, and filtration, treatment plants can now produce clean effluents, generate biogas, and extract vital agricultural nutrients like phosphorus and nitrogen. These advances align well with circular economy principles, promoting sustainability through resource recovery. However, the path to widespread adoption remains challenging. Many facilities on hold with outdated infrastructure, high modernization costs, and the absence of supportive regulatory frameworks. To unlock the full potential of these innovations, strategic actions are essential. Governments must increase research funding and foster public–private partnerships to drive technological progress. Simultaneously, regulatory systems need to be developed and modernized to enforce clear standards for energy and resource recovery. Policy initiatives should prioritize energy-positive and resource-efficient plant models, while investment in workforce training will ensure skilled operation and maintenance of these sophisticated systems. Finally, international cooperation—through shared knowledge, technology exchange, and collaborative initiatives—will be key to scaling integrated, sustainable wastewater treatment solutions globally.

Acknowledgements The authors would like to thank Chandigarh University for the infrastructural support. Support from the Usmanu Danfodiyo University is duly acknowledged.

References

Abuhasel, K., Kchaou, M., Alquraish, M., Munusamy, Y., & Jeng, Y. T. (2021). Oily wastewater treatment: Overview of conventional and modern methods, challenges, and future opportunities. *Water, 13*(7), 980.

Abuwatfa, W. H., AlSawaftah, N., Darwish, N., Pitt, W. G., & Husseini, G. A. (2023). A review on membrane fouling prediction using artificial neural networks (ANNs). *Membranes, 13*(7), 685.

Ahmed, S. F., Mehejabin, F., Momtahin, A., Tasannum, N., Faria, N. T., Mofijur, M., Hoang, A. T., Vo, D.-V.N., & Mahlia, T. M. I. (2022). Strategies to improve membrane performance in wastewater treatment. *Chemosphere, 306*, 135527.

Akhtar, N., Syakir Ishak, M. I., Bhawani, S. A., & Umar, K. (2021). Various natural and anthropogenic factors responsible for water quality degradation: A review. *Water, 13*(19), 2660.

Al-Dahidi, S., Alrbai, M., Al-Ghussain, L., & Alahmer, A. (2024). Maximizing energy efficiency in wastewater treatment plants: A data-driven approach for waste heat recovery and an economic analysis using Organic Rankine Cycle and thermal energy storage. *Applied Energy, 362*, 123008.

Alley, W. M., & Alley, R. (2022). *The water recycling revolution: Tapping into the future.* Rowman & Littlefield.

Augustyn, G., Mikulik, J., Szyba, R. R., & M. (2021). Energy self-sufficient livestock farm as the example of agricultural hybrid off-grid system. *Energies, 14*(21), 7041.

Balogun, A.-L., Marks, D., Sharma, R., Shekhar, H., Balmes, C., Maheng, D., Arshad, A., & Salehi, P. (2020). Assessing the potentials of digitalization as a tool for climate change adaptation and sustainable development in urban centres. *Sustainable Cities and Society, 53*, 101888.

Bdour, A. N., Hamdi, M. R., & Tarawneh, Z. (2009). Perspectives on sustainable wastewater treatment technologies and reuse options in the urban areas of the Mediterranean region. *Desalination, 237*(1–3), 162–174.

Bohra, V., Ahamad, K. U., Kela, A., Vaghela, G., Sharma, A., & Deka, B. J. (2022). Energy and resources recovery from wastewater treatment systems. In *Clean energy and resource recovery* (pp. 17–36). Elsevier.

Brundin, C. (2018). *Alternative energy concepts for Swedish wastewater treatment plants to meet demands of a sustainable society.*

Cheah, C. G., Chia, W. Y., Lai, S. F., Chew, K. W., Chia, S. R., & Show, P. L. (2022). Innovation designs of industry 4.0 based solid waste management: Machinery and digital circular economy. *Environmental Research, 213*, 113619.

Chen, G.-H., van Loosdrecht, M. C. M., Ekama, G. A., & Brdjanovic, D. (2020). *Biological wastewater treatment: principles, modeling and design.* IWA publishing.

Chen, Y.-F., Lin, P.-W., Chen, W.-H., Yen, F.-Y., Yang, H.-S., & Chou, C.-T. (2021). Biogas upgrading by pressure swing adsorption with design of experiments. *Processes, 9*(8), 1325.

Cheng, S., Yu, Y., Meng, F., Chen, J., Chen, Y., Liu, G., & Fan, W. (2023). Potential benefits of public–private partnerships to improve the efficiency of urban wastewater treatment. *NPJ Clean Water, 6*(1), 13.

Dalpaz, R., Konrad, O., da Silva Cyrne, C. C., Barzotto, H. P., Hasan, C., & Guerini Filho, M. (2020). Using biogas for energy cogeneration: An analysis of electric and thermal energy generation from agro-industrial waste. *Sustainable Energy Technologies and Assessments, 40*, 100774.

Das, S., Choudhury, M. R., Chatterjee, B., Das, P., Bagri, S., Paul, D., Bera, M., & Dutta, S. (2024). Unraveling the urban climate crisis: Exploring the nexus of urbanization, climate change, and their impacts on the environment and human well-being—A global perspective. *AIMS Public Health, 11*(3), 963.

Diaz-Elsayed, N., Rezaei, N., Guo, T., Mohebbi, S., & Zhang, Q. (2019). Wastewater-based resource recovery technologies across scale: A review. *Resources, Conservation and Recycling, 145*, 94–112.

Estrada-Arriaga, E. B., Zepeda-Aviles, J. A., & Garcíia-Sánchez, L. (2016). Post-treatment of real oil refinery effluent with high concentrations of phenols using photo-ferrioxalate and Fenton's reactions with membrane process step. *Chemical Engineering Journal, 285*, 508–516.

García, A. M., Gallagher, J., Chacón, M. C., & Mc Nabola, A. (2021). The environmental and economic benefits of a hybrid hydropower energy recovery and solar energy system (PAT-PV), under varying energy demands in the agricultural sector. *Journal of Cleaner Production, 303*, 127078.

Grant, C. A., & Flaten, D. N. (2019). 4R management of phosphorus fertilizer in the northern Great Plains. *Journal of Environmental Quality, 48*(5), 1356–1369.

Gu, Y., Li, Y., Yuan, F., & Yang, Q. (2023). Optimization and control strategies of aeration in WWTPs: A review. *Journal of Cleaner Production, 418*, 138008.

Guerra-Rodríguez, S., Oulego, P., Rodríguez, E., Singh, D. N., & Rodríguez-Chueca, J. (2020). Towards the implementation of circular economy in the wastewater sector: Challenges and opportunities. *Water, 12*(5), 1431.

Hakak, S., Khan, W. Z., Gilkar, G. A., Haider, N., Imran, M., & Alkatheiri, M. S. (2020). Industrial wastewater management using blockchain technology: Architecture, requirements, and future directions. *IEEE Internet of Things Magazine, 3*(2), 38–43.

Hao, X., Wu, D., Li, J., Liu, R., & Van Loosdrecht, M. (2022). Making waves: A sea change in treating wastewater–why thermodynamics supports resource recovery and recycling. *Water Research, 218*, 118516.

Hušek, M., Moško, J., & Poho\vrel\'y, M. (2022). Sewage sludge treatment methods and P-recovery possibilities: Current state-of-the-art. *Journal of Environmental Management, 315*, 115090.

Kabeyi, M. J. B., & Olanrewaju, O. A. (2022). Biogas production and applications in the sustainable energy transition. *Journal of Energy, 2022*(1), 8750221.

Kasulla, S., Malik, S. J., Zafar, S., & Saraf, A. (2021). A Retrospection of hydrogen sulphide removal technologies in biogas purification. *International Journal of Trend in Scientific Research and Development, 5*(3), 857–863.

Kehrein, P., Van Loosdrecht, M., Osseweijer, P., Garf\'\i, M., Dewulf, J., & Posada, J. (2020). A critical review of resource recovery from municipal wastewater treatment plants--market supply potentials, technologies and bottlenecks. *Environmental Science: Water Research & Technology, 6*(4), 877–910.

Klemeš, J. J., Varbanov, P. S., Walmsley, T. G., & Foley, A. (2019). Process integration and circular economy for renewable and sustainable energy systems. In *Renewable and Sustainable Energy Reviews* (Vol. 116, p. 109435). Elsevier.

Kolawole, A. S., & Iyiola, A. O. (2023). Environmental pollution: threats, impact on biodiversity, and protection strategies. In *Sustainable utilization and conservation of Africa's biological resources and environment* (pp. 377–409). Springer.

Krishnamoorthy, N., Dey, B., Unpaprom, Y., Ramaraj, R., Maniam, G. P., Govindan, N., Jayaraman, S., Arunachalam, T., & Paramasivan, B. (2021). Engineering principles and process designs for phosphorus recovery as struvite: A comprehensive review. *Journal of Environmental Chemical Engineering, 9*(5), 105579.

Kumar, P., & Choudhury, D. (2024). Innovative technologies for effective water resources management. *Water Crises and Sustainable Management in the Global South*, 555–594.

Kumar, M., Sridharan, S., Sawarkar, A. D., Shakeel, A., Anerao, P., Mannina, G., Sharma, P., & Pandey, A. (2023). Current research trends on emerging contaminants pharmaceutical and personal care products (PPCPs): A comprehensive review. *Science of the Total Environment, 859*, 160031.

Lowe, M., Qin, R., & Mao, X. (2022). A review on machine learning, artificial intelligence, and smart technology in water treatment and monitoring. *Water, 14*(9), 1384.

Mannina, G., Gulhan, H., & Ni, B.-J. (2022). Water reuse from wastewater treatment: The transition towards circular economy in the water sector. *Bioresource Technology, 363*, 127951.

Mohanty, A., Mohanty, S. K., & Mohapatra, A. G. (2024). Real-time monitoring and fault detection in AI-enhanced wastewater treatment systems. In *The AI cleanse: Transforming wastewater treatment through artificial intelligence: Harnessing data-driven solutions* (pp. 165–199). Springer.

Molinos-Senante, M., & Sala-Garrido, R. (2018). Evaluation of energy performance of drinking water treatment plants: Use of energy intensity and energy efficiency metrics. *Applied Energy, 229*, 1095–1102.

Moretti, A., Ivan, H. L., & Skvaril, J. (2024). A review of the state-of-the-art wastewater quality characterization and measurement technologies. Is the shift to real-time monitoring nowadays feasible? *Journal of Water Process Engineering, 60*, 105061.

Mulier, M. C. G. H., de Ven, F. H. M., & Kirshen, P. (2022). Circularity in the urban water-energy-nutrients-food nexus. *Energy Nexus, 7*, 100081.

Nabgan, W., Saeed, M., Jalil, A. A., Nabgan, B., Gambo, Y., Ali, M. W., Ikram, M., Fauzi, A. A., Owgi, A. H. K., Hussain, I., et al. (2022). A state of the art review on electrochemical technique for the remediation of pharmaceuticals containing wastewater. *Environmental Research, 210*, 112975.

Obaideen, K., Shehata, N., Sayed, E. T., Abdelkareem, M. A., Mahmoud, M. S., & Olabi, A. G. (2022). The role of wastewater treatment in achieving sustainable development goals (SDGs) and sustainability guideline. *Energy Nexus, 7*, 100112.

Obiora, N. K., Ujah, C. O., Asadu, C. O., Kolawole, F. O., & Ekwueme, B. N. (2024). Production of hydrogen energy from biomass: prospects and challenges. *Green Technologies and Sustainability*, 100100.

Paranjpe, A., Saxena, S., & Jain, P. (2023). A review on performance improvement of anaerobic digestion using co-digestion of food waste and sewage sludge. *Journal of Environmental Management, 338*, 117733.

Pascual, J. A., Morales, A. B., Ayuso, L. M., Segura, P., & Ros, M. (2018). Characterisation of sludge produced by the agri-food industry and recycling options for its agricultural uses in a typical Mediterranean area, the Segura River basin (Spain). *Waste Management, 82*, 118–128.

Pervez, M. N., Balakrishnan, M., Hasan, S. W., Choo, K.-H., Zhao, Y., Cai, Y., Zarra, T., Belgiorno, V., & Naddeo, V. (2020). A critical review on nanomaterials membrane bioreactor (NMs-MBR) for wastewater treatment. *NPJ Clean Water, 3*(1), 43.

Priya, A. K., Gnanasekaran, L., Dutta, K., Rajendran, S., Balakrishnan, D., & Soto-Moscoso, M. (2022). Biosorption of heavy metals by microorganisms: Evaluation of different underlying mechanisms. *Chemosphere, 307*, 135957.

Qadir, M., Drechsel, P., Jiménez Cisneros, B., Kim, Y., Pramanik, A., Mehta, P., & Olaniyan, O. (2020). Global and regional potential of wastewater as a water, nutrient and energy source. *Natural Resources Forum, 44*(1), 40–51.

Qu, J., Dai, X., Hu, H.-Y., Huang, X., Chen, Z., Li, T., Cao, Y., & Daigger, G. T. (2022). Emerging trends and prospects for municipal wastewater management in China. *ACS ES&T Engineering, 2*(3), 323–336.

Rane, N., Choudhary, S., & Rane, J. (2023). Leading-edge artificial intelligence (AI), Machine Learning (ML), Blockchain, and Internet of Things (IoT) technologies for enhanced wastewater treatment systems. *Machine Learning (ML), Blockchain, and Internet of Things (IoT) Technologies for Enhanced Wastewater Treatment Systems (October 31, 2023).*

Rani, A., Snyder, S. W., Kim, H., Lei, Z., & Pan, S.-Y. (2022). Pathways to a net-zero-carbon water sector through energy-extracting wastewater technologies. *NPJ Clean Water, 5*(1), 49.

Rani, M., Ankit, K., & Shanker, U. (2024). Efficient degradation of endocrine disruptor pesticides by biochar iron oxide-based nanocomposite: Green synthesis, kinetics, and photoactivity. *Biomass Conversion and Biorefinery, 14*(18), 22657–22672.

Rashid, R., Shafiq, I., Akhter, P., Iqbal, M. J., & Hussain, M. (2021). A state-of-the-art review on wastewater treatment techniques: The effectiveness of adsorption method. *Environmental Science and Pollution Research, 28*, 9050–9066.

Rodriguez, D. J., Serrano, H. A., Delgado, A., Nolasco, D., & Saltiel, G. (2020). *From waste to resource: Shifting paradigms for smarter wastewater interventions in Latin America and the Caribbean.*

Sánchez, A. S., & Martins, G. (2021). Nutrient recovery in wastewater treatment plants: Comparative assessment of different technological options for the metropolitan region of Buenos Aires. *Journal of Water Process Engineering, 41*, 102076.

Sarker, S., Lamb, J. J., Hjelme, D. R., & Lien, K. M. (2019). A review of the role of critical parameters in the design and operation of biogas production plants. *Applied Sciences, 9*(9), 1915.

Sarpong, G., Gude, V. G., Magbanua, B. S., & Truax, D. D. (2020). Evaluation of energy recovery potential in wastewater treatment based on codigestion and combined heat and power schemes. *Energy Conversion and Management, 222*, 113147.

Serrano, A., Siles, J. A., Martín, M. A., Chica, A. F., Estévez-Pastor, F. S., & Toro-Baptista, E. (2016). Improvement of anaerobic digestion of sewage sludge through microwave pre-treatment. *Journal of Environmental Management, 177*, 231–239.

Sewerin, T., Elshof, M. G., Matencio, S., Boerrigter, M., Yu, J., & de Grooth, J. (2021). Advances and applications of hollow fiber nanofiltration membranes: A review. *Membranes, 11*(11), 890.

Sharifian, S., & Wang, N.-H. L. (2024). Resin-based approaches for selective extraction and purification of rare earth elements: A comprehensive review. *Journal of Environmental Chemical Engineering*, 112402.

Shen, Y., Linville, J. L., Urgun-Demirtas, M., Mintz, M. M., & Snyder, S. W. (2015). An overview of biogas production and utilization at full-scale wastewater treatment plants (WWTPs) in the United States: Challenges and opportunities towards energy-neutral WWTPs. *Renewable and Sustainable Energy Reviews, 50*, 346–362.

Shi, Z., Huang, H., Chiu, Y., Zhang, B., & Zhang, C. (2021). Linkage analysis of water resources, wastewater pollution, and health for regional sustainable development—Using undesirable three-stage dynamic data envelopment analysis. *Environmental Science and Pollution Research, 28*, 19325–19350.

Song, C., Cai, F., Yang, S., Wang, L., Liu, G., & Chen, C. (2024). Machine learning-based prediction of methane production from lignocellulosic wastes. *Bioresource Technology, 393*, 129953.

Speight, J. G. (2023). *Gas engineering: Vol. 3: Uses of gas and effects*. Walter de Gruyter GmbH & Co KG.

Ungureanu, N., Vlăduţ, V., & Voicu, G. (2020). Water scarcity and wastewater reuse in crop irrigation. *Sustainability, 12*(21), 9055.

van Leeuwen, K., de Vries, E., Koop, S., & Roest, K. (2018). The energy & raw materials factory: Role and potential contribution to the circular economy of the Netherlands. *Environmental Management, 61*, 786–795.

Velenturf, A. P. M., & Jopson, J. S. (2019). Making the business case for resource recovery. *Science of the Total Environment, 648*, 1031–1041.

Yang, G., Luo, Y., Bian, Y., Chen, X., Chen, L., & Huang, X. (2025). Electro-mediated cathodic oxygen drives respiration chain electron transfer of electroactive bacteria to enhance refractory organic biological oxidation. *Water Research, 268*, 122585.

Chapter 11
A Comprehensive Analysis of Wastewater Management Challenges in India: Infrastructure and Policy Perspectives

Showkat Ahmad Wani, Vinaya Tari, and Sheikh Inam Ul Mansoor

Abstract The increasing urbanization and industrial pollution are a serious concern for future generations. Having a proper sewage system in India casts human lives and unfortunately remains the least concern for the policymakers. As per reports, almost 80% of global wastewater is released into the environment without adequate treatment, which can have disastrous health effects. In 2021, The Central Pollution Control Board (CPCB) in its report said that India's current capacity for treating sewage is 18.6% and its current capacity for treating water is 27.3%. According to government official statistics, 62.5% of urban India's wastewater is either partially or not at all treated. There are existing research studies that claim "most of the sewage treatment plants that were established under the Ganga Action Plan (GAP) and Yamuna Action Plan (YAP) are not working, and interestingly, out of the 33,000 MLD of waste generated, only 7000 MLD is collected and treated. To underscore the environmental urgency and its pitfalls, it becomes imperative that the centralized wastewater treatment demands a well-developed network of interconnected sewers and drainage for the wastewater to be collected in a central location." This chapter mainly deals with a critical review of "policies, rules, regulations, on wastewater management in India." The study will also examine the implementation challenges in enforcing wastewater laws and bye-laws in India. The chapter will also delve into the existing challenges and constraints that impede the development of current waste management practices and propose solutions. Moreover, the study investigate the possibilities for waste management, so complementing centralized treatment plants with less expensive alternatives to reduce potential effects on the aquatic environment

S. A. Wani (✉)
School of Law, Alliance University Bengaluru India, Bangalore, Karnataka, India
e-mail: showkatamu001@gmail.com

V. Tari
Department of Biology, Universitas Airlangga, Surabaya, Indonesia

S. I. U. Mansoor
Symbiosis Law School Hyderabad Campus, Symbiosis International (Deemed University), Pune, India
e-mail: sheikh.mansoor@slsh.edu.in

S. Dhiman and G. Mukherjee (eds.), *Resource Resurgence*, Springer Water,
https://doi.org/10.1007/978-3-032-02750-4_11

and suggest a more effective waste management approach. This chapter seeks to undertake a legal and policy assessment of the reuse of Wastewater in India and establish the challenges and prospects in mitigating water scarcity. Regarding the role of Wastewater reuse policies and regulations, it provides a comprehensive overview of the applied economic, social, and environmental value.

Keywords Water management · Policies · Pollution · Rules · Environmental law

11.1 Introduction

Water is a crucial resource for living, including domestic, agricultural, industrial, and recreational operations. Eventually, water availability, quality, and dependability can significantly impact business and profitability (Li et al., 2019; Siddiqui et al., 2022). Hence, effective management of water resources is influential for industries to safeguard sustainability (Colella et al., 2021; Rahman et al., 2023; Tari, 2021). However, industrial sectors face many challenges in managing this crucial resource in the present context. One of the key challenges is the disposal of industrial effluent. The discharge of industrial wastewater into the water resources can project several environmental concerns such as wetland pollution, loss of biodiversity, eutrophication, biomagnification, and saturation of non-degradable material in the resources, eventually leading to unembellished health and environmental consequences (Al-Tohamy et al., 2022; Tari et al., 2021).

To combat the challenge of wastewater discharge in India, Central Pollution Control Board (CPCB) and State Pollution Control Boards (SPCB) are supervising industries. Industries have to comply with quality standards issued by the World Health Organization (WHO) in 1991, the Bureau of Indian Standards (BIS) IS: 10,500 in 1983, the Ministry of Works and Housing (MWH) in 1975, the Indian Council for Medical Research (ICMR) in 1963 for drinking water and industrial effluent discharge standards set by Bureau of Indian Standards (BIS) IS: 2490 in 1981 (Including 35 parameters), General standards for discharge of environmental pollutants: Part—A, Effluent (Schedule-VI, issued in GSR, 422 (E), dated 19th May, 1993, MOEF (Including 40 parameters) (Maiti, 2004). Thus, industries are adopting numerous technologies for wastewater treatments including physical, chemical, and biological methods. However, compliance with spatial regulations is crucial for industries to avoid fines, loss of reputation, and legal allegations (Zan et al., 2022). Water is essential for manufacturing processes, mining, oil and gas production, power generation, institutional laboratories, drinking, cleaning, safety and hygiene, construction, agriculture, gardening, recreational activities, and many more (Gingerich et al., 2017; Tari et al., 2024). In a nutshell, the utility of water in industrial operations is crucial, on the other hand, efficient management of water resources is necessary to minimize the environmental impact for the sustainable development of business, and profitability.

Table 11.1 shows heavy water demanding industrial operations eventually producing waste water out of each process. Now, the real environmental concern is "*How are these industries managing their wastewater*?" "*What are the sustainable technologies being used to ensure economic and environmental resilience*?".

However, to address these issues, collaborative efforts are required from industries, stakeholders, and governments to protect water resources for upcoming generations.

Table 11.1 Overview of heavy water demanding industrial operations

S. No	Industry	Processes	References
1	Textile industry	Pretreatment process, dying, printing, coating, chemical finishing, etc.	Faheem Uddin (2019), Singh et al. (2019)
2	Paper and pulp	Raw material preparation, tumbling, fiber separation, ester-pulping, bleaching, paper making,	Man et al. (2018), Princeton (1989)
3	Petrochemical refining	Polymerization, natural gas treatment process, physical adsorption, Chemisorption, naphtha cracking, etc	Speight (2019)
4	Pharmaceutical industry	Emulsification, separation process, cooling, Solvent extraction, Chemical reactions, printing, coating, etc	World Bank Group, (1998)
5	Mining industry	Mineral processing, Ore extraction, material separation and processing, dust suppression, etc	Norgate and Haque, (2010)
6	Coal mining	Washing, fragmentation by the high-pressure water jet, separating impurities, etc	Balasubramanian, (2016)
7	Oil and Gas industry	Hydraulic fracturing, separation, refining, cooling, etc	Fthenakis and Kim, (2010), Håvard Devold (2008)
8	Geothermal energy production	Condensation, boiling, flash separation, etc	Ziyodulla Yusupov and Mohamed Almaktar (2021)
9	Solar panel manufacturing industries	Cleaning, cooling, materials purification, etc	Compton et al. (2018)
10	Breweries, wineries, and soft drink manufacturing	Cleaning, distillation, Ingredient preparation, fermentation, etc	Luc Fillaudeau et al. (2008)
11	Agricultural products processing units	Washing, peeling, cooking, pulp production, packaging, etc	Sodhi et al. (2020)
12	Food processing industries	Washing, rinsing, cooking, etc	Compton et al. (2018); Sodhi et al. (2020)

11.2 Existing Challenges and Scope to Adopt Effective Wastewater Management

Untreated or partially treated industrial wastewater discharge can lead to contamination of natural resources and is a major environmental concern in the present context worldwide. This concern is not only related to the contamination of surface water resources but also has significant implications on groundwater quality, aquatic flora and fauna, microbes, and public health (Ortúzar et al., 2022; Xi et al., 2022).

Thus, effective industrial effluent management includes some key practices such as wastewater treatment, recycling, reuse, efficient use of water resources (in the present arena "*Water budgeting*"), and pollution prevention principles for sustainable development (Patterson, 1985). There are several existing wastewater treatments for the management of industrial effluent (Aswini et al., 2024; Atangana et al., 2021; Palani et al., 2021).

Commonly used physical treatments where physical processes are used to remove pollutants through manual screening, grit removal, sedimentation, filtration, membrane separation, etc. (Dwivedi et al., 2022). Chemical treatment processes such as coagulation, flocculation, precipitation, oxidation, and adsorption are subjected to the use of chemicals for wastewater treatment (Zou et al., 2021). However, cost-effective and eco-friendly means of treatment are biological wastewater treatments where biological entities are used to break down pollutants and are digested aerobically or anaerobically. These biological processes include activated sludge process, up-flow anaerobic sludge blanket (UASB), trickling filters, Facultative lagoons, oxidation ditch, Membrane bioreactor (MBR), Moving bed biological reactor (MBBR), fixed bed bioreactor (FBR), sequencing batch reactor (SBR) etc. (Verstraete et al., 2022).

All the methods mentioned above are effective for wastewater management (Fig. 11.1); however, they sometimes fail to meet the established guidelines for effluent discharge into natural water bodies. This is where advanced technologies such as reverse osmosis (RO), ultrafiltration, and ion exchange come into play (Burge et al., 2020). Furthermore, nano-based advancements in wastewater treatment represent a rapidly growing field. Their effectiveness has been demonstrated through various processes, including nanofiltration, nano sensors, nanostructured catalytic membranes, nano catalysts, bioactive nanoparticles, biomimetic membranes, and nano sorbents. Examples include silver nanoparticles, copper nanoparticles, iron nanoparticles, carbon-based nanomaterials, graphene-based nanosheets, and fullerenes (Palani et al., 2023; Malakootian et al., 2019; Wahyuni et al., 2024; Khani, 2024; Tari et al., 2022; Lakkaboyana et al., 2022; Gupta et al., 2024; Aswini et al., 2022).

Table 11.2 summarizes the challenges, scope, corresponding issues, and effective management practices. This table seeks to provide a thorough overview of the complexities and possible solutions to these challenges across various industries and sectors.

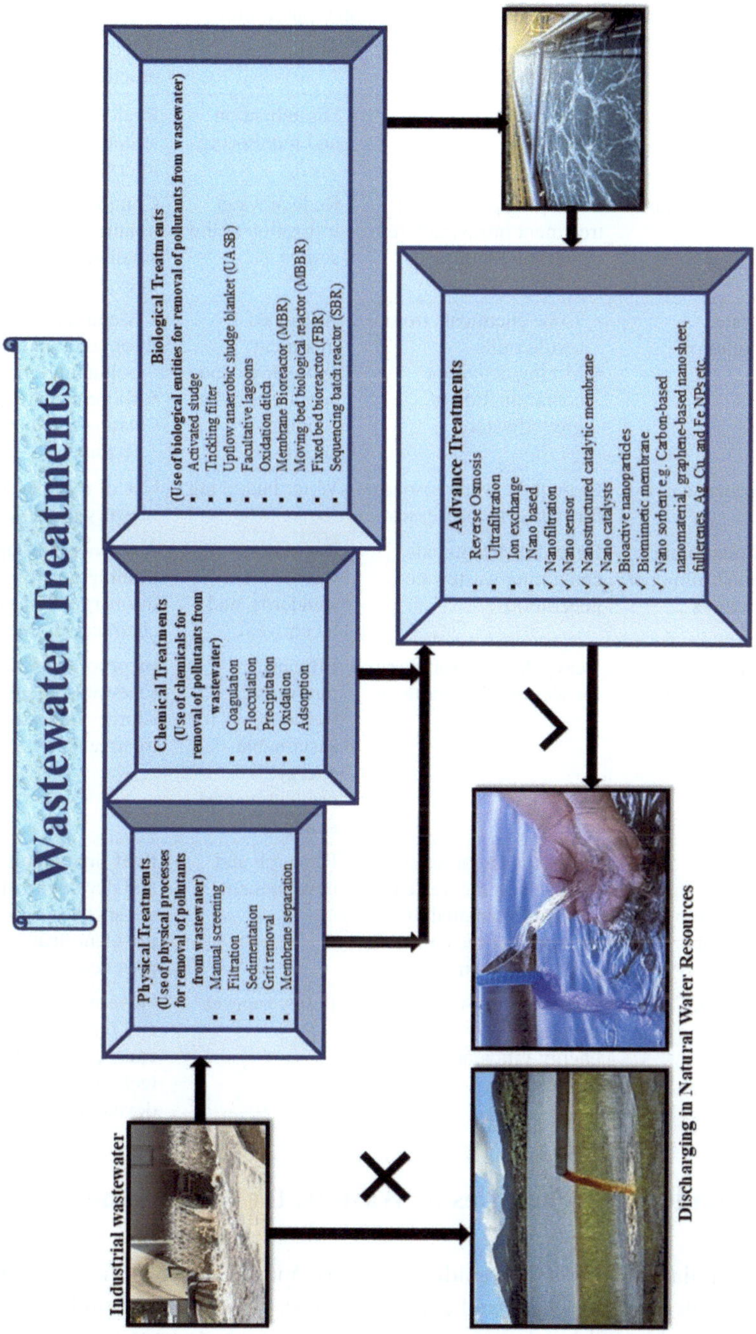

Fig. 11.1 Existing scope for effective wastewater treatment

Table 11.2 Challenges and Wastewater Management Opportunities

Sr. No	Challenges	Corresponding issues	Scope	Effective Management practices
1	Regulatory compliance	Inadequate advanced wastewater treatment access	Digitalization and monitoring	Real-time monitoring systems for continuous intervention
2	Infrastructure	Inadequate wastewater treatment infrastructure for small-scale industries	Reduce water utilization at the source	Common effluent treatment plant for small-scale industries in every industrial zone
3	Water pollution	• Toxic chemicals from textile mills • Mining activities contamination of groundwater	Advanced treatment technology and adopt a circular economy approach	• Membrane filtration for color and pollutant removal • Bioremediation or nano-remediation techniques
4	Water scarcity	Inadequate water resources for industrial processes	Water budgeting	Use of recycled water in every possible way
5	Collaboration and training of trainers	Hazardous chemicals containing wastewater generation	Regulatory frameworks, standards, and incentives	Awareness through training regarding zero discharge of hazardous chemicals frameworks
6	Energy demand and sustainability	Heavy demand for energy for wastewater treatment	Efficient innovative technologies for sustainable energy production and storage	Incorporating renewable energy sources into wastewater treatment
7	Novel contaminants e.g. nano toxicants	The presence of nano toxicants in the treated water, limited analytical methods for the detection of novel contaminants	Research and development	Real-time monitoring, and development of advanced sensors for the identification of novel contaminants
8	Funding	Financial crisis for developing energy-efficient technologies	Policy support	Government incentives and funding for energy-efficient technology development

11.3 Inadequate Facilities in Wastewater Management

India's population is growing rapidly, and consequently, there is increased demand for clean water for residential and commercial use. The failure of public services to keep up with demand has led to poor water waste management infrastructure and efforts. Few urban dwellers are covered by water supply, sanitation, solid waste, and sewage management. There is clear inequality and disparity in the public services that residents receive according to their socioeconomic status.

Shockingly, the treatment capacity for wastewater in India remains grossly inadequate relative to its production. In general, the proper treatment infrastructure for most urban centres is simply missing or woefully inadequate, with few plants having operation levels of even less than full capacity because of both inefficiency and lack of effective maintenance. Skewed towards bigger cities, much of small-town India and peri-urban areas find little or hardly any access to any treatment. This worsens environmental and public health issues emanating from this untreated wastewater. The data reflects a huge gap in the infrastructure of wastewater treatment in India. Class I cities and Class II towns together generate about 38,254 million liters per day of wastewater, while the existing municipal wastewater treatment facilities can treat only about 11,787 million liters per day (MLD), which constitutes 31% of the total wastewater generated in these urban areas. This gap means that about 69% of wastewater goes untreated, and the situation may lead to severe environmental and public health issues. Closing this gap requires huge investment in wastewater treatment infrastructure, policies for efficient water resource management, and innovative solutions for recycling and reuse. It is therefore imperative to bridge this gap which is crucial for sustainable urban development, public health and natural resources. (Kamyotra et al., 2011).

In India, waste management not only deals with water but also solid waste management and is generally done together without any segregation; either they are dumped in pits or sent through purifiers to be made barely good enough to be dumped into lakes and rivers without any legal violations, industries in India chose the river dumping option as it is cheaper and more economical for them, it also helps them reduce storage space as it is a quicker method than collecting and transporting dump. The municipal authorities cannot interfere in such matters due to the legal provisions. Still, due to improper maintenance and regular check-ups of these plants, the quality of purified wastewater gets further degraded over the years due to wear and tear of machinery, leading to unchecked waste disposal with no monitoring done by authorities.

11.3.1 Wastewater (Reuse) Management in India

Water is an inherent part of life and critical in agriculture, manufacturing, and other human uses. In the context of India, as water deficiency is growing to be one of the major problems of the country, proper utilization of wastewater is a challenge (Chaplin, 2001). Population growth, urbanization, industrialization, and increased water consumption have aggravated water stress in the country and significantly impacted water security and environmental challenges (Soni et al., 2024). Sanitation is gradually being appreciated as one of the solutions to the problems accompanying the depletion of freshwater sources (Küfeoğlu, 2022). Wastewater is used in industries, homes, commercial buildings, and agriculture. It contains compounds and microorganisms which, if not well treated, are disastrous to human health and the environment (Dutta et al., 2021). Nevertheless, development in treatment technologies and health care measures incorporating standard guidelines has ensured wastewater reuse becomes the solution to these effects (Ealias et al., 2024). By adopting water reuse and recycling, freshwater is saved and added to the advantage of economic effectiveness, reducing negative impacts on the environment and public health, making it possible in a country like India (Kumar & Goyal 2020). However, India's freshwater resources are under severe stress due to increased population, unpredictable monsoons, and water-intensive agriculture. Per capita water availability has been progressively declining over the years, and millions of people, including rural, urban and peri-urban dwellers, are struggling to access portable water sources (Prabha et al., 2020a, 2020b). Research suggests that nearly 70% of water in India is unsafe due to poor treatment of Wastewater and unsanitary standards. Due to such concerns, the Indian government has focused on wastewater management, and several steps have been incorporated to enhance the treatment and utilization of Wastewater (Bhattacharya et al., 2022). However, despite these measures, the country still has some difficulties in Wastewater management, such as treatment problems, lack of public awareness, and weak implementation of legislation. As a result, most large cities and other urban areas face the challenge of untreated sewage and the discharge of dangerous effluent into rivers and water bodies (Jadeja et al., 2022a, 2022b).

Wastewater reuse in India cannot be overemphasized because of the growing water scarcity in the country. When adequately treated, Wastewater can be used for several purposes, including agriculture, dust control, recreation, industrial cooling and non-portable domestic uses such as use in latrine systems and washing of clothes (Kesari et al., 2021). As the demand increases, the water availability in urban areas increases, and because the index of fresh water sources is low, the use of treated Wastewater has become inevitable. There are many benefits if Wastewater is to be reused. It includes pressure off the freshwater resources, pressure off the cost of water supply systems, and pressure off pollution at large (Prabha et al., 2020a, 2020b). In addition, applying Wastewater in the agriculture sector may improve the yields in areas characterized by water rationing (Mishra et al., 2023). The rich nutrient content in treated water acts as a fertilizer that helps farmers without necessarily applying more inorganic chemicals. Consequently, Wastewater reuse is an efficient solution to water shortage and environmental pollution problems (Ofori et al., 2021). Several legal and policy formulations guide Wastewater management, though the key concern is compliance with existing provisions. The statutes of water pollutants include but are not limited to the Environmental Protection Act of 1986 and the Water (Prevention and Control of Pollution) Act of 1974 (Meher, 2023). These laws require the pre-treatment of Wastewater before discharge into water bodies according to the Central Pollution Control Board (CPCB) and State Pollution Control Board (SPCB's) standards. Revised the National Water Policy periodically focuses its activities on Wastewater treatment and reuse as one of the key issues of water saving (Ballal et al., 2021). However, there is a strong contradiction regarding how policies are enacted. Although the situation has improved to some extent concerning the construction of Wastewater treatment facilities, mainly in big cities, the Wastewater treatment problem in small towns and rural areas largely remains unsolved (Varma et al., 2022). In addition, there is still no coherent national policy for wastewater reuse, which has led to different situations across the states, where some are in a better position than others (Kumar & Tortajada 2020).

11.4 Legal and Policy Framework

Due to the vast and complex geographical and socio-economic structure, India suffers from numerous problems meeting its water needs. These problems are aggravated by rapid urbanization, industrialization, and population growth (Sharma et al., 2024). The increasing need for water and decreased availability of freshwater resources makes the utilization of Wastewater a matter of concern. Sewerage has emerged as more critical as it is a determinant of public health, the state of the environment, and the availability of water (Pani et al., 2021). Legal and policy provisions have largely influenced the current state of wastewater management in India under the legal instruments for abating water pollution and efficient use of water resources (Jadeja et al., 2022a, 2022b). Nonetheless, the available regulations and efforts, such as smart water recycling policies, proper regulatory provisions, and execution mechanisms, have not been effectively implemented and enforced, making the effectiveness of the frameworks of wastewater management in India significantly unsatisfactory (Sahu, 2018). Thus, any evaluation of barriers and possibilities for wastewater reuse in the country requires a clear insight into legislation and policies guiding wastewater management in India (Morris et al., 2021). The basis of wastewater management in India is a strong black letter law, which comprises a set of several relevant rules and standards for the effective prevention of water pollution, treatment, and disposal of wastewater in the country (Kumar & Tortajada 2020). The Water (Prevention and Control of Pollution) Act is among the most important for effective prevention of water pollution enacted in 1974. This legislation was passed to contain and manage water pollution in India (Yadav, 2016). According to this Act, the CPCB and the SPCBs were also authorized to prescribe the effluent discharge standards and the extent of their enforcement. The Act states that all industries and municipal corporations must undertake Wastewater Treatment Plants to release Wastewater into rivers, lakes, and other water sources (Khanam et al., 2023). Also, the Act creates provisions for the setup of effluent treatment plants (ETPs) within industrial areas, so that treated Wastewater is treated as per the set standards.

Water pollution control is also a part of India's broader body of legislation, which includes the Environmental Protection Act of 1986. This Act extends the power of the central government to make guidelines for the protection of the environment, and it also requires specifying standards for many pollutants, such as the one in Wastewater (Jadeja et al., 2022a, 2022b). In addition, the Act also provided for corrective measures in case the stipulated environmental measures are not observed. Also, the Air (Prevention and Control of Pollution) Act of 1981 supports the Water

Act since it also deals with the air pollution problem, which is always associated with industrial effluent discharge that impacts the quality of water sources (Khanam et al., 2023). These national regulations focus on the utilization and reclamation of Wastewater as a significant strategy in addressing industrial and domestic effluent disposal. But even in these cases, some legal and policy frameworks do not address the issues concerning Wastewater reuse at the national level (Kesari et al., 2021). However, there are no regulations on the recycling and reuse of Wastewater that compel a random approach to their management. It gives rise to an appalling situation in that each state adopts its policies, rules, and regulations, hence practice disparities. Some of the above states, like Gujarat and Maharashtra, have developed structures for managing Wastewater and its reuse due to the differences in political will, and financial and institutional capabilities of other states (Dubey et al., 2024). If Wastewater is not well treated, it will have more hazardous impacts on the health and environment in general. Untreated Wastewater contains pathogens, heavy metals, chemical substances, and other dangerous substances that are lethal to human health because they pollute drinking water sources (Jan et al., 2022). To eliminate these risks, CPCB has put in place effluent discharge standards, which are procedures that determine levels of pollutants may be released in effluent. These standards strive to help verify that the Wastewater Treatment Plants efficiently remove those contaminants before discharge of the treated water into water bodies or its reuse (Rajaram et al., 2008). Key quality indicators are indicated by designators like biochemical oxygen demand (BOD), chemical oxygen demand (COD), total suspended solids (TSSs), and faecal coliform. Observance of these standards is an effective procedure for achieving adequate protection of the health of the public and the environment. However, efforts to implement the standards continue to face several hurdles because of poor monitoring and compliance, especially in remote areas (Jouanneau et al., 2014).

Besides these legal requirements, several governmental regulations and actions have been implemented and adopted regarding wastewater management and reusing of wastewater. At the national level, the National Water Policy, first developed in 1987, with periodic review, enunciates the concept of integrated water resources management, wastewater treatment, and reuse (Khara, 2023). The National Water Development Agency (NWDA) urges the establishment of wastewater treatment plants in urban and industrial areas to promote the reuse of treated wastewater for non-portable uses, such as irrigation, industrial cooling, and landscaping. It also supports efficient technologies in water use to minimize consumption and prevent waste product formation (Maura et al., 2023).

The Swachh Bharat Abhiyan (Clean India Mission) launched in 2014 adds to the country's emphasis on sanitation and effective wastewater management. At the same time, the prepared and conducted Mission is aimed at solid waste management and sanitation, but it has affected the quality of wastewater, especially sewage (Hutton et al., 2020). The Smart Cities Mission of development of 100 smart cities in India also mentions the sustainable and future wastewater treatment and reuse system along the modern structure of SMART cities. Likewise, the AMRUT scheme's goals include improving urban infrastructure, such as wastewater, by developing and upgrading Smart Treatment Plants and recycling treated sewerage (Prasad et al., 2020). In the case of Wastewater management, state governments and local authorities play a crucial role in enforcing the national policies and regulations that broadly fall at their disposal. Although, the legal and policy measures are set at the central level, state governments and local authorities are responsible for the development of supporting structures, as well as the control of compliance with established principles and the correct use of treated Wastewater (Breitenmoser et al., 2022). While some states have effectively implemented the option, several others are struggling with infrastructure, lack of funds, and a dearth of skilled professionals to implement the reuse programs effectively. Municipal corporations and panchayats are directly associated with community-level wastewater management. These are designing and managing sewage plants, regulating effluent discharge, and advocating for recycling and reuse (Varma et al., 2022). On the other hand, the actual level of effective wastewater management is hampered by inadequate resources, poor political will, and intergovernmental disunity.

11.5 Challenges in Wastewater Reuse

The prospects for using wastewater in India are enormous, but many factors restrict its utilization. Due to these factors, techno-structural, regulatory, financial and social constraints must be overcome to effectively implement wastewater reuse (Al-Hazmi et al., 2023). The mentioned and described benefits include water conservation, pollution control, and cost control. Nonetheless, numerous challenges arise that bar the realization of these gains, which can be understood from the technological, policy, awareness, and funding perspectives (Pani et al., 2021). Some of the main obstacles to wastewater reuse include poor technology and the unavailability of proper infrastructure to reclaim wastewater to a level that meets reuse requirements. Technologies implemented in the management and treatment of Wastewater are complicated and costly to install and maintain (Kumar & Goyal 2020). In India, the majority of Water Treatment Plants are constrained by obsolete or inadequate technologies that slow their ability to treat Wastewater to a sufficient standard for reuse in agricultural production, industrial processes, or urban areas. Furthermore, the treatment capacity of so many treatment facilities is still limited in accommodating the increasing inflows of wastewater, especially in metropolitan areas (Varma et al., 2022). The absence of infrastructure also contributes to low utilization of existing treatment facilities because the treated water is not properly used for reuse. Another challenge is the absence of proper pipelines, storage, and distribution systems to support the effective use of treated Wastewater in non-potable uses (Kesari et al., 2021). Furthermore, the scale and structure of Wastewater facilities in India are dispersed in a highly distributed network. Establishing and managing centralized Wastewater treatment neglects decentralized options that might benefit rural areas or small cities. It led to inequitable distribution of Wastewater Treatment that gave birth to significant disparities in the country's water, sanitation, and hygiene services with large disparities in the country (Varma et al., 2022). Moreover, the application of new treatment processes like membrane bioreactors and reverse osmosis has not yet expanded because of the high installation and operation costs. Therefore, the absence of modern treatment technologies, irrespective of their availability worldwide, afflicts large-scale utilization of Wastewater in India (El-Sheekh et al., 2021).

The other significant issue relates to policy and legal frameworks of water recycling and reuse in developed economies. Though the Water (Prevention and Control of Pollution) Act and the Environmental Protection Act are at the national and state levels to deal with water pollution, no practicable regulatory standards support wastewater reuse (Kumar & Tortajada 2020). The lack of a codified and coherent general policy on Wastewater reuse in the United States has also resulted in the phenomenon being approached piecemeal across the different states that differ in their commitment to the issue and legal authorities (Malisa-van der Walt et al., 2022). For instance, some states have already developed policies for using reclaimed water in agriculture, but such policies are lacking in some other states. These regulations and inconsistencies pose problems to industries, municipalities, and consumers desiring to invest in or practice wastewater reuse (Kumar & Goyal 2020). However, the discharge standards of treated Wastewater are often incompatible with international benchmarks, making it unclear regarding the permissible quality of water for reuse purposes. Inadequate measures in the enforcement and monitoring occasions complement the lack of a closed and stringent regulatory structure. The local pollution control boards and municipalities responsible for implementing and enforcing environmental standards are often overcome with technical inaptitude, inadequate funding, and political apathy to discharge requisite obligations in managing and reusing Wastewater proficiently (Awewomom et al., 2024). This leads to poor supervision of effluent discharges and violation of wastewater treatment regulations. Moreover, there is little reward or punishment for individual industries and municipalities to conserve water and reuse Wastewater (Kesari et al., 2021). The problem associated with it is the absence of standard practices for wastewater recycling and the absence of economic motives to invest in such systems. Again, the most significant factors that impact wastewater reuse programs in India are the knowledge, awareness and acceptance of the general society. The people, especially the rural population, view treated sewage as generally unsafe to reuse regardless of modern technological interventions that make the treated water safe and suitable for reuse (Dubey et al., 2024). Such challenges, where many people still do not understand the benefits of using treated Wastewater and feel that Wastewater reuse poses health risks to human beings, create a lot of hindrances. Even in urban areas, people have not embraced recycled water in applications not involved with drinking water. For example, there is still a tabernacle in irrigation, cooling, and agriculture (Lee et al., 2020). Addressing this level of public skepticism would, therefore, require very aggressive public enlightenment campaigns addressing issues of safety, utility, and the need to recycle Wastewater for use, especially with the increasing commodity scarcity. Besides, awareness is closely associated with trust in authorities who regulate and treat Wastewater or dispose of it respectively. Historically, the procedures of treating and monitoring water have not been clearly explained in detail, and the government has had little interaction with the public regarding the reuse of Wastewater (Bhattacharya et al., 2022). Therefore, there should be increased involvement of the people, especially in the decision-making of water conservation and wastewater management, thus making them more responsible.

Finally, financial and economic limitations impede the growth of more Wastewater reuse in the country. Some of the main challenges of Wastewater treatment include the financial costs of constructing and maintaining a wastewater treatment system as well as establishing wastewater recycling systems (Morris et al., 2021). Many of the municipalities and industrial segments in India experience challenges in financing such enormous projects, especially when competing with the business needs regarding financing options. The high cost of innovative tertiary treatments like reverse osmosis and membrane filtration makes the overall cost of Wastewater reuse even more expensive (Prasad et al., 2024). Small and medium industries will likely find the capital to invest in effluent treatment or recycling machinery, and many discharge raw or partially treated effluent. Also, wastewater reuse is another process that requires economic feasibility, whereby treated water should be marketable or appropriately used economically. Therefore, it calls for the formulation of price structures that will fit the value of treated water and encourage industries to adopt the practice of reuse of Wastewater (Kesari et al., 2021). Nevertheless, there are no established markets for treated Wastewater, let alone demand for it, and more importantly, no subsidies or financial incentives to encourage investment in Wastewater recycling. The costs of implementing Wastewater reuse projects are high, and the payback periods are long, and facts show that it may discourage public and private sector participation.

11.6 Conclusion

The management and reuse of wastewater in India continue to be critical to tackling the increasing problems of water shortages, environmental pollution, and increased needs for urban and industrial development. This chapter shows that although the application of Wastewater is highly possible, its implementation is accompanied by several complex issues. Nonetheless, it is possible to solve these challenges integrally by applying technologies, improving legal systems, offering public awareness, and popularizing Wastewater reuse practices. The results of this study stress the importance of enhancing technology acquisition and infrastructure improvement to address the rising demand for wastewater treatment and recycling. Current and existing water treatment plants are not well equipped or designed to meet the required quality level for reuse. Communities still need dialogue regarding extending its use and the perceived risk of utilizing treated Wastewater. They also include financial constraints in funding wastewater reuse projects by both the private and the public sectors.

Responding to these gaps involves a positive approach that is legal, technological, financial, and social on the policy level, seeking a well-coordinated and integrated national regulation of Wastewater treatment and its further use in different spheres. It would bring clear and consistent rules for the stakeholders to follow and work to improve the equal distribution of treated sewage water to those people who need it. Such a framework should include both the centralized as well as the decentralized management of Wastewater to satisfy the variety of demands of urban and rural structures. Minimum standards that conform with global benchmarks should be implemented to enable the safe discharge of Wastewater for activities like agricultural use, cooling in industries, and greening of cities. However, the technological agenda is equally important, which requires major investment for the necessary technological infrastructure. Governments should encourage Public–Private Partnerships (PPPs) to invest in the innovations of efficient treatment technologies, inclusive of membrane filtration and reverse osmosis, to produce good quality treated water. It was postulated that subsidies or tax incentives might be used to lower the initial expenses of Wastewater treatment and reuse projects. However, decentralized treatment can potentially decrease the load of the central plants, especially in rural and semi-urban areas, with great chances of proper wastewater reuse at a lower price. More emphasis must be laid on the enhancement of public understanding and acceptance of wastewater reuse. Public awareness is required to eliminate misconceptions that treated Wastewater is unsafe to consume or use erroneously which holds back technology and standards of water treatment. Stakeholders should be involved in Wastewater reuse regularly to pass the information on to the authorities, industries, and the public. The inclusion and engagement of all stakeholders in organizational planning and development will help in achieving trust while enhancing the implementation of water conservation and reuse projects.

For future research, there is a need for more work carried out to explore advanced treatment technologies in wastewater treatment to enhance their efficacy and reduce costs. In addition, more research is required to identify the economic benefits of using Wastewater in agriculture, industry, and urban areas in the long run. Studying the effects of using Wastewater, especially regarding reuse possibilities for soil and water, will supply future restrictions. Concerning practice, legal and institutional framework conditions must be stronger at the national and state level. For the lasting success of Wastewater management in India, a rational, easily enforceable body of rules that encourages the reuse of Wastewater, provides safety to the public, and invites investments will be crucial. Further endeavours should be made towards enhancing the capacity of local governments and agencies to enhance their utilization of these frameworks.

References

A Balasubramanian (2016). Coal and Mining Methods. Report No. 1, University of Mysore. https://doi.org/10.13140/RG.2.2.19117.08162

Al-Hazmi, H. E., Mohammadi, A., Hejna, A., Majtacz, J., Esmaeili, A., Habibzadeh, S., Saeb, M. R., Badawi, M., Lima, E. C., & Mąkinia, J. (2023). Wastewater reuse in agriculture: Prospects and challenges. *Environmental Research, 236*(Part 1), 116711. https://doi.org/10.1016/j.envres.2023.116711 , https://pubmed.ncbi.nlm.nih.gov/37487927/

Al-Tohamy, R., Ali, S. S., Li, F., Okasha, K. M., Mahmoud, Y. A. G., Elsamahy, T., Jiao, H., Fu, Y., & Sun, J. (2022). A critical review on the treatment of dye-containing wastewater: Ecotoxicological and health concerns of textile dyes and possible remediation approaches for environmental safety. *Ecotoxicology and Environmental Safety, 231*, 113–160. https://doi.org/10.1016/j.ecoenv.2021.113160

Aswini, R., & Kannan, K. (2022). *Bismuth-based materials for environmental remediation: Defects in bismuth series for the treatment of environmental pollutants.* IOP Publishing, 9(1), 1–20. https://iopscience.iop.org/book/edit/978-0-7503-5138-6/chapter/bk978-0-7503-5138-6ch9.epub

Aswini, R., Jothimani, K., & Kannan, K. (2024). *Carica papaya* leaf-infused metal oxide nanocomposite: A green approach towards water treatment and antibacterial applications. *Environmental Geochemistry and Health, 46*, 334. https://doi.org/10.1007/s10653-024-02090-4.

Atangana, E., Oberholster, P. J., & Turton, A. R. (2021). Will the extraction of COVID-19 from wastewater help flatten the curve? *Chemosphere, 271*, Article 129429. https://doi.org/10.1016/j.chemosphere.2020.129429

Awewomom, J., Dzeble, F., Takyi, Y. D., et al. (2024). Addressing global environmental pollution using environmental control techniques: A focus on environmental policy and preventive environmental management. *Discover Environmental, 2*, 8. https://doi.org/10.1007/s44274-024-00033-5

Ballal, A., Tambe, S., & Joe, E. T. (2021). The evolution of India's environmental policy. *Indian Forests, 147*(8). (PDF) The Evolution of India's Environmental Policy

Bhattacharya, A. K., & Roy, S. N. (2022). Wastewater management and treatment technologies with recycling and reuse issues in India leading to zero liquid discharge (ZLD). In S. Yadav, A.M. Negm, & R.N. Yadava (Eds.), *Wastewater assessment, treatment, reuse and development in India* (pp. 125–157). Earth and Environmental Sciences Library. Springer. https://doi.org/10.1007/978-3-030-95786-5_7

Breitenmoser, L., Cuadrado Quesada, G., N, A., Bassi, N., Dkhar, N. B., Phukan, M., Kumar, S., Naga Babu, A., Kierstein, A., Campling, P., & Hooijmans, C. M. (2022). Perceived drivers and barriers in the governance of wastewater treatment and reuse in India: Insights from a two-round Delphi study. *Resources, Conservation and Recycling, 182*, 106285. https://doi.org/10.1016/j.resconrec.2022.106285

Burge, S. R., Hristovski, K. D., Burge, R. G., Hoffman, D. A., Saboe, D., Chao, P., Taylor, E., & Koenigsberg, S. S. (2020). Microbial potentiometric sensor: A new approach to longstanding challenges. *Science of the Total Environment, 742*, Article 140528. https://doi.org/10.1016/j.scitotenv.2020.140528

Chaplin, M. F. (2001). Water: Its importance to life. *Biochemistry and Molecular Biology Education, 29*(2), 54–59. https://doi.org/10.1111/j.1539-3429.2001.tb00070.x

Colella, M., Ripa, M., Cocozza, A., Panfilo, C., & Ulgiati, S. (2021). Challenges and opportunities for more efficient water use and circular wastewater management: The case of Campania Region, Italy. *Journal of Environmental Management, 297*, 113171https://doi.org/10.1016/j.jenvman.2021.113171

Compton, M., Willis, S., Rezaie, B., & Humes, K. (2018). Food processing industry energy and water consumption in the Pacific northwest. *Innovat. Food Sci. Emerg. Technol., 47*, 371–383. https://doi.org/10.1016/j.ifset.2018.04.001

Devold, H. (2008). *Oil and gas production handbook: An introduction to oil and gas production* (1st ed.). Oil and gas production handbook ed3x0_web.pdf

Dubey, D., Kumar, S., & Dutta, V. (2024). A circular economy approach for sustainable water reuses in India: Policies, practices, and future prospects. *Environmental Sustainability, 7*, 303–323. https://doi.org/10.1007/s42398-024-00321-z

Dutta, D., Arya, S., & Kumar, S. (2021). Industrial wastewater treatment: Current trends, bottlenecks, and best practices. *Chemosphere, 285*, Article 131245. https://doi.org/10.1016/j.chemosphere.2021.131245

Dwivedi, K. A., Huang, S.-J., & Wang, C.-T. (2022). Integration of various technology-based approaches for enhancing the performance of microbial fuel cell technology: A review. *Chemosphere, 287*, 132–248. https://doi.org/10.1016/j.chemosphere.2021.132248

Ealias, M. A., Meda, G., & Tanzil, K. (2024). Recent progress in sustainable treatment technologies for the removal of emerging contaminants from Wastewater: A review on occurrence, global status, and impact on biota. *Reviews of Environmental Contamination, 262*, 16. https://doi.org/10.1007/s44169-024-00067-z

El-Sheekh, M. M., Ibrahim, H. A. H., Amer, M. S., & Ali, E. M. (2021). Wastewater treatment by membrane bioreactor as potent and advanced technology. In M. P. Shah & S. Rodriguez-Couto (Eds.), *Membrane-based hybrid processes for wastewater treatment*, (pp. 45–72). Elsevier. https://doi.org/10.1016/B978-0-12-823804-2.00012-4

Fillaudeau, L., Bories, A., Decloux, M. (2008). Brewing, winemaking and distilling: overview of wastewater treatment and utilisation schemes. Book: Handbook of Water and Energy Management in Food Processing https://doi.org/10.1533/9781845694678.6.929

Fthenakis, V., & Kim, H. C. (2010). Life-cycle uses of water in US electricity generation. *Renewable and Sustainable Energy Reviews, 14*(7), 2039–2048. https://doi.org/10.1016/j.rser.2010.03.008

Gingerich, S. B., Voss, C. I., & Johnson, A. G. (2017). Seawater-flooding events and impact on freshwater lenses of low-lying islands: Controlling factors, basic management and mitigation. *Journal of Hydrology, 551*, 676–688. https://doi.org/10.1016/j.jhydrol.2017.03.001

Gupta, V., Chandra, N., Tari, V., & Kannan, K. (2024). A sustainable approach for the green synthesis of silver nanoparticles using *Brassica oleracea* subsp. *botrytis* (L.) leaves and its antibacterial activity. *Physics and Chemistry of Solids, 25*(2), 303–310. https://doi.org/10.15330/pcss.25.2.303-310

Hutton, G., Patil, S., Kumar, A., Osbert, N., & Odhiambo, F. (2020). Comparison of the costs and benefits of the Clean India Mission. *World Development, 134*, Article 105052. https://doi.org/10.1016/j.worlddev.2020.105052

Jadeja, N. B., Banerji, T., Kapley, A., & Kumar, R. (2022a). Water pollution in India—Current scenario. *Water Security, 16*, Article 100119. https://doi.org/10.1016/j.wasec.2022.100119

Jadeja, N. B., Banerji, T., et al. (2022b). Water pollution in India—Current scenario. *Water Security, 16*, Article 100119. https://doi.org/10.1016/j.wasec.2022.100119

Speight, J. G. (2019). Handbook of petrochemical processes. CRC Press Taylor & Francis Group 6000 Broken Sound Parkway NW, Suite 300 Boca Raton, FL 33487–2742. LC record available at https://doi.org/10.1201/9780429155611

Jan, I., Ahmad, T., Wani, M. S., Dar, S. A., Wani, N. A., Malik, N. A., & Tantary, Y. R. (2022). Chapter 1—Threats and consequences of untreated Wastewater on freshwater environments. In G. H. Dar, R. A. Bhat, H. Qadri, & K. R. Hakeem (Eds.), *Advances in environmental pollution research: microbial consortium and biotransformation for pollution decontamination* (pp. 1–26). Elsevier. https://doi.org/10.1016/B978-0-323-91893-0.00009-2

Jouanneau, S., Recoules, L., Durand, M. J., Boukabache, A., Picot, V., Primault, Y., Lakel, A., Sengelin, M., Barillon, B., & Thouand, G. (2014). Methods for assessing biochemical oxygen demand (BOD): A review. *Water Research, 49*, 62–82. https://doi.org/10.1016/j.watres.2013.10.066

Kamyotra, J. S., & Bhardwaj, R. M. (2011). *Municipal wastewater management in India.* In *India infrastructure report 2011: Water: Policy and performance for sustainable development* (p. 299). Oxford University Press.

Kesari, K. K., Soni, R., Jamal, Q. M. S., et al. (2021). Wastewater treatment and reuse: A review of its applications and health implications. *Water, Air, & Soil Pollution, 232*, 208. https://doi.org/10.1007/s11270-021-05154-8

Khanam, Z., Sultana, F. M., & Mushtaq, F. (2023). Environmental pollution control measures and strategies: An overview of recent developments. In F. Mushtaq, M. Farooq, A. B. Mukherjee, &

M. Ghosh Nee Lala (Eds.), *Geospatial analytics for environmental pollution modelling* (pp. 385–414). Springer. https://doi.org/10.1007/978-3-031-45300-7_15

Khani, M. R., Mahdizadeh, H., Kannan, K., Kalankesh, L. R., Kamarehei, B., Baneshi, M. M., & Shahamat, Y. D. (2020). Olive mill wastewater (OMW) treatment by hybrid processes of electrocoagulation/catalytic ozonation and biodegradation. *Environmental Engineering and Management Journal, 19*(8), 1401–1410. https://eemj.eu/index.php/EEMJ/article/view/4155

Khara, D. S. (2023). Groundwater governance in India: A legal and institutional perspective. *Indian Journal of Public Administration, 69*(1), 204–220. https://doi.org/10.1177/00195561221128618

Küfeoğlu, S. (2022). SDG-6 clean water and sanitation. In *Emerging technologies* (pp. 289–304). Springer, Cham. https://doi.org/10.1007/978-3-031-07127-0_8

Kumar, A., & Goyal, K. (2020). Chapter two—Water reuse in India: Current perspective and future potential. In P. Verlicchi (Ed.), *Advances in Chemical Pollution, Environmental Management and Protection* (Vol. 6, pp. 33–63). Elsevier. https://doi.org/10.1016/bs.apmp.2020.07.011

Kumar, M. D., & Tortajada, C. (2020). The legal, institutional and policy regime for the control of water pollution in India. In *Assessing wastewater management in India*, 7–12. Springer Briefs in Water Science and Technology. Springer. https://doi.org/10.1007/978-981-15-2396-0_2

Lakkaboyana, S. K., Soontarapa, K., Kale, A. N., Reddy, A. S., Kumar, V., Karthik, K., Hanafiah, M. M., & Wan Zuhairi, W. Y. (2022). Synthesis of AC@CuO-NWs and removal of basic dye from wastewater. *Materials Today: Proceedings, 53*(Part 3), 336–338 https://doi.org/10.1016/j.matpr.2022.01.364

Lee, K., & Jepson, W. (2020). Drivers and barriers to urban water reuse: A systematic review. *Water Security, 11*, Article 100073. https://doi.org/10.1016/j.wasec.2020.100073

Li, J., See, K. F., & Chi, J. (2019). Water resources and water pollution emissions in China's industrial sector: A green-biased technological progress analysis. *Journal of Cleaner Production, 229*, 1412–1426. https://doi.org/10.1016/j.jclepro.2019.03.216

Maiti, S. K. (2004). *Handbook of methods in environmental studies: Vol. 1. Water and wastewater analysis* (2nd ed.). ABD Publishers.

Malakootian, M., Kannan, K., Gharaghani, M. A., Dehdarirad, A., Nasiri, A., Shahamat, Y. D., & Mahdizadeh, H. (2019). Removal of metronidazole from wastewater by Fe/charcoal micro electrolysis fluidized bed reactor. *Journal of Environmental Chemical Engineering, 7*(6), Article 103457. https://doi.org/10.1016/j.jece.2019.103457

Malisa-van der Walt, R., & Taigbenu, A. (2022). Policy, laws, and guidelines of wastewater reuse for agricultural purposes in developing countries. In M. Nasr, & A. M. Negm (Eds.), *Cost-efficient wastewater treatment technologies* (Vol. 117, pp. 1–20). The Handbook of Environmental Chemistry. Springer. https://doi.org/10.1007/698_2022_864

Man, Y., Han, Y., Wang, Y., Li, J., Chen, L., Qian, Y., & Hong, M. (2018). Woods to goods: Water consumption analysis for papermaking industry in China. *Journal of Cleaner Production, 195*, 1377–1388. https://doi.org/10.1016/j.jclepro.2018.06.026

Maura, J., Atreya, S., & Arshi, A. (2023). The treatment of Wastewater, recycling and reuse—Past, present, and in the future. *International Journal of Science and Research (IJSR), 12*(11). https://doi.org/10.21275/SR231013064713

Meher, A. (2023). Revisiting environmental policy in India: An analysis of structure, process, and institution. *International Journal of Political Science and Governance, 5*(1), 188–197. https://doi.org/10.33545/26646021.2023.v5.i1c.222

Mishra, S., Kumar, R., & Kumar, M. (2023). Use of treated sewage or Wastewater as an irrigation water for agricultural purposes—Environmental, health, and economic impacts. *Total Environment Research Themes, 6*, Article 100051. https://doi.org/10.1016/j.totert.2023.100051

Morris, J. C., Georgiou, I., Guenther, E., et al. (2021). Barriers in implementation of wastewater reuse: Identifying the way forward in closing the loop. *Circular Economy and Sustainability, 1*(3), 413–433. https://doi.org/10.1007/s43615-021-00018-z

Norgate, T., & Haque, N. (2010). Energy and greenhouse gas impacts of mining and mineral processing operations. *Journal of Cleaner Production, 18*(3), 266–274. https://doi.org/10.1016/j.jclepro.2009.09.020

Ofori, S., Puškáčová, A., Růžičková, I., & Wanner, J. (2021). Treated wastewater reuse for irrigation: Pros and cons. *Science of the Total Environment, 760*, Article 144026. https://doi.org/10.1016/j.scitotenv.2020.144026

Ortúzar, M., Esterhuizen, M., Olic´on-Hern´andez, D.R., Gonz´alez-L´opez, J., Aranda, E. (2022). Pharmaceutical pollution in aquatic environments: a concise review of environmental impacts and bioremediation systems. *Frontiers in Microbiology, 13*, 869332 https://doi.org/10.3389/fmicb.2022.869332

Palani, G., Arputhalatha, A., Kannan, K., Lakkaboyana, S. K., Hanafiah, M. M., Kumar, V., & Marella, R. K. (2021). Current trends in the application of nanomaterials for the removal of pollutants from industrial wastewater treatment—A review. *Molecules, 26*(9), 2799. https://doi.org/10.3390/molecules26092799

Palani, G., Trilaksana, H., Sujatha, R. M., Kannan, K., Rajendran, S., Korniejenko, K., Nykiel, M., & Uthayakumar, M. (2023). Silver nanoparticles for waste water management. *Molecules, 28*(8), 3520. https://doi.org/10.3390/molecules28083520

Pani, A., Ghatak, I., & Mishra, P. (2021). Understanding the water conservation and management in India: An integrated study. *Sustainable Water Resources Management, 7*, 77. https://doi.org/10.1007/s40899-021-00556-2

Patterson, J. W. (1985). *Industrial wastewater treatment technology.* U.S. Department of Energy, Office of Scientific and Technical Information. https://www.osti.gov/biblio/7253209

Prabha, S. A., Ram, A., & Begum Irfan, Z. (2020a). Exploring the relative water scarcity across the Indian million-plus urban agglomerations: An application of the Water Poverty Index. *HydroResearch, 3*, 134–145. https://doi.org/10.1016/j.hydres.2020.10.001

Prabha, A. S., Ram, A., & Irfan, Z. B. (2020b). Exploring the relative water scarcity across the Indian million-plus urban agglomerations: An application of the Water Poverty Index. *HydroResearch, 3*, 134–145. https://doi.org/10.1016/j.hydres.2020.10.001

Prasad, D., & Alizadeh, T. (2020). What makes Indian cities smart? A policy analysis of Smart Cities Mission. *Telematics and Informatics, 55*, Article 101466. https://doi.org/10.1016/j.tele.2020.101466

Prasad, D., Alizadeh, T., & Dowling, R. (2024). Smart city planning and the challenges of informality in India. *Dialogues in Human Geography, 14*(3), 385–402. https://doi.org/10.1177/20438206231156655

Princeton University. (1989). Technologies for Reducing Dioxin in the Manufacture of Bleached Wood Pulp (Part 4). Chapter 2: The Pulp and Paper Making Processes. https://www.princeton.edu/~ota/disk1/1989/8931/893104.PDF. pp. 17–25. Retrieved on 06 December 2024

Rahman, T. U., Roy, H., Islam, M. R., Tahmid, M., Fariha, A., Mazumder, A., Tasnim, N., Pervez, M. N., Cai, Y., Naddeo, V., & Islam, M. S. (2023). The advancement in membrane bioreactor (MBR) technology toward sustainable industrial wastewater management. *Membranes, 13*(2), 181. https://doi.org/10.3390/membranes13020181

Rajaram, T., & Das, A. (2008). Water pollution by industrial effluents in India: Discharge scenarios and case for participatory ecosystem specific local regulation. *Futures, 40*(1), 56–69. https://doi.org/10.1016/j.futures.2007.06.002

Sahu, G. (2018). Rethinking environmental regulation in India. In M. Sekher, S. Parasuraman, & R. Kattumuri (Eds.), *Governance and governed* (pp. 129–153). Springer. https://doi.org/10.1007/978-981-10-5963-6_8

Sharma, M., Kumar, V., & Kumar, S. (2024). A systematic review of urban sprawl and land use/land cover change studies in India. *Sustainable Environment, 10*(1). https://doi.org/10.1080/27658511.2024.2331269

Siddiqui, N., Faiyaz, T., & Tari, V. (2022). Ganga and Yamuna: Through the lens of National Green Tribunal. *Nature, Environment and Pollution Technology, 21*(4), 1657–1664. https://doi.org/10.46488/NEPT.2022.v21i04.018.

Singh, R. P., Singh, P. K., Gupta, R., & Singh, R. L. (2019). Treatment and recycling of wastewater from textile industry. In R. L. Singh (Ed.), *Advances in biological treatment of industrial waste water and their recycling for a sustainable future* (pp. [insert page numbers]). Springer. https://doi.org/10.1007/978-981-13-1468-1_8.

Sodhi, H. S., Singh, D., & Singh, B. J. (2020). A conceptual examination of lean, six sigma and lean six sigma models for managing waste in manufacturing SMEs. *World Journal of Science, Technology and Sustainable Development, 17*(1), 20–32. https://doi.org/10.1108/WJSTSD-10-2019-0073

Soni, H., Yadav, R. K., & Patra, S. K. (2024). Global impact of urbanization on ecosystems: A comprehensive bibliometric analysis. *Natural Hazards Research*. https://doi.org/10.1016/j.nhres.2024.04.001

Tari V. S. (2021). Radioactive pollution in drinking water. Short communication, *Acta Scientific Microbiology, 4*(6), 3–4 ISSN: 2581-3226. (28) Radioactive Pollution in Drinking Water

Tari, V., Gupta, R., & Siddiqui, N. (2021). Monitoring and evaluating the ecological health of wetlands. In *Impact of climate change on Upper Ganga Ramsar site, UP, India* (1st ed., pp. 92–105). IGI Global. https://www.igi-global.com/gateway/chapter/295742

Tari, V., Kannan, K., & Vishwakarma, V. (2022). Emerging antiviral technologies. In P. Nunthavarawong, S. M. Rangappa, S. Siengchin, & M. Thoppil-Mathew (Eds.), *Antimicrobial and antiviral materials: Polymers, metals, ceramics, and applications* (1st ed., pp. 137–150). CRC Press, Taylor & Francis Group. https://doi.org/10.1201/9781003143093-8.

Tari, V., Siddiqui, N., Rathi, D., Siddiqui, N. N., & Wahyuni, D. K. (2024). Need for an evolved groundwater justice in rural areas of Uttar Pradesh, India. *Nature, Environment and Pollution Technology, 23*(1), 125–137. https://doi.org/10.46488/NEPT.2024.v23i01.009

Uddin, F. (2019). Introductory chapter: Textile manufacturing processes. In F. Uddin (Ed.), Textile manufacturing processes (pp. 1–10). IntechOpen. https://doi.org/10.5772/intechopen.87968

Van Der Walt, R. M., & Taigbenu, A. (2022). Policy, laws, and guidelines of wastewater reuse for agricultural purposes in developing countries. In M. Nasr & A. M. Negm (Eds.), *Cost-efficient wastewater treatment technologies* (Vol. 117, pp. 1–20). The Handbook of Environmental Chemistry. Springer. https://doi.org/10.1007/698_2022_864

Varma, V. G., Jha, S., Raju, L. H. K., Kishore, R. L., & Ranjith, V. (2022). A review on decentralized wastewater treatment systems in India. *Chemosphere, 300*, Article 134462. https://doi.org/10.1016/j.chemosphere.2022.134462

Verstraete, W., Yanuka-Golub, K., Driesen, N., & De Vrieze, J. (2022). Engineering microbial technologies for environmental sustainability: Choices to make. *Microbial Biotechnology, 15*(1), 215–227. https://doi.org/10.1111/1751-7915.13986

Wahyuni, D. K., Atere, T. G., Kharisma, V. D., Tari, V., Rahmawati, C. T., Murtadlo, A. A. A., Syukriya, A. J., Nuha, G. A., Wacharasindhu, S., Prasongsuk, S., & Purnobasuki, H. (2024). Antimicrobial potentials of *Pandanus amaryllifolius* Roxb.: Phytochemical profiling, antioxidant, and molecular docking studies. *PLOS ONE, 19*(1). https://doi.org/10.1371/journal.pone.0305348

World Bank Group. (1998, July). *Pollution prevention and abatement handbook: Pharmaceuticals manufacturing*. World Bank. https://www.ifc.org/content/dam/ifc/doc/1990/pharma-ppah.pdf Retrieved on 06/12/2024.

Xi, H., Zhou, X., Arslan, M., Luo, Z., Wei, J., Wu, Z., & Gamal El-Din, M. (2022). Heterotrophic nitrification and aerobic denitrification process: Promising but a long way to go in the wastewater treatment. *Science of the Total Environment, 805*, Article 150212. https://doi.org/10.1016/j.scitotenv.2021.150212

Yadav, V. (2016). Measures for prevention of water pollution under the Water (Prevention and Control of Pollution) Act, 1974: A critical analysis. *International Journal of Law, 2*(4), 24–27. Available at SSRN: https://ssrn.com/abstract=3621193

Yusupov, Z., & Almaktar, M. (2021). Geothermal power generation. *Geothermal Energy*. https://doi.org/10.5772/intechopen.97423

Zan, F., Iqbal, A., Lu, X., Wu, X., & Chen, G. (2022). "Food waste–wastewater–energy/resource" nexus: Integrating food waste management with wastewater treatment towards urban sustainability. *Water Research, 211*, Article 118089. https://doi.org/10.1016/j.watres.2022.118089

Zou, Y., Li, Q., Tao, T., Ye, P., Zhang, P., & Liu, Y. (2021). Fe-Mn binary oxides activated aluminosilicate mineral and its Tl (I) removal by oxidation, precipitation and adsorption in aqueous. *Journal of Solid State Chemistry, 303*, Article 122383. https://doi.org/10.1016/j.jssc.2021.122383

Chapter 12
Economic and Social Impacts of Sustainable Water Management—Cost Benefit Analysis of Waste Water Recycling and Socio-economic Benefits

Naveen Gaurav, Pankaj Tripathi, Yunus Ali, Venus Dimri, and Love Gairola

Abstract Water is a crucial need for maintaining life which needs sustainable management practices to reduce its exploitation. Wastewater recycling is important to meet the demands of the public as well as protect waterbodies. This chapter focuses on the economic and social effects of sustainable water recycling and its socioeconomic effects. In economic scenario, sustainable water management decreases water pollution and increases agricultural production. With time, advanced technologies have promoted economically sustainable water usage and water quality which decreases the financial strain on various low-income regions. Sustainable water management and water scarcity. Access to clean and safe water is a fundamental human right. This chapter highlights the significance of wastewater recycling and sustainable water management. Communities will be more immune to climate change, prevent environmental exploitation and water shortages by following such recycling. In conclusion, recycling wastewater is an acceptable solution to financial, social and environmental issues related to water crisis. This sustainable way avoids the over-use of water resources and develop the communities in all aspects equally.

N. Gaurav (✉)
Department of Biotechnology, School of Basic and Applied Sciences, Shri Guru Ram Rai University, Dehradun, Uttarakhand, India
e-mail: naveengaurav@sgrru.ac.in

P. Tripathi
Institute of Plant Sciences, Agricultural Research Organization (ARO), Rishon LeZion, Israel
e-mail: pankajkt@volcani.agri.gov.il

Y. Ali
Department of Agricultural Sciences in the Tropics and Subtropics, University of Hohenheim, Stuttgart, Germany

V. Dimri · L. Gairola
Department of Microbiology, School of Basic and Applied Sciences, Shri Guru Ram Rai University, Dehradun, Uttarakhand, India

S. Dhiman and G. Mukherjee (eds.), *Resource Resurgence*, Springer Water,
https://doi.org/10.1007/978-3-032-02750-4_12

Keywords Wastewater recycling · Infrastructure · Sustainable management · Freshwater · Policies and strategies · Communities · ETC

12.1 Introduction

The world is encountering major water scarcity issue, contributed by several factors like climate change, population growth, unsustainable agricultural practices and increased industrial activity. It's crucial to implement sustainable water management strategies so as to lessen the exploitation of freshwater resources. Among these strategies, the recycling of wastewater is being preferred to ensure the efficient use of water resources. It has been predicted that by 2025 over 2.7 billion people will experience water shortages which will increase the need to implement sustainable water management solutions. Wastewater recycling or water reuse, suggests as a viable solution to this challenge addressing both water scarcity and water management. This problem can be countered by reducing reliance on freshwater sources, controlling population and preventing pollution. Water reuse has been practiced for over 5000 years; however, during the last 100 years efforts have been made in many regions of the world, for the production of high-quality reused water. In developed areas with intense water scarcity, water reuse is currently practiced in an efficient way. Such places include Southern USA (primarily, California, Florida, Texas and Arizona), Australia, Singapore, Israel and countries in the Persian Gulf. In Europe, there is a significant variation between the North and the South, with respect to water scarcity, and respectfully with water reuse applications (Angelakis & Gikas, 2014; Angelakis et al., 2015; Friedler, 2001; Qadir et al., 2007).

Advancements in Wastewater Recycling Technologies.

Due to the advancement in technology wastewater recycling has become more cost-effective and scalable. Current wastewater recycling technologies include:

Primary Treatment/Initial Treatment: This stage involves in elimination of big particles and solids.

Secondary Treatment/Subsequent Treatment: This stage applies natural processes to decrease organic substances.

Treatment/Third-stage Treatment: Utilizes advanced filtering (membrane filtering, reverse osmosis) and sterilization (chlorination, UV radiation) to guarantee top notch reused water.

Quaternary Treatment/ Fourth-stage Treatment: It mainly focuses on pharmaceutical contaminants and pollutants like microplastics (Angelakis & Gikas, 2014; Angelakis et al., 2015; Asano et al., 2007; Qadir et al., 2007). The water purification process has been summarized in the Fig. 12.1.

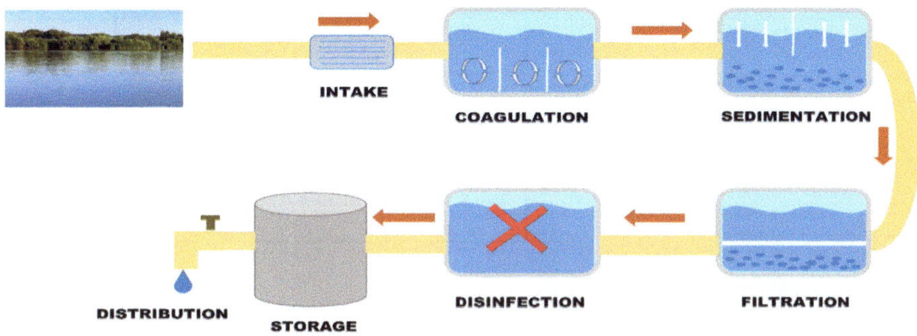

Fig. 12.1 Water purification process

12.2 The Economic Rationale for Wastewater Recycling

The following section studies emphasize about the socio-economic advantages along with economic and societal consequences related to wastewater recycling.

12.2.1 Initial Investment and Infrastructure Costs

Adequate investment is required to construct strong infrastructure and develop reliable technologies such as reverse osmosis, membrane filtration, and ultraviolet disinfection for functioning of treatment plant. It is necessary to improve water pricing policies, such as innovative rate structures and reforms of subsidies to make it cost efficient (Gleick, 2018). The long-term economic and environmental benefits outweigh the high upfront costs. Water recycling is a better alternative to reduce higher costs of sourcing of freshwater sources for industrial, agricultural and other purposes. This result in significant cost reduction on water bills and pressure on natural water bodies over time (Asano et al., 2007; Muga & Mihelcic, 2008; PUB Singapore, 2021).

12.2.2 Operational and Maintenance Costs

Operational and maintenance costs must be evaluated beforehand for successful water management. These expenses include the energy consumption, plant maintenance, and skilled labour for operation costs. The water recycling processes have become more cost-effective with technological advancements and significant investment. Operational or labour costs can be reduced by implementing automation and smart water management systems. These ongoing expenses provide better long-term financial benefits and outweigh the water acquisition expenses and pollution control

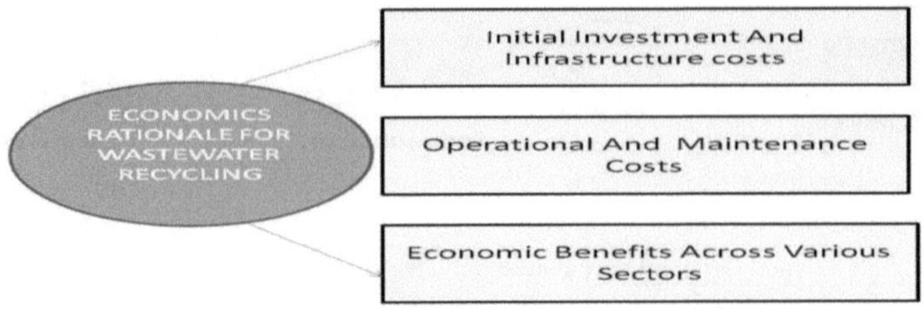

Fig. 12.2 Economics rationale for waste water recycling

costs It is important to consider the operational and maintenance costs during evaluation of wastewater recycling time (Asano et al., 2007; Muga andMihelcic, 2008; PUB Singapore, 2021). Figure 12.2 portrays the recycling and economics rationale for waste water.

12.2.3 Economic Benefits Across Sectors

The economic benefits of wastewater recycling extend across various sectors.

12.2.3.1 Agriculture

Field irrigation requires large amount of water which pressurise the natural water bodies. Using recycled water prevents over exploitation of freshwater bodies which will enhance agricultural efficiency. In water scarce areas, reusing wastewater can play a significant role in boosting agricultural productivity by maintaining crop yields and strengthening food security (Jiménez and Jiménez & Asano, 2008; Lazarova et al., 2013; Lautze et al., 2014; Mekonnen & Hoekstra, 2016).

Research has demonstrated that crops irrigated with treated wastewater yield higher outputs, attributed to the nutrient content in the recycled water. Treated wastewater poses fewer environmental risks, leading to reduced expenses related to environmental clean-up (Asano et al., 2007; Muga & Mihelcic, 2008; PUB Singapore, 2021). There are several economic benefits of wastewater recycling in agricultural sector such as lower irrigation cost, reduce dependency on freshwater, reduce need of fertilizers, more consistent supply, better yield, boost rural development, and resilience against drought conditions.

12.2.3.2 Industry

Various sectors, including textiles, chemicals, and food processing, rely on large water volumes. Reusing wastewater offers a reliable and cost-efficient water supply, mitigating operational risks related to water scarcity and enhancing the sustainability of industrial activities (Asano et al., 2007; Barton et al., 2019; Muga et al., 2008; PUB Singapore, 2021).

12.2.3.3 Energy Sector

The energy industry, especially in thermal and nuclear power generation, heavily depends on water. Reusing wastewater can decrease the need for fresh water in energy production, resulting in energy and cost savings. Nevertheless, advanced treatment methods like reverse osmosis can consume considerable energy, leading to increased operational expenses (Asano et al., 2007; Barton et al., 2019; Muga et al., 2008; PUB Singapore, 2021). The economic benefits of treated wastewater in energy industries are low cost of freshwater processing, lowers water treatment costs, support sustainable methods, minimize fines and penalties, improve public outlook and hence increase the market value.

The cost of initial capital investment in wastewater recycling systems may seem high, but technological advancements have significantly improved their long-term economic feasibility. Modern membrane technologies and energy recovery systems can now effectively lower operational expenses and enhance efficiency. Investing in wastewater recycling can reduce the dependence on freshwater sources and further reduce the freshwater extraction expenses (Asano et al., 2007; Barton et al., 2019; Muga et al., 2008; PUB Singapore, 2021; Qadir et al., 2007). The Economic Costs of Conventional Technologies, Advanced Technologies, Long-Term Financial Gains and Economic Benefits of Wastewater Recycling are as follows (Tables 12.1, 12.2, 12.3 and 12.4).

Table. 12.1 Economic costs of conventional technologies (David Young, 2014)

Cost component	Estimated range (USD/cubic meter)
Infrastructure (Capex)	0.50–1.00
Operational and maintenance	0.20–0.40
Energy (depending on technology)	0.10–0.30

Table 12.2 Economic costs of advanced technologies (Raphael et al., 2018)

Technology type	Capex (USD/cubic meter)	Operational costs (USD/cubic meter)	Energy requirement (kWh/m^3)
Conventional treatment	0.40–0.80	0.20–0.50	0.3–0.7
Membrane bioreactor (MBR)	1.00–1.50	0.50–0.80	0.6–1.2
Reverse osmosis (RO)	0.70–1.30	0.40–0.60	1.0–2.5

Table. 12.3 Long-term financial gains

Parameter	Cost of freshwater supply (USD/m^3)	Cost of recycled water (USD/m^3)	Potential savings (%)
Industrial use	1.20–2.50	0.50–1.20	30–60
Municipal use (non-potable)	1.50–3.00	0.40–1.00	40–70
Agricultural use	0.80–1.50	0.30–0.60	50–75

Table 12.4 Economic benefits of wastewater recycling (Tziakis et al., 2009)

Category	Cost/investment	Economic benefits
Infrastructure Investment	High (initial costs for treatment plants and pipelines)	Long-term savings on freshwater extraction and transport
Operational costs	Moderate (ongoing maintenance of treatment facilities)	Potential revenue from the sale of treated water, reduced costs in freshwater treatment
Energy savings	Low to moderate	Lower energy consumption for transport of freshwater over long distances
Job creation	High (employment in water treatment plants)	Local economic growth, especially in rural regions

12.3 Social Impacts of Wastewater Recycling

12.3.1 Public Health and Hygiene

Public health enhancement is the major social advantages of wastewater recycling. Utilization of treated recycled water for non-drinking purposes such as industrial processes, irrigation, and toilet flushing can decrease the need for freshwater. The freshwater resources should only be used for domestic purposes like drinking, cooking and overall hygiene. In water scarce regions, usage of treated wastewater lowers the chances of waterborne diseases spread through consumption of contaminated water. This contribute to healthier environment and reducing the medical. Recycled wastewater can uplift public health by supplying clean, safe water for non-potable and potable uses. In areas with insufficient availability of clean water, water recycling is a crucial step to provide clean and safe water, promote better hygiene and overall community development (Orange County Water District (OCWD) 2020; Pachauri et al., 2020; Ploeg et al., 2016).

12.3.2 Social Equity and Access to Water

Certain marginalized groups are unequally impacted by lack of clean water due to water shortage. One of the major benefits of recycling wastewater is upliftment of such backward communities as accessibility to clean water is a basic human right. Sustainable water management ensures equal distribution of safe and clean water among communities of all socio-economic statuses. Communities residing in remote area can have better access to clean water and reduce the reliance on freshwater sources. Furthermore, wastewater recycling encourages social upliftment of such communities and reduce socio-economic gaps. Additionally, limited availability of freshwater resources can lead to conflicts over water sources and its distribution, causing social inequality. Hence, it is necessary to implement alternative sustainable wastewater recycling to prevent such conflicts in the future (Rodriguez et al., 2009; Sapkota et al., 2007; Sato et al., 2013).

12.3.3 Employment and Economic Development

The construction and management of wastewater recycling infrastructure generates several job opportunities in field of engineering, management, construction and transportation (Table 12.5). These opportunities play a crucial role in economic growth by offering financial stability and lifting local economy by aiding the industries dependent on a consistent water supply. The presence of recycled water attracts new industries in water scarce regions, increases job scopes and supporting local economies. The construction of wastewater recycling plants demand required skills and hence, it helps in generating employment not only in construction sector but also in their working and management. Furthermore, the availability of large amount water benefits the agriculture sector with increased outputs, production and job opportunities in this field (Vij, 2018; Water Environment Federation., 2017; Wichelns, 2017).

Table 12.5 Employment opportunities and economic growth (Heintz et al., 2009)

Sector	Job opportunities (numbers) created
Infrastructure development	10,000
Operations and maintenance	2,000
Agriculture and industry expansion	15,000

12.4 Environmental Benefits and Climate Resilience

The main focus of this chapter involves the economic and social impacts, but it is important to highlight the environmental benefits of sustainable wastewater recycling. This recycling practice lowers the exploitation of freshwater sources thus, contributes to the preservation of biodiversity and aquatic habitats. The wastewater is treated efficiently before being dumped into the natural water bodies. This long-term sustainable way presents several benefits like boosting economy, protecting environment, increasing tourism and decreasing health issues. Wastewater recycling also aids in the preservation of aquatic ecosystems by reducing extraction of freshwater from rivers and aquifers (Ellen MacArthur Foundation., 2020; Farooq et al., 2020; Tortajada et al., 2021). The important environmental benefits are shown in the following diagram (Fig. 12.3):

The water recycling presents long-term benefits for the environment and economy, especially in areas related to economy, tourism, fishing, and the preservation of biodiversity. Some key benefits include:

Benefits for Agriculture and Industry: Reusing water can increase the availability of water for irrigation and industrial processes, which leads to increased productions and cost reduction.

Decreased Freshwater Usage: The recycling of water helps reduce the demand for freshwater, thus preventing its over exploitation.

Reduced Pollution: Treated wastewater reduces environmental clean-up expenses. Wastewater recycling prevents water pollution by treating it efficiently before being released into the water bodies. Moreover, with drastic climate change, droughts and water scarcity have become more frequent, recycling wastewater increases water supply that can help communities withstand water shortages (Ellen MacArthur Foundation., 2020; Farooq et al., 2020; Tortajada et al., 2021).

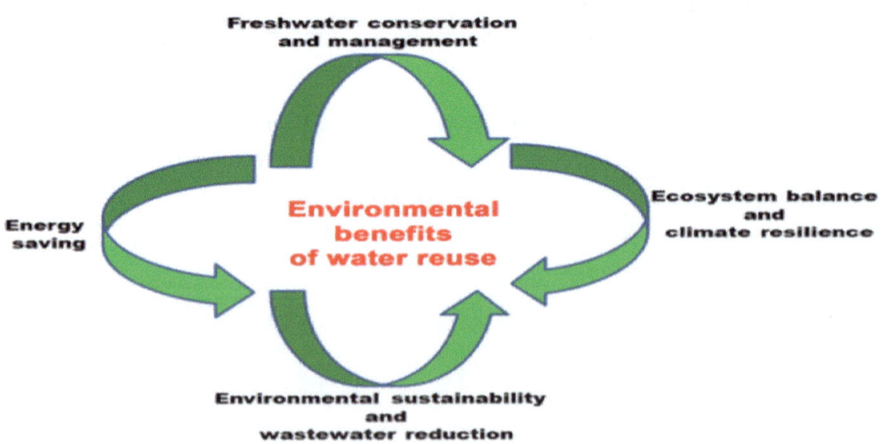

Fig. 12.3 Environmental benefits of water reuse

12.5 Challenges and Barriers to Implementation

There are several challenges linked with wastewater recycling that hinder its widespread adoption. These includes public perception, regulatory and policy frameworks and financial constraints.

12.5.1 Public Perception

There is a "yuck factor" attached with recycled water serving as a major obstacle. Several communities refuse to use the recycled water for their daily chores like drinking due to safety and health concerns. Public acceptance is a necessary factor in growth of wastewater recycling programmes.

12.5.2 Regulatory and Policy Frameworks

It is important for government to establish viable policies and regulations to improve water recycling programmes. The waste water recycling will meet the desired standard with strong legal framework.

12.5.3 Financial Constraints

The construction of expensive infrastructure demotivates the low-income communities and developing nations in joining such initiatives. It is necessary to develop cost effective strategies and technologies. However, initiatives such as public–private partnerships (PPPs) can provide an economic aid to the low-income areas in constructing recycling plants (Drechsel et al., 2010; Vij, 2018; Water Environment Federation., 2017; Wichelns, 2017).

12.6 Case Studies and Best Practices

To illustrate the practical applications of waste water recycling, this section presents case studies from various regions:

12.6.1 Singapore's NEWater Program

Singapore's innovative water recycling program called NEWater (New-Water) has provided high-quality recycled water for various purposes. This program has decreased their cost of importing water from outside. NEWater was developed in the 1990s and officially launched in 2003 is a crucial part of Singapore's "Four National Taps" strategy, alongside imported water, desalinated water, and local catchment. NEWater produces safe water through a three- step cleaning process which includes reverse osmosis, microfiltration, and ultraviolet disinfection. Singapore now recycles 40% of its water through technologies like advanced membrane technologies and ultraviolet disinfection According to its new water security strategy, Singapore, is planning to increase its fulfilment of 40% of Singapore's total water demand to 55% by 2060. Along with lowering demands of imported water, NEWater has also augmented their economy, public confidence through constructing visitor centres for educational purposes which inspires the public to learn and devise more such sustainable plans in the future. Singapore's success involves combination of both advanced technology, effective policy and public acceptance that has established NEWater as a role model for water-scarce regions. It globally demonstrates that recycled water can be a safe, sustainable, and cost-effective solution to several water related problems (PUB Singapore, 2021; Tal, 2016; Tortajada, 2006). The below flowchart mentions the procedure for wastewater treatment and management (Fig. 12.4).

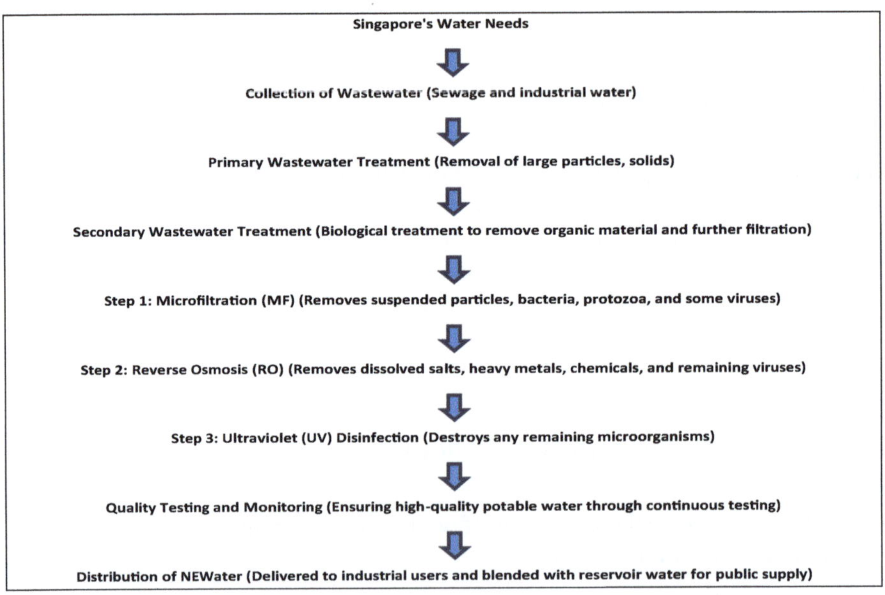

Fig. 12.4 Singapore's NEWater program (PUB Singapore, 2021; Tal, 2016; Tortajada, 2006)

12.6.2 California's Water Recycling Initiatives

California has made considerable investments in wastewater recycling to combat drought conditions and support both its urban water requirements and agricultural sector. The state has implemented cost-effective recycling initiatives to solve water scarcity. California's water recycling programs have played a significant role in addressing persistent water shortages, especially focusing on frequent droughts and population growth. The Groundwater Replenishment System (GWRS) in Orange County, one of the world's largest water recycling facility utilizes same mechanism as NEWater program that includes microfiltration, reverse osmosis, and ultraviolet light to purify wastewater before supplying it for industrial and portable use. Currently, this system generates 130 million gallons of recycled water per day to satisfy the needs of the public. Other cities, such as Los Angeles and San Diego have followed similar sustainability initiative to lower their dependence on water imported from Colorado River. California's attempt in water recycling served as a significant model in improving wastewater management and minimize negative impacts on health and the environment (Orange County Water District (OCWD) 2020; Qadir et al., 2010). The below flowchart mentions the procedure for wastewater treatment and management (Fig. 12.5).

12.6.3 Israel's Agricultural Reuse

Israel is globally leading in sustainable wastewater management, with over 85% of its treated wastewater is being recycled for agricultural purposes. It has advanced at water management strategies and recycling wastewater for agricultural irrigation. This ensures the sustainability of agricultural sector by significantly reducing the dependence on freshwater sources. Israel has built an advanced network for wastewater collection and treatment, to resolve the water scarcity issues. One of the largest facilities in the country, The Shafdan waste water treatment plant, uses developed filtration and biological processes that treats water to a tertiary level. The effluent or the recycled water is nutrient-rich, reduces the dependency on chemical fertilizers for agricultural practices. Israel has preserved the fresh water sources for domestic and ecological purposes. Reusing wastewater has also provided success in desert farming which enables the country to maintain food security despite its arid climate. Israel's agricultural reuse model is a role-model on cost-effective and sustainable wastewater treatment and resource for food production (Friedler, 2001; Smith et al., 2013). The below flowchart mentions the procedure for wastewater treatment and management (Fig. 12.6).

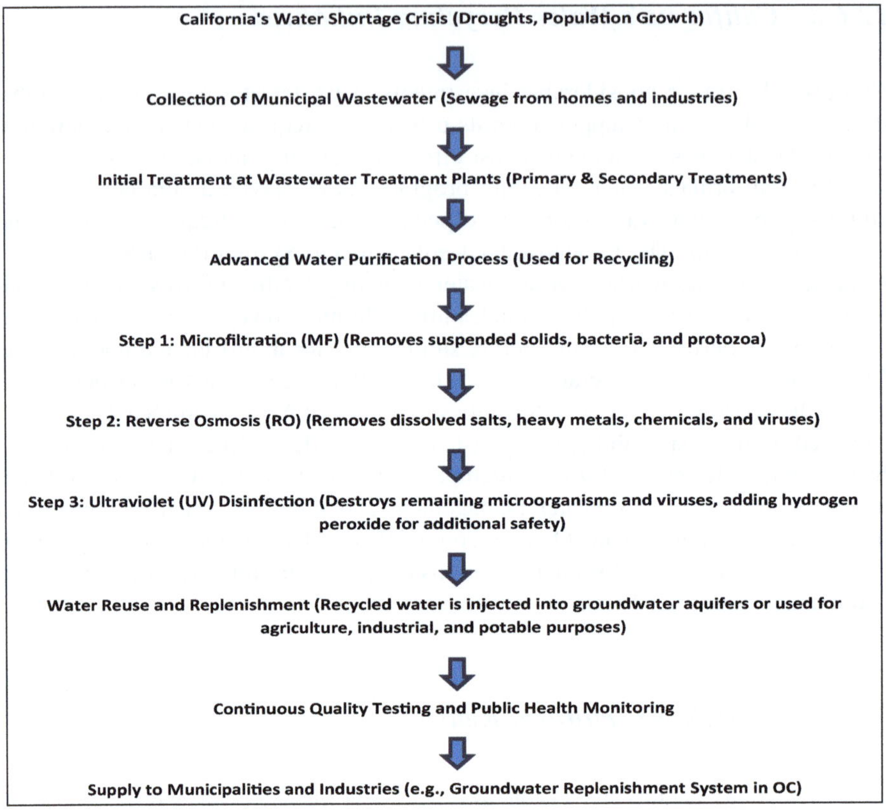

Fig. 12.5 California's water recycling initiatives (Orange County Water District (OCWD), 2020; Rodriguez et al., 2009)

12.6.4 Germany's Water Recycling Initiatives

Germany has focused on water recycle which prioritize the reuse of industrial wastewater, the management of stormwater, and the advanced wastewater treatment to promote sustainable development and efficient usage of resources. Germany has implemented water recycling initiatives due to its heavy dependence on freshwater resources such as rivers or ponds. The Emscher Project in North Rhine-Westphalia, involves treating and reusing wastewater for various domestic and industrial purposes. Further Germany has reduced urban runoffs by constructing green roof sand wetlands in urban areas which helps in purifying stormwater. These strategies align with the European Union's Water Framework Directive, which prefers sustainable water management practices. Germany is gradually implementing its water recycling practices into a broader environmental policy framework by focusing on reducing exploitation of freshwater in industrial processes and urban areas (Bixio

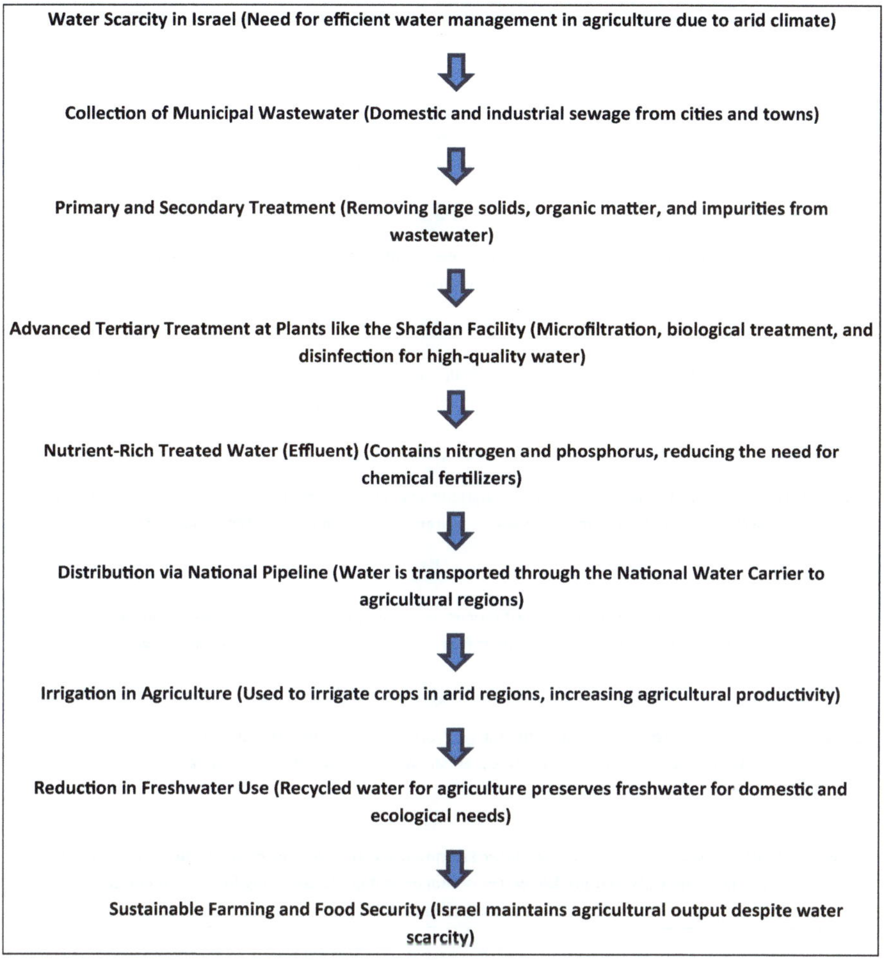

Fig. 12.6 Israel's agricultural reuse (Friedler, 2001; Smith et al., 2013)

et al., 2006; European Commission, 2020). The below flowchart mentions the procedure for wastewater treatment and management (Fig. 12.7):

12.6.5 India's Water Recycling Initiatives

India is facing severe water crisis; hence, it is increasing its strategies for recycling water as it decreases severe water scarcity and reduces excessive exploitation of groundwater resources. Various cities, including Bengaluru, Chennai, and Delhi, have

Fig. 12.7 Germany's water recycling initiatives (Bixio et al., 2006; European Commission, 2020)

initiated projects that focus on treating and reusing wastewater to decrease dependence on fresh water sources. As in case of Chennai which has commence large-scale wastewater recycling to provide industries in the water-scarce regions. In Bengaluru, groundwater replenishment is done through sustainable treatment of wastewater. Moreover, the NamamiGange program launched by the Indian government in 2014 promotes for the purification and revival of the Ganges River through treatment and reusage of wastewater. Regardless, India still confronts significant number of challenges, such as inadequate infrastructure, insufficient funds and lack of public awareness and support. Nevertheless, with increasing investments and sturdy policy backing, wastewater recycling is expected to play a crucial role in the country's upcoming water management (Drechsel et al., 2010; Ghosh et al., 2008; United

Nations Water (UN-Water), 2017). The below flowchart mentions the procedure for wastewater treatment and management (Fig. 12.8).

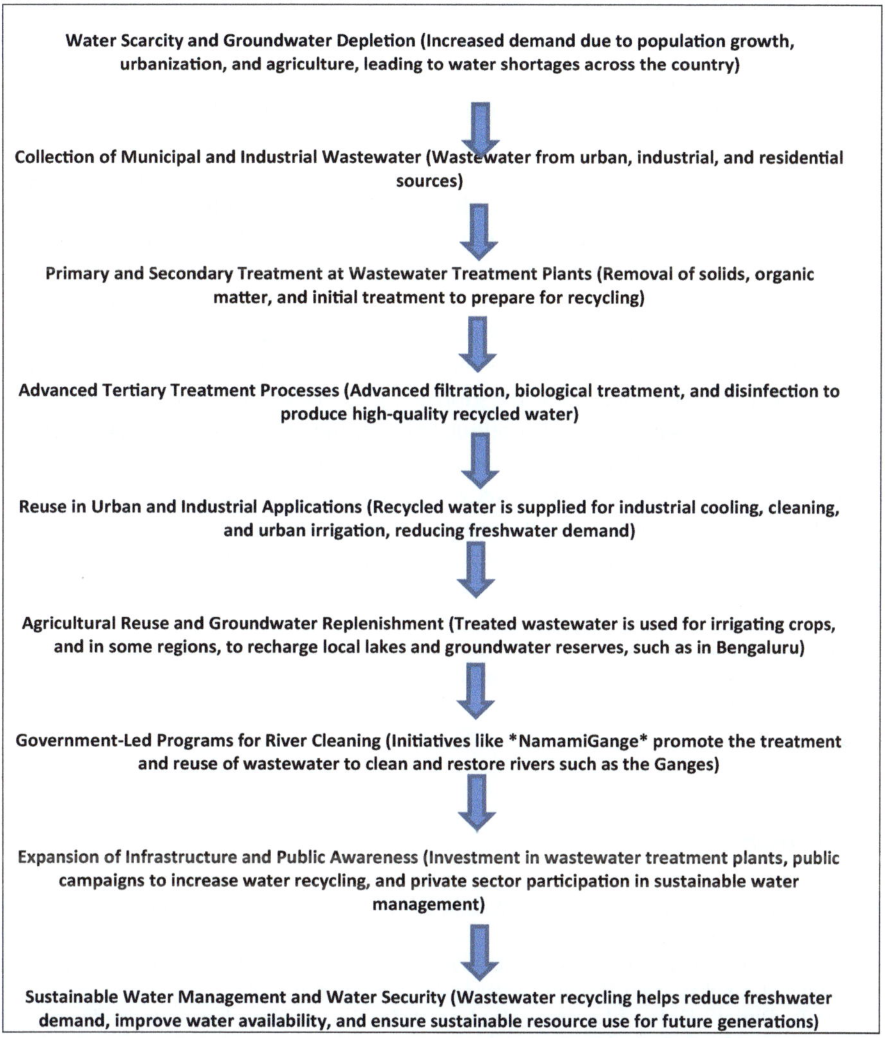

Fig. 12.8 India's water recycling initiatives (Ghosh et al., 2021; United Nations Water (UN-Water), 2017)

12.7 Conclusion and Future Directions

Wastewater recycling is a crucial element of sustainable water management. It is cost effective as long-term advantages overshadows the initial costs of developing infrastructure. Wastewater recycling promotes social equality, water security and prevent environmental degradation of natural waterbodies. Water scarcity is a global problem due to several reasons like climate change, water pollution, and droughts, therefore, it is necessary to develop sustainable infrastructure, technologies and action plans for water recycling. This will improve public perception, public acceptance and equal distribution of water among all communities. The reliable water management policies can help human society to reach sustainable and strong future. This chapter provides a detailed insight about the economic and social effects of wastewater recycling, that can assist policymakers, practitioners, and researchers in developing advanced water management technologies and policies. Sustainable water management can provide benefits in both social and economic scenarios. In conclusion, wastewater recycling is a reliable solution to the global water crisis and new laws, regulations, technologies and researches will continue to lower the production costs and increase its efficiency, availability and acceptance to all the nations regardless of their socio-economic status. At last, wastewater recycling is a strong and reliable solution to meet globalwater demands and tackle water crisis with stronger policy frameworks, rules and regulation, and increased public awareness and equality, further contributing to a sustainable future.

Acknowledgements The authors wish to express their gratitude to the authorities of Shri Guru Ram Rai University, Dehradun, for granting access to laboratory and library facilities throughout the duration of this study.

References

Angelakis, A. N., & Snyder, S. A. (2015). Wastewater treatment and reuse: Past, present, and future. *Water, 7*(9), 4887–4895.

Angelakis, A. N., & Gikas, P. (2014). Water reuse: Overview of current practices and trends in the world with emphasis on EU states. *Water Utility Journal, 8*, 67–78.

Asano, T., Burton, F. L., &Leverenz, H. L. (2007). *Water reuse: Issues, technologies, and applications*. McGraw-Hill Professional.

Barton, A. B., & Seidel, C. (2019). Innovative approaches to water reuse: Economic and environmental perspectives. *Water Research, 154*, 127–141.

Bixio, D., Thoeye, C., De Koning, J., Joksimovic, D., Savic, D., Wintgens, T., & Melin, T. (2006). Wastewater reuse in Europe. *Desalination, 187*(1–3), 89–101.

Drechsel, P., Scott, C. A., Raschid-Sally, L., Redwood, M., & Bahri, A. (Eds.). (2010). *Wastewater irrigation and health: Assessing and mitigating risk in low-income countries*. International Water Management Institute (IWMI) & Earthscan.

Ellen MacArthur Foundation. (2020). *The circular economy and the water sector: Opportunities for sustainability*.

European Commission. (2020). *Water reuse in Europe.* Retrieved from https://ec.europa.eu/environment/water/reuse.htm

Farooq, S., & Ansari, Z. I. (2020). Wastewater recycling: A cost-benefit analysis for the Middle East and North Africa (MENA) region. *Environmental Economics and Policy Studies, 22*(4), 781–799.

Friedler, E. (2001). Water reuse: An integral part of water resources management—Israel as a case study. *Water Policy, 3*(1), 29–39.

Ghosh, D., Medhi, C. R., & Purkait, M. K. (2008). Treatment of fluoride-containing drinking water by electrocoagulation using monopolar and bipolar electrode connections. *Chemosphere, 73*(9), 1393–1400.

Ghosh, S. K., Ghosh, S. K., & Baidya, R. (2021). Circular economy in India: Reduce, reuse, and recycle through policy framework. In *Circular economy: Recent trends in global perspective* (pp. 183–217). Singapore: Springer Nature Singapore.

Gleick, P. H. (2018). *The world's water volume 9: The biennial report on freshwater resources.* Island Press.

Heintz, J., Pollin, R., & Garrett-Peltier, H. (2009). How infrastructure investments support the US economy: employment, productivity and growth. *Political Economy Research Institute (PERI), University of Massachussetts Amberst.*

Jiménez, B., & Asano, T. (2008). *Water reuse: An international survey of current practice, issues, and needs.* IWA Publishing.

Lautze, J., Stander, E., Drechsel, P., da Silva, A. K., &Keraita, B. (2014). Global experiences in water reuse. *International Water Management Institute.*

Lazarova, V., Asano, T., Bahri, A., & Anderson, J. (2013).*Milestones in Water Reuse: The Best Success Stories.* IWA Publishing.

Mekonnen, M. M., & Hoekstra, A. Y. (2016). Four billion people facing severe water scarcity. *Science Advances, 2*(2), Article e1500323.

Molinos-Senante, M., Hernández-Sancho, F., & Sala-Garrido, R. (2011). Cost-benefit analysis of water-reuse projects for environmental purposes: A case study for Spanish wastewater treatment plants. *Journal of Environmental Management, 92*(12), 3091–3097.

Muga, H. E., & Mihelcic, J. R. (2008). Sustainability of wastewater treatment technologies. *Journal of Environmental Management, 88*(3), 437–447.

National Research Council. (2012). *Water reuse: Potential for expanding the nation's water supply through reuse of municipal wastewater.* National Academies Press.

Orange County Water District (OCWD). (2020). *Groundwater Replenishment System.* Retrieved from https://www.ocwd.com.

Pachauri, S., & Van Ruijven, B. J. (2020). Water-energy-food nexus: An integrated approach to sustainable water management. *Annual Review of Environment and Resources, 45*, 295–319.

Ploeg, M. J., & Eichner, M. J. (2016). Urban water management and economic impact of wastewater recycling in the European Union. *Environmental Economics and Policy Studies, 18*(3), 419–439.

PUB Singapore (2021). *NEWater: Singapore's Water Recycling Success Story.* Retrieved from https://www.pub.gov.sg.

Qadir, M., Bahri, A., Sato, T., & Al-Karadsheh, E. (2010). Wastewater production, treatment, and agricultural use in Middle Eastern and North African countries. *Food and Agriculture Organization of the United Nations (FAO).*

Qadir, M., Wichelns, D., Raschid-Sally, L., Singh, M., & Amerasinghe, P. (2007). Agricultural use of marginal-quality water—Opportunities and challenges. *Water Policy, 9*(S1), 45–55.

Tarpani, R. R. Z., & Azapagic, A. (2018). Life cycle costs of advanced treatment techniques for wastewater reuse and resource recovery from sewage sludge. *Journal of Cleaner Production, 204*, 832–847. ISSN 0959-6526.

Rodriguez, C., Van Buynder, P. G., Lugg, R., Blair, P., Devine, B., Cook, A., & Weinstein, P. (2009). Indirect potable reuse: A sustainable water supply alternative. *International Journal of Environmental Research and Public Health, 6*(3), 1174–1209.

Sapkota, A., Lefferts, L. Y., McKenzie, S., & Walker, P. (2007). What do we feed to food-production animals? A review of animal feed ingredients and their potential impacts on human health. *Environmental Health Perspectives, 115*(5), 663–670.

Sato, T., Qadir, M., Yamamoto, S., Endo, T., & Zahoor, A. (2013). Global, regional, and country-level need for data on wastewater generation, treatment, and use. *Agricultural Water Management, 130*, 1–13.

Smith, K., & Kazmierczak, R. (2013). The economic value of treated wastewater in agricultural irrigation: A case study from Israel. *Water Economics and Policy, 6*(2), 195–208.

Tal, A. (2016). *The land is full: Addressing Overpopulation in Israel.* Yale University Press.

Tortajada, C. (2006). Water management in Singapore. *Water Resources Development, 22*(2), 227–240.

Tortajada, C., & Biswas, A. K. (2021). Water reuse and desalination for sustainable water management in Singapore. *Water International, 46*(3), 304–320.

Tziakis, I., Pachiadakis, I., Moraitakis, M., Xideas, K., Theologis, G., & Tsagarakis, K. P. (2009). Valuing benefits from wastewater treatment and reuse using contingent valuation methodology. *Desalination, 237*(1–3), 117–125.

United Nations Water (UN-Water). (2017). *Wastewater: The untapped resource.* World Water Development Report.

Vij, S. (2018). India's water crisis: Need for wastewater reuse. *Economic and Political Weekly, 53*(26–27), 15–17.

Water Environment Federation. (2017). *The reuse handbook*. McGraw-Hill Education.

Wichelns, D. (2017). Economic and environmental considerations in wastewater treatment and reuse. *Water Resources and Economics, 17*, 1–13.

Winpenny, J., Heinz, I., & Koo-Oshima, S. (2010). *The wealth of waste: The economics of wastewater use in agriculture.* Food and Agriculture Organization of the United Nations (FAO).

Young, D. (2014). The development of empirical models to evaluate energy use and energy cost in wastewater collection. *All Dissertations*, 1365.

Chapter 13
Future Trends and Innovations: Emerging Technologies in Wastewater Treatment and Potential Breakthroughs in Resource Recovery

R. Dhinesh, S. S. Kirthiga, Anu Gopinath, and Mohammad Taherzadeh

Abstract The escalating need for sustainable wastewater treatment has driven the development of innovative technologies that not only address pollutant degradation but also enable resource recovery. This chapter provides an in-depth analysis of emerging technologies in wastewater treatment, highlighting their potential to optimize treatment processes and contribute to a circular economy. Key technologies such as advanced oxidation processes (AOPs), membrane bioreactors (MBRs), and forward osmosis have shown significant promise in efficiently removing complex pollutants, improving water reuse, and enhancing overall treatment efficiency in which these processes are crucial for removing a wide range of contaminants, particularly those that are difficult to remove using conventional methods. The integration of bio electrochemical systems (BES) and microbial fuel cells has emerged as a key innovation in wastewater treatment, offering the dual advantage of treating wastewater while simultaneously generating energy. Additionally, electrocoagulation is identified as an effective, chemical-free treatment method for the removal of suspended solids, heavy metals, and organic pollutants, presenting a more sustainable approach to wastewater management. The role of nanotechnology in wastewater treatment is explored in depth, particularly for the removal of emerging contaminants such as pharmaceuticals, personal care products, and endocrine-disrupting compounds. Furthermore, microalgae-based systems and algal–bacterial consortia are highlighted for their ability to recover nutrients, mitigate carbon dioxide emissions, and support the treatment of wastewater, contributing to both resource recovery and environmental sustainability. Besides, the application of artificial intelligence (AI) and machine learning (ML) in wastewater treatment has the potential to revolutionize process optimization. Despite these significant advancements, challenges

R. Dhinesh (✉) · S. S. Kirthiga · A. Gopinath
Department of Aquatic Environment Management, Faculty of Fisheries Science, Kerala University of Fisheries and Ocean Studies, Kochi, Kerala, India
e-mail: dhinesh7143@gmail.com

M. Taherzadeh
Bioprocess Technology, Swedish Centre for Resource Recovery, University of Borås, Borås, Sweden

S. Dhiman and G. Mukherjee (eds.), *Resource Resurgence*, Springer Water,
https://doi.org/10.1007/978-3-032-02750-4_13

such as membrane fouling, high energy consumption, and scalability of these technologies were also discussed in the chapter. The chapter also emphasizes the importance of evaluating the techno-economic feasibility, environmental sustainability, and establishing robust regulatory frameworks to support the widespread adoption of these emerging technologies and resource recovery in wastewater treatment.

Keywords Advanced oxidation process · Bio electrochemical systems · Nanomaterials · Resource recovery · Artificial intelligence · Emerging contaminants

13.1 Introduction

Wastewater treatment plays a pivotal role in environmental management and public health, acting as a critical safeguard against pollution and disease transmission. Untreated or inadequately treated wastewater, rich in organic matter, pathogens, nutrients, and toxic contaminants, poses significant risks to aquatic ecosystems, soil health, and human populations through the contamination of water resources (Tariq & Mushtaq, 2023). Conventional wastewater treatment systems, which rely on primary, secondary, and tertiary processes, have long been used to remove organic matter, nutrients, and pathogens. However, these traditional approaches are increasingly inadequate in addressing the challenges posed by modern waste streams particularly the presence of emerging contaminants. These contaminants, including pharmaceuticals, pesticides, endocrine-disrupting chemicals, personal care products, and microplastics, has raised significant concerns due to their persistence, bio accumulative nature, and potential toxic effects on ecosystems and human health (Shanmuganathan et al., 2023). Moreover, conventional treatment systems often have high operational costs, energy demands, and environmental footprints, making their long-term sustainability a growing concern (Peng et al., 2018a, 2018b).

A transformative shift in wastewater management is imperative, transitioning from traditional treatment and disposal practices to an integrated approach that prioritizes resource recovery, sustainability, and circular economy principles. Wastewater should be seen as a source of valuable resources, including clean water, energy, and nutrients, rather than merely a waste product. Implementing innovative technologies can enhance pollutant removal while enabling the recovery of useful materials, aligning with circular economy principles. Emerging technologies are transforming wastewater treatment by addressing the shortcomings of conventional treatment through enhanced efficiency, specificity, and multi-functionality. Advanced treatment methods, such as advanced oxidation processes (AOPs), have demonstrated exceptional potential for the degradation of recalcitrant organic pollutants by generating highly reactive hydroxyl radicals (Manna & Sen, 2022). Similarly, membrane bioreactors (MBRs) have revolutionized treatment systems by combining biological degradation with membrane filtration, offering superior effluent quality and compact designs suitable for urban areas (Qrenawi & Rabah, 2023). Forward osmosis, with its low-energy requirements and ability to concentrate wastewater for resource recovery, represents another innovative approach gaining traction in the field (Xu et al., 2022a, 2022b). Integrating these technologies with resource recovery systems not only enhances pollutant removal but also promotes circular economy principles by recovering nutrients, biogas, and water for reuse, thereby reducing environmental impact and operational costs (Zarei, 2020). The inclusion of bio electrochemical systems (BES) such as microbial fuel cells in wastewater treatment is particularly promising for energy generation alongside contaminant removal. By leveraging electroactive microbes to convert organic matter into electricity, BES offers a sustainable solution for low-carbon treatment systems (Das et al., 2024). Furthermore, the integration of algal–bacterial consortia has shown potential for simultaneous nutrient recovery and the sequestration of carbon dioxide, contributing to climate mitigation efforts (Jin et al., 2024). The development of nanotechnology-based approaches has expanded the scope of contaminant removal through tailored materials with high surface area, enhanced reactivity, and specificity (Asghar et al., 2024). Besides, Synthetic biology and bioengineering are driving the development of innovative bioremediation techniques that leverage genetically modified microorganisms to degrade specific contaminants or produce valuable byproducts (Ezeako, 2024). Particularly, the use of bio electrochemical systems in combination with synthetic biology has opened new avenues for the development of biohybrid systems with enhanced functional properties for wastewater treatment (Dong et al., 2022). Artificial intelligence (AI) and machine learning (ML) are also emerging as game-changers in optimizing wastewater treatment systems in which these tools enable predictive modeling, real-time process monitoring, and adaptive management, ensuring the precise control of treatment variables for maximum efficiency (Nguyen et al., 2022a, 2022b). Through AI-driven insights, the energy consumption of processes like aeration and chemical dosing can be minimized, leading to significant cost savings and improved environmental sustainability. However, the application of AI requires the integration of robust data acquisition systems and expertise in computational modeling, which presents an interdisciplinary challenge for researchers and practitioners.

Despite the advantages of emerging technologies, there also exists a few critical challenges. The high capital costs, complexity of operations, and technical expertise required for implementation can hinder their large-scale adoption. The techno-economic feasibility of integrating these advanced systems into existing treatment plants requires careful assessment to ensure long-term viability and accessibility. Moreover, environmental sustainability must be prioritized in the development of these technologies, ensuring that their implementation does not introduce new risks or exacerbate existing ones, such as secondary pollution from nanomaterial use (Nowack and Bucheli, 2007). This chapter explores a range of emerging and future technologies in wastewater treatment, offering a detailed analysis of their mechanisms, advantages, and limitations. It provides insights into how these innovations can be harnessed to address the growing complexity of wastewater streams while emphasizing their role in resource recovery and sustainability.

13.2 Emerging Wastewater Treatment Technologies

13.2.1 Advanced Oxidation Processes (AOPs)

Advanced oxidation processes (AOPs) have emerged as a powerful class of treatment technologies for degrading organic pollutants and emerging contaminants in wastewater, leveraging highly reactive oxygen species (ROS) to achieve near-complete mineralization of complex compounds into carbon dioxide, water, and inorganic ions (Mukherjee et al., 2023). Figure 13.1 represents the processes and methods involved in advanced oxidation processes. The fundamental mechanism of AOPs lies in the generation of hydroxyl radicals (·OH), which are among the most potent oxidants known, with a redox potential of 2.8 V (Cheng et al., 2016) and capable of attacking and breaking down a wide number of pollutants such as pharmaceuticals, pesticides, endocrine disruptors, and dyes, which are found to be highly resistant against conventional treatment methods (Khan et al., 2023).

One of the most studied AOPs is photocatalysis, which utilizes light-activated catalysts such as titanium dioxide (TiO_2) to generate ROS upon exposure to ultraviolet (UV) or visible light (Paumo et al., 2021). Photocatalysis facilitates the degradation of persistent organic pollutants by promoting electron–hole pair generation within the catalyst, followed by electron transfer to adsorbed water or oxygen molecules, resulting in the formation of •OH and superoxide radicals (Fan et al., 2024). This

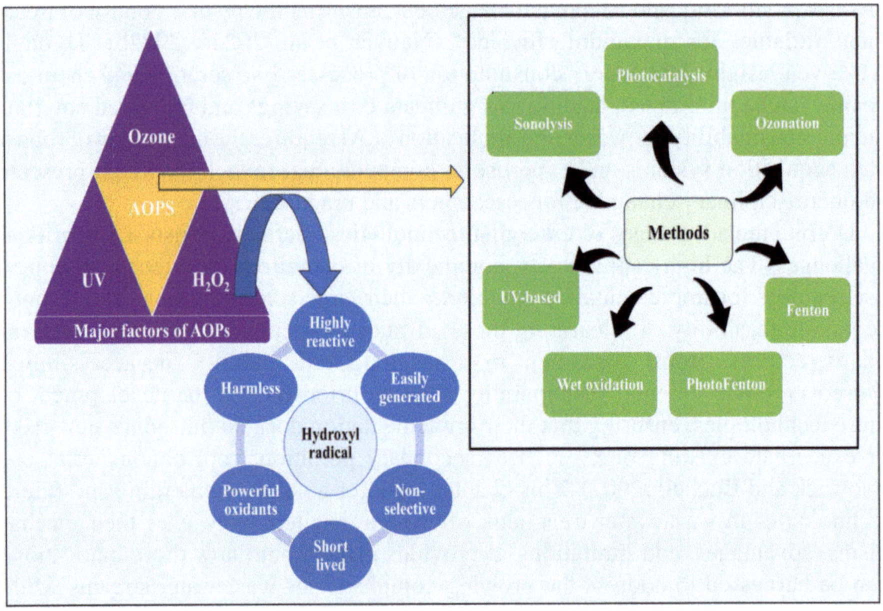

Fig. 13.1 Processes and methods involved in advanced oxidation processes (AOPs)

technology has shown remarkable efficiency in degrading dyes, pharmaceuticals, and industrial chemicals, although its practical application is hindered by the need for high-intensity light sources and the tendency for catalyst deactivation due to fouling or aggregation (Zeghioud et al., 2016). The Fenton and photo-Fenton processes represent another subset of AOPs that utilize iron-based catalysts and hydrogen peroxide (H_2O_2) to produce •OH radicals. In the classical Fenton reaction, ferrous iron (Fe^{2+}) reacts with H_2O_2 to generate •OH, while in the photo-Fenton process, UV or visible light is used to enhance the catalytic efficiency by regenerating Fe^{2+} from ferric iron (Fe^{3+}) and promoting additional ROS generation (Lai and Chun et al., 2017). These processes have been widely applied for the treatment of industrial effluents, textile wastewater, and landfill leachates, achieving significant reductions in chemical oxygen demand (COD) and total organic carbon (TOC). However, the Fenton-based processes face limitations such as the need for acidic conditions (pH ~ 3), high sludge generation due to iron hydroxide precipitation, and the cost of H_2O_2. Efforts to address these challenges include the development of heterogeneous Fenton catalysts using immobilized iron on solid supports, which reduces sludge production and extends the operational pH range (Bouzayani and Sanromán, 2024).

Ozonation and ozone-based AOPs are another prominent category, utilizing ozone (O_3) as a strong oxidant to degrade organic pollutants directly or via ROS generation. In water, ozone reacts with hydroxide ions or hydrogen peroxide to produce ·OH, enhancing its oxidative capacity. Ozone-based processes, such as peroxone (O_3/H_2O_2) and ozone/UV systems, have been successfully employed for the removal of recalcitrant contaminants like pharmaceuticals, nitrosamines, and endocrine disruptors (Rekhate & Srivastava, 2020). Ozonation has also demonstrated efficacy in disinfecting wastewater by inactivating a broad range of pathogens, including bacteria, viruses, and protozoa. Despite these advantages, the widespread adoption of ozone-based AOPs is constrained by high energy consumption for ozone generation, the potential formation of harmful by-products such as bromate (BrO_3^-) in bromide-containing waters, and the need for advanced monitoring and control systems to optimize reaction conditions (Zahrani, 2024). Recent advancements, including the integration of ozone with nanomaterials and catalytic systems, aim to enhance ROS generation efficiency while minimizing by-product formation.

13.2.2 Membrane Bioreactors (MBRs)

Membrane bioreactors (MBRs) represent an innovative and highly effective approach to wastewater treatment by integrating biological treatment processes with membrane filtration systems, offering significant advancements over conventional treatment methods. The benefits of MBR systems are multifaceted, contributing to their increasing popularity in wastewater treatment applications in which the membrane filtration unit separates treated water from suspended solids, pathogens, and other contaminants, ensuring superior effluent quality (Jijingi et al., 2024). A key advantage is their ability to produce high-quality effluent that is free from suspended solids

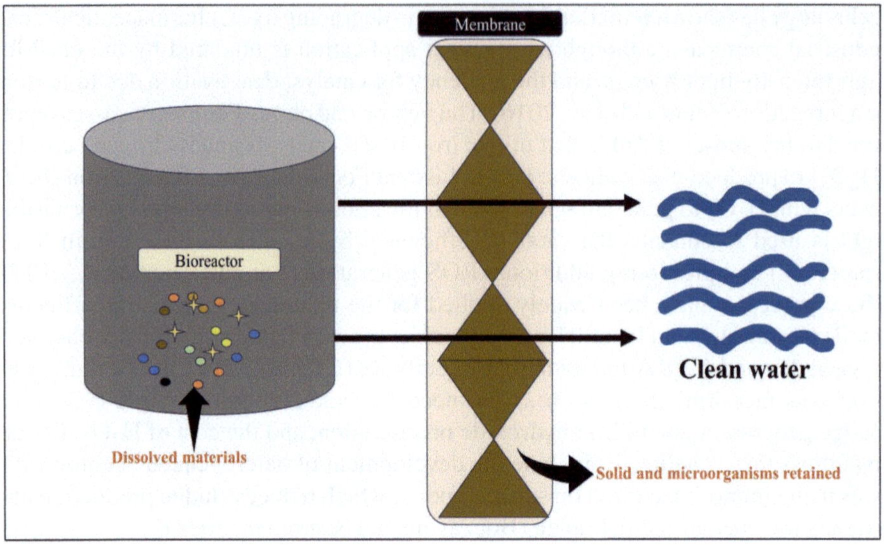

Fig. 13.2 Design of membrane bioreactors

and has significantly reduced concentrations of biochemical oxygen demand (BOD) and total nitrogen, making it suitable for direct discharge or reuse (Bera et al., 2021). MBR systems also feature a compact design, as the membrane filtration unit eliminates the need for secondary clarifiers and other large treatment components, reducing the overall footprint of the treatment plant (Qrenawi & Rabah, 2023). Furthermore, these systems minimize sludge production by retaining a higher concentration of biomass within the bioreactor, enhancing the degradation of organic matter and reducing the costs associated with sludge disposal (Fig. 13.2). The ability of MBRs to operate under varying loading conditions and treat high-strength wastewater further underscores their versatility and robustness (Fida et al., 2021).

The membranes used in MBRs typically fall into two categories: microfiltration (MF) and ultrafiltration (UF). MF membranes, with pore sizes ranging from 0.1 to 10 microns, are effective in removing suspended solids, bacteria, and some viruses. UF membranes, with smaller pore sizes of 0.01 to 0.1 microns, provide additional filtration capabilities, effectively capturing even smaller contaminants such as colloids and most viruses (Xiao et al., 2023). These membranes act as a physical barrier, preventing the passage of particles while allowing water and dissolved substances to pass through, ensuring consistent effluent quality irrespective of fluctuations in influent composition (Pendergast & Hoek, 2011). The widespread adoption of MF and UF membranes in MBR systems is driven by their robustness and ability to maintain high hydraulic flux, although challenges such as fouling and cleaning requirements remain significant concerns (Du et al., 2020). For the removal of dissolved substances and emerging contaminants, MBR systems can be coupled with nanofiltration (NF) or reverse osmosis (RO) membranes (Monachan et al., 2022).

NF membranes, with pore sizes typically in the nanometer range, are effective in removing divalent and larger monovalent ions, organic molecules, and micropollutants, making them suitable for applications requiring partial demineralization and removal of pharmaceuticals or pesticides (Bodzek et al., 2015). RO membranes, which function on the principle of rejecting nearly all dissolved solutes through a dense, non-porous membrane, are capable of removing salts, heavy metals, and a broad spectrum of emerging contaminants (El Batouti et al., 2021). When integrated into MBR systems, these advanced membranes enhance the removal efficiency of both conventional pollutants and trace-level contaminants, ensuring compliance with stringent water quality standards and enabling water reuse in industrial and municipal settings. The application of MBR systems has been extensively evaluated in various wastewater treatment settings, ranging from municipal to industrial and decentralized systems. In municipal wastewater treatment, MBRs have been successfully employed to achieve stringent effluent quality standards, particularly in regions with water scarcity where treated water reuse is essential (Yang et al., 2020a, 2020b). In industrial settings, MBRs have demonstrated effectiveness in treating complex waste streams containing high concentrations of organic pollutants, oils, and heavy metals, such as those generated by food processing, pharmaceutical manufacturing, and petrochemical industries (Jijingi et al., 2024). Decentralized wastewater treatment systems incorporating MBR technology have also gained traction, providing cost-effective solutions for small communities and remote locations by reducing infrastructure requirements and facilitating local water reuse.

13.2.3 Forward Osmosis (FO)

Forward osmosis (FO) has emerged as a promising technology in wastewater treatment and water recovery, utilizing osmotic pressure differences across a semi-permeable membrane to drive water flux. Unlike traditional pressure-driven processes such as reverse osmosis (RO), FO operates on the principle of osmosis, where water naturally flows from a low-concentration solution (feed water) to a high-concentration solution (draw solution) without requiring high-pressure pumps (Blandin et al., 2020). This unique mechanism significantly reduces energy consumption, making the process of forward osmosis an energy-efficient alternative for water extraction from wastewater, particularly when paired with low-energy draw solutions (Singh et al., 2021a, 2021b). Draw solutions, often composed of highly soluble salts or other osmotic agents, play a critical role in enhancing water flux while maintaining the osmotic gradient. These solutions can be engineered for specific applications, such as desalination, wastewater reclamation, or industrial effluent treatment, by tailoring their chemical properties to optimize performance (Ibraheem et al., 2023). Its applications in desalination have gained significant attention, particularly as a pre-treatment or hybrid system integrated with RO or other technologies. By reducing the salinity of feedwater before it enters high-pressure systems, FO decreases the energy requirements and fouling potential of downstream processes, improving the overall efficiency of desalination plants (Chia et al., 2021). Industrial effluent treatment is another critical area where FO is proving valuable, particularly in industries producing high-strength waste streams, such as textile manufacturing, pharmaceuticals, and food processing industries (Xu et al., 2022a, 2022b). One of the primary advantages of FO is its lower fouling potential compared to pressure-driven membrane processes. The absence of hydraulic pressure minimizes the compaction of fouling layers and reduces the physical stress on the membrane, prolonging its lifespan and lowering maintenance costs (Chia et al., 2021). However, challenges remain in the broader implementation of FO, with draw solution regeneration being a key technical hurdle. The separation and recovery of the draw solute from the extracted water often require additional processes, which can offset the energy savings and increase operational complexity (Chaoui et al., 2019).

13.2.4 Microbial Fuel Cells (MFCs)

Microbial fuel cells (MFCs) represent an innovative and sustainable technology that leverages the metabolic activity of microorganisms to achieve simultaneous wastewater treatment and energy recovery. MFCs operate by exploiting the ability of certain microorganisms, known as exoelectrogens, to transfer electrons generated during the oxidation of organic matter in wastewater to an anode in which this electron transfer drives an electrochemical reaction, generating electrical energy while breaking down pollutants (Gude et al., 2016). The dual functionality of MFCs in treating wastewater and producing renewable energy has garnered significant attention as a green alternative to traditional energy-intensive treatment methods, particularly in decentralized and resource-limited settings (Aghababaie et al., 2015). The construction of an MFC typically consists of two chambers separated by a proton exchange membrane (PEM): an anaerobic anode chamber where microorganisms oxidize organic matter, and a cathode chamber where electrons and protons combine with an electron acceptor, usually oxygen, to form water. The anode, often made from conductive materials such as carbon cloth, carbon felt, or graphite, serves as the site for microbial attachment and electron transfer. The cathode, similarly constructed from conductive materials, facilitates the reduction reactions (Daniel et al., 2009; He et al., 2017). Electrode materials are critical in determining the efficiency of electron transfer and power generation, with recent advancements focusing on modifying the surface properties of electrodes to enhance microbial adhesion and electron transport (Prathiba et al., 2022. In terms of wastewater treatment, MFCs effectively degrade a wide range of organic pollutants, including carbohydrates, proteins, and lipids, as well as emerging contaminants like pharmaceuticals and dyes. The use of diverse microbial communities ensures broad-spectrum degradation capabilities, with different microbial species specializing in metabolizing specific substrates (Khandaker et al., 2021). MFCs are also increasingly explored as biosensors for detecting water quality parameters, such as biochemical oxygen demand (BOD) and toxic compounds, by correlating the generated current with the concentration of specific analytes (Olias & Di Lorenzo, 2021).

One of the most significant advantages of MFCs is their low operational cost. The reliance on microbial metabolism, rather than external energy inputs, minimizes the energy footprint of the treatment process (Boas et al., 2022). Furthermore, the ability to recover energy in the form of electricity, albeit currently at low power densities, contributes to the economic feasibility of the technology (He et al., 2018). Challenges include maintaining consistent microbial activity, ensuring efficient mass transfer, and managing operational complexities in large reactors (Hassan et al., 2024).

13.2.5 Microalgae-Integrated Systems

Microalgae-based systems are increasingly recognized for their potential to provide an integrated solution for wastewater treatment and resource recovery. These systems are considered to be highly efficient as they utilize the naturally assimilating characteristic of microalgae in which these are used to absorb high amounts nutrients like nitrogen and phosphorus, which are often found in wastewater from cities, industries, and farms (Li et al., 2019). Microalgae play a critical role in capturing and transforming contaminants from wastewater, including emerging pollutants such as pharmaceuticals, heavy metals, and personal care products (Maryjoseph & Ketheesan, 2020). Their robust metabolic pathways allow them to biotransform or bioaccumulate such contaminants, thereby reducing their bioavailability in treated effluent. During photosynthesis, microalgae also sequester CO_2, making them an effective tool for reducing greenhouse gas emissions while simultaneously treating wastewater (Molazadeh et al., 2019). The nutrient uptake by microalgae occurs through a combination of biological assimilation and metabolic transformations, where nitrogen is incorporated into cellular proteins and phosphorus into nucleotides and other biomolecules (Su et al., 2021). This dual functionality of nutrient removal and carbon sequestration positions microalgae systems as an eco-friendly alternative to conventional nutrient removal technologies, which are often energy-intensive and chemically reliant. A significant advantage of microalgae systems is their ability to generate valuable biomass as a byproduct of wastewater treatment. Microalgae biomass can be utilized to meet the growing demand for renewable energy such as the production of biodiesel, biogas or bioethanol since the microalgal biomass is rich in the components like lipids, carbohydrates and proteins (Abdelfattah et al., 2023). Moreover, the residual biomass that is left over after the production can be utilized in aquaculture and production of animal feed as well as bio-based products like bioplastics and fertilizers which aid in the development of a circular economy approach. Microalgae systems have been successfully applied to a variety of wastewater streams, including municipal, industrial, and agricultural effluents. Municipal wastewater provides an excellent substrate for microalgae growth due to its balanced nutrient profile, while industrial wastewater, such as from food processing and textile industries, offers specific nutrients that support selective algal strains (Wang et al., 2017). Agricultural wastewater, characterized by high nitrogen and phosphorus content, is also ideal for microalgae-based systems, especially in addressing eutrophication concerns (Wollmann et al., 2019).

13.2.6 Algal–Bacterial Consortia

Algal–bacterial consortia represent a promising and sustainable approach in wastewater treatment, leveraging the synergistic interactions between algae and bacteria to enhance nutrient recovery and contaminant removal. These systems rely on mutualistic relationships, where algae, such as *Chlorella* and *Scenedesmus* species, provide oxygen through photosynthesis, supporting the growth of aerobic bacteria like *Nitrosomonas*, *Nitrobacter*, and *Pseudomonas*, which degrade organic pollutants and facilitate nitrogen cycling (Saravanan et al., 2021). The mutualistic relationship between algae and bacteria in these systems not only improves the overall efficiency of wastewater treatment but also supports the simultaneous production of valuable resources, such as biomass, and reduces the formation of harmful by-products. By integrating the biological processes of both microorganisms, algal–bacterial consortia offer a more efficient alternative to conventional wastewater treatment methods, which often rely on more energy-intensive or chemically intensive processes. The use of such consortia is being increasingly explored for municipal, industrial, and agricultural wastewater treatment, offering an eco-friendly, cost-effective, and renewable solution to the growing concerns over water quality and sustainability (Saravanan et al., 2021). The synergistic interactions between algae and bacteria play a pivotal role in the nutrient recovery and pollutant removal processes in wastewater treatment systems. Algae, through the process of photosynthesis, can efficiently remove excess nutrients such as nitrogen and phosphorus from wastewater (Nguyen et al., 2022a, 2022b). The growth of algae facilitates the uptake of these nutrients, transforming them into biomass and reducing their concentration in the treated water (Liu et al., 2020). Besides, bacteria also contribute to the biodegradation of organic pollutants through various metabolic pathways, such as anaerobic and aerobic processes in which the bacteria tend to break down complex organic compounds into simpler, non-toxic molecules (Cuellar-Bermudez et al., 2017). The presence of algae provides an oxygen-rich environment that supports aerobic bacterial processes, while bacteria can offer organic substrates or metabolites that promote algal growth, leading to a feedback loop that enhances the overall treatment efficiency (Ren et al., 2024). This mutualistic relationship significantly accelerates the removal of pollutants, especially in systems treating wastewater contaminated with both organic matter and nutrients. Moreover, bacteria play a key role in the degradation of micro-pollutants and emerging contaminants, such as pharmaceuticals and personal care products, which algae alone might not efficiently remove (Oruganti et al., 2022). In algal–bacterial consortia, algae primarily contribute to nutrient recovery through photosynthesis, which is driven by light energy to convert carbon dioxide (CO_2) into organic carbon compounds, releasing oxygen in the process. This oxygen facilitates the growth of aerobic bacteria, creating a balanced environment where both microorganisms thrive. Algae uptake nitrogen and phosphorus from the wastewater during their growth, storing these nutrients in their biomass and reducing their concentration in the effluent. This photosynthetic nutrient recovery is particularly advantageous in treating wastewater with high nutrient levels, such as municipal and agricultural effluents that often lead to eutrophication in receiving water bodies (Hasan et al., 2021). The combined metabolic activities of algae and bacteria, particularly in the removal of nitrogen, phosphorus, and organic matter, contribute to a more effective and holistic treatment process, allowing for improved water quality and the recovery of valuable resources. Despite the promising potential of algal–bacterial consortia in wastewater treatment, several challenges remain, particularly related to system stability, light availability, and biomass harvesting. The stability of algal–bacterial systems can be compromised by factors such as nutrient imbalances, pH fluctuations, and temperature variations, which may negatively affect the performance of both algae and bacteria (Jiang et al., 2021). Light availability is another critical factor influencing the efficiency of algal–bacterial systems. Algae require adequate light for photosynthesis, but in large-scale systems, light penetration may be limited, leading to suboptimal growth. This issue can be addressed by using artificial light sources or optimizing the system's design to improve light distribution, such as by employing photobioreactors with high surface-area-to-volume ratios (Vasumathi et al., 2012). Biomass harvesting also presents a challenge, as the separation of algal biomass from the treated wastewater can be energy-intensive and costly. Various methods, such as flocculation, centrifugation, and filtration, have been explored to improve biomass recovery efficiency. Recent advancements in harvesting technologies, including electrocoagulation and flotation techniques, are being integrated into algal–bacterial systems to streamline biomass recovery while minimizing energy consumption and operational costs (Mathimani & Mallick, 2018).

13.2.7 Nanotechnology in Wastewater Treatment

Nanotechnology has emerged as a promising field in wastewater treatment due to the unique properties of nanomaterials that enhance pollutant removal, improve treatment efficiency, and enable more effective resource recovery. Nanomaterials possess unique characteristics, such as high surface area-to-volume ratio, reactivity, which aids in the ability of nanomaterials to interact with pollutants at the molecular level (Saleh, 2024). These properties make nanomaterials highly effective in removing contaminants from wastewater, including heavy metals, organic pollutants, and microbial pathogens (Yaqoob et al., 2020). Nanomaterials play a significant role in improving the efficiency of wastewater treatment processes by enhancing pollutant removal through mechanisms such as adsorption, catalytic degradation, and photocatalysis. Their high surface area-to-volume ratio allows for efficient interaction with contaminants, resulting in higher removal efficiencies compared to conventional materials (Amin et al., 2014). Nanomaterials have shown remarkable effectiveness in removing the contaminants such as heavy metals, such as lead, arsenic, and mercury, as well as organic pollutants, including pesticides, pharmaceuticals, and industrial chemicals (Ugwu et al., 2022). Furthermore, the use of nanomaterials in membrane filtration technologies, such as nanofiltration and reverse osmosis, has led to improved separation capabilities and reduced fouling, thus enhancing treatment efficiency and lifespan of membranes (Joseph et al., 2023). Titanium dioxide (TiO_2) nanoparticles which is one of the most widely used nanomaterials for wastewater treatment, that particularly works on the mechanism of photocatalysis. Under ultraviolet (UV) light, TiO_2 nanoparticles generate reactive oxygen species (ROS) that can degrade a number of organic pollutants, including dyes, pesticides, and pharmaceutical residues. Since TiO_2-based photocatalysis has been shown to be highly effective in breaking down complex organic compounds that are resistant to traditional treatment methods, which makes it a promising technique for advanced water purification (Nematov et al., 2024). Furthermore, the silver nanoparticles (AgNPs) have been utilized in water disinfection systems owing to its antimicrobial property, which results in a rapid and efficient method for microbial removal, thus enhancing the safety of treated water (Bhardwaj et al., 2021). However, while both TiO_2 and AgNPs have shown promising results, issues such as the potential leaching of silver ions and the need for UV light activation in TiO_2 photocatalysis remain challenges for large-scale implementationCarbonbased nanomaterials, such as graphene, carbon nanotubes (CNTs), and activated carbon, are highly efficient in wastewater treatment thn the traditional treatment systems because of their high adsorption capacity and surface area in addition to their functional properties (Mohapatra et al., 2023). The effectiveness of nanomaterials in wastewater treatment relies heavily on the molecular interactions between the nanomaterial and the target pollutants (Amin et al., 2014). These interactions occur through various mechanisms, including physical adsorption, electrostatic attraction, ion exchange, and chemical bonding (Singh et al., 2021a, 2021b). Nanomaterials have thus enabled the development of more energy-efficient and environmentally friendly treatment processes, overcoming some limitations of conventional treatment methods. The application of nanotechnology in wastewater treatment encompasses a wide range of processes, including adsorption, catalysis, disinfection, and advanced oxidation, making it a multifaceted approach for tackling water pollution challenges (Sharma et al., 2022).

13.3 Wastewater as a Resource

The growing recognition of wastewater as a resource has led to the development of technologies that enable the recovery of multiple resources from wastewater streams (Fig. 13.3). Rather than viewing wastewater solely as a nuisance that needs to be treated and disposed of, the circular economy model encourages the extraction of valuable components, such as energy (through biogas production), nutrients (such as nitrogen and phosphorus), water (through reuse or desalination), and carbon (via biochar production). By integrating energy recovery, nutrient recycling, and water reuse in a single treatment process, wastewater treatment plants can operate more efficiently and reduce the overall environmental impact. Technologies such as anaerobic digestion, membrane filtration, and forward osmosis are being combined in integrated systems that allow for the recovery of multiple resources simultaneously. For instance, biogas produced from anaerobic digestion can be used to generate electricity or heat, while the resulting digestate can be processed into biochar or compost for agricultural use (Dhinesh et al., 2024a). Simultaneously, nutrients such as nitrogen and phosphorus can be recovered and reused as fertilizers, and treated water can be recycled for agricultural, industrial, or even potable uses. This integrated approach reduces the need for external resources, lowers operational costs, and minimizes the environmental footprint of wastewater treatment facilities. Zero-Waste Wastewater Treatment is one of the core concepts which balance both the treatment systems and the resource recovery. The ultimate goal of a zero-waste wastewater treatment system is to recover and reuse all available resources from wastewater, leaving no residual waste (Thabet et al., 2022). This closed-loop system is achievable through the application of multi-resource recovery technologies that target energy, water, and nutrients simultaneously. For instance, technologies like membrane bioreactors (MBRs) can treat wastewater while simultaneously recovering energy and nutrients. Advances in bio electrochemical systems (BES) and microbial fuel cells (MFCs) further contribute to this closed-loop by using microbial processes to treat wastewater, recover energy, and produce by-products that can be reused or valorized. The implementation of zero-waste systems not only helps in reducing the environmental pollution associated with traditional wastewater treatment methods but also promotes sustainability by minimizing reliance on external resources. Additionally, these systems create opportunities for resource recovery, enabling the extraction of valuable materials from wastewater, such as phosphorus for fertilizer production, and carbon for use in biochar and other applications. The adoption of zero-waste systems can also enhance the economic feasibility of wastewater treatment, as the recovered resources can be sold or used in-house, creating new revenue streams for treatment plants. The increased focus on sustainable practices and resource recovery will drive the transition to circular economy models, ensuring that wastewater treatment not only improves water quality but also contributes to a sustainable and resource-efficient future.

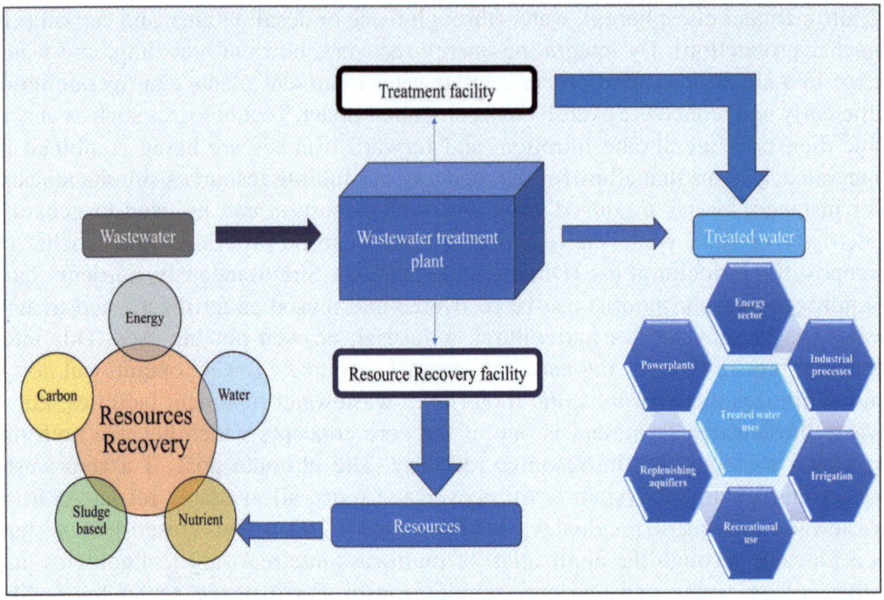

Fig. 13.3 Benefits of treated wastewater and the resources recovered from wastewater

13.4 Potential Breakthroughs in Resource Recovery

13.4.1 Energy Recovery from Wastewater

Energy recovery from wastewater treatment is a key step towards making wastewater treatment facilities more sustainable. Wastewater treatment plants typically require large amounts of energy to process and treat water. However, various emerging technologies are shifting the focus from energy consumption to energy production and recovery, making treatment more energy-efficient and environmentally friendly (Gu et al., 2017). Anaerobic digestion (AD) is a widely used biological process that breaks down organic matter in the absence of oxygen, producing biogas rich in methane (Anukam et al., 2019; Chatterjee & Mazumder, 2019). This process enhances wastewater treatment sustainability by converting waste into a renewable energy source, reducing the energy demand of treatment plants (Mao et al., 2015). However, conventional AD systems struggle with dilute wastewater due to the large reactor volumes required for long hydraulic retention times (HRT), often up to 30 days. To overcome this limitation, high-rate anaerobic systems such as the upflow anaerobic sludge blanket (UASB) reactor have been developed to improve treatment efficiency. UASB reactors enhance microbial activity by facilitating the upward flow of wastewater through a dense sludge bed, reducing HRT while maintaining effective organic matter degradation (Ngwenya et al., 2022; Sarker et al., 2019). Recent advancements in AD focus on optimizing operational parameters such as temperature, pH, and HRT to maximize biogas production and improve microbial community performance. Additionally, co-digestion which is the simultaneous digestion of multiple waste streams) has emerged as a way to enhance the process and increase biogas production (Xie et al., 2018). The integration of advanced microbial consortia, including syntrophic bacteria, has further improved the yield and stability of biogas production (Grimalt-Alemany et al., 2020). Furthermore, research is exploring the use of enzymes or genetic modification to boost the activity of microorganisms involved in methane production. The digestate, or solid by-product of the process, can also be treated and converted into valuable fertilizer, further enhancing the sustainability of the treatment process (Chojnacka & Moustakas, 2024). Microbial fuel cells (MFCs) are an emerging technology which incorporates the use of microorganisms to directlyconvert the chemical energy stored in wastewater organic pollutants into electrical energy (Rahimnejad et al., 2020). In an MFC, microorganisms oxidize organic matter, transferring electrons to an anode, which are then passed through an external circuit to the cathode, generating electricity (Kalathil et al., 2018). The latest research in MFCs focuses on improving the efficiency of electron transfer, particularly between microorganisms and the electrodes, which is often a limiting factor in MFC performance (Prathiba et al., 2022). Advanced materials for the anodes, such as carbon nanotubes or graphene, are being explored to facilitate better electron conductivity and to increase the overall power output of the system (Kothandam et al., 2023). Additionally, optimizing the microbial communities used in MFCs through bioaugmentation or selecting specific strains of bacteria with high electron transfer capabilities can improve the efficiency and scalability of the technology (Wang et al., 2022). In large-scale applications, integrating MFCs with other wastewater treatment processes like anaerobic digestion or activated sludge systems is being investigated to reduce the operational costs and improve overall energy recovery (Dwivedi et al., 2022). Hydrogen is an energy carrier with immense potential as a clean fuel. Wastewater, as an organic waste source, has the potential to be converted into hydrogen through both electrochemical and biological processes (Tak et al., 2022). Electrochemical hydrogen production typically involves electrolysis, where water is split into hydrogen and oxygen using an electric current. In wastewater, this process can be powered using the organic pollutants in the wastewater as the electricity source, providing a dual benefit of wastewater treatment and hydrogen generation (Veeramani et al., 2023). On the biological front, microorganisms such as algae or bacteria can produce hydrogen through fermentation processes, converting organic matter into hydrogen gas (Akhlaghi & Najafpour-Darzi, 2020). Advances in genetic engineering and metabolic pathway manipulation are pushing the boundaries of biological hydrogen production by enhancing the hydrogenase enzymes in microbes, thereby improving the yields (Pandit et al., 2024). Moreover, integrating electrochemical and biological methods could enhance the overall hydrogen production efficiency from wastewater, contributing to cleaner energy production from waste sources.

13.4.2 Water Recovery and Reuse

Water recovery from wastewater is an essential part of ensuring water sustainability, particularly in arid regions or areas experiencing water scarcity. Emerging technologies in water recovery are focusing on improving the efficiency of wastewater treatment and making reuse a more economically viable option. Membrane distillation (MD) is a thermal separation process that uses hydrophobic membranes to separate clean water from contaminated waste water (Julian et al., 2022). The driving force in MD is the difference in vapor pressure between the two sides of the membrane, which allows water vapor to permeate through the membrane while leaving contaminants behind. This technology is particularly useful for treating brine and high-salinity wastewater, such as desalination brine or industrial effluents (Zou et al., 2020). Recent innovations in MD are focused on improving membrane materials to increase flux rates and minimize fouling, which can significantly reduce membrane lifespan and system efficiency (Osman et al., 2024). In order to improve the efficiency of the distillation process, advanced materials such as electro spun nanofibers and composite membranes are being developed (Xu et al., 2022a, 2022b). The ability to recover clean water from high-salinity and difficult-to-treat wastewater is particularly important in regions where freshwater is in short supply (Ahmad et al., 2021). Forward Osmosis can be particularly effective for treating wastewater streams with high levels of dissolved solids and organic contaminants (Mahto et al., 2021). The recent advances in FO technology are aimed at improving the efficiency and practicality of the process. Desalination, the process of removing salts and minerals from seawater or brackish water to produce fresh water, is critical in water-scarce regions (Sabour & Ghorashi, 2024). While traditional desalination methods such as reverse osmosis (RO) are effective, they are energy-intensive, particularly when treating high-salinity water. To reduce energy consumption, research is focusing on low-energy desalination technologies. One such development is the use of low-energy RO membranes, which can operate at lower pressures, thus requiring less energy (Park et al., 2020). Another promising innovation is solar-powered desalination in which solar energy is used to run the desalination processes, particularly through reverse osmosis or distillation techniques. Solar-powered desalination systems are especially relevant in arid and remote regions with abundant sunlight but limited access to grid power (Al-Addous et al., 2024). Additionally, new hybrid systems combining multiple desalination techniques, such as FO and MD, are being developed to enhance the efficiency of water recovery while minimizing energy use.

13.4.3 Nutrient Recovery: Nitrogen and Phosphorus

Nutrient recovery from wastewater is crucial for reducing nutrient pollution in water bodies while also providing sustainable resources for agriculture. Nitrogen and phosphorus are the two primary nutrients involved in eutrophication, a process that leads to the overgrowth of algae and depletion of oxygen in aquatic environments (Ansari et al., 2010). Therefore, the recovery and reuse of these nutrients from wastewater is essential for both environmental protection and agricultural sustainability. Struvite, a crystalline compound composed of magnesium, ammonium, and phosphate (MAP), is one of the most promising methods for recovering phosphorus from wastewater. This process involves adding magnesium to wastewater, raising the pH, and precipitating struvite crystals, which can then be filtered out and used as a slow-release fertilizer (Jia et al., 2014). Innovations in struvite precipitation focus on improving the efficiency and economics of the process. The development of continuous-flow reactors, which allow for the continuous recovery of struvite, has significantly improved recover process (Ramesh Sakthivel et al, 2020). Additionally, research is focused on optimizing the operating conditions, such as pH, temperature, and the ratio of magnesium to phosphorus, to increase yield and reduce chemical consumption (Peng et al., 2018a, 2018b). Bio electrochemical systems (BES) use electricity to drive microbial processes that recover valuable nutrients from wastewater (Bajracharya et al., 2016). In BES, microorganisms in the anodic compartment of an electrochemical reactor reduce nitrogen compounds, such as nitrate or ammonium, while phosphorus is captured at the cathode or through electrochemical precipitation (Wan et al., 2023). Innovative BES technologies are focused on improving the electrochemical conditions, such as the development of more efficient electrode materials (e.g., conducting polymers or carbon composites), optimizing microbial communities to enhance nutrient recovery, and improving reactor designs (Das et al., 2024). BES can be integrated with other wastewater treatment processes, such as anaerobic digestion or activated sludge, to recover nutrients while treating organic pollutants simultaneously. This integration has the potential to create more sustainable and energy-efficient wastewater treatment systems. Adsorption-based methods for nutrient recovery utilize materials that have a high affinity for specific nutrients like nitrogen and phosphorus. Materials such as zeolites, activated carbon, and biochar are effective in adsorbing ammonium, phosphorus, and other nutrients from wastewater (Han et al., 2021) in which these adsorbents can be regenerated and reused, that aid for sustainability and cost-effectiveness. Recent developments in adsorbent materials focus on improving selectivity, adsorption capacity, and regeneration efficiency which for instance, the modified biochar have found to show improved nutrient uptake compared to pristine biochars and traditional adsorbents (Dhinesh et al., 2024b). As more cost-effective and efficient materials are developed, nutrient recovery from wastewater will become an increasingly viable solution for sustainable agricultural practices.

13.4.4 Resource Recovery from Sludge

Sludge, a by-product of wastewater treatment, contains valuable organic matter, nutrients, and trace metals that can be recovered for reuse (Tyagi & Lo, 2013). However, the recovery of these resources from sludge presents significant technical challenges due to its complex composition. Sludge contains valuable nutrients like nitrogen, phosphorus, and organic matter, which can be recovered and processed into fertilizers (Tyagi & Lo, 2013). One of the most common methods for sludge treatment is composting in which microorganisms tend to break down the organic material into humus. Advances in composting technology, such as controlled aeration and temperature regulation, have improved the quality and consistency of the resulting compost (Shi et al., 2018). Thermal treatments, such as pyrolysis and hydrothermal carbonization, can also applied to efficiently convert sludge into biochar, a form of carbon-rich material that has slow-release fertilizer properties. Biochar is particularly valuable in improving soil fertility, water retention, and reducing soil erosion (Gopinath et al., 2021). Sludge also contains trace metals such as gold, silver, copper, and rare earth elements, which are valuable in various industries in which techniques like bioleaching can be applied, where microorganisms extract metals from sludge, and hydrometallurgical processes that use chemical solutions to solubilize metals, are being explored to recover these valuable resources (Gu et al., 2018). These methods are more environmentally friendly compared to traditional mining processes, and they allow for the recycling of metals from waste streams. In addition to precious and rare metals, research is also focused on recovering metals like zinc, nickel, and copper, which are essential in electronics and other industries. As the demand for these metals increases and mining becomes less sustainable, recovering them from wastewater sludge will play an important role in reducing the environmental impact of metal extraction.

13.4.5 Carbon Recovery

The recovery of carbon from organic waste and its use in various applications has become an important strategy for both mitigating climate change and recovering valuable resources. One of the most promising materials for carbon recovery is biochar, which is produced through the pyrolysis of organic waste (Yang et al., 2020a, 2020b). This material is gaining recognition not only for its carbon sequestration potential but also for its versatile applications in wastewater treatment, soil enhancement, and environmental management. Biochar is a stable, carbon-rich material produced by heating organic material in the absence of oxygen through the respective process known as pyrolysis) (Tripathi et al., 2016). During this process, the volatile compounds of the organic material are driven off as gases, and the remaining solid material, rich in carbon, is referred to as biochar. Biochar is highly efficient for long term carbon sequestration rather than other organic components since biochar is highly resistant to further decomposition where organic matter decomposes quickly and releases carbon dioxide into the atmosphere (Leng & Huang, 2018). The ability of biochar to lock carbon in soils for extended periods (potentially thousands of years) is due to its highly stable molecular structure (Dhinesh et al., 2024b). Biochar's porous nature provides a large surface area for adsorption and retention of carbon, making it a long-term reservoir for this greenhouse gas. By integrating biochar into soil, it effectively stores carbon that would otherwise be released into the atmosphere as CO_2 (Lehmann et al., 2021). Furthermore, biochar application has additional soil benefits, such as improving soil structure, water retention, and nutrient availability, making it a win–win solution for both climate mitigation and agricultural productivity. This carbon sequestration property has prompted the incorporation of biochar into wastewater treatment processes, where it can not only capture carbon but also offset the carbon emissions associated with conventional treatment methods, such as aerobic processes, which typically consume large amounts of energy and release greenhouse gases (Dissanayake et al., 2020). Biochar's role in climate change mitigation lies in its potential for large-scale application in agricultural and land management practices. When produced from waste biomass, biochar not only locks away carbon from the atmosphere but also helps reduce the environmental impact of landfills and waste incineration (Tripathi et al., 2019). The higher adsorption property of biochar makes it an effective tool for the remediation of pollutants in which the large surface area and porous nature of biochar is highly efficient in removing a wide range of pollutants such as heavy metals, organic compounds, and other toxic substances commonly found in wastewater (Dhinesh et al., 2024a, b). Biochar has been widely used for the removal of heavy metals such as cadmium, lead, mercury, and arsenic from wastewater through the process of adsorption, in which these toxic metals, if left untreated, can accumulate in water bodies, leading to long-term environmental contamination and potential harm to human health (Qiu et al., 2021). The ability of biochar to effectively adsorb heavy metals is mainly due to the functional groups, such as carboxyl and hydroxyl groups present on the surface of the biochar which interact with the metal ions, thereby removing them from the water. Research has shown that the adsorption capacity of biochar can be influenced by factors such as feedstock type, pyrolysis temperature, and activation processes (Dhinesh et al., 2024a). By optimizing these parameters, biochar's ability to remove metals from wastewater can be enhanced, making it a viable alternative to traditional methods of heavy metal removal, such as chemical precipitation and ion exchange, which often produce secondary waste and require additional processingIn addition to heavy metals, biochar is also effective at absorbing organic pollutants, such as pesticides, pharmaceuticals, and volatile organic compounds (VOCs) which are commonly found in industrial effluents and agricultural runoff, and are also difficult to remove using conventional treatment methods due to their complex molecular structures (Lin et al., 2022). Biochar's large surface area and reactivity make it highly effective in adsorbing these organic substances, preventing them from leaching into water sources and reducing their potential impact on the environment.

13.4.6 Circular Economy in Resource Recovery

The concept of the circular economy has been transformative in rethinking how wastewater treatment is approached (Fig. 13.4). Thus, wastewater is not only seen as a waste product, but rather as a valuable resource that can be used to produce valuable

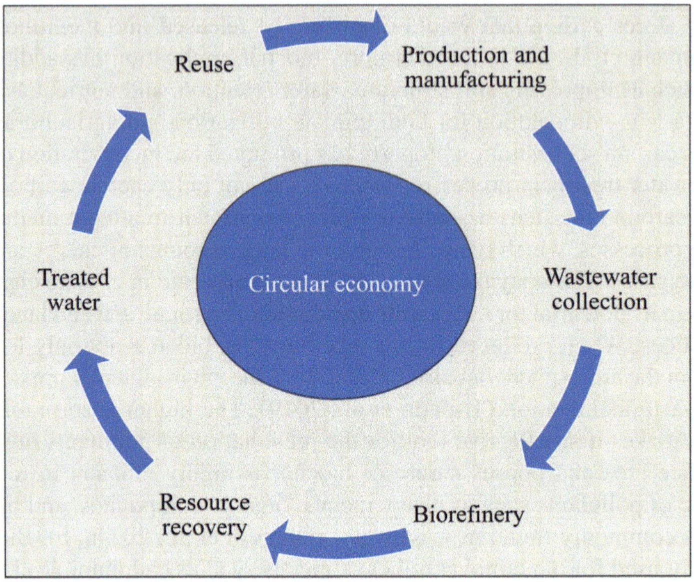

Fig. 13.4 Circular economic approach of wastewater treatment

resources such as energy, nutrients, water, and even carbon (Guerra-Rodríguez et al., 2020). By recovering these resources from wastewater, we reduce the environmental footprint of treatment processes and contribute to a more sustainable system where waste is minimized, and valuable materials are continuously reused.

13.5 Conclusion

The technologies explored in the chapter have showcased significant advancements in wastewater treatment processes, addressing the limitations of traditional methods and expanding the capabilities of modern systems. Key technologies such as Advanced Oxidation Processes, Membrane Bioreactors, Forward Osmosis, Microbial Fuel Cells, and Microalgae-Integrated Recovery have proven to be at the forefront of innovation, offering solutions to critical challenges such as the removal of emerging contaminants, improving effluent quality, and enhancing resource recovery. These technologies not only offer environmental and operational benefits but also provide opportunities for sustainable practices by turning waste into valuable resources, thereby supporting circular economy initiatives. Their application in municipal, industrial, and agricultural wastewater management systems holds the potential to revolutionize wastewater treatment, making it more efficient, sustainable, and economically viable. Advanced wastewater treatment technologies hold the promise of reshaping the landscape of environmental sustainability and resource recovery.

The ability to recover valuable resources from wastewater, such as water, energy, and nutrients, while reducing environmental impacts, positions these technologies as key enablers of sustainable water management. Technologies such as AOPs, MBRs, and FO not only enhance pollutant degradation but also support the recovery of water and nutrients that can be reused in various sectors, including agriculture and industrial processes. In terms of sustainability, these technologies contribute to reducing the overall environmental footprint of wastewater treatment by lowering energy consumption, improving effluent quality, and facilitating resource recovery. The convergence of these technologies with the principles of circular economy, where waste is minimized and resources are continuously reused, offers a pathway to a more sustainable, resilient, and resource-efficient wastewater treatment system. Besides, the integration of artificial intelligence (AI) and machine learning (ML) into wastewater management systems is a growing area of interest in which AI and ML have the potential to optimize operational parameters, predict treatment outcomes, and improve the overall efficiency of wastewater treatment plants (Kotsialos et al., 2021). Further research in this field will help refine these algorithms and ensure their applicability in real-time decision-making, potentially transforming the way wastewater treatment is managed globally. To accelerate the widespread adoption of these technologies, a concerted effort is needed from governments, regulatory bodies, and industry stakeholders to provide the necessary policy frameworks, incentives, and funding.

References

Abdelfattah, A., Ali, S. S., Ramadan, H., El-Aswar, E. I., Eltawab, R., Ho, S.-H., Elsamahy, T., Li, S., El-Sheekh, M. M., Schagerl, M., Kornaros, M., & Sun, J. (2023). Microalgae-based wastewater treatment: Mechanisms, challenges, recent advances, and future prospects. *Environmental Science and Ecotechnology, 13*, Article 100205. https://doi.org/10.1016/j.ese.2022.100205

Aghababaie, M., Farhadian, M., Jeihanipour, A., & Biria, D. (2015). Effective factors on the performance of microbial fuel cells in wastewater treatment—A review. *Environmental Technology Reviews, 4*(1), 71–89. https://doi.org/10.1080/09593330.2015.1077896

Ahmad, N. N. R., Ang, W. L., Leo, C. P., Mohammad, A. W., & Hilal, N. (2021). Current advances in membrane technologies for saline wastewater treatment: A comprehensive review. *Desalination, 517*, Article 115170. https://doi.org/10.1016/j.desal.2021.115170

Akhlaghi, N., & Najafpour-Darzi, G. (2020). A comprehensive review on biological hydrogen production. *International Journal of Hydrogen Energy, 45*(43), 22492–22512. https://doi.org/10.1016/j.ijhydene.2020.06.182

Al-Addous, M., Bdour, M., Rabaiah, S., Boubakri, A., Schweimanns, N., Barbana, N., & Wellmann, J. (2024). Innovations in solar-powered desalination: A comprehensive review of sustainable solutions for water scarcity in the middle east and North Africa (MENA) region. *Water, 16*(13), 1877. https://doi.org/10.3390/w16131877

Amin, M. T., Alazba, A. A., & Manzoor, U. (2014). A Review of removal of pollutants from water/wastewater using different types of nanomaterials. *Advances in Materials Science and Engineering, 2014*, 1–24. https://doi.org/10.1155/2014/825910

Ansari, A. A., Gill, S. S., & Khan, F. A. (2010). Eutrophication: threat to aquatic ecosystems. *Eutrophication: Causes, Consequences and Control*, 143–170. https://doi.org/10.1007/978-90-481-9625-8_7

Anukam, A., Mohammadi, A., Naqvi, M., & Granström, K. (2019). A review of the chemistry of anaerobic digestion: Methods of accelerating and optimizing process efficiency. *Processes, 7*(8), 504. https://doi.org/10.3390/pr7080504

Asghar, N., Hussain, A., Duc Anh Nguyen, Ali, S., Hussain, I., Aurangzeb Junejo, & Ali, A. (2024). Advancement in nanomaterials for environmental pollutants remediation: A systematic review on bibliometrics analysis, material types, synthesis pathways, and related mechanisms. *Journal of Nanobiotechnology, 22*(1). https://doi.org/10.1186/s12951-023-02151-3

Bajracharya, S., Sharma, M., Mohanakrishna, G., Benneton, X. D., Strik, D. P., Sarma, P. M., & Pant, D. (2016). An overview on emerging bioelectrochemical systems (BESs): Technology for sustainable electricity, waste remediation, resource recovery, chemical production and beyond. *Renewable Energy, 98*, 153–170. https://doi.org/10.1016/j.renene.2016.03.002

Bakhta, B., & Sanromán, M. Á. (2024). Polymer-supported heterogeneous Fenton catalysts for the environmental remediation of wastewater. *Molecules, 29*(10), 2188–2188. https://doi.org/10.3390/molecules29102188

Bera, S. P., Godhaniya, M., & Kothari, C. (2021). Emerging and advanced membrane technology for wastewater treatment: A review. *Journal of Basic Microbiology, 62*(3–4), 245–259. https://doi.org/10.1002/jobm.202100259

Bhardwaj, A. K., Sundaram, S., Yadav, K. K., & Srivastav, A. L. (2021). An overview of silver nano-particles as promising materials for water disinfection. *Environmental Technology and Innovation, 23*, Article 101721. https://doi.org/10.1016/j.eti.2021.101721

Blandin, G., Ferrari, F., Lesage, G., Le-Clech, P., Héran, M., & Martinez-Lladó, X. (2020). Forward osmosis as concentration process: Review of opportunities and challenges. *Membranes, 10*(10), 284. https://doi.org/10.3390/membranes10100284

Boas, J. V., Oliveira, V. B., Simões, M., & Pinto, A. M. F. R. (2022). Review on microbial fuel cells applications, developments and costs. *Journal of Environmental Management, 307*, Article 114525. https://doi.org/10.1016/j.jenvman.2022.114525

Bodzek, M. (2015). Membrane technologies for the removal of micropollutants in water treatment. In *Advances in membrane technologies for water treatment* (pp. 465–517). Woodhead Publishing. https://doi.org/10.1016/b978-1-78242-121-4.00015-0

Chaoui, I., Abderafi, S., Vaudreuil, S., & Bounahmidi, T. (2019). Water desalination by forward osmosis: Draw solutes and recovery methods—Review. *Environmental Technology Reviews, 8*(1), 25–46. https://doi.org/10.1080/21622515.2019.1623324

Chatterjee, B., & Mazumder, D. (2019). Role of stage-separation in the ubiquitous development of Anaerobic digestion of organic fraction of municipal solid waste: A critical review. *Renewable and Sustainable Energy Reviews, 104*, 439–469. https://doi.org/10.1016/j.rser.2019.01.026

Cheng, M., Zeng, G., Huang, D., Lai, C., Xu, P., Zhang, C., & Liu, Y. (2016). Hydroxyl radicals based advanced oxidation processes (AOPs) for remediation of soils contaminated with organic compounds: A review. *Chemical Engineering Journal, 284*, 582–598. https://doi.org/10.1016/j.cej.2015.09.001

Chia, W. Y., Chia, S. R., Khoo, K. S., Chew, K. W., & Show, P. L. (2021). Sustainable membrane technology for resource recovery from wastewater: Forward osmosis and pressure retarded osmosis. *Journal of Water Process Engineering, 39*, Article 101758. https://doi.org/10.1016/j.jwpe.2020.101758

Chojnacka, K., & Moustakas, K. (2024). Anaerobic digestate management for carbon neutrality and fertilizer use: A review of current practices and future opportunities. *Biomass and Bioenergy, 180*, Article 106991. https://doi.org/10.1016/j.biombioe.2023.106991

Cuellar-Bermudez, S. P., Aleman-Nava, G. S., Chandra, R., Garcia-Perez, J. S., Contreras-Angulo, J. R., Markou, G., Muylaert, K., Rittmann, B. E., & Parra-Saldivar, R. (2017). Nutrients utilization and contaminants removal. A review of two approaches of algae and cyanobacteria in wastewater. *Algal Research, 24*, 438–449. https://doi.org/10.1016/j.algal.2016.08.018

Daniel, D. K., Mankidy, B. D., Ambarish, K., & Manogari, R. (2009). Construction and operation of a microbial fuel cell for electricity generation from wastewater. *International Journal of Hydrogen Energy, 34*(17), 7555–7560. https://doi.org/10.1016/j.ijhydene.2009.06.012

Darwish, M., Aris, A., Puteh, M. H., Abideen, M. Z., & Othman, M. N. (2016). Ammonium-nitrogen recovery from wastewater by struvite crystallization technology. *Separation and Purification Reviews, 45*(4), 261–274. https://doi.org/10.1080/15422119.2015.1119699

Das, I. (2024). Bioelectrochemical systems (BES) for wastewater treatment. *Earth and Environmental Sciences Library*, 383–401. https://doi.org/10.1007/978-3-031-63046-0_16

Dhinesh, R., Aruna, S., Ravaneswaran, K., Kirthiga, S. S., Keerthivarman, S. G., Al-Ghanim, K. A., & Kamaraj, C. (2024). Aquatic weed Eichhornia crassipes as sustainable feedstocks for biochar production: A potential of ammonium adsorption and kinetic models. *Journal of Cleaner Production, 459*, Article 142566. https://doi.org/10.1016/j.jclepro.2024.142566

Dissanayake, P. D., You, S., Igalavithana, A. D., Xia, Y., Bhatnagar, A., Gupta, S., Kua, H. W., Kim, S., Kwon, J.-H., Tsang, D. C. W., & Ok, Y. S. (2020). Biochar-based adsorbents for carbon dioxide capture: A critical review. *Renewable and Sustainable Energy Reviews, 119*, Article 109582. https://doi.org/10.1016/j.rser.2019.109582

Dong, F., Simoska, O., Gaffney, E., & Minteer, S. D. (2022). Applying synthetic biology strategies to bioelectrochemical systems. *Electrochemical Science Advances*. https://doi.org/10.1002/elsa.202100197

Du, X., Shi, Y., Jegatheesan, V., & Haq, I. U. (2020). A Review on the Mechanism, Impacts and Control Methods of Membrane Fouling in MBR System. *Membranes, 10*(2). https://doi.org/10.3390/membranes10020024

Dwivedi, K. A., Huang, S.-J., & Wang, C.-T. (2022). Integration of various technology-based approaches for enhancing the performance of microbial fuel cell technology: A review. *Chemosphere, 287*, Article 132248. https://doi.org/10.1016/j.chemosphere.2021.132248

El Batouti, M., Al-Harby, N. F., & Elewa, M. M. (2021). A review on promising membrane technology approaches for heavy metal removal from water and wastewater to solve water crisis. *Water, 13*(22), 3241. https://doi.org/10.3390/w13223241

Ezeako, E. C. (2024). Harnessing synthetic biology for sustainable industrial innovation: Advances. *Challenges, and Future Direction*. https://doi.org/10.2139/ssrn.4991755

Fan, X., Fu, Q., Liu, G., Jia, H., Dong, X., Li, Y.-F., & Cui, S. (2024). Applying molecular oxygen for organic pollutant degradation: Strategies, mechanisms, and perspectives. *Environmental Science and Ecotechnology, 22*, Article 100469. https://doi.org/10.1016/j.ese.2024.100469

Fida, Z., Price, W. E., Pramanik, B. K., Dhar, B. R., Kumar, M., Jiang, G., & Hai, F. I. (2021). Reduction of excess sludge production by membrane bioreactor coupled with anoxic side-stream reactors. *Journal of Environmental Management, 281*, Article 111919. https://doi.org/10.1016/j.jenvman.2020.111919

Gopinath, A., Divyapriya, G., Srivastava, V., Laiju, A. R., Nidheesh, P. V., & Kumar, M. S. (2021). Conversion of sewage sludge into biochar: A potential resource in water and wastewater treatment. *Environmental Research, 194*, Article 110656. https://doi.org/10.1016/j.envres.2020.110656

Grimalt-Alemany, A., Łężyk, M., Kennes-Veiga, D. M., Skiadas, I. V., & Gavala, H. N. (2020). Enrichment of mesophilic and thermophilic mixed microbial consortia for syngas biomethanation: The role of kinetic and thermodynamic competition. *Waste and Biomass Valorization, 11*(2), 465–481. https://doi.org/10.1007/s12649-019-00595-z

Gu, T., Rastegar, S. O., Mousavi, S. M., Li, M., & Zhou, M. (2018). Advances in bioleaching for recovery of metals and bioremediation of fuel ash and sewage sludge. *Bioresource Technology, 261*, 428–440. https://doi.org/10.1016/j.biortech.2018.04.033

Gu, Y., Li, Y., Li, X., Luo, P., Wang, H., Robinson, Z. P., Wang, X., Wu, J., & Li, F. (2017). The feasibility and challenges of energy self-sufficient wastewater treatment plants. *Applied Energy, 204*, 1463–1475. https://doi.org/10.1016/j.apenergy.2017.02.069

Gude, V. G. (2016). Wastewater treatment in microbial fuel cells—An overview. *Journal of Cleaner Production, 122*, 287–307. https://doi.org/10.1016/j.jclepro.2016.02.022

Guerra-Rodríguez, S., Oulego, P., Rodríguez, E., Singh, D. N., & Rodríguez-Chueca, J. (2020). Towards the implementation of circular economy in the wastewater sector: Challenges and opportunities. *Water, 12*(5), 1431. https://doi.org/10.3390/w12051431

Han, B., Butterly, C., Zhang, W., He, J., & Chen, D. (2021). Adsorbent materials for ammonium and ammonia removal: A review. *Journal of Cleaner Production, 283*, Article 124611. https://doi.org/10.1016/j.jclepro.2020.124611

Hasan, M. N., Altaf, M. M., Khan, N. A., Khan, A. H., Khan, A. A., Ahmed, S., Kumar, P. S., Naushad, M., Rajapaksha, A. U., Iqbal, J., Tirth, V., & Islam, S. (2021). Recent technologies for nutrient removal and recovery from wastewaters: A review. *Chemosphere, 277*, Article 130328. https://doi.org/10.1016/j.chemosphere.2021.130328

Hassan, M., Kanwal, S., Singh, R. S., Ali SA, M., Anwar, M., & Zhao, C. (2024). Current challenges and future perspectives associated with configuration of microbial fuel cell for simultaneous energy generation and wastewater treatment. *International Journal of Hydrogen Energy*. https://doi.org/10.1016/j.ijhydene.2023.08.134

He, L., Du, P., Chen, Y., Lu, H., Cheng, X., Chang, B., & Wang, Z. (2017). Advances in microbial fuel cells for wastewater treatment. *Renewable and Sustainable Energy Reviews, 71*, 388–403. https://doi.org/10.1016/j.rser.2016.12.069

He, L., Zhang, S., Chen, Y., Ren, L., & Li, J. (2018). Techno-economic potential of a renewable energy-based microgrid system for a sustainable large-scale residential community in Beijing, China. *Renewable and Sustainable Energy Reviews, 93*, 631–641. https://doi.org/10.1016/j.rser.2018.05.053

https://doi.org/10.1016/j.chemosphere.2023.140473

Ibraheem, B. M., Aani, S. A., Alsarayreh, A. A., Alsalhy, Q. F., & Salih, I. K. (2023). Forward osmosis membrane: Review of fabrication. *Modification, Challenges and Potential, 13*(4), 379–379. https://doi.org/10.3390/membranes13040379

Jia, G. (2014). *Nutrient removal and recovery by the precipitation of magnesium ammonium phosphate* (Doctoral dissertation, University of Adelaide, School of Chemical Engineering).

Jiang, L., Li, Y., & Pei, H. (2021). Algal–bacterial consortia for bioproduct generation and wastewater treatment. *Renewable and Sustainable Energy Reviews, 149*, Article 111395. https://doi.org/10.1016/j.rser.2021.111395

Jijingi, H. E., Yazdia, S. K., Abakr, Y. A., & Etim, E. (2024). Evaluation of membrane bioreactor (MBR) technology for industrial wastewater treatment and its application in developing countries: A review. *Case Studies in Chemical and Environmental Engineering*, 100886. https://doi.org/10.1016/j.cscee.2024.100886

Jin, D., Zhang, X., Zhou, L., Zhang, X., & Wu, P. (2024). Emerging applications and mechanisms of algal-bacterial symbiosis on sustainable wastewater treatment: A comprehensive review. *Journal of Water Process Engineering, 65*, Article 105912. https://doi.org/10.1016/j.jwpe.2024.105912

Joseph, T. M., Al-Hazmi, H. E., Śniatała, B., Esmaeili, A., & Habibzadeh, S. (2023). Nanoparticles and nanofiltration for wastewater treatment: From polluted to fresh water. *Environmental Research, 238*, Article 117114. https://doi.org/10.1016/j.envres.2023.117114

Julian, H., Nurgirisia, N., Qiu, G., Ting, Y.-P., & Wenten, I. G. (2022). Membrane distillation for wastewater treatment: Current trends, challenges and prospects of dense membrane distillation. *Journal of Water Process Engineering, 46*, Article 102615. https://doi.org/10.1016/j.jwpe.2022.102615

Kalathil, S., Patil, S. A., & Pant, D. (2018). Microbial fuel cells: Electrode materials. *Encyclopedia of Interfacial Chemistry*, 309–318. https://doi.org/10.1016/b978-0-12-409547-2.13459-6

Khan, Z.U.H., Gul, N.S., Sabahat, S., Sun, J., Tahir, K., Shah, N.S., Muhammad, N., Rahim, A., Imran, M., Iqbal, J. and Khan, T.M. (2023). Removal of organic pollutants through hydroxyl radical-based advanced oxidation processes. *Ecotoxicology and Environmental Safety, 267*, 115564. https://doi.org/10.1016/j.ecoenv.2023.115564

Khandaker, S., Das, S., Hossain, Md. T., Islam, A., Miah, M. R., & Awual, Md. R. (2021). Sustainable approach for wastewater treatment using microbial fuel cells and green energy generation—A comprehensive review. *Journal of Molecular Liquids, 344*, Article 117795. https://doi.org/10.1016/j.molliq.2021.117795

Kothandam, G., Singh, G., Guan, X., Lee, J. M., Ramadass, K., Joseph, S., Benzigar, M., Karakoti, A., Yi, J., Kumar, P., & Vinu, A. (2023). Recent advances in Carbon-Based electrodes for energy storage and conversion. *Advanced Science*. https://doi.org/10.1002/advs.202301045

Lai, L., & Chun, H. U. (2017). Heterogeneous Fenton catalytic water treatment technology and mechanism. *Progress in Chemistry, 29*(9), 981. https://doi.org/10.7536/PC170552

Lehmann, J., Cowie, A., Masiello, C. A., Kammann, C., Woolf, D., Amonette, J. E., Cayuela, M. L., Camps-Arbestain, M., & Whitman, T. (2021). Biochar in climate change mitigation. *Nature Geoscience, 14*(12), 883–892. https://doi.org/10.1038/s41561-021-00852-8

Leng, L., & Huang, H. (2018). An overview of the effect of pyrolysis process parameters on biochar stability. *Bioresource Technology, 270*, 627–642. https://doi.org/10.1016/j.biortech.2018.09.030

Li, K., Liu, Q., Fang, F., Luo, R., Lu, Q., Zhou, W., Huo, S., Cheng, P., Liu, J., Addy, M., Chen, P., Chen, D., & Ruan, R. (2019). Microalgae-based wastewater treatment for nutrients recovery: A review. *Bioresource Technology, 291*, Article 121934. https://doi.org/10.1016/j.biortech.2019.121934

Lin, Q., Tan, X., Almatrafi, E., Yang, Y., Wang, W., Luo, H., Qin, F., Zhou, C., Zeng, G. & Zhang, C. (2022). Effects of biochar-based materials on the bioavailability of soil organic pollutants and their biological impacts. *Science of The Total Environment, 826*, 153956. https://doi.org/10.1016/j.scitotenv.2022.153956

Liu, J., Pemberton, B., Lewis, J., Scales, P. J., & Martin, G. J. O. (2020). Wastewater treatment using filamentous algae—A review. *Bioresource Technology, 298*, Article 122556. https://doi.org/10.1016/j.biortech.2019.122556

Mahto, A., Aruchamy, K., Meena, R., Kamali, M., Nataraj, S. K., & Aminabhavi, T. M. (2021). Forward osmosis for industrial effluents treatment—sustainability considerations. *Separation and Purification Technology, 254*, Article 117568. https://doi.org/10.1016/j.seppur.2020.117568

Manna, M., & Sen, S. (2022). Advanced oxidation process: A sustainable technology for treating refractory organic compounds present in industrial wastewater. *Environmental Science and Pollution Research*. https://doi.org/10.1007/s11356-022-19435-0

Mao, C., Feng, Y., Wang, X., & Ren, G. (2015). Review on research achievements of biogas from anaerobic digestion. *Renewable and Sustainable Energy Reviews, 45*, 540–555. https://doi.org/10.1016/j.rser.2015.02.032

Maryjoseph, S., & Ketheesan, B. (2020). Microalgae based wastewater treatment for the removal of emerging contaminants: A review of challenges and opportunities. *Case Studies in Chemical and Environmental Engineering, 2*, Article 100046. https://doi.org/10.1016/j.cscee.2020.100046

Mathimani, T., & Mallick, N. (2018). A comprehensive review on harvesting of microalgae for biodiesel – Key challenges and future directions. *Renewable and Sustainable Energy Reviews, 91*, 1103–1120. https://doi.org/10.1016/j.rser.2018.04.083

Mohapatra, L., Cheon, D., & Yoo, S. H. (2023). Carbon-based nanomaterials for catalytic wastewater treatment: A review. *Molecules, 28*(4), 1805. https://doi.org/10.3390/molecules28041805

Molazadeh, M., Ahmadzadeh, H., Pourianfar, H. R., Lyon, S., & Rampelotto, P. H. (2019). The use of microalgae for coupling wastewater treatment with CO_2 Biofixation. *Frontiers in Bioengineering and Biotechnology, 7*. https://doi.org/10.3389/fbioe.2019.00042

Monachan, M., Dixit, N., Maliyekkal, S. M., & P. Singh, S. (2022). Reverse osmosis (RO) and nanofiltration (NF) membranes for emerging contaminants (ECs) removal. *New Trends in Emerging Environmental Contaminants*, 407–425. https://doi.org/10.1007/978-981-16-8367-1_17

Mukherjee, J., Lodh, B. K., Sharma, R., Mahata, N., Shah, M. P., Mandal, S., & Bhunia, B. (2023). Advanced oxidation process for the treatment of industrial wastewater: A review on strategies, mechanisms, bottlenecks and prospects. *Chemosphere, 345*, Article 140473.

Nematov, D. (2024). Titanium dioxide and photocatalysis: A detailed overview of the synthesis, applications, challenges, advances and prospects for sustainable development. *Applications,*

Challenges, Advances and Prospects for Sustainable Development (October 02, 2024). https://doi.org/10.20944/preprints202408.1733.v1

Nguyen, L. N., Aditya, L., Vu, H. P., Johir, A. H., Bennar, L., Ralph, P., Hoang, N. B., Zdarta, J., & Nghiem, L. D. (2022). Nutrient removal by algae-based wastewater treatment. *Current Pollution Reports, 8*. https://doi.org/10.1007/s40726-022-00230-x

Nguyen, X. C., Nguyen, T. T. H., Tran, Q. B., Bui, X. T., Ngo, H. H., & Nguyen, D. D. (2022). Artificial intelligence for wastewater treatment. In *Current developments in biotechnology and bioengineering* (pp. 587–608). Elsevier. https://doi.org/10.1016/B978-0-323-99874-1.00008-7

Ngwenya, N., Gaszynski, C., & Ikumi, D. (2022). A review of winery wastewater treatment: A focus on UASB biotechnology optimisation and recovery strategies. *Journal of Environmental Chemical Engineering, 10*(4), Article 108172. https://doi.org/10.1016/j.jece.2022.108172

Olias, L. G., & Di Lorenzo, M. (2021). Microbial fuel cells for in-field water quality monitoring. *RSC Advances, 11*(27), 16307–16317. https://doi.org/10.1039/d1ra01138c

Oruganti, R. K., Katam, K., Show, P. L., Gadhamshetty, V., Upadhyayula, V. K. K., & Bhattacharyya, D. (2022). A comprehensive review on the use of algal-bacterial systems for wastewater treatment with emphasis on nutrient and micropollutant removal. *Bioengineered, 13*(4), 10412–10453. https://doi.org/10.1080/21655979.2022.2056823

Osman, A. I., Chen, Z., Elgarahy, A. M., Farghali, M., Mohamed, A. K. P., Hawash, H. B., & Yap, P. (2024). Membrane technology for energy saving: Principles, techniques, applications, challenges, and prospects. *Advanced Energy and Sustainability Research*. https://doi.org/10.1002/aesr.202400011

Pandit, C., Kumar, M., Chander Kuhad, R., Sharma, K., Roy, A., Shukla, R., Pandit, S., Ranjan, N., Mishra, S. K., & Prasad, R. (2024). Exploring the performance of biocatalysts for biohydrogen production. *Biocatalysis and Agricultural Biotechnology, 60*, Article 103290. https://doi.org/10.1016/j.bcab.2024.103290

Park, K., Kim, J., Yang, D. R., & Hong, S. (2020). Towards a low-energy seawater reverse osmosis desalination plant: A review and theoretical analysis for future directions. *Journal of Membrane Science, 595*, Article 117607. https://doi.org/10.1016/j.memsci.2019.117607

Paumo, H. K., Dalhatou, S., Katata-Seru, L. M., Kamdem, B. P., Tijani, J. O., Vishwanathan, V., Kane, A., & Bahadur, I. (2021). TiO_2 assisted photocatalysts for degradation of emerging organic pollutants in water and wastewater. *Journal of Molecular Liquids, 331*, Article 115458. https://doi.org/10.1016/j.molliq.2021.115458

Pendergast, M. M., & Hoek, E. M. V. (2011). A review of water treatment membrane nanotechnologies. *Energy and Environmental Science, 4*(6), 1946. https://doi.org/10.1039/c0ee00541j

Peng, L., Dai, H., Wu, Y., Peng, Y., & Lu, X. (2018a). A comprehensive review of phosphorus recovery from wastewater by crystallization processes. *Chemosphere, 197*, 768–781. https://doi.org/10.1016/j.chemosphere.2018.01.098

Peng, T., Kellens, K., Tang, R., Chen, C., & Chen, G. (2018b). Sustainability of additive manufacturing: An overview on its energy demand and environmental impact. *Additive Manufacturing, 21*, 694–704. https://doi.org/10.1016/j.addma.2018.04.022

Prathiba, S., Kumar, P. S., & Vo, D.-V.N. (2022). Recent advancements in microbial fuel cells: A review on its electron transfer mechanisms, microbial community, types of substrates and design for bio-electrochemical treatment. *Chemosphere, 286*, Article 131856. https://doi.org/10.1016/j.chemosphere.2021.131856

Qiu, B., Tao, X., Wang, H., Li, W., Ding, X., & Chu, H. (2021). Biochar as a low-cost adsorbent for aqueous heavy metal removal: A review. *Journal of Analytical and Applied Pyrolysis, 155*, Article 105081. https://doi.org/10.1016/j.jaap.2021.105081

Qrenawi, L. I., & Rabah, F. K. (2023). Membrane bioreactor (MBR) as a reliable Technology for Wastewater Treatment. *Journal of Membrane Science and Research, 9*(1). https://doi.org/10.22079/jmsr.2022.548826.1532

Rahimnejad, M., Asghary, M., & Fallah, M. (2020). Microbial fuel cell (MFC): An innovative technology for wastewater treatment and power generation. *Bioremediation of Industrial Waste for Environmental Safety: Volume II: Biological Agents and Methods for Industrial Waste Management*, 215–235. https://doi.org/10.1007/978-981-13-3426-9_9

Ramesh Sakthivel, S., Azizurrahaman, M., Ganesh Prabhu, V., & Chariar, V. M. (2020). Recovery of phosphorus from stored urine using continuous flow reactor in decentralised level operations. *Blue-Green Systems, 2*(1), 237–249. https://doi.org/10.2166/bgs.2020.005

Rekhate, C. V., & Srivastava, J. K. (2020). Recent advances in ozone-based advanced oxidation processes for treatment of wastewater—A review. *Chemical Engineering Journal Advances, 3*, Article 100031. https://doi.org/10.1016/j.ceja.2020.100031

Ren, Z., Li, H., Sun, P., Fu, R., Bai, Z., Zhang, G., Sun, L., & Wei, Y. (2024). Development and challenges of emerging biological technologies for algal-bacterial symbiosis systems: A review. *Bioresource Technology, 413*, 131459–131459. https://doi.org/10.1016/j.biortech.2024.131459

Sabour, S. M. J. S., & Ghorashi, B. (2024). A comprehensive review of major water desalination techniques and mineral extraction from saline water. *Separation and Purification Technology*, 127913. https://doi.org/10.1016/j.seppur.2024.127913

Saleh, T. A. (2024). Materials, nanomaterials, nanocomposites, and methods used for the treatment and removal of hazardous pollutants from wastewater: Treatment technologies for water recycling and sustainability. *Nano-Structures and Nano-Objects, 39*, 101231–101231. https://doi.org/10.1016/j.nanoso.2024.101231

Saravanan, A., Kumar, P. S., Varjani, S., Jeevanantham, S., Yaashikaa, P. R., Thamarai, P., Abirami, B., & George, C. S. (2021). A review on algal-bacterial symbiotic system for effective treatment of wastewater. *Chemosphere, 271*, Article 129540. https://doi.org/10.1016/j.chemosphere.2021.129540

Sarker, S., Lamb, J. J., Hjelme, D. R., & Lien, K. M. (2019). A review of the role of critical parameters in the design and operation of biogas production plants. *Applied Sciences, 9*(9), 1915. https://doi.org/10.3390/app9091915

Shanmuganathan, R., Kadri, M. S., Mathimani, T., Le, Q. H., & Pugazhendhi, A. (2023). Recent innovations and challenges in the eradication of emerging contaminants from aquatic systems. *Chemosphere, 332*, Article 138812. https://doi.org/10.1016/j.chemosphere.2023.138812

Sharma, S., Dhingra, P., & Jain, S. (2022). Advancements and challenges of nanotechnology in waste water treatment. *Materials Today: Proceedings*. https://doi.org/10.1016/j.matpr.2022.09.481

Shi, S., Xu, G., Yu, H., & Zhang, Z. (2018). Strategies of valorization of sludge from wastewater treatment. *Journal of Chemical Technology and Biotechnology, 93*(4), 936–944. https://doi.org/10.1002/jctb.5548

Singh, S. K., Sharma, C., & Maiti, A. (2021a). A comprehensive review of standalone and hybrid forward osmosis for water treatment: Membranes and recovery strategies of draw solutions. *Journal of Environmental Chemical Engineering, 9*(4), Article 105473. https://doi.org/10.1016/j.jece.2021.105473

Singh, S., Kapoor, D., Khasnabis, S., Singh, J., & Ramamurthy, P. C. (2021b). Mechanism and kinetics of adsorption and removal of heavy metals from wastewater using nanomaterials. *Environmental Chemistry Letters, 19*(3), 2351–2381. https://doi.org/10.1007/s10311-021-01196-w

Su, Y. (2021). Revisiting carbon, nitrogen, and phosphorus metabolisms in microalgae for wastewater treatment. *Science of the Total Environment, 762*, Article 144590. https://doi.org/10.1016/j.scitotenv.2020.144590

Tak, S. S., Shetye, O., Muley, O., Jaiswal, H., & Malik, S. N. (2022). Emerging technologies for hydrogen production from wastewater. *International Journal of Hydrogen Energy*. https://doi.org/10.1016/j.ijhydene.2022.06.225

Tariq, A., & Mushtaq, A. (2023). Untreated wastewater reasons and causes: A review of most affected areas and cities. *International Journal of Chemical and Biochemical Science, 23*(1), 121–143. (ISSN 2226–9614)

Thabet, R. H., Fouad, M. K., Sherbiny, S. A. E., & Tony, M. A. (2022). Zero-waste approach: Assessment of aluminum-based waste as a photocatalyst for industrial wastewater treatment

ecology. *International Journal of Environmental Research, 16*(3). https://doi.org/10.1007/s41742-022-00414-9
Tripathi, M., Sahu, J. N., & Ganesan, P. (2016). Effect of process parameters on production of biochar from biomass waste through pyrolysis: A review. *Renewable and Sustainable Energy Reviews, 55*, 467–481. https://doi.org/10.1016/j.rser.2015.10.122
Tripathi, N., Hills, C. D., Singh, R. S., & Atkinson, C. J. (2019). Biomass waste utilisation in low-carbon products: harnessing a major potential resource. *Npj Climate and Atmospheric Science, 2*(1). https://doi.org/10.1038/s41612-019-0093-5
Tyagi, V. K., & Lo, S.-L. (2013). Sludge: A waste or renewable source for energy and resources recovery? *Renewable and Sustainable Energy Reviews, 25*, 708–728. https://doi.org/10.1016/j.rser.2013.05.029
Ugwu, E. I., Karri, R. R., Nnaji, C. C., John, J., Padmanaban, V. C., Othmani, A., Ikechukwu, E. L., & Helal, W. M. K. (2022). Application of green nanocomposites in removal of toxic chemicals, heavy metals, radioactive materials, and pesticides from aquatic water bodies. *Sustainable Nanotechnology for Environmental Remediation*, 321–346. https://doi.org/10.1016/b978-0-12-824547-7.00018-7
Vasumathi, K. K., Premalatha, M., & Subramanian, P. (2012). Parameters influencing the design of photobioreactor for the growth of microalgae. *Renewable and Sustainable Energy Reviews, 16*(7), 5443–5450. https://doi.org/10.1016/j.rser.2012.06.013
Veeramani, K., Janani, G., Kim, J., Surendran, S., Lim, J., Jesudass, S. C., & Sim, U. (2023). Hydrogen and value-added products yield from hybrid water electrolysis: A critical review on recent developments. *Renewable and Sustainable Energy Reviews, 177*, Article 113227. https://doi.org/10.1016/j.rser.2023.113227
Wan, Y., Li, R., Wang, X., & Liao, C. (2023). Recovery of reactive nitrogen from wastewater using bioelectrochemical systems. *Separation and Purification Technology, 327*, Article 125002. https://doi.org/10.1016/j.seppur.2023.125002
Wang, J., Ren, K., Zhu, Y., Huang, J., & Liu, S. (2022). A review of recent advances in microbial fuel cells: Preparation, operation, and application. *Biotech, 11*(4), 44. https://doi.org/10.3390/biotech11040044
Wang, J.-H., Zhang, T.-Y., Dao, G.-H., Xu, X.-Q., Wang, X.-X., & Hu, H.-Y. (2017). Microalgae-based advanced municipal wastewater treatment for reuse in water bodies. *Applied Microbiology and Biotechnology, 101*(7), 2659–2675. https://doi.org/10.1007/s00253-017-8184-x
Wollmann, F., Dietze, S., Ackermann, J., Bley, T., Walther, T., Steingroewer, J., & Krujatz, F. (2019). Microalgae wastewater treatment: Biological and technological approaches. *Engineering in Life Sciences, 19*(12), 860–871. https://doi.org/10.1002/elsc.201900071
Wu, D., Yang, L., Gan, L., Chen, Q., Li, L., Chen, X., Wang, X., Guo, L., & Miao, A. (2015). Potential of novel wastewater treatment system featuring microbial fuel cell to generate electricity and remove pollutants. *Ecological Engineering, 84*, 624–631. https://doi.org/10.1016/j.ecoleng.2015.09.068
Xiao, T., Zhu, Z., Li, L., Shi, J., Li, Z., & Zuo, X. (2023). Membrane fouling and cleaning strategies in microfiltration/ultrafiltration and dynamic membrane. *Separation and Purification Technology, 318*, Article 123977. https://doi.org/10.1016/j.seppur.2023.123977
Xie, S., Higgins, M. J., Bustamante, H., Galway, B., & Nghiem, L. D. (2018). Current status and perspectives on anaerobic co-digestion and associated downstream processes. *Environmental Science: Water Research & Technology, 4*(11), 1759–1770. https://doi.org/10.1039/c8ew00356d
Xu, D., Zhu, Z., & Li, J. (2022a). Recent progress in electrospun nanofibers for the membrane distillation of hypersaline wastewaters. *Advanced Fiber Materials, 4*(6), 1357–1374. https://doi.org/10.1007/s42765-022-00193-0
Xu, Y., Zhu, Y., Chen, Z., Zhu, J., & Chen, G. (2022b). A comprehensive review on forward osmosis water treatment: Recent advances and prospects of membranes and draw solutes. *International Journal of Environmental Research and Public Health, 19*(13), 8215. https://doi.org/10.3390/ijerph19138215

Yang, H., Ye, S., Zeng, Z., Zeng, G., Tan, X., Xiao, R., Wang, J., Song, B., Du, L., Qin, M., Yang, Y., & Xu, F. (2020a). Utilization of biochar for resource recovery from water: A review. *Chemical Engineering Journal, 397*, Article 125502. https://doi.org/10.1016/j.cej.2020.125502

Yang, J., Monnot, M., Ercolei, L., & Moulin, P. (2020b). Membrane-based processes used in municipal wastewater treatment for water reuse: state-of-the-art and performance analysis. *Membranes, 10*(6), 131. https://doi.org/10.3390/membranes10060131

Yaqoob, A. A., Parveen, T., Umar, K., & Mohamad Ibrahim, M. N. (2020). Role of nanomaterials in the treatment of wastewater: A review. *Water, 12*(2), 495. https://doi.org/10.3390/w12020495

Zahrani, A. A. (2024). *Unveiling the mechanisms of bromate formation and NOM treatment in drinking water via plasmon-enhanced catalytic ozonation.* The University of Alabama in Huntsville.

Zarei, M. (2020). Wastewater resources management for energy recovery from circular economy perspective. *Water-Energy Nexus, 3*, 170–185. https://doi.org/10.1016/j.wen.2020.11.001

Zeghioud, H., Khellaf, N., Djelal, H., Amrane, A., & Bouhelassa, M. (2016). Photocatalytic reactors dedicated to the degradation of hazardous organic pollutants: Kinetics, mechanistic aspects, and design—A review. *Chemical Engineering Communications, 203*(11), 1415–1431. https://doi.org/10.1080/00986445.2016.1202243

Zou, Q., Zhang, Z., Li, H., Pei, W., Ding, M., Xie, Z., & Li, H. (2020). Synergistic removal of organic pollutant and metal ions in photocatalysis-membrane distillation system. *Applied Catalysis B: Environmental, 264*, Article 118463. https://doi.org/10.1016/j.apcatb.2019.118463

MIX
Papier aus verantwortungsvollen Quellen
Paper from responsible sources
FSC® C105338

If you have any concerns about our products,
you can contact us on
ProductSafety@springernature.com

In case Publisher is established outside the EU,
the EU authorized representative is:
Springer Nature Customer Service Center GmbH
Europaplatz 3, 69115 Heidelberg, Germany

Printed by Libri Plureos GmbH
in Hamburg, Germany